STRATHCLYDE UNIVERSITY LIBRARY

ml

30125 00373511

KV-577-219

ANDERSONIAN LIBRARY
★
WITHDRAWN
FROM
LIBRARY
STOCK
★
UNIVERSITY OF STRATHCLYDE

Welding Metallurgy of Structural Steels

ANDERSONIAN LIBRARY
★
WITHDRAWN
FROM
LIBRARY
STOCK
★
UNIVERSITY OF STRATHCLYDE

Welding Metallurgy of Structural Steels

Proceedings of an International Symposium on Welding Metallurgy of Structural Steels, sponsored by the TMS Ferrous Metallurgy Committee and the Heat Treatment Committee, and co-sponsored by the Edison Welding Institute, held at the Annual Meeting of the Metallurgical Society, Inc. in Denver, Colorado, February 22-26, 1987.

Edited by:

Jay Y. Koo
Exxon Research and Engineering Co.
Route 22 East
Annandale, New Jersey 08801

A Publication of The Metallurgical Society, Inc.

UNIVERSITY OF STRATHCLYDE
2 9 SEP 1989
UNIVERSITY LIBRARY

A Publication of The Metallurgical Society, Inc.
420 Commonwealth Drive
Warrendale, Pennsylvania 15086
(412) 776-9000

The Metallurgical Society, Inc. is not responsible for statements or opinions and absolved of liability due to misuse of information contained in this publication.

Printed in the United States of America.
Library of Congress Catalogue Number 87-42882
ISBN NUMBER 0-87339-025-3

Authorization to photocopy items for internal or personal use, or the internal or personal use of specific clients, is granted by The Metallurgical Society, Inc. for users registered with the Copyright Clearance Center (CCC) Transactional Reporting Service, provided that the base fee of $3.00 per copy is paid directly to Copyright Clearance Center, 27 Congress Street, Salem, Massachusetts 01970. For those organizations that have been granted a photocopy license by Copyright Clearance Center, a separate system of payment has been arranged.

© 1987

D
672.52
INT

TABLE OF CONTENTS

TOUGHNESS EVALUATION

FACTORS CONTROLLING PROPERTIES OF WELDMENTS

MICROSTRUCTURAL EVOLUTION IN WELDMENTS

STEEL DEVELOPMENTS FOR IMPROVED WELDABILITY

PREFACE

This book presents the proceedings of an International Symposium on "Welding Metallurgy of Structural Steels" which was held during the Annual Meeting of The Metallurgical Society of AIME in Denver, Colorado, February 1987. Co-sponsored by the TMS Ferrous Metallurgy Committee and the Heat Treatment Committee, together with the Edison Welding Institute, the Symposium included sessions on Weldability, Microstructural Evolution in Weldments, Microstructure and Toughness of Weldments, and Factors Controlling Properties of Weldments. Thirty eight papers were presented by authors from eleven countries including Australia, Belgium, Canada, England, Finland, Holland, Japan, Korea, Norway, Sweden, and the USA.

With the advent of the elastic-plastic fracture mechanics and the modern probe forming analytical tools, significant advances have been made in the past decade in our understanding of microstructure and property relationship in welded steel joints. The purpose of the symposium was to provide a comprehensive review of the subject and to bring up to date the progress made in improving weldability and service performance of the structural steel weldments.

Major topics in the book include weldability, microstructure characterization, fracture toughness test methods, local brittle zones, factors controlling the properties of weldments, and the structural integrity of the welded joints. Special topics are devoted to fundamentals of the formation and control of heat affected zone and weld metal microstructures, fabricator's and user's experience, and steel manufacturing technologies of modern structural steels. All the papers in this volume underwent peer reviews and I wish to acknowledge the efforts of the anonymous reviewers, many of whom were conferees.

Thanks are due to the members of the TMS/AIME Ferrous Metallurgy Committee/Heat Treatment Committee and the Edison Welding Institute for their support, in particular to Dr. J. M. Gray of Microalloying International Inc. and Dr. P. L. Threadgill of Edison Welding Institute for their assistance in organizing the symposium. The help of the TMS/AIME staff, including Mr. Kevin Marsden, Ms. Judy Parker, Ms. Shirley Miller, and Ms. Kristen Rothert is also greatly appreciated. Finally, I would like to thank Mrs. B. Bice of Exxon Research and Engineering Company for secretarial assistance.

Jay Y. Koo
Exxon Research and Engineering
Company
Annandale, New Jersey

CONFERENCE ORGANIZER

Jay Y. Koo
Exxon Research and Engineering Company
Annandale, New Jersey

SESSION CHAIRMEN/CO-CHAIRMEN

S. A. David	Oak Ridge National Laboratory
R. H. Frost	Colorado School of Mines
J. M. Gray	Microalloying Internationl Inc.
J. Y. Koo	Exxon Research and Engineering Co.
K. W. Mahin	Sandia National Laboratory
T. H. North	Welding Institute of Canada
D. T. Read	National Bureau of Standards
C. P. Royer	Exxon Production Research Company
P. L. Threadgill	Edison Welding Institute
G. A. Vaughn	Exxon Production Research Company

Weldability

A HISTORICAL VIEW OF WELDABILITY

Warren F. Savage

Professor Emeritus
Materials Division
Rensselaer Polytechnic Institute
Troy, New York   12180-3590

Abstract

This review is based upon articles pertaining to weldability which
have appeared in the Welding Journal over the past 55 years. Prior to
World War II, most arc welding in this country was performed using bare
electrodes, and the early cellulosic-coated electrodes were not suitable
for welding the alloy steels used in armor and aircraft. No until the mid
40's when the role of hydrogen in causing cold-cracking was recognized were
the so-called "controlled-hydrogen", lime-coated ferritic electrodes developed which facilitated the welding of hardenable steels. This, in turn,
expanded the application of welded fabrication and stimulated research in
weldability that led to the development of numerous tests to evaluate both
fabricability of materials and the serviceability of weldments. Since then,
extensive fundamental research has been conducted in order to understand
the mechanisms responsible for the creation of welding defects, to develop
new, more weldable alloys, and to aid in the selection of optimum welding
conditions. 84 references.

Introduction

The American Welding Society and I were born in 1922, making us both 65 this year. In preparation for this talk, I decided to review those articles published in the Welding Journal pertaining to the subject of weldability, and, since little was written on the subject during the first ten years, I elected to cover only the 55 years from 1931 to date.

It should be remembered that in 1931, arc welding was only 23 years old, and that the arc welding electrodes used in this country were either bare wire or sull-coated (i.e. bare wire from which the lime-slurry used as a drawing lubricant was not removed). Thus, since no protection from atmospheric gases was afforded, the oxygen reacted vigorously with the carbon in steels to form gaseous porosity, and with iron and other elements to form non-metallic inclusions. Meanwhile, atmospheric nitrogen reacted to form embrittling nitrides. In fact, the resulting quality of the weld deposits was so poor that apologists published statements like the following, (1):

"Ductility greater than that of the parent metal would ruin welding but luckily it can be obtained only in advertisements. ... In arc welding, ductility equal to the parent metal can only be obtained in test pieces where it does no harm. ... We believe that it is about time that the public learned the truth about welding and the art is just about old enough to be acquainted with the fact that there is no Santa Claus. Welding is now twenty-four years old and it is about time the fact was recognized that an arc weld is porous, always has been and always will be."

Therefore, it is not surprising that much of both the production and repair welding was performed using oxy-acetylene as the heat source. Since the flame could be made reducing, protection from the atmospheric gases was provided. In addition, since few alloy steels were welded, the cooling rates associated with oxy-acetylene welding were, for the most part, slow enough to prevent the formation of hard heat-affected-zone microstructures, and thus weldability was not a major concern in those days.

It was not until the beginning of World War II that coated electrodes began to be used extensively in the United States, and the majority of the early coated electrodes for welding steels had cellulosic coatings. Thus, during the war when the necessity arose for welding the alloy steels used in aircraft and armor, hydrogen-induced cracking, or, as it was then called, "Hard-Cracking", became a serious problem.

Weldability Literature (1931 to 1940)

Speller and Texter in an article entitled "The Quality of Materials for Fusion Welding" (2), were among the first to write about weldability. They state:

"The ideal welded joint obviously is one in which the properties of the weld metal and the base metal are similar both chemically and physically. Although at the present time this objective does not seem commercially attainable, such an ideal should be kept in mind and investigation pointed in that direction."

Then in the next paragraph:

"It is not possible to definitely define all the factors which influence weldability and the statement in some specifications that steel shall be of good weldable quality leaves the manufacturer in somewhat of a quandary."

4

They then quote from a motion passed by A.S.T.M. Subcommittee XXI on Steel for Welding, with reference to a list of ASTM specifications for steels:

"In connection with carbon limitations in these specifications, the committee was of the opinion that, while higher carbon steels are successfully welded, commercial practice at present often limits the carbon (by ladle test) to 0.30 per cent in open-hearth steel and 0.15 per cent in bessemer steel. In the case of seamless steel tubes covered by Specifications A 53, the carbon limit was placed at 0.35 per cent."

This motion, therefore, more or less represents a fore-runner of the carbon equivalent.

In 1936, Bob Aborn published what is probably the first comprehensive treatice (3) on the welding metallurgy of steels. This, and a subsequent publication (4), did much to acquaint the welding community with the complexity of the weld heat-affected zone ("WHAZ") and explained the great influence of heat-affected-zone microstructures on the weldability of steels.

In 1939, Werner (5), a German scientist, proposed that hydrogen derived from the dissociation of water vapor was responsible for cracking in chrome-moly aircraft steels. This hypothesis met with considerable criticism from reviewers and was not widely accepted at the time. It is interesting that in the same paper he reported that copper from the copper-plated welding rods, when abraded on the surface of his weldability test specimens caused hot-cracks. After showing that when welding on sheet steel which had been coated with copper by dipping in a copper-sulfate solution, cracks appeared whether or not the welding rod was coated with copper, he concluded:

"The cracking caused by copper is another instance of intergranular embrittlement of steel by molten metals."

Thus, in a single paper only a few pages long, he recognized not only the role of moisture and hydrogen in cold-cracking, but also the phenomenon of copper-contamination cracking (the latter thirty-two years before Steve Matthews, Matt Mattock, and I did (6)).

In 1940, Spraragen and Clausen published a literature review (7) of papers which discussed the effects of hydrogen (together with arsenic, titanium, and "miscellaneous" elements) on the welding of steel. This, as near as I have been able to ascertain, was the first comprehensive review of the role of hydrogen in the welding of steel. Later that same year, Zappfe and Sims published their paper setting forth the Planar-Pressure Theory of hydrogen-induced cracking (8).

The War Years

One of the most active periods in weldability research began with the onset of World War II, and continued for about a decade. During this period the development of improved heavily coated welding electrodes made shielded metal arc welding (SMA) the work-horse of the ship-building, armament, aircraft and construction industries.

However, recognition of the importance of hydrogen was slow in coming. Witness these statements taken from a 1941 article (9) by an author associated with the "Mechanical Department" of a major oil refinery:

"There has been in recent articles on the welding of steel a growing emphasis on the effect of hydrogen in the deposited weld metal. While

5

undoubtedly this gas does have an effect on the formation of fissures, the defects are to some degree only of academic interest. "Cat" or "Fish" eyes are defects have never assumed more than laboratory interests. While they may to a limited extent lower the tensile strength and the ductility, their influence is so slight that the strength and plasticity of the weld metal remains well above the minimum requirements if they are the only contributing factor."

Then near the end of the article, the author states:

"In welding the plate the dendritic impurities (He earlier cites phosphorus, sulfur and silicon as the offending elements.) were entrapped again by the rapidly cooling weld metal. These areas of entrapment were the ultimate cause of failure. Hydrogen pressure areas caused by the change from the atomic to the molecular state may increase the size of the primary cavities, but it would only be a contributing factor."

At least he read Zapffe!

The role of hydrogen in cold-cracking continued to be a subject of debate among welding engineers throughout World War II. Between 1938 and 1944, most reports on cracking in arc-welded steels blamed transformation stresses and thermally-induced deformation and did not consider hydrogen as a significant contributing factor.

However, as early as 1934, it had been reported that 18Cr-8Ni austenitic stainless steel electrodes were used successfully in Germany and Japan for welding alloy steels. Although it was postulated that the weld metal yielded enough under the thermally induced stresses to prevent cracking in the base metal, this failed to explain why base metal cracking failed to occur when welding was done with bare wire or sull-coated electrodes.

Although it is not entirely clear who first replaced the cellulosic coatings then popular for ferritic electrodes with the heavy, lime-type coatings which were usually employed on austenitic electrodes, the first of such so-called "controlled-hydrogen electrodes" were developed by a wartime consortium of electrode manufacturers and fabricators of steel weldments, working under the auspices of the Office of Scientific Research and Development. A related project at Battelle Memorial Institute (10) provided fundamental information on the characteristics of ferritic electrodes with lime-type coatings. The results of this research, as reported by Mallett and Rieppel (11), indicated that electrodes which produced arc atmospheres containing 30% or more hydrogen by volume caused cold-cracking in hardenable steels, whereas no cracking occurred with electrodes which produced low-hydrogen atmospheres, such as stainless steel and lime-coated ferritic electrodes.

In 1947, Voldrich (12), in a paper dealing with the nature of cold cracking, presented data showing the effect of various electrode coatings on the arc atmosphere and cold-cracking in hardenable steels. He also investigated the effects of base metal composition, homogeneity, carbide size, composition, and heat-affected zone hardness on cold-cracking and concluded:

"The results of a number of investigations lead to the conclusion that cold cracking in the heat-affected zone of metallic-arc-welded joints can occur whenever a critical combination of four factors is present. These factors are hydrogen, rate of heating and cooling, chemical composition and structure, and stress."

He went on to state that, since the magnitude of the stresses and the rate of heating cannot be controlled to any useful degree by choice of welding conditions, the governing factors are "hydrogen, the steel, and the cooling rate", a conclusion equally valid today, 40 years later.

In 1946, the Welding Research Council, in trying to agree upon a definition for the term "Weldability", came up with two definitions, neither of which appeared to satisfy Bill Warner, who wrote the following in an extensive critique of both (13):

"However, if welding engineering is to take its place along side other established engineering professions welding engineers must learn how to prescribe welding procedures to satisfy the requirements for fabricating a design so that the service properties of the base metal after welding are not reduced below the values required for adequate serviceability of the weldment."

He then concludes:

"If the engineer makes adequate laboratory tests he can determine weldability of metals without fabricating production weldments. The objective of welding research and development work is determination of such data together with proper welding procedures. The rule of thumb procedure in production welding should be on its way out. When it becomes the exception rather than the rule welding engineering may then be regarded more as a science than an art."

Systems For Choosing Welding Conditions

In 1941, Hess (14), after describing the results of longitudinal tensile testing procedures used to evaluate the effect of welding conditions on the properties of welded joints, stated:

"It is apparent that in order to completely study a single type of steel, it is necessary to find for a range of thicknesses the welding conditions which will satisfy the above series of tests. In finding the most satisfactory conditions for a given thickness it will be necessary to study several sets of welding conditions. In addition to studying the effects of the ordinary welding conditions of voltage, current and travel speed, for many steels it would be necessary to study a variety of preheat temperatures in order to find the optimum welding conditions. The task of covering the complete range of medium carbon and low-alloy steels would obviously be extremely laborious."

"In order to simplify the determination of optimum welding conditions for the above wide range of steels, it is proposed to make use of fundamental metallurgical information concerning the different steels. The information which is most important is that concerning the transformation characteristics upon rapid cooling. If the structure and properties of a particular steel are known for various rates of cooling, a satisfactory rate of cooling for this steel may be selected, and welding conditions which will give this rate of cooling may be specified."

In October of 1941, upon recommendation of N.D.R.C., Dr. A.B. Kinzel was appointed Supervisor of a cooperative program between Rensselaer and Lehigh, financed by O.S.R.D., which was undertaken to measure weld cooling rates for various combinations of welding conditions, weld geometries, and initial plate temperatures. Ultimately the two universities were to utilize completely different methods for determination of weld cooling rates, and their results were to be published in different forms.

At Rensselaer, thermocouples, percussion welded to the bottom of small
diameter holes strategically located with reference to the weld interface,
were used to measure the actual weld thermal cycles directly and, from these
measurements, tables of weld cooling rates as a function of weld energy input and initial plate temperature were assembled for various plate thicknesses and weld geometries. In all, measurements were made for welds in
four plate thicknesses ($\frac{1}{4}$, $\frac{1}{2}$, 1, 1$\frac{1}{2}$ inch) using at least four different
energy inputs at plate temperatures of 3, 22, 100 and 200° C. In order to
extend the experimental data, modifications of the general mathematical solutions for heat flow in weldments published by Rosenthal (15) were developed
and employed. The published results (16) contain easily used tables, each
of which lists the weld cooling rates which can be achieved in butt welds
in a particular plate thickness using practical welding conditions and permits combinations of energy input and preheat temperatures to be chosen which
will produce each of the cooling rates listed. A second publication (17)
contains similar tabular data for fillet welds.

At Lehigh, bead-on-plate, fillet, and first pass single and double-vee
butt welds were made with four plate thicknesses ($\frac{1}{4}$, $\frac{1}{2}$, 1, and 1$\frac{1}{2}$ inch)
using four to six energy inputs and four different plate temperatures (32,
75, 200 and 400° F) for each type of joint. Transverse sections $\frac{1}{2}$ inch
thick were sawed under coolant from the center of each weld, polished and
lightly etched to reveal the weld cross section. An average of six to eight
Vickers hardness indentations were then made in the hardest region of the
heat-affected zone immediately adjacent to the fusion zone using a 10kg load.

The cooling rates experienced at various distances from the water-
quenched end of a Jominy bar during the quenching operation were measured
using thermocouples attached to standard Jominy bars. Triplicate Jominy
bars machined from the steels used in the welding experiments were aus-
tenitized for 30 minutes at 2100°F to duplicate the grain size produced in
the coarse grained heat-affected zone where the maximum hardness was found
on the weld cross-sections. Hardness measurements were made at intervals
of 1/32-in from the water quenched end on all three bars after grinding a
$\frac{1}{2}$-in longitudinal flat. The cooling rates in the welds were then determined
by assuming that the weld cooling rate was equal to that measured at the location in the Jominy bar made from the same steel where the hardness corresponded to the maximum hardness observed in the weld heat-affected zone.
Cooling rates corresponding to those in the weld heat-affected zone were
found at Jominy distances between 4/32 and 24/32-in. where the cooling rates
at 1300°F were 130°F/sec and 20°F/sec, respectively.

The Lehigh system ultimately employed a novel method for reducing the
influence of weld geometry to a single arithmetic factor. It was determined
empirically that the influence of the surrounding base metal on the weld
cooling rate could be related to the mass of metal contained within 3-in. of
the center of the molten pool. The area of base metal intercepted by the
surface of an imaginary sphere with a 3-in. radius centered on a bead-on-
plate weld pool on a $\frac{1}{2}$-in. plate was arbitrarily chosen as a geometric factor of unity, and the intercept areas for other plate thicknesses and weld
geometries were calculated as ratios of this standard. Tables of values for
"Geometric Factors" calculated in this fashion for commonly encountered
plate thicknesses and weld geometries were included as part of the system
for choosing welding conditions.

A second set of tables listed the values of arc power as a function of
welding speed and plate temperature which had been found to give a maximum
heat-affected-zone hardness in a bead-on-plate weld made on a $\frac{1}{2}$-in. plate of
a given steel equal to that produced at a specific location on a Jominy bar
of the same steel. According to the Lehigh system, this quantity, called

the "Volt-Ampere Factor", was chosen such that the resulting weld cooling rate would produce the desired maximum heat-affected-zone hardness, and then multiplied by the Geometric Factor to obtain the proper arc power to be used for obtaining the desired weld cooling rate.

In addition to providing a method for predicting the maximum hardness in the heat-affected zone, the research at Lehigh included longitudinal-bead-weld notch-bend tests and attempted to correlate the ductility of the weld heat-affected zone to the maximum heat-affected zone hardness in a number of carbon and low alloy steels. The results of the Lehigh research were published in three papers (18,19,20). Reference 19, entitled "Guide to the Weldability of Steels", gives a detailed explanation of the use of the Lehigh system together with specific examples.

It is interesting to note that the cooling rates predicted by the Lehigh method agree with those measured at Rensselaer for comparable weld geometries within a few percent. However, both the Rensselaer and the Lehigh methods for predicting the maximum hardness in the weld heat-affected zone rely upon accurate Jominy hardenability data in order to relate the hardness of the austenite transformation products to the cooling rates associated with a particular set of welding conditions, weld geometry and pre-heat temperature. None-the-less, I have found that Jominy hardenability data calculated from chemical composition using methods contained in a recent AIME publication (21) can be used with the Rensselaer cooling rate data to estimate the maximum hardness in the heat-affected zone within a few Rockwell C points for many carbon and low-alloy constructional steels.

Unfortunately, neither system met with wide acceptance and few welding engineers are familiar with them today. The concept of the carbon-equivalent with its many divergent formulae seems to have displaced the more fundamental approach to weldability afforded by the Rensselaer and Lehigh systems. I find this retreat from the fundamental approach disheartening, and it never fails to amaze me that so many different formulae can purport to describe the same quantity. In a recent IIW document (22), Suzuki compares seven different formulae and on the basis of two methods of testing for cold-cracking sensitivity concludes that no single formula applies to all steels. As an exercise, I used the seven formulae to calculate carbon equivalents for a typical 1¼Cr-½Mo steel containing 0.15%C, 0.40%Mn, 0.75%Si, 1.05%Cr, and 0.55%Mo and obtained values ranging from 0.285 to 0.586, with a mean of 0.45 and a standard deviation of 0.13! These data do not inspire my confidence in the concept of the carbon equivalent.

Furthermore, the value of the carbon equivalent is limited since it considers only the nominal composition of the base metal and ignores section thickness, joint geometry, and restraint, all of which may be of greater significance than the calculated carbon equivalent. Therefore, since I have but little faith in any of the published formulae, I make no recommendations as to their use and can only hope that a better system will ultimately evolve when we have a better fundamental understanding of the problem.

Some Weldability Tests

During the War years, numerous tests were devised to evaluate the weldability of base metals, and they can be divided into two groups as tests of fabricability or of serviceability. Among those designed to test base metals for fabricability were the Longitudinal-Bead-Weld Underbead-Cracking Test (23), and the Circular-Bead-Weld Underbead Cracking Test (24), developed at Battelle, the Reeve Fillet Test (25) (which was a fore-runner of the Controlled-Thermal-Severity Test that was subsequently developed in 1953 (26)), the Lehigh Restraint Test (27), and the N.R.L. Restraint

Specimen (28). These tests were all designed to evaluate either hot-cracking or cold-cracking tendencies, but only the Lehigh Restraint Test and the Controlled-Termal-Severity Test are widely used today.

Tests for serviceability were mainly concerned with techniques for evaluating toughness and ductility of weldments. In addition to the Charpy V-Notch test, both notched and unnotched bend specimens were popular. The Longitudinal-Bead-Weld Notch-Bend Specimen developed at Lehigh (29) and the Kinzel Test Specimen (30), a modification of the Lehigh specimen, were the most widely used. Bend test data were recorded in many forms, including bend angle at failure, deflection at failure, energy absorbed at failure and, in the Kinzel test, the lateral contraction at the root of the notch after failure.

During the fifties and sixties additional fabricability tests were developed which attempted to generate reproducible conditions of restraint through various complex specimen designs. Included in these were the Cruci- form Test (31), tapered-fin type restraint-cracking tests, such as the Key- hole Slotted-Plate Restraint Specimen developed at Battelle (32), the Hould- craft Cracking Test (33) which attempted to design varying degrees of re- straint into a single specimen, and circular-weld restraint cracking tests, such as the Circular Patch Test (34), the Segmented-Groove Circular Patch Test (35) and the Navy Circular Patch Test (36).

In 1954, the Murex Hot Cracking Test, the first of a number of restraint cracking tests utilizing an externally applied load, was described in the British literature (37). in 1956, Boudreau developed a hot-cracking sus- ceptibility Test (38) which applied a transverse weld to a sheet metal spec- imen as it was held under a longitudinal dead load. In 1965, Carl Lundin and I developed the Varestraint Test (39,40) which applied a reproducible augmented strain to a weld by bending the specimen around a radiused die- block as the weld was being deposited on the top surface of the specimen. In 1968 Gene Goodwin developed what we called the "TIG-A-MA-JIG Test" (41), which is now more commonly known as the Spot Varestraint Test, as part of his doctoral thesis. This test also generated a controlled augmented strain by bending the specimen around a radiused die-block during welding, but used a GTA spot weld instead of a stringer bead. Versions of both of these test- ing procedures, I am pleased to say, are being widely used in evaluating hot-cracking susceptibility today.

With regard to new tests for serviceability, in 1951, Hartbower and Pellini developed the Explosion Bulge Test (42), which was designed to evaluate the resistance of materials to the high dynamic loading imposed in military service. By conducting this test over a range of temperatures, they showed that three useful transition temperatures can be determined:

1. The Nil Ductility Temperature (NDT) - below which all ability to deform plastically is lost.

2. The Fracture Transition for Elastic Loading (FTE) - below which a fracture will run into the elastically loaded region at the edge of the specimen.

3. The Fracture Transition for Plastic Loading (FTP) - above which the fracture is arrested in the Plastically deformed region.

Subsequently, the Drop Weight Tear Test (DWTT), later modified and renamed the Dynamic Tear Test (DT), was developed at NRL as a less expensive means for determining the NDT. These tests provided the basis for the sys- tem developed by Pellini which he has termed "Fracture Safe Design" (43,44).

Briefly, according to this system, limiting minimum service temperature to slightly above NDT protects against fracture initiation from small cracks in regions of high local stress. Restricting service temperatures to above FTE, which he showed to be about 60°F above NDT, provides fracture-arrest protection if nominal stresses do not exceed the yield strength. Restricting service temperatures to above FTP, which he showed to be approximately 120°F above NDT, ensures that only fully ductile failure can occur. These concepts have been applied with a high degree of success in the design and fabrication of military hardware, but have met with considerable resistance from designers and fabricators of buildings and bridges as being unnecessarily conservative.

The Evolution of the Gleeble

In 1946, Nippes and I attempted to evaluate the notch toughness of various regions of the heat-affected zone of ship steel weldments by careful placement of the notches in Charpy V-Notch specimens machined from the weldments (45). However, the results were biased by the sloping nature of the heat-affected zone which caused the fracture to traverse more than one portion of the zone. Consequently, we decided to develop a thermal simulator and utilize temperature measurement data accumulated during the earlier cooling-rate studies at Rensselaer (46) to generate the various heat-affected-zone microstructures in Charpy specimens. In 1949, after several years of abortive effort, the first thermal simulator capable of reproducing programmed thermal cycles duplicating actual measured weld thermal cycles was developed (46) and utilized to generate heat-affected-zone microstructures for impact testing (48).

This thermal simulator underwent a series of successive improvements over the next few years and was used to study knife-line attack in Type 347 stainless steel (49,50). It was also used to determine the minimum cooling rate which would avoid the formation of proeutectoid transformation products in T-1 steel and, from this information and the cooling rate data previously obtained at R.P.I., to establish the maximum permissable energy inputs for welding T-1 Steel (51,52). Owczarski, in studying the toughness of heat-affected-zone microstructures in NAXtra (53), showed that the partially transformed region which were exposed to peak temperatures in the intercritical temperature region could exhibit the most severe embrittlement of any region in the weld heat-affected-zone because of the formation of pockets of high carbon martensite. He also used the Gleeble to study the weld heat-affected zones in ductile iron (54) and Paez used the apparatus to study the heat-affected zones in Mn-Mo armor steels (55,56).

In 1955, a high speed tensile-testing capability was added to the thermal simulator (57) and the Gleeble was born. During the late fifties and early sixties, the Gleeble was used extensively for Hot-Ductility testing of candidate heats of austenitic stainless steels for steam power plants. The findings of tests on a variety of high temperature alloys were documented in a Welding Research Council Bulletin (58).

Studies of the elevated temperature properties of the heat-affected zone of 18Ni Maraging steels by Pepe (59,60) led to the discovery of the phenomenon of Constitutional Liquation which has since been found to be an important mechanism contributing to hot-cracking in many alloys.

Since 1957, when the first commercial unit was produced, the Gleeble has experienced several metamorphoses, and, at last count, over 135 were in use throughout the world. The current model features closed-loop, computer-programmed control of both thermal cycles and mechanical testing operations with heating rates up to 10,000 °C/sec and loads up to 10,000 kg with cross-

11

head speeds up to 120 cm/sec in both tension and compression.

Some Recent Tests For Hydrogen-Induced Cracking

Granjon, in his 1962 Adams Memorial Lecture (61), described a novel
system for photographing the nucleation and growth of hydrogen-induced
cracking in an alloy steel. He first polished through-thickness surfaces on
two specimens, and after clamping the prepolished surfaces together, deposited a stringer bead across the top surface in a direction perpendicular
to the mating interface, using a moistened SMA electrode to introduce hydrogen. He then broke the two specimens apart and took time-lapse motion pictures of the prepolished surfaces as the hydrogen-induced under-bead cracks
nucleated and grew. By applying a drop of microscope immersion oil on the
surface, he was also able to photograph hydrogen bubbles as they evolved
from the growing crack.

Szekeres, (62) modified this technique by placing the prepolished
specimens in a three-point loading frame after they were separated and
applying a sufficient load to cause a small amount of plastic bending, thus
producing yield strength stresses in the outer fibers in a direction transverse to the weld bead to simulate the restraint stresses present in a full-
scale weldment. In this fashion, he was able to observe the nucleation of
hydrogen-induced cracks in HY-80 and to document their growth over a period
of several hours. He noted that the hydrogen evolution always occurred
gradually after the crack had propagated past a given point, thus adding
support to the Troiano Model (63) and discrediting the Zappfe Planar Pressure Theory (8). Szekeres also established the existence of an unmixed zone,
within which the base metal at the edge of the fusion zone melted and re-
solidified without experiencing mechanical mixing with the filler metal, and
showed that crack initiation tended to occur in this region and in the
adjacent regions where partial melting had been experienced either as a result of exposure to temperatures between the solidus and liquidus or constitutional liquation of sulfides (64).

Homma improved upon Szekeres's technique by instrumenting the loading
system (65) and found that while stresses in the elastic region could
nucleate microcracks, yield strength stresses were required for their propagation. He then elected to force the specimens to conform to a radiused
die-block in order to produce a controlled augmented strain in the outer
fibers. Using this technique, he demonstrated that if sufficient diffusible
hdyrogen were present, anelastic rumpling of the prepolished surface by dislocation motion continued for several hours after the initial loading. However, in the absence of a critical level of diffusible hydrogen, no such
anelastic deformation occurred. This behavior not only supported the hypothesis advanced by Beachem (66) that hydrogen interacts with dislocations to
activate dislocation sources, but also indicated that dislocation motion
aids in the transport of hydrogen in the crack tip.

A simplified version of this test was subsequently named the Augmented
Strain Cracking Test (ASCT). Results obtained using this test on rare earth
treated HY-80 steels showed that although proper additions of rare earth
metals significantly reduced the sensitivity to hydrogen-induced cracking,
additions in excess of an optimum amount caused the susceptibility to increase (67).

In 1969, Granjon developed the "Implants Test" for studying hydrogen-
induced cracking in steels (68). Sawhill modified the original Implants
Test in 1974 (69) by incorporating a spiral notch in place of the original
single circumferential notch used by Granjon. The Implants Test has as its
major advantage the ability to investigate the effect of welding procedures

on the cold-cracking susceptibility of an alloy using a minimum amount of material, since the individual specimens are only ¼-in. in diameter and less than 4-in. long.

Although the Augmented Strain Cracking Test has been used successfully by investigators at Rensselaer (70,71,72), it does not appear to have been accepted elsewhere. The Implants test, the Controlled Thermal Severity Test and the Lehigh Restraint Test are probably the most widely used tests for studying hydrogen-induced cracking today.

Some Miscellaneous Contributions To The State Of The Art

In 1968, during research on hot-cracking of Inconel welds sponsored by ORNL, Goodwin observed major variations in the geometry of the weld pool caused both by oxygen and nitrogen as contaminants in the argon shielding gas (73), and by variations in the amount of minor elements well within the allowable composition of the alloy. This research, ultimately reported in 1977 (74), was among the first to document the influence of minor elements on weld penetration. However, the now classic paper by Heiple et.al (75), was the first to establish the enormous influence of survice active elements such as sulfur, known as the marangoni Effect, on weld penetration.

Aronson showed that epitaxial nucleation and competitive growth were the major factors controlling the grain size in the weld fusion zone (76) and Hrubec demonstrated that the transition from an elliptical to a teardrop shaped weld pool caused by excessively high welding speeds influenced the competitive growth process and promoted the formation of center-line hot-cracks (77). Zanner developed an impact decanting technique which was useful in studying the effect of welding conditions on the shape of the weld pool (78), and Walsh has used a modification of this technique to study the effect of minor elements on the geometry of the molten pool (79).

Chase showed that the composition of the anode (base metal) has a major influence on the characteristics of a gas tungsten arc (80), thus proving the studies of arc physics performed on water-cooled copper anodes to be of questionable value. Agusa measured the arc force during GTA welding and showed the effect of arc force on defect formation (81).

In the days of rivetted construction lamellar tearing was unknown since it is impossible to load plate material in the short-transverse direction with rivets. under such conditions, the two false assumptions (that materials are homogeneous and isotropic) and the fallacy (that stress is proportional to strain within the elastic limit) on which design formulae are based did no harm to the integrity of structures. Unfortunately, with the advent of welding fabrication, designers ignored the fact that steels are far from homogeneous and that the mechanical properties in the short-transverse direction are far inferior to those parallel to the plate surface, and designed weldments which forced major loads to be borne in the through-thickness direction. As a result, lamellar tearing became a serious problem in welded bridges and buildings. Until 1973, when Bob Stout invented his ingeneous test (82) no method of studying lamellar tearing with a laboratory scale specimen was available. In my opinion this test has contributed more to our understanding of the phenomenon than all other studies combined. However, it is unfortunate that we can't make metallurgy a required course for designers, since if they recognized the inherent inferiority of the short transverse properties of plate material, the problem would disappear.

Another of Bob Stout's many contributions to the state of the arc is the simple Slot-Weld Specimen for field weldability testing of pipeline steels (83). These specimens can be field fabricated from plate by welding and permit establishing a schedule of preheat and time between root and

13

hot-pass welds to ensure pipe welds free of cracking. This Slot Weldability test is reported to correlate well with pipes girth welded in the shop for a variety of welding conditions.

One of the major limitations of both the Varestraint and the Spot Varestaint Tests is their inability to test sheet material. The Sigamjig Test recently developed by Goodwin (84) fills that gap by providing a method of obtaining a quantitative, rating of the hot-cracking sensitivity of sheet materials.

And Of The Future?

The high strength low alloy (HSLA) steels so popular today are an enigma, since their low carbon content would lead one to believe them to be immune to hydrogen-induced cracking. I believe the major problem lies in the fact that we ignore that very important but little understood peritectic reaction. This reaction is virtually impossible to equilibrate for the simple reason that the reaction product, the peritectic alloy, forms as a wall which separates the reactants, one of which is a liquid containing some 0.5% carbon. I'm sure that how this carbon-enriched liquid ultimately solidifies and how the carbon, other alloy alloying elements and hydrogen it contains are redistributed by solid state diffusion has a major influence on the sensitivity to hydrogen-induced cracking. I understand that Dave Howden and his students are tackling this problem at Ohio State. I wish him well, but it is a formidable task and one that I hope others in the welding research community will address in the near future.

As a final thought, the data acquisition systems currently available are so far superior ot those available in 1943 when the temperature measurements were made at Rensseler that it would be appropriate to initiate a concerted effort on the part of university laboratories to generate a new data base of measured weld thermal cycles. With the sophisticated computer technology we have today, it should then be possible to establish computer codes which would accurately generate weld thermal cycles for virtually an combination of welding conditions and weld geometry. Using the currently available Gleeble technology, these data could then be employed to synthesize the various heat-affected-zone microstructures for property testing. With the aid of systems for calculating hardenability, a truly fundamental approach to the evaluation of weldability and the selection of optimum welding conditions should be possible.

References

1. C.J. Holslag, "Ductility and Penetration - Two Fallacies", Weld. J., 11(4) (1931), 47-48.

2. F.N. Speller and C.R. Texter, "The Quality of Materials for Fusion Welding", Ibid., 19(6) (1931), 23-24.

3. R.H. Aborn, "Metallurgical Aspects of Welding Steel", Ibid., 15(10) (1936), 21-31.

4. R.H. Aborn, "Metallurgical Changes at Welded Joints and the Weldability of Steels, "Weld. J., Res. Suppl.", 19(10) (1940), 345s-351s.

5. O. Werner, "The Causes of Cracks in Welded Aircraft Steels", Ibid., 18(11) (1939), 425s-428s.

6. S.J. Matthews, M.O. Maddock, and W.F. Savage, "How Copper Surface Contamination Affects Weldability of Cobalt Superalloys", Weld. J., 51(5) (1971), 326-328.

7. W. Spraragen and G.E. Clausen, "The Effect of Hydrogen, Arsenic Titanium and Miscellaneous Elements in the Welding of Steel", Weld. J., Res. Suppl., 19(1) (1940), 24s-30s.

8. C.A. Zappfe and C.E. Sims, "Defects in Weld Metal and Hydrogen in Steel", Ibid., 19(10) (1940), 377s-394s.

9. A.G. Ford, "Weldability of Steel", Ibid., 20(7) (1941), 287s-288s.

10. "Development of Improved Electrode Coatings", Report on N.D.R.C. Research Project NRC-76.

11. M.W. Mallett and P.J. Rieppel, "Arc Atmospheres and Underbead Cracking", Weld. J., Res. Suppl., 25(11) (1946), 748s-759s.

12. C.B. Voldrich, "Cold Cracking in the Heat-Affected-Zone", Ibid., 26(3) (1947), 153s-169s.

13. W.L. Warner, "This Elusive Character Called Weldability", Ibid., 18(11) (1939), 185s-188s.

14. W.F. Hess, "Evaluating Welded Joints", Ibid., 20(10) (1941), 453s-458s.

15. D. Rosenthal, "Mathematical Theory of Heat Distribution During Welding and Cutting", Ibid., 20(5) (1941), 220s-234s.

16. W.F. Hess et.al., "The Measurement of Cooling Rates Associated with Arc Welding and Their Application to the Selection of Optimum Welding Conditions", Ibid., 22(9) (1943), 377s-422s.

17. W.F. Hess et.al., "Determination of Cooling Rates of Butt and Fillet Welds as a Result of Arc Welding with Various Types of Electrodes on Plain Carbon Steel", Ibid., 23(8) (1944), 376s-391s.

18. G.E. Doan, R.D. Stout, and J.H. Frye, "A Tentative System for Preserving Ductility in Weldments: Parts I and II", Ibid., 22(7) (1943), 278s-299s.

19. Advisory Committee on Weldability Projects of the War Metallurgy Committee, "Guide to Weldability of Steels", Ibid., 22(8) (1943), 333s-352s.

20. R.D. Stout, S.S. Tor, and G.E. Doan, "A Tentative System for Preserving Ductility in Weldments: Part III", Ibid., 22(9) (1943), 423s-437s.

21. Douglas Doane, "A Critical Review of Hardenability Predictors", Hardenability Concepts with Applications to Steel, The Metallurgical Society of AIME, 1978, 351-396.

22. Haruyoshi Suzuki, "Comparison of Carbon Equivalents for Steel Weldability", IIW Doc. IX-1306-84. March 1984.

23. G.G. Luther et.al., "Weldability of Managenese-Silicon High-Tensile Steels", Weld. J., Res. Suppl., 24(4) (1945), 77s-90s.

24. S.L. Hoyt et.al., "Metallurgical Factors of Underbead Cracking", Ibid., 24(9) (1945), 433s-445s.

25. L. Reeve, "A Summary of Reports of Investigations on Selected Types of High-Tensile Steels", Transactions of the Institute of Welding (Brit.), 3(4) (1940), 177.

26. C.L.M. Cottrell, "Controlled Thermal Severity Test Simulates Practical Practical Joints", Weld. J., Res. Suppl., 32(6) (1953), 257s-272s.

27. R.D. Stout et.al., "Quantitative Measurements of the Cracking Tendancy in Welds", Ibid., 26(9) (1947), 522s-531s.

28. G.G. Luther et.al., "A Review and Summary of Weldability Testing Carbon and Low Alloy Steels", Ibid., 25(7) (1946), 376s-396s.

29. R.D. Stout et.al., "Effect of Welding on Ductility and Notch Sensitivity of Some Ship Steels", Ibid., 26(6) (1947), 335s-357s.

30. A.B. Kinzel, "Ductility of Steels for Welded Structures", Ibid., 27(3) (1948), 217s-234s.

31. S. Weiss, J.N. Ramsey, and H. Udin, "Evaluation of Weld Cracking Tests on Armor Steel", Ibid., 35(7) (1956), 348s-356s.

32. H.W. Mishler et.al., WADC Technical Report 59-531, August 1959.

33. P.T. Houldcroft, "A Simple Cracking Test for Use with Argon Arc Welding", British Welding Journal, 2(10) (1955), 471-475.

34. J.E. Hockett and L.O. Seaborn, "Evaluation of Circular Weld-Patch Test", Weld. J., Res. Suppl., 31(8) (1952), 387s-392s.

35. A. Hoerl and T.J. Moore, "The Welding of Type 347 Steels", Ibid., 36(10) (1957), 442s-448s.

36. Military MIL-E-986.

37. E.C. Rollason and D.F.T. Roberts, "Operating Data for the Murex Hot-Crack Testing Machines", British Welding Journal, 1(10) (1954), 441-447.

38. J.S. Boudreau, "A Weld Cracking Susceptibility Test for Sheet Ma-
     terials", Weld. J., Res. Suppl., 35(4) (1956), 164s-168s.

39. W.F. Savage and C.D. Lundin, "The Varestraint Test", Ibid., 44(10)
     (1965), 433s-442s.

40. W.F. Savage and C.D. Lundin, "Application of the Varestraint Test to
     the Study of Weldability", Ibid., 45(11) (1966), 497s-503s.

41. G.M. Goodwin, E.F. Nippes, and W.F. Savage, "Effect of Minor Elements
     on Hot-Cracking Tendencies of Inconel 600", Ibid., 56(8) (1977),
     245s-253s.

42. C.E. Hartbower and W.S. Pellini, "Explosion Bulge Test Studies of
     the Deformation of Weldments", Ibid., 30(6) (1951), 307s-318s.

43. W.S. Pellini, "Principles of Fracture Safe Design - Part I", Ibid.,
     50(3) (1971), 91s-109s.

44. W.S. Pellini, "Principles of Fracture Safe Design - Part II", Ibid.,
     50(4) (1971), 147s-162s.

45. E.F. Nippes and W.F. Savage, "The Weldability of Ship Steels - A
     Study of the Effect of Travel Speed, Preheat Temperature and Arc
     Power Level on the Notch Toughness of Weld Metal and the Heat-
     Affected Zone", Ibid., 25(11) (1946), 776s-787s.

46. E.F. Nippes, L.L. Merrill, and W.F. Savage, "Cooling Rates in Arc
     Welds in ½-Inch Plate', Ibid., 28(11) (1949), 556s-564s.

47. E.F. Nippes and W.F. Savage, "Development of Specimen Simulating
     Weld Heat-Affected Zones", Ibid., 28(11) (1949), 534s-546s.

48. E.F. Nippes and W.F. Savage, "Tests of Specimens Simulating Weld
     Heat-Affected Zones", Ibid., 28(12) (1949), 599s-616s.

49. E.F. Nippes, H. Wawrousek, and W.L. Fleischmann, "Some Properties
     of of the Heat-Affected Zone in Arc Welded Type 347 Stainless Steel",
     Ibid., 34(4) (1955), 169s-182s.

50. E.F. Nippes et.al., "Time-Temperature Effect on Properties of Weld
     Heat-Affected Zone in Type 347 Stainless Steel", Ibid., 36(6) (1957),
     265s-270s.

51. E.F. Nippes and C.R. Sibley, "Impact Performance of Synthetically
     Reproduced Heat-Affected-Zone Microstructures in "T-1" Steel",
     Ibid., 35(10) (1956), 473s-480s.

52. E.F. Nippes, W.F. Savage, and R.J. Allio, "Studies of the Weld Heat-
     Affected Zone of T-1 Steel, Ibid., 36(12) (1957), 531s-540s.

53. W.F. Savage and W.A. Owczarski, "Microstructure and Notch Impact
     Behavior of a Welded Structural Steel", Ibid., 45(2) (1966), 55s-65s.

54. E.F. Nippes, W.F. Savage, and W.A. Owczarski, "The Heat-Affected Zone
     of Arc-Welded Ductile Iron", Ibid., 39(11) (1960), 465s-472s.

55. E.F. Nippes, W.F. Savage, and J.M. Paez, "Transformational Behavior
     of Mn-Mo Armor Steels , Ibid., 38(12) (1959), 475s-481s.

17

56.   E.F. Nippes, W.F. Savage, and J.M. Paez, "Impact Characteristics of Heat-Affected Zones in Mn-Mo Armor Steels", Ibid., 39(1) (1960), 31s-39s.

57.   W.F. Savage et.al., "An Investigation of the Hot Ductility of High Temperature Alloys", Ibid., 34(4) (1955), 183s-196s.

58.   E.F. Nippes, W.F. Savage and G. Grotke, "Further Studies of the Hot-Ductility of High-Temperature Alloys", Welding Research Council Bulletin No. 33, February 1957.

59.   J.J. Pepe and W.F. Savage, "Effects of Constitutional Liquation in Maraging Steel Weldments", Weld. J., Res. Suppl. 46(9) (1967), 411s-422s.

60.   J.J. Pepe and W.F. Savage, "The Weld Heat-Affected Zone of the 18 Ni Maraging Steels", Ibid., 49(12) (1970), 545s-553s.

61.   H.P. Granjon, "Studies on Cracking of and Transformation in Steels During Welding", Ibid., 41(1) (1962), 1s-11s.

62.   E.S. Szekeres, E.F. Nippes, and W.F. Savage, "A Study of Hydrogen-Induced Cracking in a Low-Alloy Steel", Ibid., 55(9) (1976), 276s-283s.

63.   A.R. Troiano, "The Role of Hydrogen and Other Interstitials on the Mechanical Behavior of Metals", ASM Trans., 52(1) (1960), 54-80.

64.   E.S. Szekeres, E.F. Nippes, and W.F. Savage, "A Study of Weld Inter-face Phenomena in a Low-Alloy Steel", Weld. J., Res. Suppl., 55(9) (1976), 260s-268s.

65.   H. Homma, E.F. Nippes, and W.F. Savage, "Hydrogen-Induced Cracking in HY-80 Weldments", Ibid., 55(11) (1976), 368s-376s.

66.   C.D. Beachem, "A New Model for Hydrogen-Assisted Cracking (Hydrogen Embrittlement)", Met. Trans., 3(2) (1972), 437-451.

67.   W.F. Savage, "The Effect of Rare-Earth Additions on Hydrogen-Induced Cracking in HY-80 Weldments", Sulfide Inclusions in Steel, No. 6 in Materials Technology Series, ASM, 1975.

68.   H. Granjon, "The Implants Method for Studying the Weldability of High Strength Steels", British Welding Journal, 1(11) (1969), 509-515.

69.   J.M. Sawhill, A.W. Dix, and W.F. Savage, "Modified Implant Test for Studying Delayed Cracking", Weld. J., Res. Suppl., 53(12) (1974), 554s-560s.

70.   Y. Tokunaga, E.F. Nippes, and W.F. Savage, "Hydrogen-Induced Cracking in HY-130 Steel Weldments", Ibid., 57(4) (1978), 118s-128s.

71.   M.A. Pellman, E.F. Nippes, and W.F. Savage, "Hydrogen-Induced Cracking in Soviet Steels Involved in USA/USSR Exchange Program", Ibid., 58(11) (1979), 315s-323s.

72.   E.L. Husa, E.F. Nippes, and W.F. Savage, "Hydrogen Assisted Cracking in HY-130 Weldments", Ibid., 61(8) (1982), 233s-241s.

73.   G.M. Goodwin, C.D. Lundin, and W.F. Savage, "An Effect of Shielding Gas on Penetration in Inconel Weldments", Ibid., 47(12) (1968), 313s & 322s.

74. G.M. Goodwin, E.F. Nippes, and W.F. Savage, "Effect of Minor Elements on Fusion Zone Dimensions of Inconel 600", Ibid., 6(4) (1977), 126s-132s.

75. C.R. Heiple, J. Roper, R.T. Stagner, and R.J. Aden, "Surface Element Effects on the Shape of GTA, Laser and Electron Beam Welds", Ibid., 62(3) (1983), 72s-77s.

76. A.H. Aronson and W.F. Savage, "Preferred Orientation in the Weld Fusion Zone", Ibid., 45(2) (1966), 85s-89s.

77. R.J. Hrubec, C.D. Lundin, and W.F. Savage, "Segregation and Hot Cracking in Low-Alloy Quench and Tempered Steels", Ibid., 47(9) (1968), 420s-425s.

78. F.J. Zanner, E.F. Nippes, and W.F. Savage, "Determination of Weld Puddle Configuration by Impact Decanting", Ibid., 57(7) (1978), 201s-210s.

79. "Minor Element Effects on Gas Tungsten Arc Weld Penetration", Final Report NSF Contract No. DMC-8208950, December 1986, pp.3-1 to 3-44, National Science Foundation, Division of Design, Manufacturing, and Computer-Integrated Engineering, Washington, DC 20550.

80 T.F. Chase, Jr. and W.F. Savage, "The Effect of Anode Composition on Tungsten Arc Characteristics", Weld. J., Res. Suppl., 50(11) (1971), 467s-473s.

81. K. Agusa, E.F. Nippes, and W.F. Savage, "Effect of Arc Force on Defect Formation in GTA Welding", Ibid., 58(7) (1979), 212s-224s.

82. R.D. Stout and R.P. Oates, "A Quantitative Test for Susceptibility to Lamellar Tearing", Ibid., 52(11) (1973), 481s-491s.

83. R.D. Vasudevan, A.W. Pense, and R.D. Stout, "A Field Weldability Test for Pipeline Steels, Part II", Ibid., 59(3) (1980), 76s-84s.

84. G.M. Goodwin, "Development of a New Hot-Cracking Test - The Sigmajig", Ibid., 66(2) (1987), 33s-38s.

A VIEWPOINT ON THE WELDABILITY OF MODERN STRUCTURAL STEELS

P R Kirkwood

British Gas plc
Engineering Research Station
P O Box 1LH Killingworth
Newcastle upon Tyne NE99 1LH
England

Abstract

The quest for increased toughness and improved weldability has resulted in lower carbon equivalents and cleaner steel making processes.

Such changes have progressively shifted attention from the common welding problems of earlier decades, for example, lamellar tearing and hydrogen induced HAZ cold cracking, to HAZ hardness and toughness problems.

In the context of offshore structural steels of the C-Si-Mn-Nb-Al type there is no doubt that initial reductions in CE from the 0.45/0.48 range to 0.41/0.44 have been beneficial, but with large tonnages now produced in the 0.35/0.40 CE range, direct observations by fabricators are causing certain fundamental assumptions to be rigorously re-examined.

This paper discusses the current controversy which has led, in some quarters, to the belief that 'modern' structural steels are more difficult to weld than their counterparts of the early seventies. Drawing on the results of extensive investigations, the author provides a personal viewpoint and interpretation of the available evidence, and concludes that lower CE steels are not always the panacea they appear to be at first sight. Many more subtle metallurgical factors are involved, and a tentative hypothesis, to explain the apparent paradox of recent observations, is forwarded.

Introduction

In the late 1960's the majority of European higher strength steels were of the carbon-manganese-niobium semi-killed or silicon-killed varieties, with occasional aluminium additions for enhanced toughness though usually in the absence of niobium. However, from around 1970 onwards, progressively increasing toughness requirements in thick section high strength steels led to the development of C-Si-Mn-Nb-Al compositions, particularly for offshore structure jacket fabrication.

The evolution and development of specifications for such steels has been adequately covered elsewhere (1), and it suffices to note here that yield strengths in excess of 345 N/mm² (50 Ksi) combined with longitudinal Charpy toughness requirements of about 40 Joules (30 ft-lbf) at -20°C were common. Additionally, transverse impact demands of the order of 27 Joules at -50°C soon became commonplace.

For various reasons there was strong preference for normalised steels, the aluminium additions being essential to lower impact transition temperatures at the slight expense of strength. Furthermore, the increased homogeneity and decreased anisotropy of properties could only be achieved with much lower sulphur levels. The necessary steelmaking improvements were readily at hand, since steels of similar chemistries with improved cleanliness had already been required by the Gas Industry for steels at the API 5LX 60 and 65 strength levels.

Perhaps the most important aspect of a structural steel's performance is its weldability, and while weldability of course means many things, the ability to resist hydrogen induced cold cracking in the weld heat affected zone is of paramount interest. Classically, a guide to such resistance is the material's carbon equivalent. In Europe, the most common carbon equivalent formula is the IIW one, namely

$$CE = C + \frac{Mn}{6} + \frac{Cr + Mo + V}{5} + \frac{Ni + Cu}{15}$$

and the British Standard (2) which is commonly used by fabricators to predict safe welding preheat levels relies on well established correlations between IIW CE and welding procedure variables. The predictive diagrams in the latter Standard are based on extensive work carried out at the British Welding Institute in the 1960's and, while the Standard has been updated since original publication, the predictive nomograms have remained unchanged, and rely heavily on information derived from steels with higher carbon equivalents, often in the range .46 to .58.

Although by 1970, carbon equivalent levels of steels in the strength range of interest had decreased dramatically, the increased use of aluminimum for improved toughness actually, temporarily, reversed this trend. It is a fact not widely appreciated that the C-Si-Mn-Nb-Al steels first used for major offshore structures in the North Sea were higher in carbon equivalent (up to 0.46) than the equivalent British Standard, lower toughness, structural steels at the same strength level. The precise reasons for this are adequately covered in other publications (1). Equally, it is often not appreciated that low sulphur steels were fairly readily available for the offshore projects, even in 1972, and as shown in Table 1, node steels with sulphur levels less than .01% were being specified even then.

However, the remainder of the '70's decade and the early 1980's have seen considerable advances resulting in improved steelmaking controls and process developments and the introduction of a wide range of new techniques such as vacuum degassing, double and triple slagging, secondary refinement techniques, ladle processing, etc. Furthermore, there has been a steady replacement of ingot cast steel with the continuously cast product, and significant changes in rolling schedules, heat treatment, etc, all of which have contributed to a much improved situation at the present time.

Sulphur levels, as indicated in Table 1 and Figure 1, have now further decreased and the required strength levels are now readily developed at much lower carbon equivalent values (see Figure 2). In fact, much of the decrease in carbon equivalent is attributable to decreases in carbon itself, and this point may be of considerable significance, as discussed later. There are, of course, other significant changes, many of which will be referred to later in the paper, but it is fair to say that nitrogen levels today are much lower and, of course, oxygen is closely controlled to remarkably low values.

While the weldability predictive approach referred to earlier adequately stood the test of time through the 1970's, more recent years have brought to the fore considerable controversy concerning the performance of the so called "new steels" from advanced steelmaking processes. It has, for many years, been suspected that lower sulphur steels are perhaps slightly harder to weld, both in respect of their propensity to develop higher hardnesses and with respect to their resistance to hydrogen induced cold cracking. The evidence in the literature is conflicting, but a relatively recent review of the subject(3) has concluded that the effect is a real one and that, for example, decreasing sulphur level from .025% to .005% may be equivalent to an increase in carbon equivalent of about .03, which in turn could lead to increased preheat levels of between 50 and 100°C (depending on which precise welding circumstances prevailed). However, the situation is by no means clear cut, since some individual references refer solely to increased hardness levels with lower sulphur, while others claim significant effects on resistance to hydrogen induced cracking in the heat affected zone. Both of these factors are important, since while the primary objective must be to produce crack-free fabrications, there is a trend to require ever decreasing maximum hardness levels in fabricated steel work, and there are well known examples of recent major offshore specifications where levels of Vickers Hardness in excess of 300HV 10 were not permitted. Such levels of hardness in steels of the generic type under consideration are extremely difficult to achieve under practically relevant welding conditions, and it is this trend which is primarily responsible for the move towards lower carbon and carbon equivalent materials in the hope that such changes will lead to the desired welding performance.

Within the last two years, the welding literature has abounded with claims and counter claims concerning the weldability of so called modern cleaner steels (4,5) and, more specifically, there have been direct accusations that lower sulphur steels are significantly more difficult to fabricate than the materials commonly being encountered by fabricators in the early 1970's (6).

British Gas has made its own contribution to this subject area and indeed the results of a fairly extensive investigation were published in 1984 and 1985 (7,8). That work did not appear to support the contention that the newer steels were more difficult to weld, but did implicate lower carbon equivalent materials in general, suggesting that the British Standard weldability prediction method became increasingly inaccurate as carbon equivalent decreased.

The present paper reviews the salient feature of earlier studies, both by British Gas and other workers, and analyses in more detail the various metallurgical observations made on the steels previously studied by Boothby and Rodgerson (7).

Steel Compositions Studied

The steels considered in the present paper are detailed in Table 2; they are all materials supplied to major offshore platform fabrication contracts between 1975 and 1982. The steels have a range of carbon contents from .10 to .17, varying degrees of microalloying additions, and carbon equivalents ranging from .36 to .43. The highest sulphur level studied is .016 and the majority of the materials would be classed, on most standards, as low sulphur with values in the range .002 to .009.

The experimental work carried out by British Gas on these materials consisted of comprehensive series of controlled thermal severity (CTS) tests used to define critical heat inputs to avoid HAZ hydrogen induced cold cracking, and high speed dilatometry work necessary to establish continuous cooling transformation diagrams and thus key information on martensite start (Ms) temperatures and critical times such as TC_1 (the cooling time from 1300°C below which a fully martensitic microstructure was always developed).

This paper does not describe any of that work in great detail, but draws on the important features of it to develop the author's own viewpoint on the subject, with the aim of highlighting additional factors which have, perhaps, been overlooked by earlier workers.

Hardenability

The hardenability of a steel can be defined in several ways. The classical definition is based on the 'critical cooling rate' from a continuous cooling transformation (CCT) diagram. The latter parameter is the slowest cooling rate at which transformation to a fully martensitic structure is obtained. Thus steels with high hardenability have slow critical cooling rates. However, in the welding context, it is more appropriate to think of hardenability in relation to the steel's ability to develop specific hardness levels for a given set of thermal conditions. Thus, for example, a steel which more readily develops hardnesses in excess of 375HV 10 in the heat affected zone at a given welding heat input would be considered to be of high hardenability.

Boothby and Rodgerson's data (7) for the steels studied is shown for their CTS tests in Figure 3. It is immediately obvious that a wide range of so called hardenability has been exhibited. This point is emphasised by Figure 4 which reveals that the ability of the different steels to produce martensite varies very significantly. Hardenability is classically considered to be controlled by chemistry and austenite grain size, and thus these factors will be examined more thoroughly later in this paper.

The Welding Institute predictive method relies in part on the assumption that there exists a clear relationship between thermal parameters required to produce certain hardness levels and IIW carbon equivalent (modified by the addition of a factor Si/6 to fully describe the steel's hardening "chemistry"). This information has been available for many years and was recently reviewed by Boothby (9). It was therefore prudent to assess whether or not the data from the steels in this study followed the expected pattern of behaviour. Figure 5, from one of the earlier publications (8),

24

shows clearly that this is not the case; indeed there are significant deviations apparently unrelated to steel vintage but perhaps slightly accentuated at lower carbon equivalent.

The continuous cooling transformation studies provided further valuable information. In addition to providing full CCT diagrams, it was possible to derive data on the cooling rates necessary to achieve a whole range of hardness levels for each steel examined. It was shown (7) that the CCT approach could be readily compared with the CTS data and that it was therefore a very valuable technique for the study of steel weldability. The extensive data can be interrogated in a number of different ways but, for the purposes of this paper, I propose initially to address the question of steel cleanliness.

Sulphur and Hardenability

The earlier review by Hart (3) suggested that the so called low sulphur phenomenon depended more on a decrease in the number of inclusions than the actual levels of the elements sulphur and oxygen. Thus, Boothby and Rodgerson (7) attempted to measure, using optical metallography, the numbers of inclusions per unit area within the matrix of the various steel microstructures studied. Using their measurements, which admittedly were unable to adequately detect inclusions much less than 0.3 µm in size, they attempted to derive various correlations between the cooling rates necessary to achieve a whole range of different hardness levels and a range of chemistry variables such as carbon, carbon equivalent and Pcm. Figure 6 shows one of their diagrams from which it would certainly be impossible to conclude that higher numbers of inclusions meant lower hardenability. If anything, the reverse appeared to apply.

However, the volume fraction of inclusions in a steel can be predicted by the relationship $Vf = 5.5$ (Sulphur + Oxygen) (10) and since in the context of the steels currently under study, oxygen level is consistently low, this means that higher sulphur levels suggest (or predict) higher volume fractions of inclusions. Figure 7 shows that there is no clear correlation between calculated Vf and the measured numbers of inclusions per unit area, and it is the present author's opinion that this probably simply results from the extreme experimental difficulty in accurately counting numbers of inclusions, particularly when these are very fine.

Thus it was considered relevant to re-examine the earlier data and to examine the possible effects of calculated Vf. The results of this re-analysis for the 375 hardness level are shown in Figures 8, 9 and 10. The calculated inclusion volume fractions are given in Table 3, and for the purposes of the graphical presentations in Figures 8, 9 and 10, dirty steels are taken to be those with Vf greater than or equal to .08, and clean steels those with Vf less than .03. The others are naturally classed as medium cleanliness. Figures 8-10 clearly suggest that steels with greater than .01% sulphur, ie those with Vf greater than .08, are in general significantly less hardenable, ie they require much faster cooling rates to achieve the 375 hardness level. This apparent effect is clearly seen whether the abscissa is taken as carbon equivalent, Pcm or carbon content itself. The diverging behaviour of the higher sulphur steels appears to be markedly accentuated at lower carbon or lower carbon equivalents. Interestingly, considering the medium and low cleanliness steels, it does not appear to be possible to separate these out. A comprehensive study of all the data has revealed that the "sulphur effect" on hardness is greater at higher hardness levels and of much less significance when considering hardnesses at the 300 level, see for example Figure 11.

Summarising these data, it appears that there is a sulphur effect on hardenability, or at least there is a significant and measurable difference between the performance of steels in the .002 to .009% range and those with sulphur levels above .01% approximately. However, the extent and importance of the effect may vary considerably depending on the hardness level of interest to a particular fabrication specification and, of even greater importance, the effect appears to be negligible at higher chemistries of the type frequently encountered in the early to mid 1970's.

But what about cold cracking susceptibility?

Hydrogen Induced Cold Cracking

Boothby and Rodgerson's data from the controlled thermal severity tests provided a rather interesting pattern of results, see Figure 12. The critical heat input required to avoid hydrogen induced cracking appeared to increase as IIW carbon equivalent decreased. When the predictive diagram lines from the British Welding Institute and British Standard methods are superimposed upon the data, see Figure 13, it is immediately apparent that the cold cracking behaviour of the lower carbon equivalent steels is entirely inadequately predicted by either method. Clearly and significantly, deviation from prediction is not restricted to new steels, low sulphur steels or any immediately apparent steelmaking variable, such as for example nitrogen level. However, once again we see that lower carbon equivalent materials are exhibiting maximum deviation from predicted behaviour.

Why is the prediction failing?

The British Standard method relies on an assumed relationship between carbon equivalent and hardenability which we have already seen does not exist for the steels being studied, and secondly, it relies on the concept that cracking occurs at a critical hardness level which is assumed to be independent of carbon equivalent.

What about this latter point?

Figure 14 shows that for the steels in the present investigation, there appears to be a decrease in critical hardness for cracking at lower carbon equivalent levels. Figure 15 emphasises this point by drawing on data from the open literature to extrapolate the plot to higher carbon equivalent levels. This same trend has been previously reported by other workers (11).

Thus both foundation stones of the predictive methodology have been damaged. There is ample evidence that, as carbon equivalent decreases, hardenability does not always drop as anticipated, and is not therefore readily predicted from any obvious chemistry based relationship. Furthermore, for whatever reason, the critical hardness for cracking is not independent of carbon equivalent and clearly decreases quite dramatically as carbon equivalent in turn drops. This latter point is, of course, highly significant since it implies that at very low carbon equivalents, towards the bottom end of the spectrum currently being produced by some steelmakers, the critical hardness for cracking may be, in some circumstances, as low as 300HV 10.

There is no evidence from this data that low sulphur steels are, in themselves, more difficult to weld with respect to the avoidance of hydrogen induced cold cracking. But, since we have already observed that as sulphur levels decrease so hardness levels increase, particularly at lower carbon equivalents, and since the critical hardness for cracking also decreases with reduction in the latter parameter, it is perhaps not unreasonable to assume,

26

as suggested by Hart (3), that the sulphur effect could be significant in some circumstances.

Clearly, however, there is something still missing in relation to possible reasons for the deviant behaviour of lower carbon equivalent steels, and this paper now proceeds to look for clues within the information generated in Boothby and Rodgerson's extensive work.

Metallurgical Factors Affecting
Transformation Characteristics and Hardenability

During the course of their work, Boothby and Rodgerson made various measurements of austenite gamma grain size, both in the dilatometer samples used for the derivation of the continuous cooling transformation diagrams and at selected heat input levels in the controlled thermal severity test blocks. They observed a considerable degree of correlation between the two sets of data (when comparable cooling rates were considered) but, for the purposes of my own overview, I will concentrate on data derived from the more slowly cooled CCT dilatometer samples. The gamma grain sizes measured are given in Table 3 beside the calculated inclusion volume fractions referred to earlier. It is immediately apparent that a significant range of gamma grain size values was observed. In the subsequent analysis, I have initially discounted Steel K since it contains .026 titanium, though it will be clearly seen later that its behaviour is, in fact, consistent with the final conclusions drawn rather than contradictory.

As indicated earlier, hardenability is controlled by two important variables; composition and austenite grain size. Hardenability usually increases with increasing austenite grain size since this latter change results in a decrease in grain boundary nucleating area. Alloying elements, on the other hand, slow down transformation kinetics and influence the rate of austenite decomposition into ferrite, pearlite or bainite. Thus the range of austenite grain sizes recorded is considered to be very significant and indeed, as shown in Figure 16, the critical cooling time below which transformation to 100% martensite always occurred, appears to be loosely correlatable with gamma grain size. Higher gamma grain sizes give increased TC_1 values.

Now, in the welding context, it has been well documented that the level of microalloying elements can be quite important in relation to grain coarsening behaviour, and indeed Hannerz has shown (12), see Figure 17, that, for example, niobium level can be of considerable significance, particularly for low heat input welding processes. He observed that higher niobium levels led to significantly finer austenite grain sizes, presumably via a grain pinning process. Easterling (13) has developed these ideas and derived various predictive models, concluding that precipitate distributions of various nitride and carbo-nitride particles can have greater or lesser efficiencies as grain size controllers, see Figure 18. With the exception of titanium-nitride, it is clear that, as observed by Hannerz, niobium carbo-nitride is perhaps the most efficient precipitate to restrict austenite grain growth. It occurred to me that the introduction, in the mid Seventies, of restrictions on niobium and vanadium levels in offshore structural steel specifications, might well have been more significant than first met the eye. Figure 19 shows typical data from 1972 and 1981 production of niobium steels, the latter being fairly representative even of today's practice. Clearly there has indeed been a major and significant change in niobium levels. Perhaps even more importantly in the modern day production, where the steelmaker is working to maxima in the .04 to .05 range, the typical product with a mean around .03 will inevitably have a tail of very low niobium casts where, perhaps, levels even less than .01

will occasionally be encountered. Work carried out many years ago by Irvine et al (14) has shown the possible important effects of different combinations of carbon, nitrogen and niobium level in respect of precipitate dissolution temperature (for niobium carbo–nitrides) and it is quite clear from Figure 20 that changes in niobium level of the type which have been experienced since the early Seventies, combined with a trend towards lower carbon content, could indeed have very significant effects on dissolution temperature, and thus gamma grain size. In Figures 21, 22 and 23, I have therefore looked at the influence of niobium, nitrogen and carbon on gamma grain size in the steels studied by Boothby and Rodgerson. While there is considerable scatter, it is apparent that higher niobium levels do lead to smaller gamma grain size, as observed by Hannerz. Superimposed on these three latter graphs are the calculated aluminium/nitrogen ratios for the steels studied, and it is again significant that higher aluminium nitrogen ratios, even at low niobium levels, appear to be associated with larger gamma grain sizes.

This latter observation is perhaps not too surprising since it is well known that higher Al/N ratios can lead to excessive sized aluminium nitride particles which themselves may partially destroy the grain size control in niobium aluminium steels, and thus while the current trend in offshore steel specifications is to stipulate minimum levels of aluminium–nitrogen ratio which must be achieved, it may be necessary in future to avoid excessively high levels, for example, 8/1 or 10/1.

It is now worth looking briefly at the behaviour of steel K, ie the titanium treated product. As expected, this material exhibited the smallest austenite grain sizes in the heat affected zone of the CTS tests and, as shown in Table 3, also in the CCT samples. Referring back to Figure 16, steel K's low TC_1 seems to be directly related to its ability to resist austenite grain growth during welding. Thus, by classical definition, this steel is of low hardenability. However, its carbon equivalent, at .43. is the highest of all the materials studied, and in terms of its ability to generate high hardnesses, its behaviour is dominated by this latter factor. Thus, in Figure 8 we note that the cooling rate required to generate 375HV is fairly low. This observation is very significant since, on this basis, the steel would clearly be classed as having high harden-ability. We must therefore be very careful how we define hardenability when we are discussing weldability, and in any case, there is certainly no correlation between 'hardenability', as defined by either criteria, and cold cracking susceptibility. Indeed, this high titanium steel's behaviour is, as shown in Figure 13, very adequately predicted by the existing British Standard technique, ie its propensity to harden is relatively unimportant.

Overview

On the basis of the information presented in this paper, the following important observations can be made. While there is indeed evidence that lower sulphur steels may have an increased tendency to produce higher hardness levels, this does not necessarily imply increased hardenability in the classical context. The 'sulphur effect' appears to increase in magnitude as carbon equivalent decreases, and to be of maximum importance when considering the ability of each steel to generate hardness levels above 375HV. There is no evidence that the lower sulphur steels in this study are more susceptible to hydrogen induced cold cracking.

Lower carbon equivalent levels are, however, implicated in two ways. They clearly show maximum deviation from the predicted behaviour in respect of cold cracking susceptibility, and while it is not possible to precisely define the reason for this, it is obviously associated with lower critical hardnesses required for cracking as carbon equivalent decreases, and with the nature of the transformation products produced in certain low carbon equivalent materials. Furthermore, lower carbon equivalent steels exhibit a considerable scatter in hardenability. It has been suggested that this latter variability is probably related to subtle metallurgical changes such as lower niobium levels and variations in aluminium nitrogen ratio. More specifically, the data suggests that the slightly higher niobium levels encountered in the mid to early Seventies may have provided better heat affected zone austenite grain size control, particularly at lower heat inputs, and that Al/N ratios, which are now often excessively high, for example 8/1 or greater, may be very detrimental.

In summary, the author's viewpoint is as follows.

The higher carbon equivalent steels of the late Sixties and very early Seventies possessed metallurgical characteristics and weldabilities totally dictated by their carbon and carbon equivalent values. Naturally, gamma grain size must have played some role, but this was effectively minimised by the dominant effect of overall chemistry. As carbon and carbon equivalent levels have been progressively decreased, the importance of other metallurgical factors has increased. Thus, effects of cleanliness (ie sulphur and/or oxygen levels) and gamma grain size variations become more and more important. In turn, since gamma grain size is strongly influenced by the presence of particular microalloying elements, if we continue to press towards lower CE values, it will be necessary to tailor steel compositions with extreme care. In this context, the trend towards lower niobium levels must be seriously questioned, as must the higher aluminium-nitrogen levels often encountered in modern offshore structural steels. We may have been too preoccupied with possible sulphur effects and have, in consequence, turned a blind eye to the increasingly dominant role of gamma grain size on transformation behaviour as carbon equivalent decreases below 0.39.

Niobium - Transformation and HAZ Toughness

The proposals above to consider increasing niobium levels will, of course, be met with considerable resistance in certain quarters. Indeed, as indicated earlier, many current offshore steel specifications do not permit niobium levels greater than .04 or .05%. The logic and detailed reasoning behind such restrictions is often unclear, but in general, appears to result from adverse literature which was prevalent in the early 1970's. During that period niobium and vanadium were often accused of being detrimental to weld metal and, more specifically, HAZ toughness, especially in the stress relieved condition. While there is no doubt that some genuine effects in support of this contention were recorded in high carbon and high carbon equivalent materials, the literature portrayed an imbalanced view of the situation.

Interestingly, the modern structural steels have also been suspected of exhibiting variable, and in some instances, unacceptably low HAZ toughness values (15, 16). Since the timescale over which such observations have been made coincides with that relevant to the hardenability and cold cracking controversy with which this paper has been primarily concerned, then speculation that the various phenomena are interrelated is inevitable. HAZ toughness, at least in the as-welded condition, is primarily dictated by microstructure and thus if, as suspected, the role of gamma grain size becomes increasingly important as carbon equivalent decreases, then it is perhaps not too surprising that 'toughness variability' is also observed in certain modern offshore structural steels. Variable HAZ toughness is currently the subject of major collaborative research work at the

29

UK Welding Institute, and such programmes may yet throw considerable light on the phenomenon.

The potential importance of changes in niobium level over the past decade has been adequately highlighted earlier, but there may be more subtle factors to unravel as our knowledge and understanding continue to grow. However, it is the present author's opinion, based on a detailed review published in 1981 (17), that there would be no insurmountable difficulties associated with a move back to higher niobium levels in steels of the class currently under discussion. Indeed, if the overall steel chemistry is tailored correctly, with a view to the appropriate end application in each instance, there is no reason why much higher levels of microalloying elements cannot be considered.

The author would like to thank British Gas plc for permission to publish this paper and his many colleagues at the Engineering Research Station for their valued help in the preparation of this contribution.

References

(1) J G Garland and P R Kirkwood.
The Selection and Fabrication of Steels for Offshore Structures.
Proceedings of Financial Times/Petroleum Times, International
Conference, London, October 1974.

(2) BS 5135: 1984: Specification for Process of Arc Welding of Carbon and
Carbon Manganese Steels, British Standards Institution, London.

(3) P H M Hart.
The Influence of Steel Cleanliness on HAZ Hydrogen Cracking: The
Present Position.
IIW Doc IX-1308-84 (Also the Welding Institute Research Bulletin,
March 1986).

(4) McKeown et al.
The Weldability of Low Sulphur Steels.
Metal Construction, November 1983.

(5) Boniszewski, McKeown and Patchett.
An Exchange of Open Letters to the Editor of Metal Construction,
published in Metal Construction in January 1984, February 1984,
January 1985 and May 1985.

(6) T Boniszewski and T Keeler.
HAZ Hardness Control in C-Mn Microalloyed Structural Steels.
Metal Construction, October 1984.

(7) P J Boothby and P Rodgerson.
Offshore Structural Steels: Hardenability and Weldability.
Proceedings of Welding Institute Conference "Towards Rational and
Economic Fabrication of Offshore Structures - Overcoming the Obstacles".
London, England, 22-23 November 1984.

(8) P J Boothby.
Weldability of Offshore Structural Steels.
Metal Construction, August and September 1985.

(9) P J Boothby.
Predicting Hardness in Steel HAZ's.
Metal Construction, June 1985.

(10) D J Widgery.
Toughness of Weld Metal.
Welding Institute Seminar, Newcastle, England, October 1977.

(11) P H M Hart.
Low Sulphur Levels in C-Mn Steels and their Effects on HAZ Hardenability
and Hydrogen Cracking.
International Conference, Trends in Steels and Consumables for Welding.
London, November 1978.

(12) N E Hannerz.
Effect of Niobium on HAZ Ductility in Constructional HT Steels.
Welding Journal, Volume 54, No 5, 1975 Research Supplement.

(13) K Easterling.
Introduction to the Physical Metallurgy of Welding - Chapter 3, The
Heat Affected Zone.
Butterworths Monographs in Metals, Butterworths and Co Ltd, 1983.

(14) K Irvine et al.
      Journal of the Iron and Steel Institute, 2, 1967.

(15) R E Dolby.
      Steels for Offshore Construction.
      Proceedings of Welding Institute Conference "Towards Rational and
      Economic Fabrication of Offshore Structures – Overcoming the Obstacles".
      London, England, 22-23 November 1986.

(16) O Grong and O M Akselsen.
      HAZ Toughness of Microalloyed Steels for Offshore.
      Metal Construction, September 1986.

(17) P R Kirkwood.
      Welding of Niobium Containing Microalloyed Steels.
      Proceedings of International Symposium "Niobium '81", San Francisco,
      November 8-11 1981. A publication of the Metallurgical Society of
      AIME.

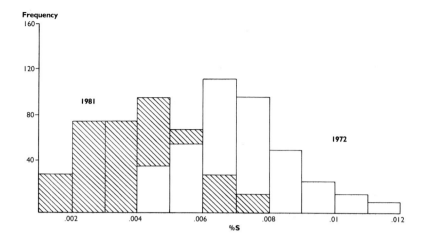

Figure 1 - Change in sulphur levels in
production of C-Si-Mn-Nb-Al steels

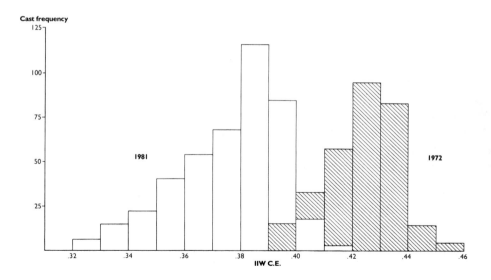

Figure 2 - Changes in carbon equivalent level
in production of C-Si-Mn-Nb-Al steels.

33

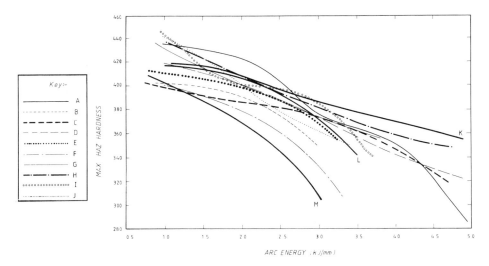

Figure 3 - Hardenability data from
CTS tests made without preheat.

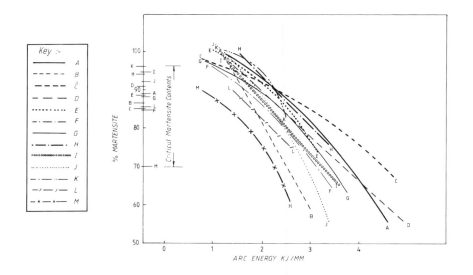

Figure 4 - Martensite content v arc energy
for CTS test without preheat.

34

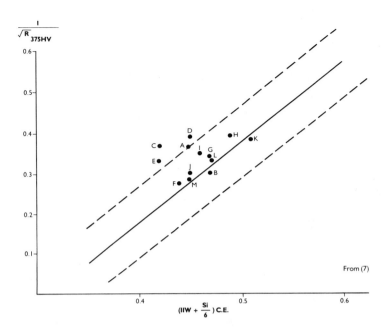

Figure 5 – Comparison of CTS hardness data with
scatter band of published results at 375 HV.

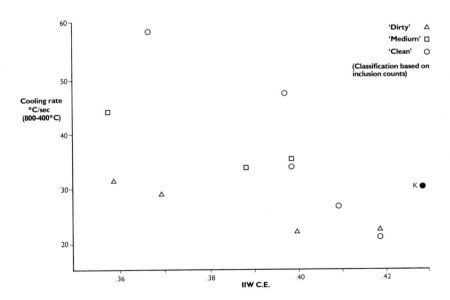

Figure 6 – CCT cooling rate required to achieve
375 HV10, versus carbon equivalent.

35

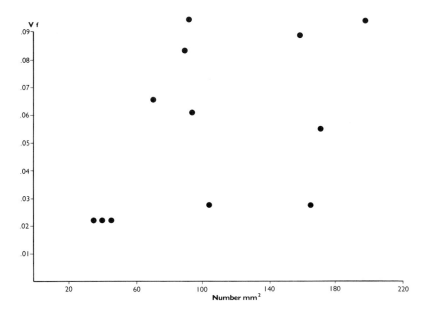

Figure 7 - Calculated volume fraction of inclusions
versus measured number/mm² (⩾ 0.3 μm).

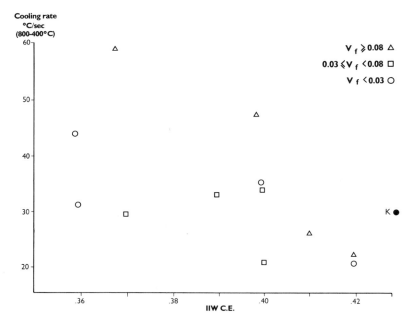

Figure 8 - CCT cooling rate required to achieve
375 HV10, versus carbon equivalent.

36

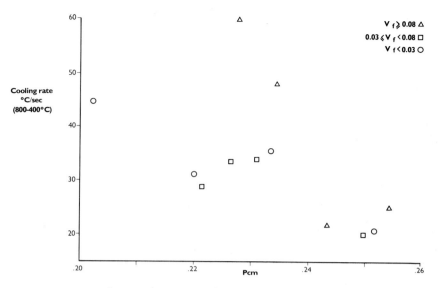

Figure 9 - CCT cooling rate required to achieve
375 HV10, versus Ito Bessyo carbon equivalent.

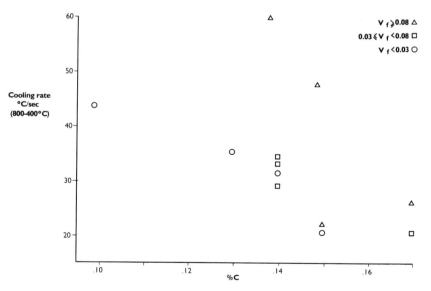

Figure 10 - CCT cooling rate required to achieve
375 HV10, versus carbon content.

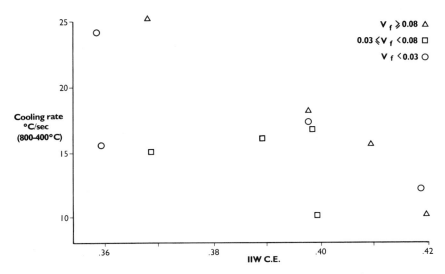

Figure 11 - CCT cooling rate required to achieve
300 HV10, versus carbon equivalent.

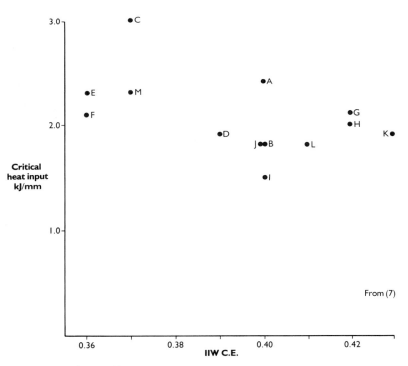

Figure 12 - CTS results - 90mm joint combined thickness.
Electrode hydrogen equivalent to Scale 'B'
conditions with reference to BS 5135.

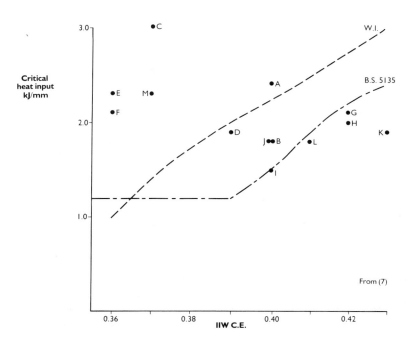

Figure 13 - CTS test results from Figure 12 with
Welding Institute and BS 5135 predictions superimposed.

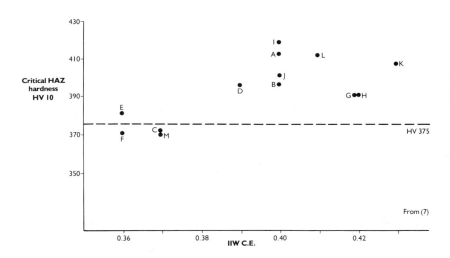

Figure 14 - Critical hardness from CTS tests
versus IIW carbon equivalent.

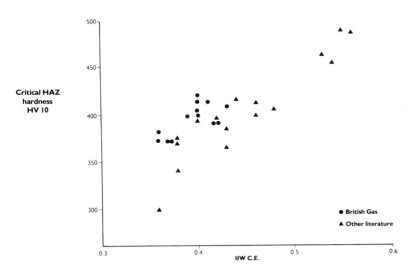

Figure 15 - CTS, critical hardness data from
Figure 14 with open literature information
superimposed (BS 5135: Scale 'B' hydrogen levels).

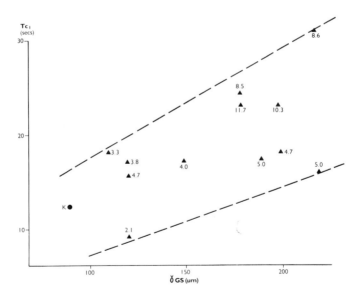

Figure 16 - The critical cooling time (Tc_1)
below which a 100% martensitic microstructure
is observed, versus gamma grain size.
(Al/N ratios sumperimposed)

40

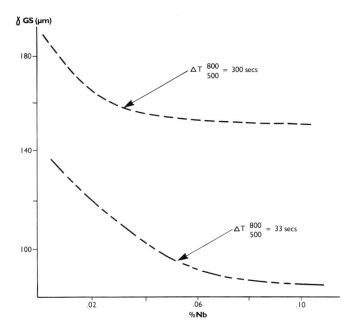

Figure 17 - The effect of niobium on heat affected zone gamma grain size after Hannerz (12).

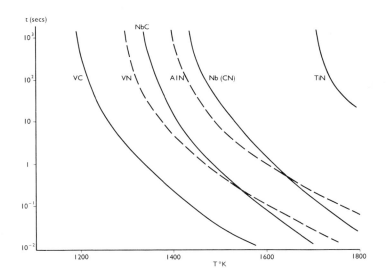

Figure 18 - Time/temperature precipitate dissolution data, after Easterling (13).

41

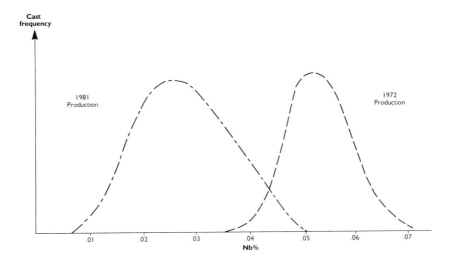

Figure 19 - Changes in niobium level in the production of C-Si-Mn-Nb-Al steels.

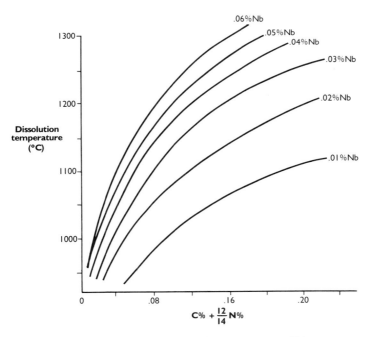

Figure 20 - Niobium-carbonitride solubility curves in austenite, after Irvine et al (14).

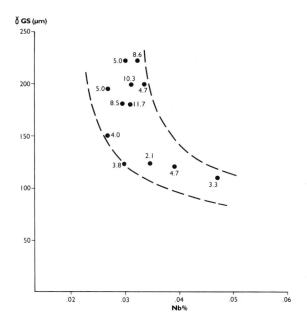

Figure 21 - CCT, gamma grain size versus
niobium content (Al/N ratios superimposed).

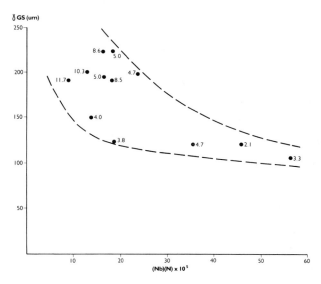

Figure 22 - CCT, gamma grain size versus
a function of niobium and nitrogen
(Al/N ratios superimposed).

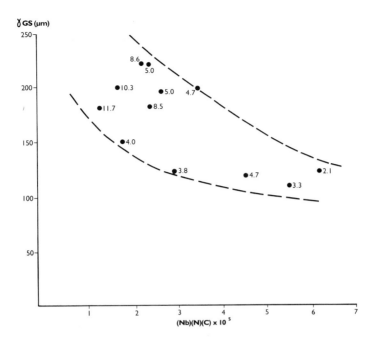

Figure 23 – CCT, gamma grain size versus
a function of niobium, nitrogen and carbon
(Al/N ratios superimposed).

MODELLING THE WELDABILITY OF STRUCTURAL STEELS

K. E. Easterling

School of Materials Science & Engineering
University of New South Wales
Sydney, Australia.

Abstract

Software, suitable for use on an IBM PC/AC microcomputer, has been developed capable of giving fairly good prediction of the changes in microstructure and properties when fusion welding various steels including the microalloyed grades. In this paper, details of this program, including the equations used for heat flow from the welding arc, phase transformation, precipitate dissolution and hardness changes are discussed with emphasis on the dissolution of complex (multi-component) carbonitrides. In addition, the effect of hydrogen in the weld is briefly discussed and a simple model for predicting hydrogen distribution is proposed.

Introduction

This paper is concerned with mathematical modelling of the changes in microstructure and properties that occur when fusion welding structural steels. Figure 1 shows a computer simulated map of the microstructural changes that occur in a Cr-Mo steel, based on the modelling procedure detailed in references 1 and 2. The changes in grain size observed in Figure 1(a) are plotted out in Figure 1(b) in a graphical form that can be compared with a relevant micrograph of the HAZ.

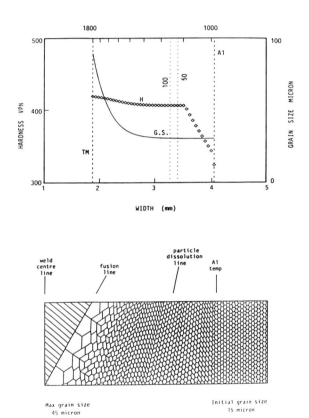

Figure 1 - *Computed grain size/hardness data for a TIG butt weld of a Cr-Mo-steel. Heat input:7,5 kW, no pre-heat. Note the uniform high hardness in the grain growth region due to martensite/bainite products. The lower diagram, also computer-generated, provides a useful "physical picture" of the HAZ.*

46

The changes in grain size and hardness are computed by taking into consideration a number of input data, including the initial grain size of the steel, the electrode size and type, the welding process and its efficiency, the composition of the steel, and the type of weld joint. In the first place it is essential to obtain a good model for the heat flow due to the moving welding arc. In the present work a simplified equation modified from that of Rosenthal(3) is used. If the temperature changes around the moving arc are known, then phenomena such as precipitate dissolution, grain growth, eutectoidal phase transformations, and reprecipitation can all be calculated, provided good equations are available in turn for each of these mechanisms. In practical steels it is difficult to obtain kinetic constants that are accurate and in the present work a solution is adopted in which equations describing the various grain growth and precipitation phenomena can be modified as necessary on the basis of a suitable experiment. In most cases a single bead-on-plate weld is sufficient to measure, e.g. grain growth and distance of the fusion line and A_1 temperatures from the centre line of the weld, etc., in order to check the predictions made by the theoretical equations. Changes in hardness as a function of microstructural constituent, are estimated semi-empirically on the basis of a number of measurements carried out by reference(4), in which the volume fraction of the constituent is related to the cooling time, T, of the weld thermal cycle. An example of this relationship is shown in Figure 2 and it is seen that the constituents concerned are martensite, bainite, pearlite and ferrite. By estimating, from the weld thermal cycle, the volume fractions of the various constituents, and knowing their average hardnesses, the final hardness can be estimated by a rule of mixtures. This works fairly well in practice and an example of experimental measurements made on a niobium microalloyed steel of hardness and grain growth for two heat inputs are shown in Figure 3. The shaded vertical bands in this Figure refer to the 50% and 100% dissolution temperatures of niobium carbide for the steel composition given.

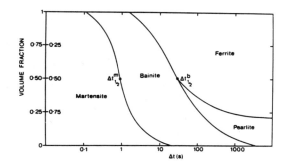

Figure 2 - *This diagram gives the relationship between the amount of the various eutectoidal decomposition products in the HAZ and the cooling rate of the weld thermal cycle. It is based on experiments of a large number of different steels (after Inagaki and Sekiguchi, ref. 4).*

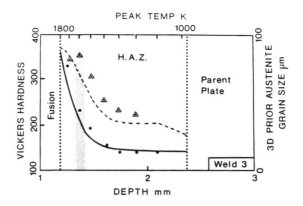

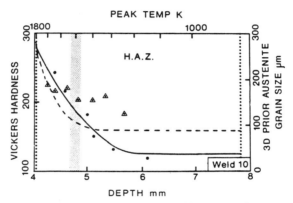

Figure 3 - *Computed grain size/hardness diagrams of a Nb-microalloyed steel for two different heat input welds. The experimental measurements of grain size and hardness for the two welds are shown and are seen to agree well with the predictions. (After Ion et al., ref. 2).*

In the present state of the art, computer programs for use on IBM PC type machines have been developed and these can be used such as to allow interaction with the welding engineer. For example by giving details of the various weld input data such as electrode size, welding process, composition of steel, etc., he can obtain a number of output data such as those shown above. The program in addition contains data of a number of different types of steels, together with appropriate experimental input data. On this basis the welding engineer can in principle obtain HAZ microstructural and hardness data from a large number of different types of welding processes, steels and welding joints.

A limitation of the present model is that it is mainly drawn up on the basis of simple binary carbonitride precipitate systems. However, most structural steels contain a number of different alloying elements which result in fairly complex carbonitride precipitates. In such cases, the present model may not be sufficiently accurate. Furthermore, it is desirable to be able to use the present output data in terms of grain size, hardness, type of weld, and electrode, etc., in order to estimate the amount of hydrogen segregation and hence susceptibility to cold cracking of a given weld. If the initial hydrogen amount allowed into the weld metal is estimated, then in principle, knowing the grain size, the cooling cycle of the weld, the type of microstructure, type of precipitates present, etc., it should be possible to estimate the approximate hydrogen segregation levels in the heat affected zone. On this basis a warning could be built into the system such as to give information on e.g. maximum hardness allowed in the weld if hydrogen cracking is to be avoided. Alternatively a suggestion might be given in terms of the type of electrode or the amount of preheat recommended. The main objective of the present paper is to discuss in more detail the effect of having complex carbonitrides present in the steel, and how the weld thermal cycle is going to affect the stability, and hence control grain growth in, the heat affected zone. Some pointers as to how the hydrogen segregation model could be set up will also be formulated.

Stability of Complex Carbonitrides in Structural Steels during Weld Thermal Cycling

High strength low alloy steels typically contain alloying elements such as aluminium, niobium, vanadium and titanium, which can combine with interstitial alloying elements such as carbon and nitrogen, to precipitate a very fine dispersion of complex carbonitrides. At any given temperature, if the carbonitride is in thermodynamical equalibrium, this composition can be as follows:

$$M^m C^n = [Al^m Nb^n V^o Ti^p][C^q N^r]$$

In this particle composition, the suffix's m,n,o,p,q,r, give the relative proportions of these constituents in the particle. In practice, the composition of the particle in the steel depends on the steel's thermal history, including preheating, hot rolling, normalising treatment, and welding procedures. The compositions of these particles determines their stability and temperatures of dissolution and precipitation during any given heat treatment, so that if these compositions can be estimated theoretically, for example from thermodynamic and kinetic modelling, then in principle it would be possible to estimate their role in allowing grain growth during a weld thermal cycle. In the case of a simple binary system, the particle solubility equation is given as:

$$\log [M^m C^n] = C_1 - \frac{C_2}{T} \tag{1}$$

where C_1 and C_2 are experimentally determined constants, and T refers to temperature. Thus the solubility temperature of a particle is given;

$$T_s = \frac{C_2}{C_1 - \log[M^m C^n]} \tag{2}$$

49

T$_s$ refers t) the temperature at which these particles will dissolve or reprecipitate during cooling. It is seen from equation 1 that the larger the amount of metallic and interstitial constituents present, the higher the solubility temperature will be. This can be understood if it is imagined that all of the alloying constituents are in solution at a given temperature in the austenite field. On cooling the steel slowly so as to maintain thermodynamic equalibrium at any given temperature, a temperature will be reached at which precipitation occurs. This is T$_s$. If there were larger amounts of alloying constituents present in solution in the austenite, then the temperature of solubility would be higher on the basis that the austenite would be more supersaturated, and this is the basis of equations (1) and (2). The thermodynamics and kinetics of complex carbonitride systems are complicated by the fact that the fusion distances of the different alloying elements are also different. This means in practice that if, for example, a titanium-niobium carbonitride is going into solution in the austenite then the temperature of dissolution is determined by how quickly individual elements can diffuse to and from the particle, while at the same time maintaining thermodynamic equalibrium within the particle and adjusting compositional changes as the temperature changes. This situation is particularly difficult to determine theoretically in welding in view of the very rapid nature of the thermal cycle involved. However, attempts to estimate the compositions and behaviour of complex carbonitrides during welding have been made(5,6), and results have been fairly encouraging.

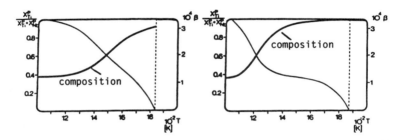

Figure 4 - *Theoretical plots of composition and molar fraction precipitated of carbonitride particles containing Ti and Nb. Figure (a) refers to a low nitrogen content and (b) to a high nitrogen content. Note the stabilising effect on Ti-rich particles if there is enough nitrogen present.*

Figure 4 shows how the composition and molar fraction of titanium-niobium-nitrogen particles varies as a function of temperature, from the work of Strid(5). Figure 4a refers to an alloy with low nitrogen content and 4b to a high nitrogen particle. It is seen that the effect of nitrogen is to hold the titanium content of the particles high until at a temperature of about 1100°C, a sharp drop in Ti composition occurs to a final amount of under 0.4. The explanation for this is that a high nitrogen content increases the stability of the titanium within the particles, the affinity between titanium and nitrogen being higher than between niobium and nitrogen. An alternative way of presenting the thermodynamic data is shown in Figure 5, after the work of Suzuki et.al(6), which in this case concerns a quarternary system of titanium-niobium-carbon-nitrogen. It is seen that the molar fraction of titanium decreases from about 0.93 at 1300°C to 0.67 at 1000°C. At the same time the carbon content in the particles increases substantially, while the nitrogen content appears to remain fairly constant. There also appears to be a dramatic increase in niobium content of the particles. Another way of displaying the data in a quarternary system of titanium-niobium-carbon-nitrogen, this time holding the carbon content constant at 0.011%, is shown in Figure 6. In this case, the contours depict solubility limits and compositional variations between titanium carbonitride at high temperatures, and niobium carbonitrides at lower temperatures. It is observed that the effect of increasing nitrogen content in the system is to increase the solubility of the titanium carbonitrides.

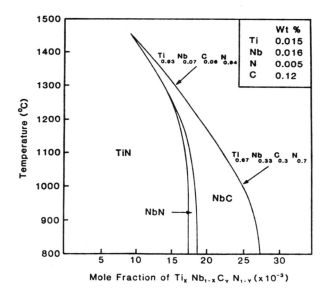

Figure 5 - *Computed molar fractions of the various compounds formed during cooling for alloys of Ti-Nb-C-N, after Suzuki et al.(6).*

51

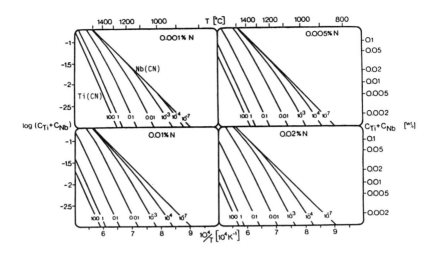

Figure 6 - *Computed dissolution curves of carbonitride particles containing Ti and Nb, as a function of different nitrogen contents. Note the effect of higher nitrogen levels in pushing the curves to higher dissolution temperatures.*

Concerning the kinetic effects of particle dissolution, Suzuki et.al(6) have shown that the effect of increasing carbon content in the niobium carbide system appears more dramatic than the effect of increasing niobium content. As seen in Figure 7, a relatively large increase in niobium content has caused only a small increase in dissolution time whereas a relatively large increase has resulted when increasing the carbon content of the alloy. A likely explanation for this is that it is easier for the interstitial elements to move out of the particles during dissolution, but that this causes a build up of the substitutional element (Nb), at the particle/matrix interface. This results in turn in a thermodynamic unbalance at the interface resulting in the particles being less stable. In complex systems, as discussed earlier, these effects are also more complex in that different substitional elements diffuse over different distances.

The composition of particles following weld thermal cycles depends very much on the welding process used. In a low energy weld process, the cooling rate is fast and the high temperature composition of particles as predicated from thermodynamic considerations, is likely to be retained. This is shown in Figure 9(a) and (b), in which (a) refers to a titanium-niobium carbonitride system with low nitrogen content and (b) refers to the same system with a relatively high nitrogen content. It is seen that the higher nitrogen content has resulted in a high titanium composition whereas the lower nitrogen content, in agreement with the thermodynamic predictions, has resulted in a spread in particle composition as well as a lower mean composition of titanium in the particles. However, if the same systems are given a high energy weld cycle, as illustrated in Figure 10(a,b), particle compositions are considerably more spread out in both cases.

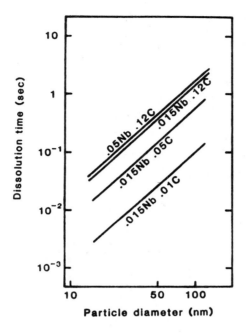

Figure 7 - *Experimental isothermal dissolution times of Nb-C particles in austenite, of various Nb and carbon contents. Note that carbon is more effective than niobium in slowing down the rate of dissolution. After Suzuki et al.(6).*

The much larger spread in composition for the smaller particles in Figure 10(b) is not considered to be due to experimental error, but is due to the fact that the smaller particles can more easily adjust their compositions during the slower cooling cycle. There is considerable debate about this, however, since it has been found that in these stable compounds the diffusion rate of the substitutional elements is usually considered to be very slow. The results could be interpreted on the basis that the smaller particles are able to adjust their composition, whereas the larger particles can not, due to differences in the diffusion distances concerned.

An important consideration in high strength structural steels is to maintain a fine particle size dispersion, even following welding. Measurements of particle size distributions are shown in Figure 11 and it is seen that in the case of titanium microalloyed steels, the high nitrogen content has a stabilising effect on the small particles and the mean particle size remains small. The effect of having other alloying elements present, such as niobium or aluminium, is to decrease their stability for reasons discussed above, and results show that the mean particle sizes are larger in these cases. As expected, the effect of welding is to cause

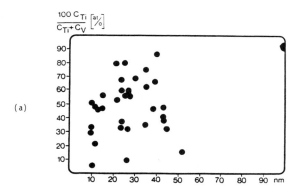

(a)

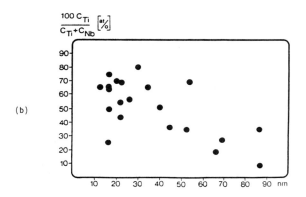

(b)

Figure 8 – Measured particle compositions in steels after
(a) normalising a Ti-V steel, (b) hot rolling a TiNb
steel. Note the wide spread of compositions.

54

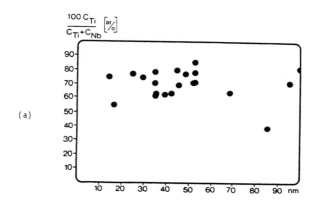

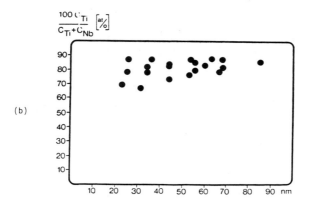

Figure 9 – Measured particle compositions of a TiNb steel with (a) low nitrogen and (b) high nitrogen content, following low energy weld thermal cycles (Δt = 17 sec). Compare with Figure 8.

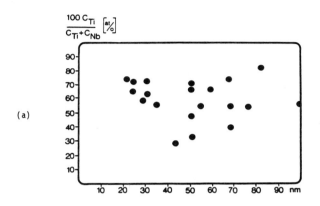

(a)

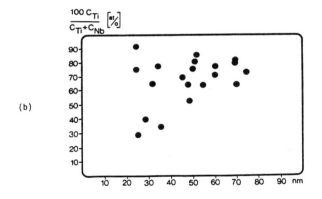

(b)

Figure 10 – *Measured particle compositions of a TiNb steel with (a) low, (b) high N-content, following high energy weld thermal cycles (Δt = 100s). Compare with Figure 9.*

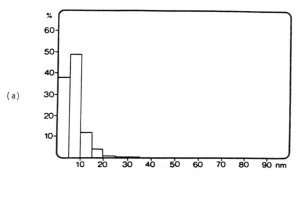

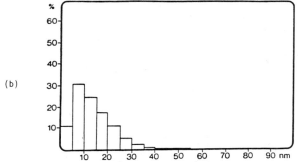

Figure 11 - Particle size distributions of (a) a normalised Ti-V-steel and (b) a normalised Ti-V-Al steel.
Note the detremental effect of Al on particle size.

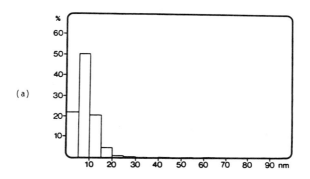

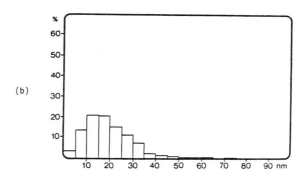

Figure 12 - *Particle size distributions of a Ti-Nb-N steel*
(a)   *before welding*
(b)   *following a rapid weld thermal cycle* (Δt = 13s).

58

particles to increase in size in all cases, and the amount of coarsening that takes place depends on the system concerned, as illustrated in Figure 12. In the case of titanium-based particles, coarsening is highly dependent upon the amount of nitrogen present. In the case of particles which dissolve during the weld thermal cycle, the effect of welding is obviously more effective, allowing considerably more grain growth to occur in the HAZ. In high energy welding processes, where the cooling rate is relatively slow, it has also been found that reprecipitation can occur(6).

Grain Growth In The Heat Affected Zone

Grain growth in the heat affected zone can occur either because particles coarsen or dissolve. According to the Zener equation, the mean grain size reached as a function of particle size and volume fraction of particles is as follows:

$$G = \frac{4\bar{r}}{3V_f} = \frac{\lambda^2}{\pi\bar{r}} \qquad (3)$$

where G refers to the grain size, $\bar{r}$ is the mean particle size, V_f is the volume fraction of precipitate, and λ is the mean free path between particles in a given cross section of the steel. From the Zener equation, it can be seen that if the particle size increases, or the volume fraction of particles decreases, then the mean grain size increases. In practice it is fairly difficult to apply this equation in the familiar form shown to the left of equation (3) because of the difficulty in measuring particle volume fraction. This is particularly difficult in the case of the titanium microalloyed steels because of the fact that large titanium nitrides often form at the melt stage. A better way of expressing the Zener equation is as shown on the right hand side of equation (3). In this case the parameter, λ , can be measured, e.g. from extraction replicas, and the mean grain size correspondingly estimated.

It is an interesting fact that in welding, the rate of grain boundary migration approaches the upper theoretical limit in which there is time exponent of n = 0.5 in the grain growth equation. However, in measurements of grain growth based on isothermal treatments of practical steels, the time exponent is invariably less than 0.5 indicating a much slower rate of grain growth (except in very pure metals). Apart from the high driving force for grain growth that is present during the severe weld thermal cycle, there appears to be little time for long range diffusion of alloying elements to take place, which implies that grain boundary segregation and grain boundary drag may be avoided in the weld situation. This has been confirmed in a study of Akselsen et.al(7) in which it is shown that in the case of low energy weld cycles the time exponent for grain growth is near 0.5, whereas in high energy weld thermocycles where the cooling rate is slow, an exponent well below 0.5 better fits the data. Nevertheless, for most practical welding cases, except for the very high energy processes such as electroslag and possibly submerged arc, the use of 0.5 for the grain growth exponent in welding appears to give satisfactory results.

Calculations of grain growth, taking into consideration changes in particle composition, and whether particles become dissolved or not for example, that complete dissolution occurs at a lower temperature in Nb-steels than in Ti-Nb steels. This in turn results in more "unrestricted" grain growth in the case of the niobium steel than in the titanium niobium

59

steel. The affect of dissolution of particles in the titanium niobium carbonitride steels, is to allow grain growth (see equation 3), and this grain growth is larger for lower nitrogen contents in the steel.

Hydrogen Distribution in the HAZ

Since the grain size variation in the HAZ as a function of any input conditions is known, it should be possible to estimate the distribution of hydrogen after a weld thermal cycle provided certain simplifying assumptions are made. To begin with, it is assumed that during cooling of the weld, hydrogen is first able to distribute itself by volume diffusion within the weld metal and HAZ. Following this, it is assumed that all hydrogen in the HAZ segregates to prior austenitic grain boundaries. There are obviously other important sinks for hydrogen, such as carbonitride particles, ferrite/martensite boundaries, etc., but as a first approximation, sinks other than grain boundaries are neglected here.

Following Coe and Chano(8), the distribution of hydrogen in a weld can be estimated by adopting a factor which accounts for the non-isothermal removal of hydrogen from a single run bead-on-plate weld during cooling. This factor, τ, is a dimensionless paramter within which the variables of time, distance and diffusivity are collected and correlated to the actual cooling rate calculated from the heat flow equations(1,3). Thus the influence of changing temperature during cooling of the weld is accounted for by noting its effect on diffusivity of hydrogen, D, and summing terms in the expression:

$$\tau = \frac{Dt}{L^2} \qquad (4)$$

where L refers to the effective diffusion distance of hydrogen in the lattice and corresponds to the distance from the top of the bead to the bottom of the base plate. Since from our modified Rosenthal equation the cooling rate of the weld is known, it is possible to estimate for any given weld process. Hence, knowing the grain size distribution, an estimation of the amount of hydrogen segregated per unit grain boundary area can now be made. Results of these calculations, showing hydrogen distribution before and after segregation, assuming an initial amount of 30p.p.m. in the weld, are illustrated in Figure 13. From these data of hydrogen segregation, and knowing the corresponding hardness distributions, the possible influence of hydrogen and hardness of the HAZ on cold cracking may now be assessed.

Main Conclusions

On the basis of thermodynamic and kinetic considerations, reasonably good qualitative prediction of what is likely to occur during weld thermal cycles on the composition and stability of carbonitrides in practical steels can be made. In particular it is possible to explain the large differences in composition of particles following a low or high energy weld thermal cycle.

HYDROGEN CONTENT

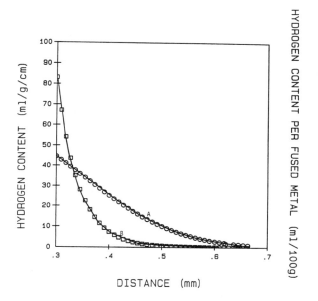

Figure 13 - *Computed hydrogen distributions in a steel weld HAZ following a TIG weld. Curve "A" refers to unsegregated hydrogen and "B" refers to hydrogen concentration assuming complete segregation to grain boundaries. Initial hydrogen level in weld assumed to be 30ml/100g.*

   The present results are promising, but in order to obtain more accurate modelling of the stability of complex carbonitrides in structural steels, and hence to make better predictions of the effect of weld cycles on the microstructure and properties of these steels, better compositional measurements across the particle/matrix interface are needed. In particular it is important to differentiate between particles that are growing or dissolving since distribution of the substitutional alloying elements is critical to the stability of the particles.

   An approximate distribution of hydrogen can be estimated and plotted as a function of grain size in the heat affected zone, for any given weld thermal cycle or weld process, assuming grain boundary segregation of hydrogen occurs.

References

(1)   M.F. Ashby and K.E. Easterling, Acta Met., 30, (1982) 1969.

(2)   J.C. Ion, K.E. Easterling and M.F. Ashby IBID, 32, (1984) 1949.

(3) M.F. Ashby and K.E. Easterling, IBID, 32, (1984) 1935.

(4) M. Inagaki and H. Sekiguchi, Trans.Natn.Res.Inst.Metals, Japan, 2, (1960) 102.

(5) J. Strid and K.E. Easterling, Acta Met., 33, (1985) 2057.

(6) S. Suzuki, G.C. Weatherly, D.C. Houghton, "The response of carbonitride particles in HSLA steels to weld thermal cycles", Acta Met., to be published.

(7) O.M. Akselsen, O. Grong, N. Ryum, N. Christensen, Acta Met., 34, (1986) 1807.

(8) F.R. Coe, Z Chano, Weld.Res.Int., 5(1), (1975) p. 33.

THERMOMECHANICALLY PROCESSED STEELS

WITH IMPROVED WELDABILITY

Isao Kozasu

Steel Research Center
Nippon Kokan K.K.
Kawasaki-ku, Kawasaki 210
Japan

Abstract

Current requirements for structural steels are enhanced strength and toughness both in base steel and weldment, and economy in fabrication. These are met by the introduction of Thermomechanical Control Process (TMCP). TMCP includes intensified controlled rolling and accelerated cooling after controlled rolling. This paper first reviews the requirement of chemical composition for improvement in cold cracking susceptibility and toughness in HAZ, and the importance of low carbon equivalent is pointed out in both respects. Secondly, outline of TMCP and its operating mechanisms are described. Then, the effect of chemical composition on TMCP base steel properties and how it is optimized in consideration of weldability are presented. Finally, examples of TMCP steels with improved weldability are presented. The benefit of TMCP is derived from the fact that it can lower CE. Recent trend for closer examination of the local brittle zones (LBZ) in HAZ seems to present new problems. If the future research can define the physical metallurgy in HAZ related to LBZ, TMCP is expected to resolve them with its potential in interchangeability with chemical composition.

Introduction

For structural steels, demands have been enhanced strength and toughness, and economy in fabrication especially in such applications as shipbuilding, offshore structures and pipelines. Specifically, typical requirements are low temperature toughness consistent with Arctic marine environment, strength higher than 500 MPa tensile strength, and heavy gauge of 50-100 mm thickness. These steels are required to be highly weldable and properties of their weldment to be comparable to those of base steels. Traditional 500 MPa tensile strength steels are normalized or controlled rolled (CR) C-Mn type with or without microalloying elements (Nb and/or V). The traditional steels are difficult to meet the current requirement in some respects and the introduction of TMCP (Thermomechanical Control Process) combined with optimized alloy design has resolved these difficulties.

The definition of TMCP has not been well established yet, but it is understood to mean three types of thermomechanical rolling process: 1) intensified controlled rolling above Ar_3, 2) intercritical rolling between Ar_3 are Ar_1 in addition to 1), and 3) interrupted accelerated cooling (ACC) in the transformation temperature range after controlled rolling. Although the mechanisms of improvement are somewhat different among them, they all give rise to strength increase which in turn results in lower carbon equivalent (CE). TMCP has been applied in commercial production and the mechanisms involved and examples of products were discussed extensively elsewhere (1-4). Weldability in its broad sense means cold cracking susceptibility and toughness of weldment, and it is mainly dependent on chemical composition or CE. Since TMCP is complementary to or interchangeable with chemical composition to achieve a specified strength level, it contributes to lowering CE, which appears to be the main benefit of TMCP steels with regard to weldability. Recent trends towards a closer examination of the local brittle zone (LBZ) revealed by validated CTOD test (5) seems to present new problems. However, if the problems are defined clearly, TMCP will be also helpful to resolve them.

The scope of this paper is to illustrate from industrial viewpoint how requirements of weldability is accommodated in actual product by the application of TMCP. First, guideline for improvement of weldability is discussed and required considerations in chemical composition are summarized. Secondly, recent overview of the mechanisms of TMCP and its process development are presented and the metallurgical design of TMCP steels is discussed in relation to the requirement of weldability. Finally, pertinent examples of TMCP steels with improved weldability are shown. Since there are already many literatures related to the present subject, the discussion here will be generic and simplified, and mainly based on recent literatures.

Considerations of Weldability

Cold cracking susceptibility in low heat input welding and low temperature toughness of heat affected zone (HAZ) in medium to high heat input welding depend largely on the chemical composition of base steel since these are influenced primarily by the reheated and transformed microstructure due to welding thermal cycle. Small second phase particles also have some effect on these properties because they influence the gamma-to-alpha transformation as gamma grain growth inhibitors and nuclei of alpha phase. In the following, the guideline of chemistry for improved weldability to be considered in the metallurgical design of TMCP steels are discussed.

64

Cold cracking is a type of delayed cracking resulting from cooperative action of the microstructures from shear type low temperature transformation, hydrogen and weld restraint. Therefore, hardenability or hardness at high cooling rate is known as a good measure of cold cracking susceptibility of base steel. As parameters to predict the susceptibility, various empirical CE's such as CE_{IIW}, CE_N, CE_{WES} and P_{cm}, which are directly or indirectly related to hardness, are currently in use. In low carbon system (<0.10%), the relative role of carbon to other elements seems to be more important in terms of resulting hardness than in medium carbon system (0.15-0.20%). For the improvement of cold cracking, the CE's in the above indicate that, among major alloying elements, carbon content should be kept as low as possible. Mn, Ni and Cu are preferred elements in comparison to Cr and Mo as explained later. Microalloying elements are considered to exert little influence on hardness at high cooling rate. The influence of CE on the maximum hardness and on the minimum preheating temperature is shown in Figs. 1 and 2 (6). It has been often pointed out that, in modern clean steels low in sulfur and oxygen, cold cracking susceptibility is not so low as predicted by CE, and this is explained by the absence of nonmetallic inclusions which act as nuclei of polygonal ferrite in transformation (7). Because the effect of these particles on transformation is influenced by the number and size of particles but not by the their total contents, there exist a wide range of scatters in the experimental results. However, the improved cold cracking susceptibility in low carbon steels appears to be not wholly offset by this inclusion effect. In addition, another type of cracking, lamellar tearing is greatly improved in clean steels. In low nitrogen steels, experiences show that residual boron must be kept as low as possible because of unwanted contribution of boron to hardenability. In actual welding operation, the improvement of cold cracking susceptibility by the above considerations permits alleviation or elimination of preheating, relaxation on the use of short bead, or increase in hydrogen content in stick electrode.

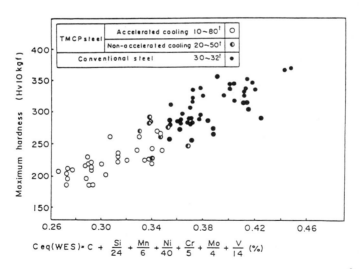

Figure 1   Effect of CE_{WES} on maximum hardness in the HAZ of 50 kg/mm^2 class steel
(Bead on plate test. "t" indicates thickness in mm)

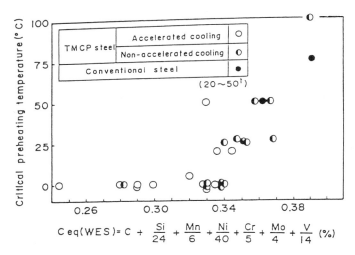

Figure 2   Relation between CE$_{WES}$ and critical preheating temperature to prevent cold cracking in JIS Y-groove cracking test on 50 kg/mm^2 class steel

$$C eq(WES)= C + \frac{Si}{24} + \frac{Mn}{6} + \frac{Ni}{40} + \frac{Cr}{5} + \frac{Mo}{4} + \frac{V}{14} \; (\%)$$

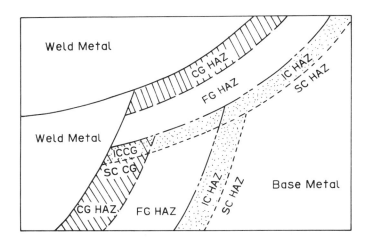

Figure 3   Schematic diagram illustrating local embrittled zones in HAZ of multi-pass welding
            CGHAZ : coarse grain HAZ
            FGHAZ : fine grain HAZ
            ICCGHAZ: intercritically reheated CGHAZ
            ICHAZ : intercritically reheated HAZ
            SCHAZ : subcritically reheated HAZ

Low temperature toughness of HAZ is determined intrinsically by the microstructures altered by welding thermal cycles. Even the microstructure resulting from single-pass welding presents an atlas of structures ranging from hardened structure from coarse grained austenite near fusion boundary to subcritically reheated ferritic structure. Subsequent welding passes further modify the microstructures altered by the preceding pass mainly causing changes in second phase such as carbides in addition to the microstructure change similar to that of the preceding pass. Usually, the worst toughness has been located at the coarse grained HAZ (CGHAZ) whose transformed microstructure is upper bainite. Such upper bainite is characterized by needle-shaped ferrite laths with cementite and martensite-austenite (MA) islands between them, and this type of morphology is known to be detrimental to toughness. Recently, closer examinations by CTOD test where the fracture initiation sites were identified indicated that the worst location was intercritically heated CGHAZ (ICCGHAZ), and another location with poor toughness was intercritically heated or subritically heated HAZ (ICHAZ or SCHAZ) as shown schematically in Fig. 3 (5,8,9). Figure 4 is an example of the toughness of ICHAZ evaluated by surface notched CTOD (9). These results were mostly from new low carbon steels and it is not clear whether similar embrittled zones can be found in traditional medium carbon steels when they are subjected to this rigorous testing. At present, generalization of LBZ situations seems premature since the effect of chemical composition of base steel, heat input and welding procedure on LBZ have not been completely clarified.

With regard to the toughness in CGHAZ including ICCGHAZ, improvement can be realized by avoiding the formation of upper bainite structure. Then, there are two alternatives: going to lower bainite or to mixed structure of bainite and polygonal ferrite. In 500 MPa grade under a given heat input, the only alternative may be the latter mixed structure by decreasing carbon or CE since increased CE which can produce lower bainite may not be tolerated from the reason of cold cracking susceptibility. Other measures to give rise to polygonal ferrite can be summarized as

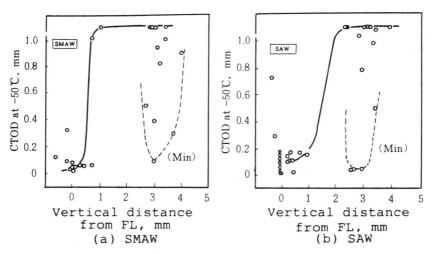

Figure 4   Critical CTOD of HAZ in relation to the vertical distance of crack tip from fusion line (FL) in surface notch CTOD test (Steel 0.10C-0.3Si-1.55Mn, 25 mm thickness, normalized)

follows: 1) prevention of austenite grain growth by dispersion of fine TiN and other second phase particles (B, Ca, REM compound) to decrease hardenability and general size of fracture units and to facilitate polygonal ferrite nucleation (10-13), 2) improvement of toughness of ferrite matrix by reducing dissociated free nitrogen (13) and by addition of Ni. Beneficial effect of decreasing carbon and niobium content to decrease CGHAZ hardenability has been documented in many literatures. Recently, it was reported that decomposition of MA islands in ICCGHAZ was retarded in the presence of Nb or V and these elements were considered detrimental to toughness (8). Poor toughness in IC or SCHAZ is an only recent finding by the use of CTOD test and the cause is inferred to be the formation of MA islands, modified carbide morphology and strain aging due to welding thermal cycles. However, the exact mechanisms of embrittlement and its remedy are yet to be investigated.

Deteriorated toughness in HAZ can be restored to acceptable levels in most cases by post weld heat treatment (PWHT). However, there are instances where this restoration is not complete in the case of steels with Nb and/or V probably because of the precipitation phenomena of these carbonitrides. Apart from toughness, lowered tensile strength of weldment sometimes becomes problem especially in high heat input welding. Usually, inevitable softening in HAZ is not a problem in actual welded structures because of the plastic constraint effect from neighboring matrix (14). When a softened zone in HAZ under an imposed heat input is too wide, CE must be raised to guarantee tensile strength in spite of the disadvantages in cold cracking.

The general effect of chemistry on weldability is summarized in Fig. 5. The actual composition has to be determined considering required properties of both base steel and HAZ, and in the following it is shown how these two requirements can be reconciled by the help of TMCP.

TMCP and Metallurgical Design

Almost all structural steels of 500-600 MPa grade can respond to TMCP and result in improvement of strength and low temperature toughness whose extent is dependent on particular alloy systems. TMCP refines the transformed microstructures in general, and is discussed first independently of chemistry. Then, the alloy system suitable for TMCP under

INCREASE IN:	COLD CRACKING SUSCEPTIBILITY	TOUGHNESS IN HAZ
CARBON	−	−
CE	−	−
MICROALLOY	+ *	−
FINE 2ND PHASE (TIN , ETC)	+	+
SULFUR , OXYGEN	(+)	(−)
NITROGEN	**	−

* FOR REDUCED CE
** AFFECTED THROUGH BORON

Figure 5   Effect of chemical composition on cold cracking susceptibility and toughness in HAZ
+: beneficial
−: detrimental

68

limitations from weldability is examined. The combination of these two constitutes the metallurgical design of weldable TMCP steels.

TMCP without water cooling is a natural extension of controlled rolling where beneficial effect of CR is exploited to a full extent (15). The basic mechanism of CR is the refinement of transformed ferrite grains by purposeful introduction of heterogeneities which act as ferrite nucleation sites in the austenite. Such heterogeneities are deformed and unrecrystallized austenitic grain boundaries and twin boundaries (also called as deformation bands) which are introduced by rolling below recrystallization temperature. Other structural features such as subgrain boundaries were also suggested as possible sites (16,17). The features of non-water cooled TMCP are very low reheating temperature (for example 1050-950°C) and heavy deformation of austenite immediately above Ar_3. In actual rolling, controlled rolling deformation is given in two or more controlled steps: for example, the total reduction more than 40% and 15% for controlling points below 910°C and 850°C respectively. However, these figures are varied according to the requirement of toughness. The low reheating temperature refines the initial austenitic grain size and the latter introduces further activated ferrite nucleation sites in increased density, both of which contribute to further refined ferrite grains. Rolling in the intercritical temperature range produces partially transformed and warm worked ferrite with fine substructures, which contributes to strengthening. An excessive rolling in this range, however, causes deterioration of toughness due to an excessive work hardening of ferrite. These situations are illustrated in Fig. 6 (15).

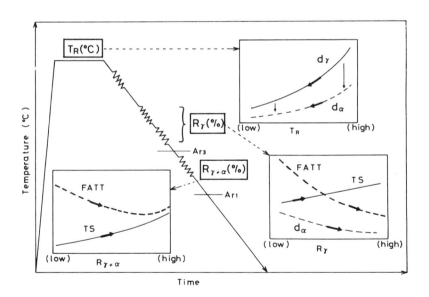

Figure 6   Schematic representation of the effect of process variables in controlled rolling (non water cooled TMCP)
T_r : reheating temperature
R^r : cumulative reduction in austenite region
$R^\gamma_{\alpha+\gamma}$: cumulative reduction in intercritical region
$d^{\alpha+\gamma}$: grain diameter

In TMCP with water cooling, accelerated cooling (ACC) is applied always in combination with various degrees of controlled rolling. Typically ACC is carried out at cooling rate of 5-15 degC/s in a temperature range of 800-500°C followed by air cooling below 500°C. ACC brings about further refinement of ferrite and produces microstructure consisting of fine ferrite and uniformly distributed fine bainite in place of lamellar pearlite in air cooled steel. ACC gives rise to additional increase in strength without deterioration of toughness while toughness is mainly determined by controlled rolling. This strengthening effect is interpreted to be derived from refinement of ferrite grain, substructure strengthening of ferrite, increased fraction of the second phase (bainite), and enhanced precipitation hardening in the case of microalloyed steels. The effect of cooling variables on properties is schematically illustrated in Fig. 7 (18).

The refinement of ferrite grain in ACC may be explained in one way by increase in the number of ferrite nuclei and suppression of ferrite grain growth as a result of supercooling of austenite (15,19). Recent experimental results suggest that the number of ferrite nuclei is drastically increased especially in grain interior in CR-ACC steel in comparison to CR-air cooled steel. Therefore, the effect of ACC on microstructure is explained by the mechanism that, in addition to the general supercooling effects on nucleation and growth of ferrite, ACC or supercooling in deformed austenite activates numerous potential nuclei in grain interior which are otherwise inactive (18,20,21).

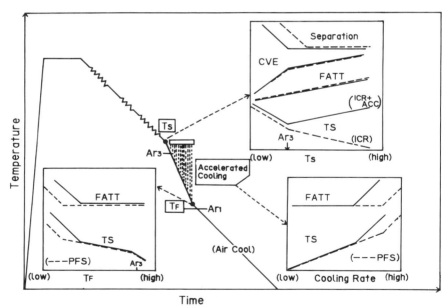

Figure 7   Shematic representation of the effect of processing variables in accelerated cooling
ICR: intensified controlled rolling
PFS: pearlite free steel
T_S and T_F: starting and finishing temperature of water cooling

70

ACC facility was first developed by NKK in 1980 and installed in the plate mill of its Fukuyama Works. All major steel manufacturers in Japan now have ACC facilities in their plate mills as shown in Table 1. In Europe, ACC facilities in Mannesmann and Nuovo Italsider are in operation. The technology of ACC was discussed in details at a recent conference (2).

The basic chemical composition of 500 MPa grade steel has been traditionally C-Mn type whether it is normalized or as rolled. Although carbon is effective in strengthening by grain refinement due to lowering of Ar_3 and increasing pearlite fraction, it is limited from consideration of weldability. Solution type alloying elements such as Mn, Ni, Cu and Cr contribute to strengthening by solid solution hardening. However, they produce more improvement in strength and toughness from grain refinement by lowering Ar_3 temperature. For example, variation of Ar_3 temperature of CR steels is shown by the equation below (22).

$$Ar_3(°C) = 910 - 310C - 80Mn - 20Cu - 15Cr - 55Ni - 80Mo \qquad Eq.(1)$$

Favorable alloying elements to replace carbon are the elements that cause the maximum Ar_3 decrease with the minimum increase in CE. Such elements are Mn and Ni, and Cu is also preferred from cost viewpoint. Ni is known to improve toughness of ferrite matrix besides the above effects. A typical alloy system of TMCP steel may be low C (<0.10) – Mn(1.2-1.5) – Ni(<0.5) – Cu(<0.5). When toughness below -60°C is required, Mn is partly replaced by Ni(23). When Cu of about 1% or above is added and tempered between 500-600°C, the steel is precipitation hardened by epsilon-Cu as in the case of A710 steel.

Microalloying elements, Nb and V, are important in high strength line pipe steels. Nb has been used in normalized steels as austenite grain refiner in the amount of 0.03%. In as rolled steels, Nb produces retardation effect of recrystallization of austenite in hot rolling and precipitation hardening during and after the gamma-to-alpha transformation. This retardation can enhance the CR effect in microstructural refinement.

Table 1   Accelerated cooling equipment in Japan

Company	NSC	NKK		KSC	SMI	KSL
Name	CLC	OLAC	OLAC II	MACS	DAC-I	KCL
Type	progressive closed	simultaneous open	progressive open	simultaneous open	progressive open	simultaneous open
Dimension (m)	4.7 x 19.8	4.5 x 44	5.3 x 20	5.35 x 40	4.7 x 27	4.7 x 39.1
Max. Water (T/min)	190	140	191	200	135	195
Nozzle Top	slit jet or flat spray	laminar	slit laminar	rod like nozzle	slit laminar	jet or pipe laminar
Nozzle Bottom	ditto	spray	spray	jet nozzle	spray	spray
Edge Mask	top & bottom	top	top	top	top	top
Distance From Finishing Mill (m)	79.8	26	53	19	25	53
Location	after HL	before HL	before HL	before HL	before HL	before HL
Cooling Rate For 25mmt (°C/s)	3-35	3-10	8-32	4-14	5-13	3-20
Direct Quench	possible	—	possible	separate equipment after MACS	separate equipment after DAC-1	possible

When Nb is added in plate steel, currently preferred amount is 0.005-0.015% from the limitation of toughness of HAZ, and even this less-than-normal addition is recognized to be useful for strength and toughness as shown in Fig. 8 (15,21). V is also precipitation hardening element but not widely in use in heavy structural plates. In V-containing plate steel, deterioration of toughness in HAZ was reportedly observed (5,8). Discussions on chemical composition so far are mainly on non-water cooled TMCP, but its effect is almost same in TMCP with ACC excepting higher base strength in ACC steels.

Besides the steels in the above which transform predominantly by diffusional process, there is a new type of steels transforming by shear mechanism. They are ultra low carbon bainite (ULCB) and low carbon bainite (LCB) steels characterized by low C-high Mn-Nb-B-Ti system and low cold cracking susceptibility (24-28). They can be favorably applied to pipelines especially from viewpoint of good field weldability. In this type of steels, the role of ACC is to suppress polygonal ferrite formation either to achieve higher strength or to obtain fully bainitic structure at lower CE.

Some precaution must be exercised especially in the fabrication of ACC steels. Because some part of strengthening is derived from the substructure strengthening of ferrite and the second phase (bainite), PWHT lowers this part of strengthening to varying extent depending on heat treatment condition (6). When PWHT is necessary, generally recommended temperature is below 600°C. Hot forming above the transformation temperature results in a large loss in strength since the strength after such thermal cycle is determined only by chemical composition, and warm working below Ac_1 is a more preferred process (29).

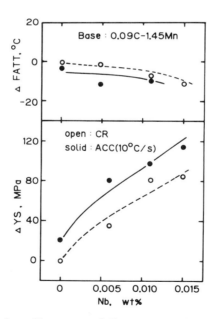

Figure 8 Effect of small amount of Nb on strength and toughness of CR and CR-ACC steel

72

Thermomechanically Processed Steels

Traditionally, for a given structural steel, optimized welding condition is determined by a compromise between welding efficiency and the resulting properties in the weldment. As was described so far, TMCP can be considered as a manufacturing parameter independent of chemical composition to achieve required properties of base steel. If the requirement of chemical composition is given based on the requirement of weldability which includes cold cracking susceptibility and toughness of HAZ, the specified properties of base steel can be met by suitable application of TMCP almost regardless of chemical composition. Examples of TMCP steels with their weldability are found in numerous technical bulletins and literatures (1-3). Although details are not reported here, there are useful informations from cooperative research and survey in Japan on the various properties of TMCP steels such as CE, required preheating, toughness of HAZ and wide plate test results (6,14).

In the below, only illustrative examples of the author's company are presented. Figure 9 is a summary of production record of structural steel of 32 mm thick plate shipped for application to an Arctic caisson (30). More detailed data of the toughness of weldment in terms of CTOD test is shown in Fig. 10. Recent results of heavier gauge plate of 63 mm are shown in Table 2. Table 2a and b show chemical composition and manufacturing conditions which are essentially similar to those of the previous case excepting that the TMCP conditions are more intensified. Table 2c shows properties of base plate, and Table 2d shows the CTOD test result of weldment with validation check. A detailed discussion on CTOD test result on similar plates is given elsewhere (31). Copper age-hardening steels are

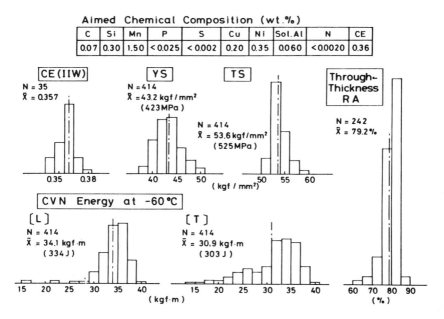

Aimed Chemical Composition (wt.%)

C	Si	Mn	P	S	Cu	Ni	Sol.Al	N	CE
0.07	0.30	1.50	< 0.025	< 0.002	0.20	0.35	0.060	< 0.0020	0.36

Figure 9   Statistical data in the production of EH36-Mod. for Arctic marine use by TMCP(CR-ACC)
Plate thickness   32 mm

73

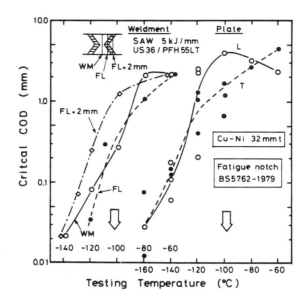

Figure 10   CTOD test result on base plate and weldment of EH36-Mod. by
TMCP(CR-ACC)

Table 2   API 2H Gr. 50 equivalent steel (63 mm^t)

(A)   Chemical composition   (Wt%)

C	Si	Mn	P	S	Cu	Ni	Nb	Ti	S.Al	N	CE^{IIW}
0.07	0.31	1.60	0.004	0.001	0.33	0.49	0.013	0.009	0.038	0.0036	0.39

(B)   Manufacturing conditions

Slab reheating temp. (°C)	CR finishing temp. (°C)	OLAC finishing temp. (°C)
950	740	470

(C)   Mechanical properties   (1/4t)

Y.S. (kgf/mm^2)	T.S. (kgf/mm^2)	vE-40 (kgf·m)	vTrs (°C)
46.9	54.2	42.6	-116

(D)   CTOD properties

Heat input (kJ/mm)	PWHT	Critical CTOD value at -10°C (mm)*	Notch location
5.0	None	>2.23 , >2.25	Coarse grain HAZ
		>2.05 , >2.25	IC/SC HAZ

* Specimen : BS5762 preferred type, through thickness notch

74

the recent promising development, where Cu exerts no harmful effect on weldability (32). Such an example is shown in Table 3 and Fig. 11 (33). The examples of ULCB and LCB steels were reported elsewhere (24,25). For practical application of TMCP steels, the adoption of them into material specification of code writing societies is important, and they have been adopted or are being considered for adoption in ASTM (A841), JIS (G3115, G3126) and API (2W).

Table 3   Chemical composition and properties of thermo-mechanically processed (CR-AcC-T) Cu bearing steel (75 mm^t)

Chemical composition , wt%												
C	Si	Mn	P	S	Cu	Ni	Nb	Ti	Sol.Al	T.N	CE (IIW)	Pcm
0.030	0.24	1.48	0.004	0.001	1.05	0.59	0.010	0.009	0.020	0.0024	0.386	0.174

Mechanical properties (600°C Aging : 1/4 t)					HAZ toughness (FL : 5kJ/mm)	
YS,MPa	TS,MPa	vE-40,J	vTrs,°C	δx-10°c,mm	vE-40,J	δx-10°c,mm
511	582	346	-92	>1.2	≥65	≥0.44

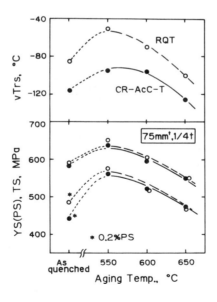

Figure 11   Aging response of TMCP (CR-ACC) and reheat-quenched Cu-bearing steel (Same steel as in Table 3)

Concluding Remarks

Low carbon TMCP steels have been recognized to possess good combination of strength and toughness in both base steel and weldment combined with low cold cracking susceptibility, and they have been finding wider applications. In these steels, the requirements from welding are accommodated by optimized chemical composition for welding and by TMCP as a manufacturing factor independent of chemical composition which refines microstructures by modification of the transformation. The production of these steels naturally requires a stable and accurate control of TMCP, which can be met adequately by the present level of steel manufacturers' technology. The drawback of TMCP may be a heavy penalty in steel manufacturers' rolling efficiency and the limitation of PWHT or hot forming in fabricators' operation.

Recently closer examination of HAZ toughness has brought up the problem of LBZ. Further research appears to be necessary to establish the extent of embrittlement, industrially viable test methods and the significance of LBZ to structural safety. To alleviate the LBZ problem, detailed investigation into the physical metallurgy of HAZ is necessary and this is expected to be able to set up the guidelines for improved HAZ toughness in terms of chemical compositions. Thus established reasonable guidelines will be implemented by TMCP and the associated alloy design.

References

1. HSLA Steels Technology and Applications, Conference Proceedings organized by M. Korchynsky; ASM, 1984
2. Accelerated Cooling of Steel; P.D. Southwick, ed. TMS-AIME, 1986
3. Proceedings of the Fifth International Offshore Mechanics and Arctic Engineering Symposium; ASME, 1986
4. A.C. de Koning, Metal Construction, Nov. 1985, pp.727-734
5. H.G. Pisarski and R.J. Pargeter, Metal Construction, July 1984, pp.412-417
6. Review on Properties of Thermo-mechanically Controlled Processed Steels; Japan Pressure Vessel Research Council, ISIJ, Sept. 1986
7. P.H.M. Hart, in Inclusions and Residuals in Steels: Effect on Fabrication and Service Behaviour, pp.435-454; J.D. Boyd and C.S. Champion, ed., CANMET, 1985
8. K. Uchino, et al, Ref. (3), pp.373-380
9. Metallurgical Investigation on the Scatter of Toughness in the Weldment of Pressure Vessel Steels (Part II Cooperative Research); Japan Pressure Vessel Research Council, ISIJ, Oct. 1985
10. S. Kanazawa, A. Nakajima, K. Okamoto and K. Kanaya, Tetsu-to-Hanage, 61 (1975), pp.2589-2603
11. S. Ueda, et al, ibid., 60 (1974), p.S549
12. S. Mukai, et al, ibid., 66 (1980), p.S1257
13. M. Suzuki, K. Tsukada and I. Watanabe, 2nd Int. Conf. Offshore Weld. Struct., P16-1-12, TWI, London, 1982
14. Report on Effective Application of 50 kg/mm^2 Grade High Strength Steels Processed by TMCP, Report No. 100; SR 193 Committee, ed., The Shipbuilding Research Association of Japan, 1985
15. I. Kozasu: Ref. (1), pp.593-607
16. R.K. Amin and F.B. Pickering, in Thermomechanical Processing of Microalloyed Austenite, pp.377-403; A.J. DeArdo, et al, ed., AIME, 1982
17. M. Umemoto, H. Ohtsuka, H. Kato and I. Tamura, in Proceedings of International Conference on High-Strength Low-Alloy Steels, pp.107-112; D.P. Dunne and T. Chandra, ed. Wollongong, Australia, 1985

18. I. Kozasu, Ref. (2), pp.15-31
19. A.J. DeArdo, in Proceedings of International Conference on High-Strength Low-Alloy Steels, pp.70-78; D.P. Dunne and T. Chandra, ed., Wollongong, Australia, 1985
20. K. Amano, T. Hatomura, M. Koda, C. Shiga and T. Tanaka, Ref. (2), pp.349-365
21. T. Abe, K. Tsukada and I. Kozasu, "Role of Interrupted Accelerated Cooling and Microalloying on Weldable HSLA Steels", paper presented at International Conference on HSLA Steels, Beijing, Nov. 1985
22. C. Ouchi, T. Sampei and I. Kozasu, Trans. ISIJ, 22 (1982), pp.214-222
23. K. Matsumoto, et al, Ref. (2), pp.151-164
24. H. Nakasugi, H. Matsuda and H. Tamehiro, in Alloys for the Eighties, pp.213-224; R.Q. Barr, ed., Climax Molybdenum Co., 1981
25. T. Taira, et al, Ref. (1), pp.723-731
26. H. Ohtani, et al, Ref. (1), pp.843-854
27. M.K. Graef, F.K. Lorenz, P.A. Peters and P. Schwaab, Ref. (1), pp.801-807
28. H. Tamehiro, N. Yamada and H. Matsuda, Trans. ISIJ, 25 (1985), pp.54-61
29. S. Tsuyama, et al, "Effect of Warm Forming on Mechanical Properties of High Strength Steel Plates Produced by On-line Accelerated Cooling for Arctic Use", paper presented at Annual General Meeting of CIM, May 1986
30. I. Kozasu, in Fracture Toughness Evaluation of Steels for Arctic Marine Use, pp.11-1-11-28; R. Thomson and C.S. Champion, CANMET, 1984
31. I. Kozasu, I. Watanabe and M. Suzuki, "Significance of Coarse Grain HAZ Toughness against Brittle Fracture", paper presented at 6th International Symposium on OMAE, Houston, March 1987
32. Y. Tomita, R. Yamaba, T. Haze, K. Ito and K. Okamoto, Ref. (3), pp.381-388
33. T. Abe, et al, Nippon Kokan Technical Report, No. 117, Apr. 1987, in press

WELDABILITY STUDIES IN

CR—MO—V TURBINE ROTOR STEEL

G. S. Kim and J. E. Indacochea[1]

and

T. D. Spry[2]

Abstract

This study was undertaken to establish the weldability of an HP-steam turbine rotor (1.0 Cr - 1.0 Mo - 0.25V steel). Characterization of the as-received material and subsequent evaluation of a de-embrittlement procedure, suggests that such heat treatment should be carried out prior to welding. But the improvement in toughness experienced by the heat affected zone (HAZ) of welds produced on both as-received and de-embrittled steel coupons suggests that a preweld de-embrittlement treatment is unnecessary. Further, a postweld heat treatment at the optimum temperature was found to improve the impact properties of both the HAZ and the base metal. The welds were automatically made on steel coupons with the welding parameters held constant, controlled by a microprocessor.

In terms of short-term mechanical properties the turbine rotor can be weld repaired. Studies are presently being conducted under simulated service conditions for long-term testing.

[1]University of Illinois at Chicago, Department of Civil Engineering, Mechanics and Metallurgy, Chicago, Illinois 60680.

[2]Commonwealth Edison Co., System Materials Analysis Department, Maywood, Illinois 60153.

Introduction

Modern electric power stations require equipment such as turbine rotors, discs and other heavy section structural alloy components to be efficient in operation and very reliable in service. Currently the designers of power plant equipment for service up to 1100° F want an optimum combination of high temperature strength and reliability. The material selection for service temperatures of about 900°F to 1100°F rests mainly between Chromium-Molybdenum type steels and Chromium-Molybdenum-Vanadium type steels (1.0-3.0 Cr 0.5 - 1.0 Mo or 1.0 Cr 1.0 Mo 0.25V). The Chromium-Molybdenum steels are time-tested compositions that are relatively simple to fabricate and offer no unexpected variations in properties. The Chromium-Molybdenum-Vanadium type steels offer higher allowable stresses at elevated temperatures up to 1100°F and perhaps higher.

Many articles dealing with these steels have left a confusing picture as to the possibility of low toughness due to temper embrittlement and the need for close control in fabrication. This lack of toughness can reduce the service life of power plants or temporarily put them out of service. In such cases, everything possible must be done to keep the plant running and to have it available at all times to meet peak load requirements. Welding would be preferred as a repair method, because the price and time of weld repair of any failed component is generally small compared with the cost of replacement power. However, defective welds in Cr-Mo and Cr-Mo-V type steels are often blamed for failures associated with welded joints. Thus there is a need for very close control of material selection, welding process, heat treatment, and inspection procedures.

A number of reports have been published regarding cases in which cracks have developed in the heat affected zone (HAZ) of joints in cast and forged assemblies for electric power equipment made of chromium-molybdenum-vanadium steels [1-3]. These cracks may show up immediately after welded assemblies have been tempered [4-8], but in other cases they have been found after 2000 to 10,000 hours of service at high temperatures [9-11]. Stress-relief cracking in thick section welds usually occurs in the coarse grained region of an HAZ and often extends right across it and terminates in the fine-grained region adjacent to the unaffected parent material.

It is recognized that when steel is welded in heavy sections, it is often necessary to perform a post-weld stress relief heat treatment to prevent cracking of the weld zone during subsequent exposure to the environment and to restore its notch toughness. This heat treatment, if not done properly, can cause a loss in toughness in both the weld metal and HAZ. This stress relief embrittlement is characterized by a decrease in the upper shelf energy as well as an increase in the Charpy V-notch transition temperature. This form of embrittlement has been attributed to the presence of partially-coherent and coherent carbide precipitates [12]. This precipitation occurs during heat treatment in the 500°F to 1100°F temperature range. If the heat treatment is sufficiently long, the semi-coherent precipitates grow and lose coherency, thus reducing the strain on the lattice and reducing the embrittlement [13].

The 0.5 Cr-Mo-V steels have good rupture strengths at high temperatures up to about 1050°F (565°C), but are known to be susceptible to HAZ cracking during stress relief [14-17]. When heat treated at high temperatures, it is thought that vanadium carbide precipitates easily on cooling as a fine, coherent, matrix dispersion causing considerable strengthening [18]. This reduces the amount of matrix deformation possible and concentrates the strain at the grain boundaries leading to a loss in overall thereby weakening the grain boundaries and promoting grain-boundary deformation.

The purpose of this study is to evaluate the weldability of a 1.0 Cr 1.0 Mo 0.25 V turbine rotor steel by characterizing the microstructure of the weld joint and evaluating its mechanical properties after different schedules of preweld and postweld heat treatments. This work looks into the possible existence of service-induced temper embrittlement in the different stages of the turbine rotor, since they are exposed to different temperature ranges. It assesses the benefit of the grain refinement in the HAZ due to multi-pass welding in terms of the improvement in the Charpy notch toughness of the weldments. Finally, it emphasizes the phenomenological changes occurring in the weld metal, heat affected zone and parent metal as a result of postweld stress relief heat treatment.

Evaluation of As-Received Rotor

This evaluation was done in terms of chemical analysis, metallographic characterization, grain size measurement, hardness measurement, tensile testing and Charpy impact testing. These tests were performed at different locations in the rotor to determine if there was uniformity in terms of mechanical properties, composition and microstructure. Samples were obtained from stages 5 and 9 and from areas next to the bore and adjacent to the periphery of the rotor. A schematic of the rotor is included in Figure 1.

Results of the chemical analyses are shown in Table I. These chemistries were compared with those specified by ASTM A293 (Carbon and Alloy Steel Forgings for Turbine Rotors and Shafts) and A470 (Vacuum Treated Carbon and Alloy Steels Forgings for Turbine Rotors and Shafts), Table II. It can be observed that the as-received rotor has a composition specified by ASTM A470, class 8. These chemistries show a homogeneous composition throughout the rotor. Most of the elements are within ASTM specifications except for phosphorus, which is slightly on the high side.

For the analysis of the microstructure of the as-received rotor, samples were extracted from stages 5 and 9 and from regions next to the bore through the periphery of the rotor. Microhardness measurements of the microstructural constituents were performed to support the qualitative metallography. The microstructure of the as-received rotor was found to be fully bainitic as shown in Figure 2. The light portions of the micrographs were observed to be a spheroidal structure at higher magnifications. The microhardness indentation tests confirm the existence of a uniform microstructure consisting of tempered bainite. The microstructural uniformity was also checked in terms of the prior austenite grain size. The samples were etched with saturated picric acid and sodium tridecylbenzene sulfonate which is known to preferentially attack regions which contain segregated phosphorous [19]. The prior austenite grain boundaries were etched strongly, Figure 3, indicating significant phosphorous segregation. Comparative measurements were conducted between stages 5 and 9, from the bore to the periphery. The results showed that the prior austenite grain size varied somewhat between the bore and the periphery from about ASTM No. 2 to No. 3 in stage 5 and between ASTM No. 2.5 and No. 3.5 in stage 9 as seen in Figure 4.

The tensile tests were conducted according to ASTM E-8. Three samples were extracted radially and longitudinally per location. The summary of the tensile data of the as-received rotor is shown in Table III along with values cited by ASTM A-470, Class 8. It is found that the values for tensile and yield strengths are above those specified by the standard and are uniform throughout the rotor. However, the elongation values in the longitudinal direction for stage 5 were observed to be below the minimum values specified by the standard, which gives an indication of probable service-induced temper embrittlement.

Charpy impact tests were performed according to ASTM E-23 type A at five different temperatures: room temperature, $100^{\circ}C$ ($212^{\circ}F$), $140^{\circ}C$ ($284^{\circ}F$), $180^{\circ}C$ ($356^{\circ}F$), and $220^{\circ}C$ ($428^{\circ}F$). Four samples per testing temperature were impacted to determine the transition temperature ($FATT_{50}$). The sample selection was made following ASTM standard A-370 for forged turbine rotors, and they were machined in the longitudinal direction with the notch facing the outer surface of the rotor.

The Charpy results for absorbed fracture energy and fracture appearance values in the as-received rotor are shown in Figure 5. Such results show that the toughness remains uniform from the bore to the periphery of the rotor in stage 5. The fracture energy at the transition temperature is about 32 ft-lb (44 joules) and the $FATT_{50}$ is approximately $195^{\circ}C$ ($383^{\circ}F$) for both regions of stage 5. The toughness results of stage 9 are better than those of stage 5. The transition temperature in stage 9 ($FATT_{50}$) dropped to about $150^{\circ}C$ ($302^{\circ}F$) and the fracture energy values at $FATT_{50}$ are also higher than those obtained in stage 5. However, for both stages the Charpy impact results are inferior to those specified by ASTM A-470 class 8.

The fracture morphologies of the Charpy specimens of stage 5 examined with the SEM are primarily brittle with intergranular failure and grain boundary separation, as shown in Figure 6. Some localized dimple type failure was observed in spots of the Charpy samples of stage 5. The fracture morphologies for the Charpy samples of stage 9, were found to contain mostly ductile fracture appearances. Figure 7 is a fractograph of a Charpy sample of stage 9 tested at $180^{\circ}C$.

<u>Temper Embrittlement in the As-Received Rotor</u>

Based on the Charpy impact test results, it was learned that toughness is not homogeneous throughout the different stages of the rotor. Further, these results are found to be inferior to those specified by ASTM A470. Assuming that the mechanical properties of this rotor prior to service met the specifications of the standard, it is apparent that the rotor did experience service-induced temper embrittlement. Also, the fact that the prior austenite grain boundaries were revealed so sharply, Figure 3, when using the etchant that brings out regions of high phosphorus segregation, gives further proof of such embrittlement.

The difference in Charpy V-notch properties between stages 5 and 9 can be attributed to their variance in service temperature, such that the hotter stage would undergo greater temper embrittlement. This result is consistent with the findings of other investigations [20,21]. Furthermore, the difference in grain size found between stages 5 and 9 could be another factor responsible for this discrepancy between the two stages. Note that stage 5 has a coarser grain size and consequently the local phosphorus enrichment of the grain boundary would be expected to be higher than the finer grain structure of stage 9. It is also observed that the sulfur level in stage 9 is double the one found in stage 5, which would imply inferior impact properties for stage 9. But this inconsistency, could be explained by a possible interaction between sulfur and phosphorus. In fact it has been reported [22], that sulfur appears to slow down phosphorus segregation.

The mechanism of temper embrittlement in Cr–Mo–V steel due to phosphorus segregation can be explained in connection with the extent of Mo carbide precipitation. In fact the inhibition of P segregation by Mo can be anticipated in a theoretical basis [23]. It is known that Mo is an effective scavenger for P and Sn when it is not tied up in carbides. But on formation of Mo-rich carbides, P and Sn are released and these impurities segregate to the grain boundaries causing embrittlement. Hence, the rate of such temper embrittlement is dependent on the kinetics of carbide precipitation. In fact, Zhe and Kuo [24] found that the rate of temper embrittlement in a 2 1/4 Cr – 1 Mo – 0.4V steel is an order of magnitude slower than the rate found by Yu and McMahon [25] in 2 1/4 Cr – 1 Mo steel. Consequently, it is believed that the presence of V in the Cr–Mo–V steel slows down the formation of Mo-rich carbides and thus lessens the release of P for segregation and embrittlement.

De-Embrittlement Heat Treatment

In view of the results of the evaluation of the as-received rotor, it was decided that due to its inferior Charpy notch toughness, stage 5 would benefit from a de-embrittlement heat treatment. This led to the extraction of weld coupons and Charpy specimens from regions next to stage 5. The selection of the time and temperature ranges for heat treatment was done by means of hardness measurements and metallographic evaluation. The heat treatment was conducted at temperatures of 1150, 1200, 1250, 1300 and 1350°F. A number of investigators [25,26] suggested that reversible temper embrittlement can be eliminated by short time heating above 1200°F (649°C) followed by rapid cooling through the embrittling temperature range.

A very extensive Charpy impact testing program was undertaken for each of the heat treating temperatures and corresponding times. These data are plotted in Figure 8 with all the significant information summarized in Table IV. It is observed that de-embrittlement for 3 and 6 hours at 1250°F resulted in better impact properties than at 1300°F. In terms of toughness, heat treatments at 1250°F for 12 hours and at 1350°F for 3 hours gave the best properties. However this was attained at the expense of a loss in strength and hardness, as shown in Figure 9. Therefore based on the hardness and Charpy data it was decided that a de-embrittlement heat treatment at 1250°F for 6 hours be followed in this investigation.

Light microscopy studies revealed no structural changes as a result of this de-embrittlement heat treatment. However, it was noted that prior austenite grain boundaries of the de-embrittled samples were not as apparent as in the as-received rotor, despite the longer etching time. This is a possible indication that the phosphorus previously segregated at the grain boundaries was driven into solution. Charpy specimens were examined in the SEM; a predominantly transgranular fracture was observed in the samples impacted at room temperature, with scattered dimples, as shown in Figure 10. The Charpy samples tested at 120°C showed a typical ductile fracture, as seen in Figure 11.

Weld Evaluation

Submerged arc welds were produced on coupons in the as-received and de-embrittled condition (6 hours at 1250°F). The welds were automatically processed with a constant potential dc power supply controlled by a Miller System 9 microprocessor. The welding parameters, shown in Table V, were held constant during weld operations. The filler wire used in the production of welds was 2 1/4 Cr–1 Mo (AWS type A5.28 class ER-90S-B3) 0.045' (1.14 mm) in diameter and the flux was a Lincoln neutral flux.

Following welding, the weldments were submitted to three different post-weld heat treatments for 2 hours at temperatures of 566, 621 and 677°C (1050, 1150 and 1250°F). One set of welds was not stress relieved. The evaluation of the postweld heat treatments was conducted in terms of Charpy V-notch impact tests of the HAZs and microstructural analysis of the weld metal, heat-affected zone and base metal for each preweld and postweld heat treat condition. The microstructural evaluation was done using light and scanning electron microscopy.

Microstructural Characterization

Examination of the unaffected parent metal after stress relieving treatments, shows microstructures consistent with those encountered in the as-received and de-embrittled samples, that is, a fully bainitic tempered microstructure with an approximate grain size of ASTM No. 3. No carbide precipitation was observed, not even in those specimens that were de-embrittled prior to welding. However, it was also noted that even in those samples that were not previously heat treated, the prior austenite grain boundaries were not as apparent when etched with picric acid and sodium tridecylbenzene sulfonate. Apparently the stress relieving treatment following welding, was sufficient to reduce or eliminate the phosphorus segregation.

Because of the extent of the heat affected zone, a selective microstructural evaluation of the HAZ was carried out at the passes circled in the schematic of the weld shown in Figure 13. The microstructures encountered varied from martensite to tempered martensite and bainite at the HAZ/base metal boundary. Figure 13 is a photo composite of micrographs from the fusion line to the unaffected parent metal, for the 21st pass, showing the above mentioned microstructures along with changes in grain size. There is no difference in the microstructures found in as-received and de-embrittled coupons. It was noted, however, that the as-welded samples show extensive carbide precipitation along the grain boundaries in the heat affected zone near the fusion line. This localized carbide precipitation was eliminated when the welds were postweld heat treated at 677°C (1250°F), as seen in Figures 13 and 14. In addition, spherodization of the microstructure was observed during the stress relieving treatment, becoming noticeable at 621°C (1150°F) and much more pronounced at 677°C (1250°F). This phenomena was the same for the two preweld conditions. This spherodization is expected due to the discrete carbide precipitation resulting from the martensitic and bainitic microstructures along with the heat treatment at these high temperatures and at relatively long times. Grain refinement of the HAZ's was observed in all the weld samples with respect to the as received material, even in the coarse grain region.

Charpy Impact Test

The region of primary interest at this point in the study has been the heat-affected zone, because of the susceptibility of this area to cracking during postweld heat treatment. In addition to a microstructural characterization, an evaluation of the impact properties was performed of all the weld specimens in each of their preweld and postweld heat treatment conditions. The Charpy impact tests were conducted at -20, 20, 60, 100 and 140°C and these results are shown in Figures 15A-D. Both the absorbed energy and the fracture appearance were plotted as a function of the testing temperature.

In general, the results show slightly better impact properties for the weld specimens that were submitted to de-embrittlement prior to welding. However this difference became less significant as the postweld heat treatment (PWHT) temperatures increased above 566°C (1050°F). It appears that the heating cycle experienced by the HAZs does not influence de-embrittlement as observed in the results of the samples that did not undergo any PWHT (Table VI). Further, the PWHT at 566°C (1050°F) deteriorated the impact properties even more. There was an increase of the $FATT_{50}$ for both of the preweld conditions. It appears that at this temperature no de-embrittlement can be obtained, as observed earlier in this study. And the fact that even the de-embrittled weld samples had a decline in their toughness, leads to speculation that some carbide precipitation may have occurred. PWHT at 677°C (1250°F) resulted in improvements for both preweld conditions. It is important to recall that this temperature was the best for the de-embrittlement procedure. It is true that the time of 2 hours at 677°C for PWHT is far shorter than the 6 hours found in the de-embrittlement studies; however, the refinement of the grains resulting from welding was another factor that contributed further to the improvement of the impact properties of the heat affected zone. Note that the transition temperature is 30-35°C lower than that of the de-embrittled rotor. Table 6 shows the results of the impact tests of the HAZs of all the weld specimens.

The fracture surfaces of several Charpy samples were examined with the scanning electron microscope. The ductile fracture morphology encountered in the specimens supports the high absorbed energy values obtained in the HAZs. Figure 16A shows the fractograph of a Charpy sample with the notch placed in the middle of the HAZ; a dimple ductile fracture is observed. Some samples gave lower impact energy values for the same conditions; the fracture surface shows some portions of brittle fracture appearance, as seen in Figure 16B. This last figure corresponds to a location between the HAZ and the unaffected parent metal. This occasional discrepancy supported our concern for the precise location of the notch due to the narrowness of the HAZ. However, examination of a set of fractured subsize Charpy samples with the notch carefully located at the weld metal heat affected zone and unaffected parent metal confirms the dimple ductile fracture found in the HAZs of the standard Charpy specimens, as shown in Figure 16.

Conclusions

The conclusions from the present study are as follows:

1. The as-received rotor was temper embrittled due to phosphorus segregation.

2. Since the degree of temper embrittlement of the as received rotor was different between stages 5 and 9, it appears that the embrittlement was service induced, rather than a result of the initial heat treatment the rotor received before going into service.

3. It was found that by submitting the as-received rotor to a de-embrittlement heat treatment, the Charpy impact toughness improved. The best de-embrittlement was heat treating at 677°C (1250°F) for 6 hours and this was selected for our study.

4. Postweld heat treatment at 677°C (1250°F) produced the best impact toughness of the HAZ. This treatment, coupled with the grain refinement produced in the heat affected zone, makes de-embrittlement prior to welding unnecessary. This PWHT also produced a substantial improvement in the base metal toughness.

References

1. N. T. Burguess, 'Some Weld Failures in Power Stations,' <u>Welding and Metal Fabrication,</u> 30, (1962), 105.

2. K. P. Bentley, 'Precipitation During Stress Relief of Welds in Cr-Mo-V Steels,' <u>British Welding Journal,</u> 11, (1964), 507.

3. V. N. Zemzin and R. Z. Shron, 'Local Breakdown of Joints in Heat-Resistant Cr-Mo-V Steels at High Temperatures,' <u>Avt. Svarka,</u> 21, (1968), No. 6:1.

4. R. Z. Shron and N. I. Nikanorova, 'The Precipitation Hardening of Welded Joints in the 15kh1M1F and 12Kh1MF Heat-Resisting Steel During Ageing at High Temperatures,' <u>Avt. Svarka,</u> 26 (1973), No. 4:5.

5. R. G. Baker and J. Nutting, 'The Tempering of 2 1/4 Cr-1 Mo Steel After Quenching and Normalizing.' <u>Journal of the Iron and Steel Institute,</u> 192 (1959), 257.

6. J. C. Lippold, 'Transformation and Tempering Behavior of 12 Cr-1 Mo-0.3V Martensitic Stainless Steel Weldments,' <u>Journal of Nuclear Materials,</u> 103, (1981), 1127.

7. T. Boniszewski, 'Metallurgical Aspects of Reheat Cracking of Weldments in Ferritic Steels.' <u>Welding Journal,</u> 51, (1972), 29.

8. J. Billy, T. Johansson, B. Loberg and K. E. Easterling: 'Stress-Relief Heat Treatment of Submerged-Arc Welded Microalloyed Steels,' <u>Metals Technology,</u> 2,(F) (1980), 67.

9. G. E. Lien, F. Eberle and R. D. Wylie, 'Results of Service Test Program on Transition Welds Between Austenitic and Ferritic Steels at the Phillip Sporn and Twin Branch Plants,' <u>Transactions of AIME,</u> 76, (1954), 1075.

10. R. W. Emerson, R. W. Jackson and C. A. Dauber, 'Transition Joints Between Austenitic and Ferritic Steel Piping for High Temperature Steam Service,' <u>Welding Journal,</u> 41, (1962), 385S.

11. R. L. Klueh, and J. F. King, 'Austenitic-Ferritic Weld-Joint Failures.' <u>Welding Journal,</u> 61 (1982), 302S.

12. R. A. Swift, and H. C. Rogers, 'Embrittlement of 2 1/4 Cr-1 Mo Steel Weld Metal by Postweld Heat Treatment.' <u>Welding Journal,</u> 52 (1973), 145S.

13. R. A. Swift, 'The Mechanism of Stress Relief Cracking in 2 1/4 Cr-1 Mo Steel,' (Paper presented at the 52nd Annual Welding Meeting, San Francisco, April 1971).

14. K. P. Bentley, 'Precipitation during Stress Relief Welds in Cr-Co-V Steels,' <u>British Welding Journal,</u> 11 (1964), 507.

15. J. D. Murry, <u>International Bibliography Information and Document,</u> 14, (1967), 447.

16. K. P. Bentley and R. G. Baker, 'Studies of Welding Metallurgy of Steels,' (British Welding Research Association, Rep. C57/11/63, (1963)).

17. C. J. McMahon Jr., Proc. 4th Bolton Landing Conference, 525 June 1974, N.Y., USA.

18. P. J. Alberry and W. K. C. Jones, 'Comparison of Mechanical Properties of 2 Cr-Mo and 0.5 Cr-Mo-V Simulated Heat-Affected Zones,' Metals Technology, (J) (1977), 45.

19. M. P. Seah, 'Interface Absorption, Embrittlement and Fracture in Metallurgy,' Sur. Sci., 53, (1975), 168.

20. Electric Power Research Institute, 'Temper Embrittlement of CrMoV Turbine Rotor Steels.' (Interim Report/EPRI CS-2242, Project 559).

21. Brown Boveri, Progress Report No. 1, Oct., (1985).

22. Y. Im, 'The Surface Segregation in Type 304 Stainless Steels by Auger Electron Spectroscopy', (M.S. Thesis, University of Illinois at Chicago, 1985), 43.

23. M. Guttmann, 'The Link Between Equilibrium Segregation and Precipitation in Ternary Solutions Exhibiting Temper Embrittlement,' Met. Sci., 10, (1976), 337.

24. Q. Zhe and K. S. Kuo, 'Embrittlement of 2 1/4 Cr-1 Mo-V Steel Bolts After Long Exposure at 540 C', J. Iron and Steel Inst., 12A, (1981).

25. J. Yu and C. J. McMahon, 'The Effect of Composition and Carbide Precipitation on Temper Embrittlement 2 1/4 Cr-1 Mo Steel,' Met. Trans., 11A, (1980), 277.

26. R. A. Swift and J. A. Gulya, 'Temper Embrittlement of Pressure Vessel Steels,' Welding Journal, 52, (1973), 57.

<u>Table I. Chemical Analysis of As-Received Rotor(w/o)</u>

CHEMICAL ELEMENT	SAMPLE				
	$9-K_a$	$9-X_a$	$9-Y_a$	$9-Z_b$	$5-Z_b$
Carbon	0.368	0.323	0.335	0.37	0.36
Manganese	0.85	0.85	0.85	0.88	0.85
Phosphorous	0.019	0.017	0.020	0.016	0.015
Sulfur	0.018	0.018	0.021	0.013	0.006
Silicon	0.249	0.248	0.248	0.22	0.22
Nickel	0.137	0.140	0.139	0.11	0.10
Chromium	1.04	1.10	1.07	1.09	1.05
Molybdenum	1.18	1.15	1.17	1.11	1.08
Vanadium	0.28	0.27	0.28	0.25	0.25
Tin	0.024	0.023	0.025	0.019	0.021
Arsenic	0.030	0.030	0.030	0.019	0.018

(a) Emission Spectroscopy   (b) Wet Chemical Analysis

<u>Table II. ASTM Chemistries of Turbine Rotor Steels</u>

Type	C	Mn	P	S	Si	Ni	Cr	Mo	V
A470 Class 8	0.25 0.35	1.00 max.	0.015	0.018	0.15/ 0.35	0.75 max.	0.90/ 1.50	1.00/ 1.50	0.20/ 0.30
A296 Class 6	0.35	1.00	0.035	0.035	0.15/ 0.35	0.5 max.	0.85 1.25	1.00/ 1.50	0.20/ 0.30

<u>Table III. Summary of the Tensile Data of As-Received Rotor</u>

Stage	Position	Sampling	Yield Strength (KSI)	Tensile Strength (KSI)	Elongation %	Reduction of Area %
	Out-side Dia.	Longitudinal	97.6	124.1	16.9	41.3
		Radial	98.6	123.7	17.2	38.8
	In-side Dia.	Longitudinal	96.6	120.7	15.5	41.7
		Radial	95.5	120.2	16.2	39.0
	Outside Dia. Radial		99.3	123.4	17.6	42.8
	·Inside Dia Longitudinal		96.0	120.7	16.7	46.6
ASTM A-470 Class 8			min 85.0	105.0 - 125.0	long. 17.0 radial 14.0	long. 43.0 radial 38.0

Table IV. CVN Test Results for De-Embrittled Rotor

CONDITION		$FATT_{50}$	FRACTURE ENERGY at $FATT_{50}$	
		C(F)	FT-LB	JOULES
AS-RECEIVED	STAGE 5 O.D.	195(383)	31.0	43
	STAGE 5 I.D.	197(387)	33.6	46
	STAGE 9 O.D.	148(298)	33.0	45
DE-EMBRITTLEMENT at 1200 F (649 C)	3 hr	134(273)	36.9	50
	6 hr	129(264)	44.4	61
	12 hr	138(280)	50.2	67
	24 hr	126(259)	46.9	64
DE-EMBRITTLEMENT at 1250 F (677 C)	3 hr	121(250)	44.0	60
	6 hr	114(237)	46.0	63
	12 hr	108(226)	49.0	67
	24 hr	116(241)	60.0	82
DE-EMBRITTLEMENT at 1300 F (704 C)	3 hr	128(262)	40.0	54
	6 hr	125(257)	42.0	57
DE-EMBRITTLEMENT at 1350 F(732 C)	3 hr	116(241)	44.0	60

Table V. Welding Parameters (Submerged Arc Process)

Voltage	34 volts
Current	230 amps
Travel Speed	11.8 in/min(30 cm/min)
Heat Input	1.56 KJ/mm (39.7 KJ/in)
Preheat Temperature	400 F (204 C)
Interpass Temperature	400 - 450 F(204-232 C)
No. of Passes	22 passes

Table VI. Summary of CVN Results for the HAZ

Condition	Transition Temperature ($FATT_{50}$)	Fracture Energies ft-lb(Joules)
AS - NO PWHT	110 C(230 F)	43 (59)
DE - NO PWHT	96 C(205 F)	51 (70)
AS - 1050 F (566 C)	114 C(237 F)	42 (57)
DE - 1050 F (566 C)	100 C(212 F)	47 (64)
AS - 1150 F (621 C)	108 C(226 F)	44 (60)
DE - 1150 F (621 C)	92 C(198 F)	48 (66)
AS - 1250 F (677 C)	79 C(174 F)	50 (68)
DE - 1250 F (677 C)	86 C(187 F)	65 (89)

ROTOR

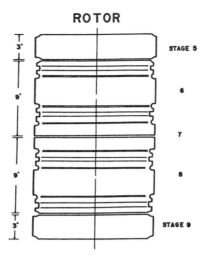

STAGE 5

6

7

8

STAGE 9

Figure 1 - Schematic drawing of the as-received rotor.

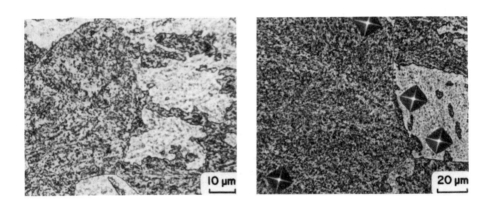

Figure 2 - (A)Microstructure of as-received rotor at high
magnification. (B)Microhardness Indentations of as-received
rotor, made with a 15gm load. Etched with 2% nital.

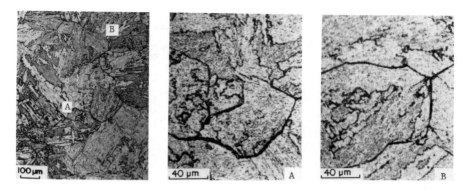

Figure 3 - Microstructures of the rotor in as-received condition
indicating the prior austenite grain boundaries. Etched with
aqueous picric acid with sodium tridecylbenzine sulfonate.

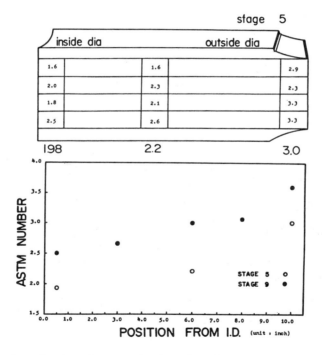

Figure 4 - Grain size variation in the as-received rotor
between the bore and the periphery for stages 5 and 9.

91

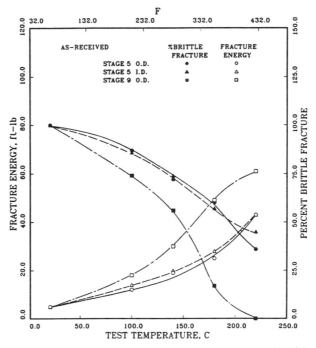

Figure 5 - CVN tests results for the as-received rotor.

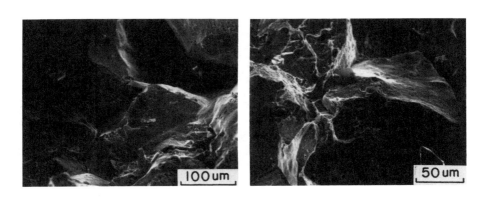

Figure 6 - Fracture surfaces of the as-received rotor stage 5, I.D. CVN sample tested at 180 C.

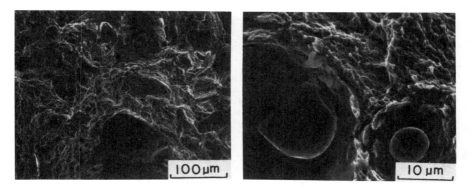

Figure 7 - Fracture surfaces of the as-received rotor stage 9
O.D. CVN sample tested at 180 C.

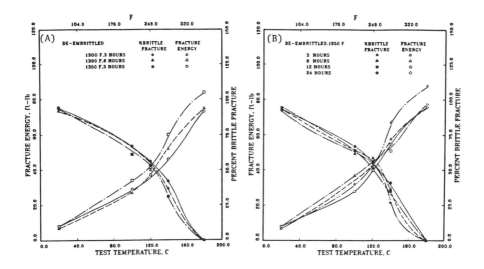

Figure 8 - Charpy test results for de-embrittled coupons
(A)1300 F. (B)1250 F. (C) 1200 F.

- Figure 8 continue -

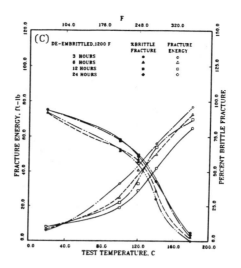

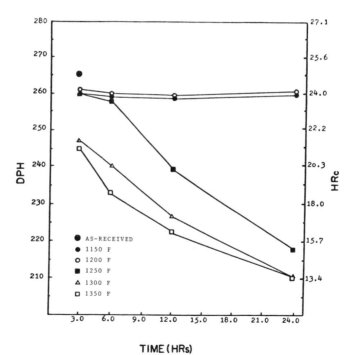

Figure 9 - Results of the preliminary heat treatment.(Hardness vs. Time and Temperature)

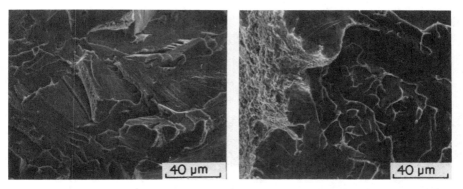

Figure 10 - Fracture surface of de-embrittled(1250 F, 6hr)
CVN samples tested at room temperature.

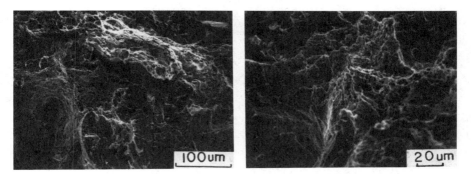

Figure 11 - Fracture surface of de-embrittled(1250 F, 6hr)
CVN samples tested at 120 C.

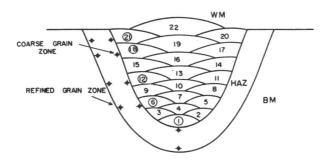

Figure 12 - Schematic diagram of the selected locations
for metallography and microhardness examination in HAZ.

21st Pass (Coarse Grained HAZ)

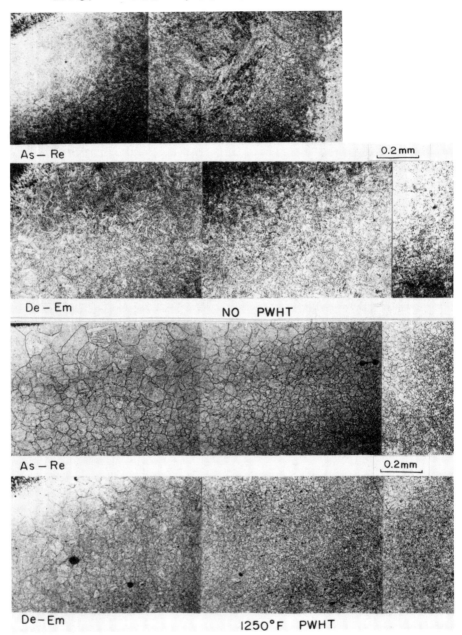

Figure 13 - Composite micrographs of the HAZ. (final pass)

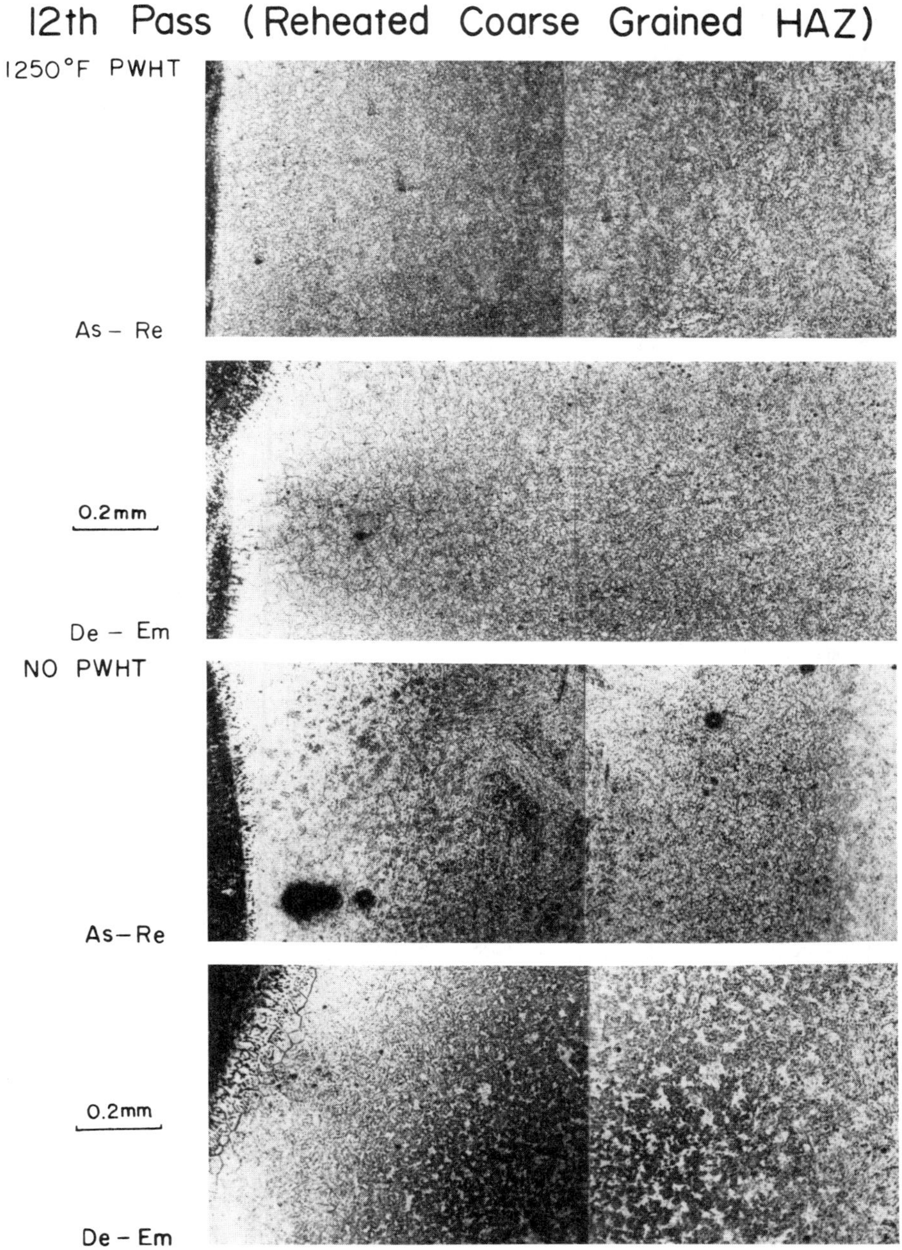

12th Pass (Reheated Coarse Grained HAZ)

1250°F PWHT

As – Re

0.2mm

De – Em

NO PWHT

As–Re

0.2mm

De – Em

Figure 14 - Composite micrographs of the HAZ. (middle pass)

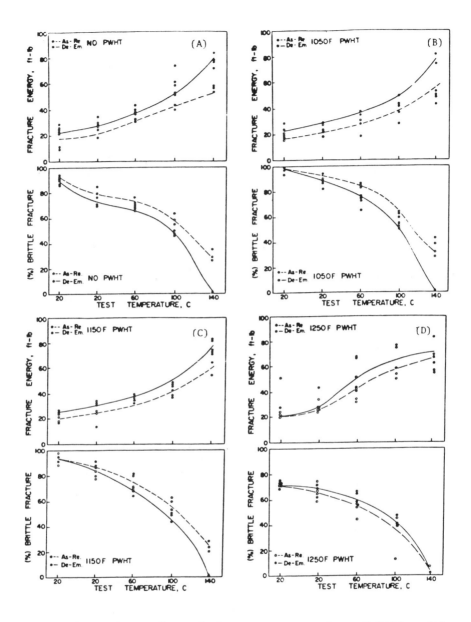

Figure 15 - Charpy V-notch impact test results of HAZ's. (A)
No PWHT. (B)1050F PWHT. (C)1150F PWHT. (D)1250F PWHT.

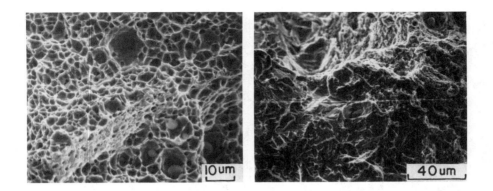

Figure 16 - Fracture of surface of broken CVN samples (A)HAZ
(B)boundary between base metal and HAZ.

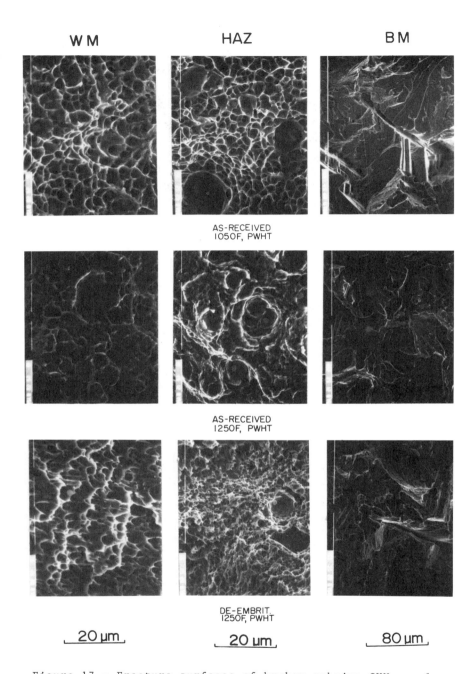

Figure 17 — Fracture surfaces of broken subsize CVN samples.

APPLICATION OF AUGER ELECTRON SPECTROSCOPY TO THE STUDY

OF TRACE ELEMENT EFFECTS ON WELDABILITY

C. L. White*, S. A. David**, and M. W. Richey***

*Department of Metallurgical Engineering   **Metals and Ceramics Division
  Michigan Technological University   Oak Ridge National Laboratory
      Houghton, MI 49931   Oak Ridge, TN 37831

***Martin Marietta Energy Systems
Y-12 Plant
Oak Ridge, TN 37831

Abstract

Auger electron spectroscopy (AES) is a surface analytical technique
that is particularly amenable to the study of interfacial segregation,
when the interfaces of interest are (or can be produced as) external free
surfaces. In several instances, AES has been applied to the study of
trace element segregation associated with fusion zone (FZ) cracking and
heat affected zone (HAZ) cracking in welds. Such studies have provided
qualitatively new and direct information regarding the relationship
between impurity segregation and the cracking processes. In general,
impurities such as S, B, and N have been observed to segregate strongly to
the surfaces of weld cracks, and the thickness of impurity enriched zones
tends to be only a few atomic layers thick. The experimental details
associated with these weld cracking studies can be somewhat different from
those for more conventional AES studies, however, and require extra care
during the acquisition and analysis of experimental data. In this paper,
we briefly overview the application of AES to the study of impurity
segregation in weld cracking problems. Although these impurities have
long been implicated in weld cracking, AES results thus far are not
sufficiently extensive to demonstrate a clear correlation between their
segregation, and the severity of weld cracking.

Introduction

Grain and interphase boundary cracking in the fusion and heat affected zones are complex phenomena, influenced by welding parameters (preheat, heat input, travel speed, heat source), environment, geometry and constraint in the workpiece, and the physical and metallurgical properties of the workpiece. Given the large number of variables that can influence the cracking phenomenon, great care must be exercised to isolate the effects associated with any one variable (or class of variables).

Among the metallurgical variables that are believed to influence cracking, the role of trace impurities is among the least understood. There is much circumstantial evidence associating certain impurities (including S, P, N, and B) with solidification and HAZ cracking in a variety of base metals (1-3). These impurities tend to have low solubilities and often form low melting eutectics with weld-metal or base-metal constituents, thus leading to the possibility of their enrichment in the interdendritic regions of solidifying welds and formation of weak or brittle low melting phases along the boundaries of the weld metal (4). Evidence for such phases are sometimes, but by no means always, observed. Failure to directly identify a second phase film may simply reflect a limitation of the microanalytical techniques employed, or it may indicate an alternate impurity related mechanism for the weld cracking.

The same elements that often form low melting phases at grain boundaries also tend to segregate strongly to grain boundaries and/or free surfaces, leading naturally to the supposition that their detrimental effect might also result from such segregation. In the following, we describe the manner in which AES has been used to provide direct information regarding the enrichment of impurities at the surfaces of weld cracks. Although such applications of AES have not been numerous, nor sufficiently systematic to clearly identify the relationships between segregation and cracking, they clearly demonstrate the feasibility and desirability of such studies.

Interfacial Segregation

Interfacial segregation can be defined as the enrichment of a solute species (intentional addition or impurity) at an interface without the formation of a second phase. Very large equilibrium concentrations of solute often occur within a few atomic distances of an interface, when overall solute concentrations are low and well within the solubility limits. A detailed discussion of this interesting phenomenon is beyond the scope of the present paper, however a number of well accepted points should be stated explicitly:

1. Solute enriched regions tend to extend only a few atomic distances from the interface in metals, primarily because of the short range nature of solute interactions in metallic bonding systems (5,6).

2. Localized enrichment by three or four orders of magnitude are fairly commonly observed at grain boundaries and free surfaces. Even though solute concentrations at the interface may be far above the bulk solid solubility, bulk precipitates are not stable (7).

3. Bulk chemical analyses are often not very reliable indicators of what impurities may or may not be segregating. With the large enrichment ratios that are often observed, the difference between 1 and 5 wt. ppm may be very significant in terms of segregation behavior, yet not reliably detectable by bulk chemical analyses.

4. Except for extreme cases where the enrichment, and the interfacial area per unit volume, are very large, the total inventory of solute at the interface will be a small fraction of the total solute in the solid. For this reason, variations in total interfacial area, or variations in the level of segregation via thermal treatment generally have little observable effect on solute concentrations in the defect free lattice (7).

5. Segregation that is equilibrium in nature will necessarily decrease interfacial energy. This is a consequence of the Gibbs Adsorption Equation, and not dependent upon any specific model of interfacial segregation (5-7).

Beyond the effect of segregation on interfacial energy, which is soundly based in classical thermodynamics, other effects of segregation on interfacial behavior have been observed empirically (e.g. segregation induced effects on interfacial diffusion, sliding, cohesion, mobility, etc.) that may also affect the tendency of an alloy to exhibit weld cracking. Effects of interfacial segregation on the related phenomena of low temperature intergranular failure and creep cavitation have been the subject of considerable research during the past 15 years (9), however comparable investigations of weld cracking are relatively rare. This situation is, no doubt, due to both experimental difficulties and the complex thermal, mechanical, and solidification histories associated with weld cracks.

Second Phase Films at Grain Boundaries

Rejection of solute from the solid to the liquid during solidification is common for low melting solutes associated with weld cracking. If this solute rejection leads to an overall concentration in the liquid, that exceeds the solid solubility limit in a given alloy, a second phase may form along the boundaries of solidifying grains. Factors governing the morphology and distribution of such phases throughout the weld microstructure are complex, however low interfacial energy between the solute rich phase and the alloy will promote formation of a thin film which may coat a large fraction of the boundaries in the fusion zone. Low interfacial energies are in turn favored by much the same chemical factors that favor the formation of low melting phases (e.g. weak bonding between the impurity and the metallic components). Such films may be studied using a number of microanalytical techniques including TEM, STEM, SEM, and AP-FIM; however if the films are only a few nanometers thick, AES may provide a useful and relatively easy way to identify or confirm their compositions.

Auger Electron Spectroscopy

The thinness of many impurity enriched regions at interfaces often makes them extremely difficult to detect and study. Several modern microanalytical techniques are capable of detecting and analyzing such regions subject to a variety of individual experimental constraints. One of the most widely used microanalytical techniques for the study of interfacial segregation is Auger electron spectroscopy (AES). Although this technique has been used extensively for study of impurity segregation at grain boundaries and free surfaces, and for characterization of thin films, it has been applied relatively rarely in studies where impurity segregation has been suspected to cause weld cracking (5).

AES is a surface sensitive microanalytical technique, that probes the top 2 or 3 atomic layers on an external surface. This surface sensitivity derives from the short mean free path of the characteristic Auger electrons, causing those Auger electrons originating more than 2 or 3 layers below the surface to be inelastically scattered, thus loosing their characteristic energy. For metallurgical studies, a finely focussed electron beam is often used for excitation of Auger spectra, as well as for imaging in much the same manner as a scanning electron microscope. Because of the surface sensitivity of AES, analysis must usually be conducted under ultra-high-vacuum (UHV) conditions (typically 10^{-8} Pa) in order to minimize the effects of environmental contamination. In the case of weld crack analyses, specimens must be fractured in-situ, and the analyzed cracks must be completely enveloped by the weld metal or base metal prior to that fracture, in order that contamination from the laboratory environment will not obscure any segregation that exists prior to sectioning the specimen. Once a surface is analyzed using AES, it is often useful to sputter ion etch the surface to remove a few atomic layers, then re-analyze it to study the underlying metal. By alternately analyzing and sputter etching, information related to the composition profile normal to the exposed surface can be obtained.

Weld Cracking in Austenitic Stainless Steels

AES analysis of fusion zone (FZ) and heat-affected-zone (HAZ) cracks in austenitic steels can be significantly complicated by the deformation of uncracked material surrounding the cracks. If the cracks are small and the surrounding uncracked material is fairly ductile, the cracks can be distorted significantly, causing difficulty in identifying the cracked regions for AES analysis. This leads to a situation where in-situ fracture through a cracked weld may or may not result in a fracture surface containing an easily identifiable weld crack. Furthermore, if the crack is large enough to be connected to the external surface of the specimen, it is likely to be heavily contaminated during specimen preparation. Such contamination effects were present in early attempts to apply AES to weld cracking phenomena (10), and severely limit the usefulness of the results.

The probability of a specimen gage section containing an internal (uncontaminated) crack can be significantly increased by knowledge of typical crack sizes and distributions, and careful specimen preparation. Metallographic preparation of representative specimens to characterize the location of extensively cracked regions in the weld, along with careful sectioning of AES fracture specimens and location of the reduced gage section are important. It is also helpful if the fracture device is capable of fracturing large gage sections, since the probability of it containing a previously unexposed crack increases with cross-sectional area.

Figure 1 shows weld solidification cracks in an autogenous arc spot weld (11). Figure 2 shows schematically how the AES tensile fracture specimen was prepared from this weld. Because solidification cracks were known to be concentrated near the transition region between the fusion

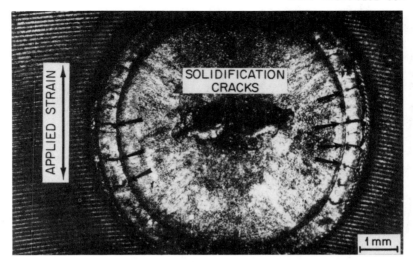

Fig. 1   Photograph showing the location of solidification cracks in an
autogenous spot arc weld on a stainless steel spot varistraint
specimen.

zone (FZ) and the heat affected zone (HAZ), the reduced gage section was
prepared by filing opposite corners containing the HAZ and the center of
the FZ, leaving the cracked transition region.   Upon fracturing the
specimen, secondary electron images obtained in the AES analysis chamber
permitted the dendritic features on the weld crack surface to be
identified (see Fig. 3-a and -b).   An Auger spectrum obtained from the
crack surface (Curve a in Fig. 4) contained large peaks from Fe, Ni, Cr,
and S, but none from carbon or oxygen.   Semiquantitative analysis of this
spectrum using published standard spectra of pure elements and compounds
for relative elemental sensitivities indicates 20-30 atom% sulfur in this
region.   We note the absence of manganese in this spectrum, suggesting
that the sulfur in this spectrum is not a manganese sulfide.   Absence of
carbon and oxygen are reliable indicators that the crack had been
completely contained, and not contaminated during specimen preparation.
An Auger spectrum taken from a transgranular portion of the fracture
surface (Curve c in Fig. 4) indicates primarily Fe, Ni, and Cr.   The very
small carbon and sulfur peaks in Curve c are probably from fine carbides,
sulfides, or carbosulfides that form nuclei for the plastically induced
voids that lead to ductile rupture.

        Figure 3-c through -f shows Auger maps of S, Fe, Cr, and Ni over the
field of view in Fig. 3-b.   Comparison of the S-map (Fig. 3-d) with the
secondary electron image indicates that the sulfur is indeed
preferentially distributed over the weld crack as indicated by the
spectrum in Fig. 4-a.   Careful comparison of the Cr-map (Fig. 3-e) with
the spectra in Fig. 4-a and -c illustrates a hazard associated with
interpretation of elemental maps in the absence of spectra.   While the
Cr-map indicates that the crack surface may be slightly Cr-poor, the Auger
spectra indicate that relative to Fe and Ni, Cr is enriched on the
fracture surface.   The reason that the maps indicate Cr depletion is that
the large sulfur signal attenuates peaks from all of the metallic
constituents.

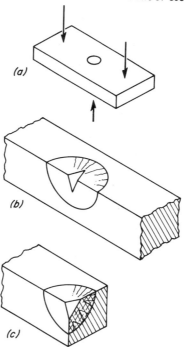

Fig. 2 Schematic illustration showing the location of fusion zone
cracks in spot varistraint specimens, and the method for
sectioning and notching AES tensile specimens cut from them.

Following analysis of the "as fractured" surface, the specimen was
sputter etched to remove 5-10 atom layers from the surface, then the
location of Curve a in Fig. 4 was re-analyzed. Curve b shows that the
sulfur peak present on the as fractured surface is almost completely
removed by the sputtering, indicating that the sulfur rich layer on the
hot crack surface is quite thin. Figure 5 shows a scanning electron
micrograph (SEM) of the fractured tensile specimen after AES analysis. In
this case, the roughly circular shape of the transition region is clearly
visible, and at higher magnification, the dendritic nature of the FZ
cracks is evident.

Another factor that significantly affects the ease (or possibility)
of analyzing weld cracks using AES is the spatial resolution of the
electron gun. State-of-the-art SAM instruments generally permit
examination of a fracture surface at resolutions of a few tens of
nanometers, permitting small areas of weld crack to be identified in
predominantly ductile transgranular fractures. Examples presented here
were obtained using an older instrument having a spatial resolution of
approximately 200 nm. While this instrument provided most valuable
results, several specimens sometimes had to be fractured in order to
obtain sufficiently large uncontaminated weld cracks for useful analysis.

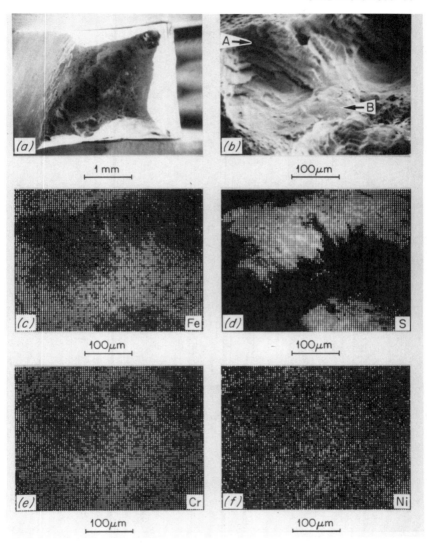

Fig. 3 Secondary electron images and Auger maps from the "as fractured"
surface of a 310 stainless steel spot varistraint specimen
containing fusion zone cracks.

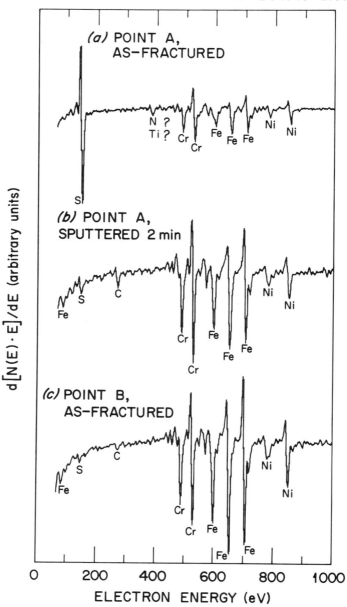

Fig. 4 Auger spectra from the "as fractured" surface of a 310 stainless steel spot varistraint specimen containing fusion zone
cracks.

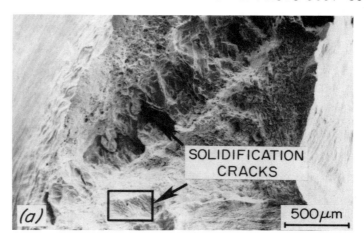

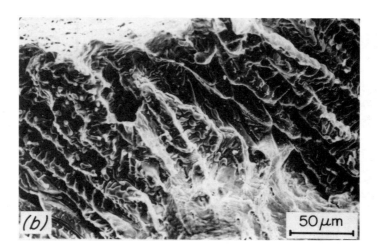

Fig. 5 Scanning electron micrographs of the fusion zone crack
analyzed in Figs. 7 and 8.

Free Surface Segregation

In many cases, one may wish to compare segregation behavior directly in two (nominally identical) welds that exhibit differing cracking tendencies. If weld cracks can be detected in only one of the welds, AES analysis of crack surfaces cannot provide a direct comparison between them. One way to achieve such a comparison in such cases is to devise a free surface segregation experiment on each material, and compare free surface segregation in the cracked weld with free surface segregation in the uncracked weld, and with crack surface analyses in the cracked welds.

Such an approach was adopted in a recent study of FZ cracking in a transition joint between a high strength Ni-Co steel [HP 9-4-20] and a nitrogen strengthened austenitic steel [21-6-9] (8). One heat of HP 9-4-20 weld wire resulted in consistently sound welds, while a second heat produced extensively cracked welds. SAM analysis of tensile fracture surfaces containing uncontaminated FZ cracks indicated extensive enrichment in B, N, and S (Fig. 6). Elemental mapping indicated that the B and N tended to be cosegregated, while S tended to segregate to regions where B and N were less enriched (see Fig. 7). Sulfur is a common impurity in many alloys, and the 21-6-9 stainless steel would provide a source for nitrogen. Bulk chemical analyses of both the weld wires and resulting weld metals failed to give any explanation for the boron segregation, however. The non-cracking weld wire actually indicated the highest boron level (20 wt. ppm) while the crack producing wire and weld metals had 10 wt. ppm boron.

In order to directly compare segregation behavior in cracked and uncracked welds, specimens were metallographically polished and sputter cleaned, then heated in-situ at 800 C for 180 min. then at 1000 for an additional 80 min. Figure 8 shows the B, N, and S Auger peak intensities (normalized to the iron Auger peak) plotted versus heating time for cracked and uncracked weld metals. The top frames show the temperature versus time plots, which were essentially the same for the two materials. While the sulfur segregation to the free surface is nearly the same for the two materials, the nitrogen and boron segregation are much more extensive in the cracked welds than in the uncracked ones. This direct comparison suggests (results of bulk analyses not withstanding) that boron and nitrogen segregation to free surfaces is strongly associated with the weld cracking. A cause-effect relationship cannot be established, however, without a systematic study in which boron and nitrogen contents are controlled and segregation levels in all alloys are monitored.

Summary and Conclusions

The application of AES and the SAM to studies of trace impurity effects on weld cracking have been briefly reviewed. Experimental details affecting the success in exposing uncontaminated weld cracks for analysis have been discussed, and applications in two weld cracking studies have been described. While weld cracks nearly always appear to be extensively enriched in impurities such as S, B, and N, a clear correlation between their segregation and the cracking has not been established in all cases. Indeed, while interfacially active elements such as these are broadly associated with cracking phenomena, they have been cited as essential for good weld penetration in other studies (12). Clear definition of the role of interfacially active impurities will require extensive systematic investigation of segregation behavior in a series of alloys exhibiting a range of cracking tendencies and impurity contents.

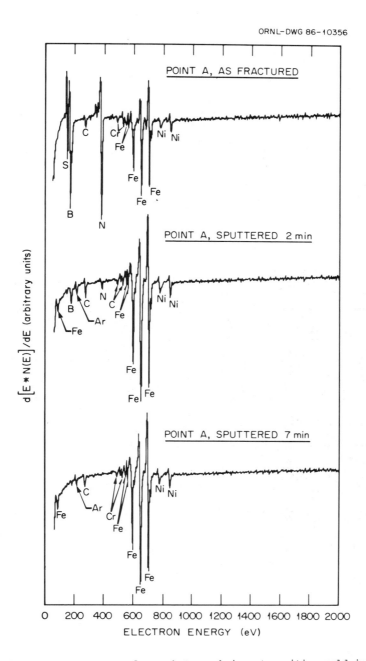

Fig. 6 Auger spectra from a hot crack in a transition weld joint
between HP 9-4-20 and 21-6-9.

(a)

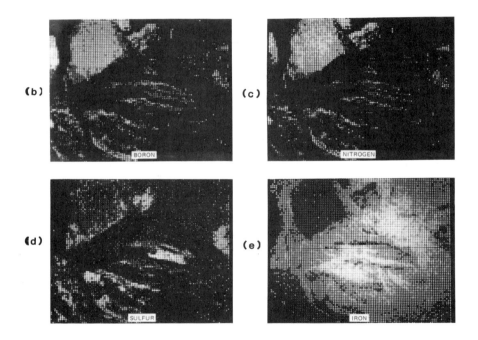

(b)   BORON

(c)   NITROGEN

(d)   SULFUR

(e)   IRON

Fig. 7   Scanning electron micrograph (a) and Auger maps [(b) through
(e)] of the "as fractured" hot cracked transition weld between
HP 9-4-20 and 21-6-9.

112

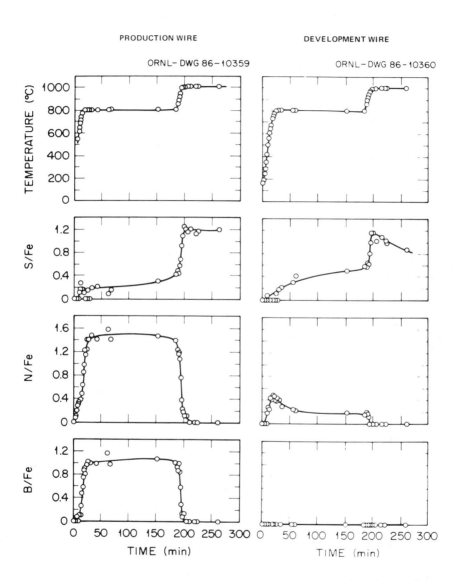

PRODUCTION WIRE

DEVELOPMENT WIRE

ORNL- DWG 86-10359

ORNL- DWG 86-10360

Fig. 8 Plots of temperature and Auger signal versus time for surface
segregation experiments on cracked (production) and uncracked
(development) transition weld joints between HP 9-4-20 and
21-6-9.

113

References

1. G. E. Linnert, "Weldability of Austenitic Stainless Steels as Affected by Residual Elements", pp. 1-5-199 in Effects of Residual Elements on Properties of Austenitic Stainless Steels, ASTM STP 418, ASTM, 1967.

2. R. D. Thomas, Jr., "HAZ Cracking in Thick Sections of Austenitic Stainless Steels-Part I", Welding Journal, vol. 63(12) pp. 24-32, Dec. 1984.

3. R. D. Thomas, Jr., "HAZ Cracking in Thick Sections of Austenitic Stainless Steels-Part II", Welding Journal - Research Supplement, vol. 63(12), pp. 355s-368s, Dec. 1984.

4. K. Easterling, Introduction to the Physical Metallurgy of Welding, Butterworths, London, pp. 164-172 (1983).

5. C. L. White, "Study of Solute Segregation at Interfaces Using Auger Electron Spectroscopy", Ceramic Bull., vol. 64(12), pp. 1571-80 (1985).

6. C. L. White, "Recent Developments Concerning Segregation and Fracture at Grain Boundaries", J. Vac. Sci. Technol., vol. 4(3), pp. 1633-47 (1986).

7. D. McLean, Grain Boundaries in Metals, pp. 116-149, Oxford (Clarendon Press) 1957.

8. M. W. Richey, M. W. Doughty, and C. L. White, "Examination of Weld Hot Cracking in a Transition joint Between HP 9-4-20 Steel and 21-6-9 Stainless Steel", in Proc. Conf. on Trends in Welding Research, edited by S. A. David, ASM.

9. M. H. Yoo, C. L. White, and H. Trinkaus, "Interfacial Segregation and Fracture", pp. 349-382 in Flow and Fracture at Elevated Temperatures, edited by R. Raj, ASM, 1985.

10. J. A. Brooks, A. W. Thompson, and J. C. Williams, "Weld Cracking of Austenitic Stainless Steels - Effects of Impurities and Minor Elements", pp. 117-136 in Physical Metallurgy of Metal Joining, edited by R. Kossowsky and M. E. Glicksman, TMS-AIME, Warrendale, PA (1980).

11. V. P. Kujanpaa, S. A. David, and C. L. White, "Formation of Hot Cracks in Austenitic Stainless Steel Welds-Solidification Cracking", Welding Research Supplement, August 1986, pp. 203s-212s.

12. B. J. Keen, K. C. Mills, and R. F. Brooks, "Surface Properties of Liquid Metals and Their Effects on Weldability", Materials Science and Technology, vol. 1, pp. 568-571 (1985).

Microstructure and Toughness of Weldments

Heat-Affected-Zone

LOCAL BRITTLE ZONE MICROSTRUCTURE AND

TOUGHNESS IN STRUCTURAL STEEL WELDMENTS

J. Y. Koo and A. Ozekcin

Exxon Research and Engineering Company
Route 22 East, Annandale, NJ 08801 USA

ABSTRACT

The impact toughness of heat affected zone(HAZ) microstructures has been investigated in experimental high strength, low alloy steels. The HAZ microstructures of interest include prior austenite grain size, martensite, upper bainite, dual phase (or MA constituent), and microalloy precipitates. These microstructural features are present in the local brittle zones (LBZ) within the HAZ. The individual LBZ microstructures were produced in a controlled laboratory simulation and were tested in both as-welded and post weld heat treated conditions. Refinement in prior austenite grain size decreases the ductile-to-brittle temperatures of as-welded martensite, but has no significant influence on the toughness of martensite after post weld heat treatment (PWHT). Among all the LBZ microstructures investigated, dual phase microstructures exhibit the lowest toughness in as-welded condition. However, the relative ranking of the LBZ microstructures in terms of embrittlement potential critically depends on the chemical compositions and the local thermal cycles including PWHT. Microalloys deteriorate toughness of martensite after PWHT at 600°C and this embrittlement occurs by strain aging phenomenon. This paper also describes detailed analytical electron microscopy characterizations of LBZ microstructures and discusses the microstructure and toughness relationship.

INTRODUCTION

A heat affected zone (HAZ) in a single-pass weld of structural steels consists of four distinct zones: coarse grain zone, fine grain zone, intercritical zone, and subcritical zone. Each zone is formed by the local thermal cycle experienced during welding. In a multipass weld, some regions in these zones undergo multiple thermal cycles, resulting in partial or complete alterations in the single-pass HAZ microstructures(1,2). A typical multipass weld HAZ consists of localized and discontinuous zones with a mixture of numerous microstructural constituents. Some areas in the HAZ contain tough microstructures while some other regions are comprised of unfavorable microstructural features which have lower toughness than the steel base metal.

Recently very low HAZ toughness values have been found in a number of high strength, low alloy steels welded with multipass submerged arc procedures(2-6). Embrittlement relative to the base steel has been observed by crack tip opening displacement (CTOD) tests in both as-welded and post weld heat treated conditions. There are two distinct local zones which show low CTOD toughness: coarse grain HAZ and intercritical/subcritical HAZ. More significant embrittlement has been found in the coarse grain HAZ adjacent to the fusion line, and this embrittled zone will be the main subject of this paper.

Detailed fractography and microstructural analysis showed that the local brittle zones (LBZ) located in the coarse grain HAZ are the frequent sites for CTOD crack initiation, and that the wide scatter band in CTOD test data is related to the linear fraction of LBZ's intersected by the fatigue crack front(2,3,7). The analysis also showed that LBZ's in a typical multipass weld are extremely narrow (e.g. 0.5 mm wide) and discontinuous, and consist of one or a combination of the following major microstructural features depending on the steel chemistry and weld thermal cycles:

(1) Large prior austenite grain size (e.g. 120 μm at 3 KJ/mm heat input)

(2) Dissolution of microalloy precipitates and subsequent precipitation hardening and/or strain aging by the dissolved atomic species

(3) Unfavorable substructures such as upper bainite, untempered martensite, and dual phase

These metallurgical parameters in concert contribute to the reduction in the CTOD toughness relative to base metal. However, there is a need to know which parameter is more crucial than the others in determining the LBZ toughness. Thus present investigation was undertaken to establish the relative significance of the LBZ constituents in terms of embrittlement potential. Such a ranking can be used for improved HAZ toughness, and also may provide a rational guideline for developing desirable steel chemistries and welding procedure. The research program is still in progress. This paper describes some of the preliminary results obtained to date.

EXPERIMENTAL DETAILS

MATERIALS

Model alloys with a systematic variation in compositions were used in order to eliminate other complicating variables such as initial microstructures and compositions, impurity and residual elements, inclusions, and so on. The chemical compositions of the model alloys are given in Table I. The base steel compositions, steel Al, meet the API 2H Grade 50 steel specification. The alloys were vacuum induction melted and cast in the form of 75 lb. cylindrical ingots. The ingots were homogenized in argon at 1100°C for 24 hrs, upset forged at 1100°C, and then hot rolled at 1100°C to approximately 12.5 mm thick plate. All of the above treatments were followed by air cooling. Oversized Charpy blanks were cut from the flat plate with the specimen length parallel to the rolling direction.

Table I. Chemical Compositions of Model Alloys(wt%)

Alloy ID	Fe	C	Mn	Si	Nb	Ti	Cu+Ni	N
Al	Bal.	0.16	1.3	0.23	-	-	-	-
A2	Bal.	0.16	1.3	0.23	0.03	-	-	-
A3	Bal.	0.16	1.3	0.22	-	0.015	-	0.006
A4	Bal.	0.16	1.3	0.23	-	-	0.2Cu 0.3Ni	

THERMAL SIMULATIONS

One of the critical phases in the experimental procedures is to isolate individual LBZ microstructure and to determine its toughness values while the other parameters are kept constant. Actual welding experiments or Gleeble simulations were not suitable for this purpose. Therefore, new thermal simulation methods were employed to produce the various LBZ microstructures.

The Charpy blanks were heat treated in a vertical tube furnace under a flowing argon atmosphere and/or in a salt pot where the molten neutral salt was maintained at a predetermined temperature. All the specimens were subjected to the solution heat treatment at 1300°C for 7 min. in a vertical furnace, followed by water quenching. This initial heat treatment was carried out to dissolve microalloy carbonitride precipitates and/or other types of carbides which may be present in the as-hot rolled steel plates.

Variations in prior austenite grain size. The austenite grain sizes of 40, 120 and 200 μm were obtained by holding Al steel specimens at 1000°C for 7, 20 and 35 minutes, respectively. After the austenization treatments, the specimens were either water quenched to produce martensite, or directly quenched into a salt pot to obtain upper bainite. The isothermal heat treatment temperatures to produce upper bainite were determined to be 260, 270 and 280°C for the prior austenite grain sizes of 40, 120 and 200 μm, respectively. The specimens were held at the temperatures for five minutes followed by water quenching. Steel Al was selected for this study since this alloy does not contain microalloys which can introduce additional variables, especially after subsequent post weld heat treatments.

LBZ Microstructures. In the simulation of LBZ microstructures, heat treatments were designed to keep the prior austenite grain size constant at about 120 μm. Simulated martensite for both Al and A4 steels was obtained by austenitization at 1000°C for 20 minutes followed by water quenching. Subsequent intercritical heat treatment at 780°C for 30 sec in the salt pot

121

and direct warm oil quenching produced dual phase microstructure, which simulates the intercritically reheated coarse grain zone. Volume fraction of the second phase (martensite and retained austenite) in the dual phase microstructure was about 10%.

The simulated martensite for A2 and A3 steels was produced by austenitization at 1200°C for 3 and 3.5 minutes, respectively. The higher temperature and shorter holding time relative to those of steel A1 were experimentally determined in order to prevent the dissolved microalloy elements from reprecipitating during the reaustenitization treatment.

Post weld heat treatment (PWHT) was carried out at 600°C for 30 minutes in the vertical furnace. After PWHT, the specimens were subjected to water quench to avoid the possibility of temper embrittlement which may occur during slow cooling to ambient temperature.

MECHANICAL TESTS

Charpy V-notch(CVN) specimens were machined from the oversized Charpy blanks after thermal simulation treatments. The impact tests were performed at the temperature range of -140°C to 100°C using a 350 ft-lb capacity testing machine. At least eight specimens for each thermal simulation were tested to construct a complete energy curve including upper shelf and lower shelf energies.

MICROSCOPY

Microstructures were examined by optical, scanning, electron and transmission electron microscopy. Thin foils for transmission electron microscopy (TEM) were prepared by the twin-jet electropolishing technique using a solution containing 75 grm of Cr_2O_3, 400 ml of CH_3COOH, and 20 ml of H_2O. The TEM analyses were carried out using Philips 400T-FEG and Philips 430T microscopes equipped with an EDAX detector and a 9900 EDAX analyzer.

RESULTS AND DISCUSSION

In all phases of thermal simulation studies, extensive TEM characterization of microstructures was carried out to ensure that the synthetic LBZ microstructures precisely simulate the weld LBZ microstructures. This was done by first examining the LBZ microstructures of single-pass, double-pass and multipass welds in commercial AP1 2H Grade 50 steels which have similar compositions to those of the model alloys used in this study. Emphasis was placed on the identification of the physical and chemical states of microalloys and their precipitates, cementite morphology, substructure morphology, and dislocation density. In addition, prior austenite grain size and hardness levels were measured in both weld and simulated LBZ microstructures. The simulation thermal cycles were determined based on the experimental verification that the LBZ microstructural features in the simulated specimens are essentially identical to those observed in the welds.

GRAIN SIZE EFFECT

In general, grain size has a significant effect on the toughness of steels by controlling ductile to brittle transition temperatures. This dependency of transition temperature on grain size is of considerable interest in the coarse grain HAZ since the prior austenite grain size can vary significantly depending on the heat inputs and steel chemistry.

Optical micrographs in Figure 1 show three different sizes of prior austenite grains, namely 40, 120 and 200 μm produced from the C-Mn steel (Al). The microstructure is simulated martensite in the coarse grain HAZ of a single-pass weld. The grain sizes of 120 μm and 200 μm, respectively, correspond to those observed in the coarse grain HAZ of C-Mn steels welded with 3 KJ/mm and 5 KJ/mm heat input submerged arc procedures.

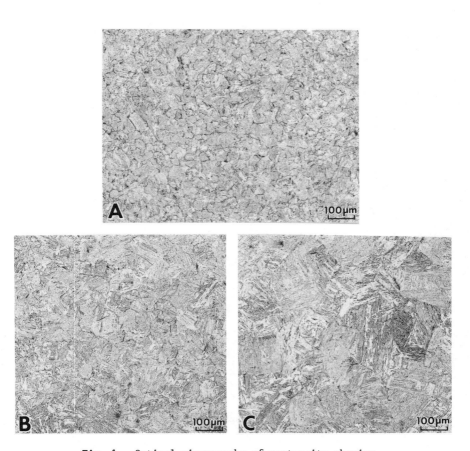

Fig. 1 - Optical micrographs of martensite showing a range of prior austenite grain sizes in Al steel (a) 40μm, (b) 120μm, (c) 200μm.

This range of heat inputs is typical of fabrication welding conditions used for offshore platform constructions. The grain size of 40 μm can be realized in a lower heat input (<1KJ/mm) welding, or in the steel compositions containing a dispersion of thermally stable particles (e.g., rare earth oxysulfides) for grain growth inhibition. CVN testing was conducted on both as-welded (as-quenched) and post weld heat treated

123

synthetic martensite at a range of temperatures to determine energy curves. The test results are plotted in Figure 2, where T_{30} is the testing temperature to obtain 30 Joule impact energy. In the specifications of offshore platform steels, 30 Joule is commonly used as an adequate energy value at a specified testing temperature. It is evident from Fig. 2 that T_{30} decreases with decreasing prior austenite grain size in the case of as-welded simulations. This finding is consistent with the numerous literature data regarding the beneficial role of finer grain size on ductile to brittle transition temperature(8-13). Upon PWHT at 600°C, however, the prior austenite grain sizes in the range of 40 to 200 μm have little influence on T_{30}.

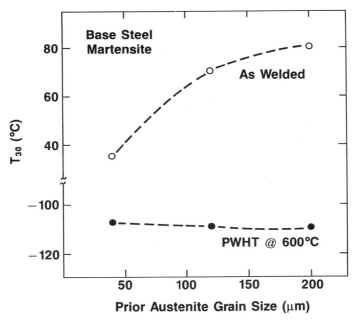

Fig. 2 - Effect of prior austenite grain size on T_{30} for as-welded(open circle) and PWHT (solid circle) martensite in A1 steel.

The reason for this may be qualitatively understood by examining the microstructures before and after PWHT. Figure 3(a) is a TEM micrograph showing as-welded martensite. The lath boundaries are low angles in nature and thus, a cleavage crack will propagate until arrested by an obstacle such as prior austenite grain boundaries. On the other hand, PWHT results in the formation of many sub-boundaries, Figure 3(b), within the prior austenite grains as a part of tempering processes. The sub-boundaries may act as barriers to the cleavage crack propagation, and thus effectively refines the prior austenite grain size. It is this effective grain size, not the prior austenite grain size, which may play an important role in determining the ductile-to-brittle transition temperatures. The effective grain size was essentially identical in the three different prior austenite grain sizes. Therefore, it is reasonable to expect that the T_{30} values after PWHT should remain constant in the range of the prior austenite grain sizes investigated.

It should be emphasized that the results and discussion of the toughness data in Figure 2 are, in a strict sense, valid only for the case of martensite in the C-Mn steel composition (A1). Different microstructure and/or modifications in steel chemistry such as microalloy additions may produce different effects of prior austenite grain size on transition temperature.

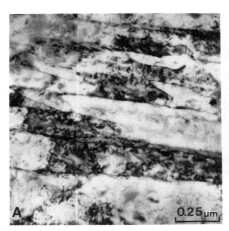 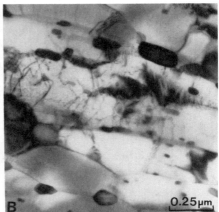

Fig. 3 - Simulated martensite in A1 steel
(a) as-welded (b) PWHT at 600°C.

EFFECT OF MICROALLOYING ELEMENTS ON TOUGHNESS

The beneficial role of the microalloying elements on base metal properties is well known(8,9). They provide precipitation strengthening, grain refinement in normalizing, and effective controlled rolling through retardation of recrystallization. The strengthening due to grain refinement and precipitation hardening by microalloying allow a reduction in carbon equivalent which results in the improvement in cold cracking susceptibility.

Such a beneficial role, however, must be weighed against possible detrimental effects on HAZ toughness. Much controversy exists in the literature concerning the specific role of microalloying elements on the deterioration of HAZ toughness(14-23). The discrepancy appears to be largely due to the lack of consistent characterizations of microstructure, especially the type and size of microalloy carbo-nitride precipitates. Furthermore, the microalloying effects appear to vary significantly depending on the local thermal cycles, type of microstructures, e.g., martensite and upper bainite, and the concentration of other elements present, such as carbon, nitrogen and aluminum.

In this study, therefore, experiments were designed to keep the other variables constant so that microalloy effect can be isolated. All the steels were given the initial solution heat treatment at 1300°C for 7 minutes with the primary purpose of dissolving microalloy precipitates which may be present in the as-received steels. TEM examinations of the solution treated steels confirmed the absence of any visible microalloy precipitates within the detection limit(about 7Å) of the microscopes used. Subsequent thermal simulations were adopted for each steel composition in order to produce martensite with the prior austenite grain size of about 120 μm. The specimens were then subjected to PWHT at 600°C. The results of impact tests are plotted in Figure 4 showing T_{30} vs. type of microalloys studied. In terms of T_{30} values, the base steel (A1) shows the best toughness followed by steel A4 (Cu + Ni), A2(Nb), and A3 (Ti + N). This study clearly demonstrates that the three types of microalloys are all detrimental to the toughness of martensite after 600°C PWHT, and that Cu + Ni is the preferred alloy addition to Nb or Ti from the toughness point of view. However, it should be recognized that the results in Figure 4 are valid for the specific set of conditions used in these experiments, namely, alloy compositions given in Table 1, martensite with PWHT at 600°C, etc. The extent of toughness deterioration due to microalloys may vary significantly, depending on the type of HAZ microstructures and the concentration of microalloys and/or other alloying elements present in the steel compositions.

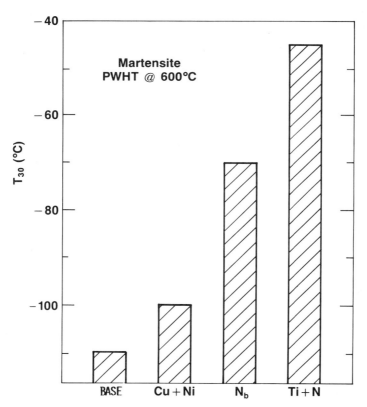

Fig. 4 - Effect of microalloying elements on T_{30} for martensite subjected to PWHT at 600°C.

Analytical electron microscopy analysis was conducted on the steel A2 to understand the mechanism by which Nb reduces toughness of martensite after PWHT. Transmission electron micrographs in Figures 5(a) and (b) show dislocation morphologies observed in the martensite before and after PWHT, respectively. Due to the role of Nb in the retardation of softening during tempering, dislocation density remained fairly high after the PWHT-related recovery process. X-ray microanalysis was carried out to determine the Cottrell atmosphere around the dislocation cores. The electron beam was intentionally astigmated to collect the fluorescent X-ray signals from the length of a dislocation. The extremely intense electron source available from the field emission gun in the Philips 400T-FEG microscope allowed the detection of Nb segregation at the dislocation cores in the post weld heat treated specimens. In the as-quenched martensite, N_b atoms were in solution and the Nb segregation at dislocations was not dectected by the microanalysis procedure outlined above.

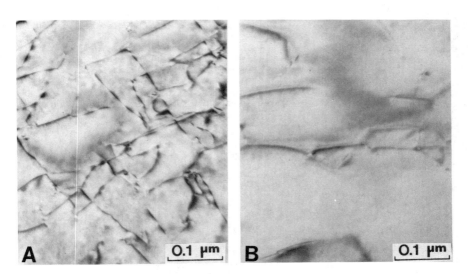

Fig. 5 - TEM micrograph showing dislocation morphology in the coarse grain HAZ near fusion line, A2 steel. (a) As-welded, Nb atoms are in solution (b) PWHT at 600°C, Nb atoms segregate to dislocations.

For the PWHT specimen, careful TEM examinations including microdiffraction and trace analysis were conducted to observe possible formation of Nb(C,N) precipitates which are found in the quenched and tempered martensite of steels containing a higher Nb content (e.g., 0.1 wt%) than that in A2 steel. The results of this study showed that Nb(C,N) precipitation was not observed in the steel A2 after PWHT, which suggests that the majority of Nb atoms may be present at dislocations as clusters. This finding indicates that after PWHT the dislocations are rendered immobile by Nb atoms or clusters that create a pinning effect. This phenomenon, strain aging, is known to deteriorate toughness of steels(24,25). The strain aging mechanism may also be responsible for the toughness degradation of the Ti-containing steel (A3) after PWHT. In the steel A3, the Ti concentration may be too low to be detected by the TEM energy dispersive analysis and, in such a case, field ion microscopy may be employed to examine the Ti segregation at dislocation cores. This work is in progress, and will be reported elsewhere. It should be pointed out that the most common strain aging in steels(both static and dynamic) refers to the interaction between interstitial element(C, N, O,...)and dislocations, and this becomes pronounced at much lower temperatures than 600°C. In this lower temperature regime, microalloying elements may be beneficial in reducing the severity of strain aging problem. In this paper, strain aging refers to dislocation segregation processes which involve substitutional elements, and hence is a higher temperature process.

RELATIVE TOUGHNESS OF LBZ CONSTITUENT MICROSTRUCTURES.

Microstructure and toughness of three types of LBZ microstructures are compared in this section. The three microstructures--martensite, upper bainite, and dual phase--were produced from the Al steel and had the prior austenite grain size of 120 μm. In the present study, the martensite and upper bainite in the as-welded condition represent the simulation of LBZ microstructures in a single-pass weld and, therefore, have not experienced subsequent thermal cycles which are typically encountered in a multipass as-welded condition. The as-welded dual phase simulates the intercritically reheated coarse grain region in a double-pass weld.

LBZ Microstructures: The morphology of the simulated martensite in the as-welded condition, Figure 3, was predominantly lath type with no evidence of twinned substructure. Regions of autotempered martensite were occasionally observed. A characteristic feature of upper bainite is shown in Figures 6(a) and (b).

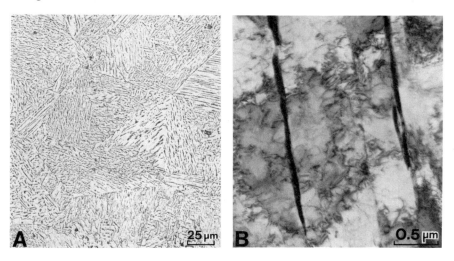

Fig. 6 - Synthetic upper bainite in Al steel. (a) Optical micrograph and (b) TEM image of a region in (a).

The presence of the long cementite precipitates along the bainitic lath boundaries, Figure 6(b), is typical of the upper bainite morphology found in both simulated microstructure and the actual weld HAZ. The as-welded dual phase microstructures are illustrated in Figures 7 and 8. The microstructure is formed by heat treatment of martensite in the $(\alpha + \gamma)$ two

128

phase region followed by rapid cooling to ambient temperature. During the intercritical heat treatment, austenite nucleates and grows preferentially along the prior austenite grain boundaries and martensite lath boundaries. Concomitant with the austenite growth, carbon partitioning occurs, resulting in the carbon enrichment in austenite and carbon depletion in the surrounding matrix. Partitioning of the substitutional alloying elements is unlikely due to the short duration at the intercritical temperature regime and the sluggish diffusion at the temperature(26,27). Carbon concentration in the austenite is determined by the intercritical temperature and holding time at the temperature, which in turn determines the volume percent of austenite in the (α + γ) region.

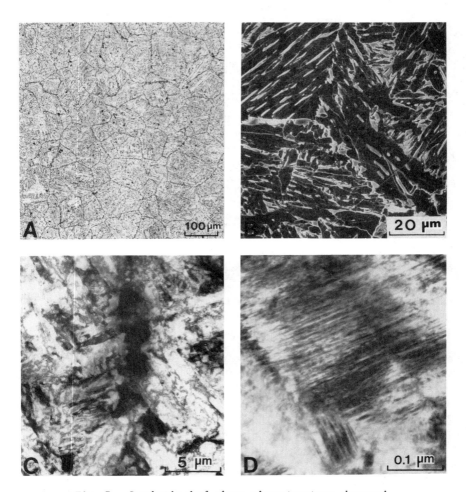

Fig. 7 - Synthetic dual phase microstructure observed in Al steel. (a) Optical micrograph, (b) SEM image, (c) TEM image, and (d) a higher mignification TEM image of the martensite islands located at the prior austenite grain boundaries.

129

Upon cooling from the intercritical temperature, the austenite will either be stabilized to room temperature, or transforms to martensite or diffusional products, depending on the austenite particle size, hardenability, and cooling rate. Such a dual phase microstructure is also called "M-A constituents" in the literature.

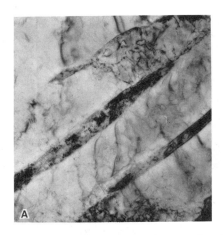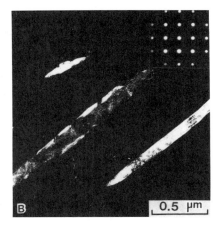

Fig. 8 - Retained austenite in dual phase microstructure (Fig. 7), (a) bright field and (b) dark field images. The inset microdiffraction pattern [(100)$_\gamma$] identifies the austenite phase.

Electron microscopy analysis showed that the austenite islands nucleated at the prior austenite grain boundaries transformed to high carbon martensite while those formed at the prior martensite lath boundaries did not undergo phase transformations and are stabilized at room temperature. In the SEM micrograph, Figure 7(b), the second phase (martensite and austenite) shows a smooth and featureless appearance. However, TEM examination reveals that the second phase islands at the prior austenite grain boundaries are primarily plate martensite with a twinned substructure as shown in Figure 8(b). This morphology is indicative of a high carbon content (>0.4 w% C) in the martensite. The second phase islands located within the prior austenite grains are elongated as shown in Figure 7(d), reflecting the austenite growth morphology along the lath boundaries. Microdiffraction analysis identified them as retained austenite whose morphology is shown in a pair of bright field and dark field images (Figures 8 a and b). Figure 9 is a high resolution lattice image of the martensite/matrix interface showing the (011) lattice fringes across the interface(28). It is clear that the interface is highly coherent except for the interfacial dislocations which are introduced to accommodate the misfits between (011) matrix and (011) martensite planes.

Toughness Comparison of LBZ Microstructures: Figure 10 compares the Charpy impact energy curves of the three simulated LBZ microstructures for both as-welded and PWHT conditions. In the as-welded simulation, dual phase exhibits the least toughness followed by martensite and upper bainite. Post weld heat treatment at 600°C results in a dramatic improvement in impact toughness of martensite and dual phase, and to a lesser extent of upper bainite. After PWHT, upper bainite shows the least toughness, while martensite shows the best toughness in terms of T30 values. The two energy curves of the as-welded and PWHT conditions for each LBZ microstructure represent the extreme upperbound and lowerbound energy values in the actual multipass weld HAZ since some portion of the LBZ microstructures will experience multiple thermal cycles from the adjacent weld passes. It must be emphasized that the toughness ranking of the three LBZ microstructures may vary substantially depending on the steel compositions. Indeed, preliminary results from our research in progress indicate that the microalloys have a strong influence on the energy curves of the LBZ microstructures after PWHT.

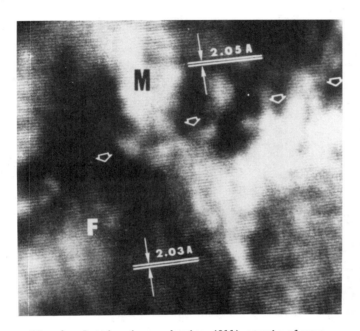

Fig. 9 - Lattice image showing (011) atomic planes across the martensite/ferrite interface. Arrows indicate the interface. Note the continuity of the (011) planes except for a few dislocations.

It is interesting to note that the as-weld dual phase is the LBZ microstructural constituent which has the lowest toughness in the as-weld HAZ. This is unexpected from the hardness point of view, since Rockwell hardness measurements showed that the hardness of dual phase is comparable to that of upper bainite, but about 1/3 that of martensite. The dual phase microstructure consisting of ferrite and martensite (with some retained austenite) has received much attention in the past decade, and this unique microstructure possesses an excellent combination of strength and formability. However, it has been shown that the microstructure is prone to embrittlement in the impact tests when the martensite islands are well connected, especially along the prior austenite grain boundaries. As shown in Figures 7 and 8, the high carbon(_ 0.5 wt.%) martensite islands are nearly continuous along the prior austenite grain boundaries. Such a grain boundary martensite can have an important influence on the cleavage fracture strength(20,32).

131

The plate martensite with a high carbon content has limited toughness, and thus, it may render an easy microcrack formation upon impact testing. This situation is analogous to the grain boundary Fe_3C films in ferrite-pearlite microstructures where a microcrack formation in the Fe_3C films can trigger a cleavage crack in the adjacent ferrite(33). Once initiated at the grain boundary martensite, a microcrack can propagate through the adjacent matrix with little resistance, since the martensite/matrix interface is highly coherent (Figure 10) and the plastic constraints provided by the continuous grain boundary martensite will limit local deformation of the matrix at the martensite/matrix interface.

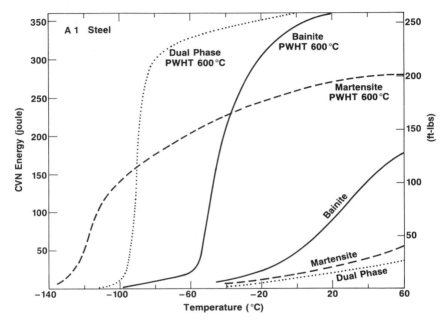

Fig. 10 - Impact energy curves of three LBZ microstructures in both as-welded and PWHT simulations, A1 steel.

The intercritically reheated coarse grain zone, where the dual phase microstructure with connected boundary martensite islands is located, occupies a small volume fraction in the coarse grain HAZ. Nevertheless, recent reports in the literature clearly demonstrate that the dual phase regions are frequent sites for crack initiation in the CTOD tests of multipass as-welded HAZ in a number of HSLA steels. The findings from the present simulation study provide a fundamental insight as to the origin of the low toughness in the intercritically reheated coarse grain zone.

SUMMARY

Local brittle zone microstructures have been produced by thermal simulation methods and their relative toughness values determined by Charpy V-notch impact test. The investigation was conducted using model alloys with a systematic variation in compositions. Based on this study, the following conclusions may be drawn:

1. Prior austenite grain size has no significant influence on the ductile-to-brittle transition temperature of tempered(PWHT) martensite in a C-Mn steel

2. Microalloying elements investigated in this study degrade toughness of PWHT martensite relative to that in C-Mn steels. The embrittlement occurs by strain aging phenomenon

3. Dual phase microstructure containing connected martensite islands along the prior austenite grain boundaries exhibits the least toughness in the as-welded condition

REFERENCES

1. Kenneth Eastering, Introduction to the Physical Metallurgy of Welding(Butterworths,1983), pp 150-153.

2. D. P. Fairchild,"Local Brittle Zones in Steel Weldments", in this proceedings.

3. C. Thaulow, A. J. Paauw, A. Gunleiksrud, and O. J. Naess, "Heat Affected Zone Toughness of a Low Carbon Microalloyed Steel," Metal Construction, 17(2)(1985),pp 94-99.

4. K. Uchino et al., "Development of 50 Kgf/mm Class Steel for Offshore Structural Use with Superior HAZ Critical CTOD," Proceedings of the Fifth International Offshore Mechanics and Arctic Engineering Conference, ASME Publication, New York, NY, Vol.2, 1986, pp 373-380.

5. H. G. Pisarski and R. J. Pargeter, "Fracture Toughness of HAZs in Steels for Offshore Platforms," Metal Construction, Vol.16, 1984, p.412.

6. C. P. Royer, "A User's Perspective on Heat Affected Zone Toughness," in this proceedings.

7. R. Ayer and J. Y. Koo, unpublished research, Exxon Research and Engineering Co.

8. F. B. Pickering,"The Spectrum of Microalloyed High Strength Low Alloy Steels," in HSLA STEELS-Technology & Applications(ASM Publication, 1984), pp.1-31.

9. W. Roberts, "Recent Innovations in Alloy Design and Processing of Microalloyed Steels," in HSLA STEELS-Technology & Applications(ASM Publication, 1984), pp.33-65.

10. A. Plumtree, "Microstructural Effects on the Ductile-Brittle Transition Temperature," Proc. of 5th int'l OMAE Symposium(ASME Publication, 1986), Vol.2, pp.228-235.

11. R. E. Smallman, Modern Physical Metallurgy(Butterworth & Co, 4th Edition,1985), pp.477-484.

12. R. W. Armstrong, "The Influence of Polycrystal Grain Size on Several Mechanical Properties of Materials," Metallurgical Transactions, Vol.1, 1970, pp.1169-1176.

13. S. Matsuda, T. Inoue, H. Mimura, and Y. Okamura, "Toughness and Effective Grain Size in Heat-Treated Low-Alloy High-Strength Steels," in Toward Improved Ductility and Toughness(Climax Molybdenum Campany Publication), 1971, pp.45-67.

14. P. L. Threadgill, "Titanium Treated Steels for High Heat Input Welding," The Welding Institute Research Bulletin, Vol.22, 1981, pp.189-196.

15. B. A. Graville, M. A. MWeldl and A. B. Rothwell, "Effect of Niobium on HAZ Toughness Using the Instrumented Charpy Test," Metal Construction, Oct.1977, pp.455-456.

16. R. E. Dolby and G. G. Saunders, "Metallurgical Factors Controlling the HAZ Fracture Toughness of Carbon-Manganese and Low Alloy Steels," in The Toughness of Weld Heat-Affected Zones(The Welding Institute Publication,1975), pp.43-66.

17. M. W. F. Cane and R. E. Dolby, "Metallurgical Factors Controlling the HAZ Fracture Toughness of Submerged-Arc Welded C-Mn Steels," ibid, pp.93-114.

18. N. E. Hannerz, "Effect of Cb on HAZ Ductility in Constructional H.T. Steels," Welding Journal, 54(5)(1975),pp.162s-168s.

19. N. E. Hannerz and B. M. Jonsson-Holmquist, "Influence of Vanadium on the Heat-Affected-Zone Properties of Mild Steel," Materials Science Journal, Vol.8, 1974, pp.228-234.

20. D. P. Fairchild and J. Y. Koo, "The Effect of Post-Weld Heat Treatment on the Microstructure and Toughness of Offshore Platform Steels," Proc. of the 6th Int'l OMAE Symposium(ASME Publication, 1987), Vol.3, pp.121-130.

21. A. B. Rothwell, J. T. McGrath, A. G. Glover, B. A. Graville and G. C. Weatherly, "Heat-Affected-Zone Toughness of Welded Jounts in Micro-Alloy Steels," Int. Inst. of Welding, DOC IX-1147(1980).

22. E. Levine and D. C. Hill, Met.Trans.A, Vol.8A, 1977, p.1453.

23. L. J. Cuddy, J. S. Lally and L. F. Porter, "Improvement of Toughness in the HAZ of High-Heat-Input Welds in Ship Steels," in HSLA STEELS-Technology & Applications(ASM Publication, 1984), pp.697-703.

24. G. Jun and W. F. Hosford, "Flow Behavior of an Aluminum-Killed Steel after Tensile Prestraining and Strain-Aging," Met.Trans.A, Vol.17A, 1986, pp.1573-1575.

25. A. K. Sachdev, "Dynamic Strain Aging of Various Steels," Met. Trans.A, Vol 13A, 1982, pp.1793-1797.

26. P. Wycliffe, "Microanalysis of Dual Phase Steels," Scripta Metallurgica, Vol.18, 1984, pp.327-332.

27. J. Y. Koo, M. Raghavan and G. Thomas, "Compositional Analysis of Dual Phase Steels by Transmission Electron Microscopy," Met.Trans.A, Vol.11A, 1980, pp.852-854.

28. J. Y. Koo and G. Thomas, "Design of Duplex Low Carbon Steels for Improved Strength : Weight Applications," in Formable HSLA and Dual-Phase Steels(AIME Publication, 1979), pp.40-55.

29. Formable HSLA and Dual-Phase Steels(AIME Publication, New York, NY, 1979). A. T. Davenport, ed.

30. Structure and Properties of Dual-Phase Steels(AIME Publication, New York, NY, 1979). R. A. Kot and J. W. Morris, ed.

31. Fundamentals of Dual-Phase Steels(AIME Publication, New York, NY, 1981). R. A. Kot and B. L. Bramfitt, ed.

32. J. Y. Koo and G. Thomas, "Metallurgical Factors Controlling Impact Properties of Two Phase Steels," Scripta Matallurgica, Vol.13, 1979, pp.1141-1145.

33. C. J. McMahon and M. Cohen, "Initiation of Cleavage in Polycrystalline." Acta Met., Vol.13, 1965, p.991.

HAZ MICROSTRUCTURE AND TOUGHNESS

OF PRECIPITATION STRENGTHENED HSLA STEELS

AND CONVENTIONAL HY-GRADE HIGH STRENGTH STEELS

by

M.R. Krishnadev*, J.T. McGrath**, J.T. Bowker** and S. Dionne*

* Department of Mining and Metallurgy,
Laval University, Quebec G1K 7P4, Canada

** Welding Section, PMRL, CANMET,
568, Booth St., Ottawa Ont. K1A 0G1
Canada

Abstract

Influence of heat input, sulphur level and inclusion morphology on the HAZ toughness and microstructure of copper precipitation strengthened HSLA steel A-710, designated HSLA-80 for naval structural applications, has been studied and compared with that of classical high strength steels HY-80 and HY-100. Variations in toughness have been related to the changes in the HAZ microstructure and fracture characteristics. Potentialities for developing higher grades (HSLA-100) of HSLA steel via alloying modifications to the basic copper-containing composition have been demonstrated.

137

Table 1: Chemical Composition of High Strength Base Metals

Steel			Weight Percent										
No	Code	Type	C	Mn	Si	S	P	Ni	Cr	Mo	Cu	V	Nb
1	A	HSLA-80	0.04	0.48	0.28	0.009	0.010	0.900	0.700	0.19	1.16	-	0.038
2	B	HSLA-80	0.03	0.40	0.24	0.012	0.008	0.800	0.700	0.22	1.18	-	0.040
3	C	HSLA-80	0.06	0.50	0.27	0.004	0.007	0.920	0.660	0.25	1.02	0.005	0.044
4	D	HY-80	0.17	0.30	0.19	0.014	0.007	2.590	1.530	0.42	0.03	0.010	0.005
5	E	QIN	0.17	0.29	0.25	0.005	0.010	2.660	1.390	0.36	0.07	0.010	0.015
6	F	HY-100	0.17	0.33	0.23	0.008	0.010	2.620	1.520	0.48	0.04	0.005	0.012
7	J	HSLA-100	0.050	1.77	0.36	0.005	0.025	0.975	0.635	0.23	1.45	-	0.050

Note: 1) Steels A to F are commercial high strength steels whereas steel J is an experimental HSLA steel under development.
2) Steels B and C have been treated with calcium for shape control.

138

Table 2. Heat Treatment of High Strength Steels

Steel			Heat Treatment
No	Code	Type	
1	A	HSLA-80	900°C, 60 min, W.Q., temper 600°C, 60 min, W.Q.
2	B	HSLA-80	900°C, 60 min, W.Q., temper 650°C, 75 min, A.C.
3	C	HSLA-80	904°C, 60 min, W.Q., temper 604°C, 60 min, W.Q.
4	D	HY-80	904°C, 70 min, W.Q., temper 715°C, 70 min
5	E	Q1N	904°C, 70 min, W.Q., temper 715°C, 70 min
6	F	HY-100	904°C, 70 min, W.Q., temper 660°C, 70 min
7	J	HSLA-100	Direct-Quench after rolling, temper 600°C, 60 min

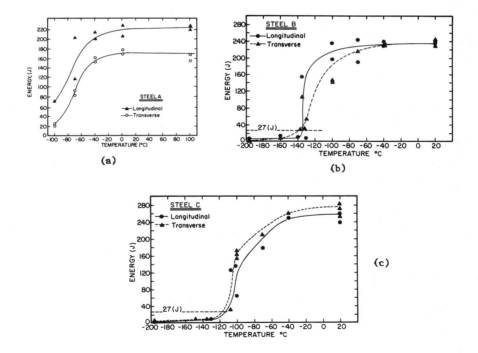

Fig. 1 Charpy impact energy transition curves
for HSLA-80 steels
a) Steel A   b) Steel B   c) Steel C

Introduction

For the past several years the authors have been involved in the development and characterization of high strength low alloy steels for artic and naval structural applications (1-5). The work has spanned the characterization of commercial as well as experimental high strength steel compositions, in particular, those containing copper as a major precipitation hardening element. The present paper is an extension of the earlier studies and examines (i) the structure and notch toughness properties of the base metal and heat affected zone (HAZ) of commercial copper-containing HSLA steel A710 (designated HSLA-80 for naval structural applications) and (ii) how the HAZ toughness of HSLA-80 steel compares with that of classical HY-80 and HY-100 grades of steel under varying weld process energy input conditions (2 kJ/mm and 4 kJ/mm). Some results of a preliminary study of the development of an HSLA-100 grade experimental steel with improved HAZ notch toughness are also presented.

Experimental Procedure

Materials

Compositions of the steels used in the study are shown in Table 1. The first six compositions are commercial steels and are generally used for naval structural applications. The HY-series and Q1N series are representative of high carbon, hardenable quenched and tempered steels with the Q1N steel having the lowest sulphur content. HSLA-80 steels have a much lower carbon level than conventional high strength steels and contain around 1.2% copper to provide substantial precipitation strengthening after heat treatment. Heat treatments of the steels are included in Table 2. An experimental HSLA-100 steel is included in Table 1. This steel contains a higher level of manganese, copper and niobium than normally used in commercial HSLA-80 steels.

Simulation of the HAZ

When the steels are welded in the fabrication of engineering structures, the heat affected zone (HAZ) at the welds will have notch toughness properties which will be dependent upon the microstructure developed during the heating and cooling cycles associated with welding. The Gleeble 1500 thermal/mechanical simulator was used to simulate the coarse grained HAZ (CGHAZ) in the high strength steels. A peak temperature of 1350°C and cooling times of 11.5 and 45 seconds between 800 and 500°C represented the cooling rates for welding of a 25 mm plate at energy inputs of 2 and 4 kJ/mm, respectively.

The Gleeble specimens, 10 mm x 10 mm x 75 mm, were machined from the base materials with the long axis transverse to the plate rolling direction. The simulated CGHAZ was contained in approximately 5 mm of the central portion of the specimen. Following the weld thermal cycling, specimens were notched through-thickness for Charpy-V notch toughness testing.

Mechanical Testing

For hardness testing, a Vickers hardness testing machine with 10 kg load was used. For impact testing, full size standard Charpy specimens have been used.

Metallography

For optical microscopy, the polished surface of the specimen was etched in 2% nital. Fractured surfaces were examined by scanning electron micro-

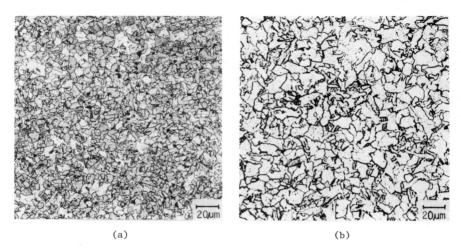

(a)   (b)

Figure 2 - Optical Micrographs a) Steel B, b) Steel C.

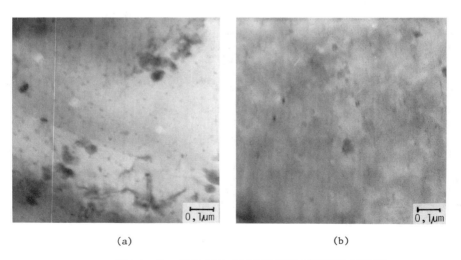

(a)   (b)

Figure 3 - Electron micrographs showing copper
precipitation. a) Steel B, b) Steel C.

141

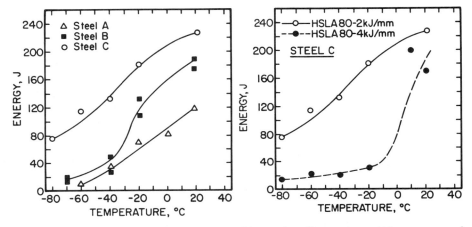

Figure 4 – Charpy transition curves of simulated CGHAZ in HSLA-80 steels (2 kJ/mm).

Figure 5 – Charpy transition curves of simulated CGHAZ in HSLA-80 steel (Steel C, 2 and 4 kJ/mm).

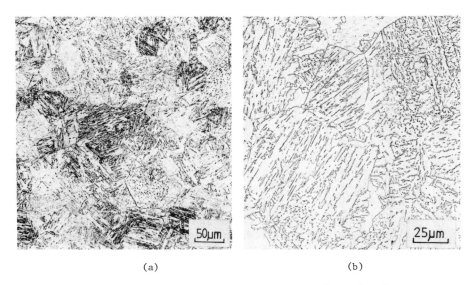

(a)   (b)

Figure 6 – Optical micrograph of simulated CGHAZ of HSLA-80 (Steel C) a) 2 kJ/mm b) 4 kJ/mm

Metallography

For optical microscopy, the polished surface of the specimen was etched in 2% nital. Fractured surfaces were examined by scanning electron microscope (SEM). For transmission microscopy, slices with a thickness 0.3 mm were cut from bulk material with an Isomet low speed saw. They were mechanically thinned by grinding and chemically polished in a solution of $HF-H_2O-H_2O_2$. The final thinning was carried out by means of electrochemical polishing in a solution of $Na_2CrO_4-CH_3COOH$ with a jet thinning apparatus (South Bay Technology Inc.). The foils were examined in a Philips EM 420 transmission electron microscope at 120 kV.

Results and Discussion

HSLA-80 Steels

Structure and Notch Toughness of Base Plate. Figure 1 shows Charpy impact energy transition curves for three commercial HSLA-80 grade steels. In steel A (see Fig. 1a), which was not subjected to any inclusion shape control, anisotropy in the impact properties is very clear, i.e. the upper shelf energy values are much higher for the longitudinal specimens than the transverse specimens. It is interesting to note that although steel B has a slightly higher sulphur level than steel A, its impact properties (Figure 1b) are much superior and there are no significant differences between the longitudinal and transverse impact properties. This is due to the inclusion shape control. Comparing steels C and B, steel C had the lowest sulphur level and the highest upper shelf energy (see Fig. 1c) (270 J); however, steel B had a lower transition temperature. This is mainly related to the fact that steel B has a lower carbon level and a finer grain size than steel C (for example see Fig. 2). Also, although steel B was in the class 3 condition (i.e. quenched and tempered) it had been aged at a higher temperature (650°C for 75 min.) than steel C (600°C for 1 hour), and had a large number of copper particles precipitated in the matrix (see Fig. 3). Finely distributed copper particles are effective in distributing slip uniformly and improving toughness (6).

HAZ Microstructure and Toughness

The Charpy impact transition curves of the Gleeble simulated CGHAZ for the HSLA-80 steels are shown in Figures 4 and 5 as a function of energy input. At both energy inputs of 2 and 4 kJ/mm the CGHAZ of steels C and B had an overall higher notch toughness than that of steel A. This is illustrated in Figure 4 for the energy input of 2 kJ/mm. This ranking of toughness for the CGHAZ, which is similar to that exhibited by the base metals particularly in the upper shelf region, can be related to sulphide inclusion characteristics. The CGHAZ of steel A, which does not have inclusion shape control, has the lowest upper shelf energy, while steel C with inclusion shape control in addition to having the lowest sulphur level has the highest upper shelf energy. All these steels have been subjected to the same weld thermal cycle which has resulted in similar microstructures, i.e. low carbon martensite for the 2 kJ/mm simulation and coarse upper bainite after 4 kJ/mm energy input.

The most important observation for HSLA-80 steels was the fact that the low temperature notch toughness decreased sharply for specimens subjected to the 4 kJ/mm simulation. This is shown typically for steel C in Figure 5. This difference in low temperature toughness between the 2 and 4 kJ/mm simulations for the HSLA-80 steels can be explained on the basis of microstructure. The microstructure of the CGHAZ of the HSLA-80 steel was composed predominantly of low carbon martensite with some bainitic regions after simula-

143

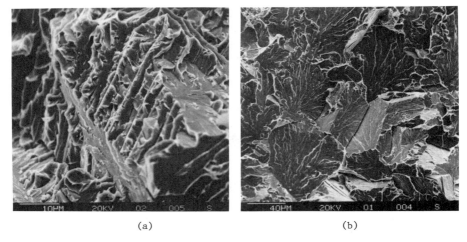

(a)   (b)

Figure 7 - SEM fractograph showing cleavage facets
for simulated CGHAZ of HSLA-80 steel (Steel C)
a) 2 kJ/mm, b) 4 kJ/mm.

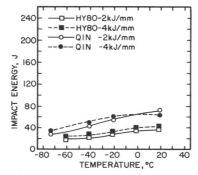

Figure 8 - Charpy transition curves
of simulated CGHAZ in HY-80 and
Q1N steels.

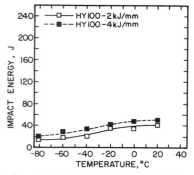

Figure 9 - Charpy transition curves
of simulated CGHAZ in HY-10 steel.

Figure 10 - Microstructure of simulated CGHAZ
in HY-80 cycled at 4 kJ/mm.

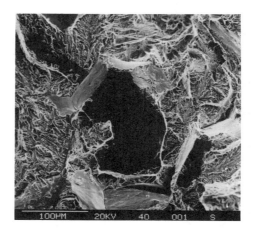

Figure 11 - SEM fractograph showing intercrystalline facets on fracture surface of HY-80.

tion at 2 kJ/mm. Using the terminology of Torronen (7) the martensitic structure in Figure 6a consisted of elongated packets which formed in bundles within prior austenite grains. Although transmission electron microscopy was not employed to measure the orientation relationship between adjacent packets it was assumed, based upon other studies of low carbon martensite (7,8) and from SEM observations of fractured surfaces, that the boundaries between packets were mainly high angled. The results of SEM examination of the fractured surface, shown in Figure 7a, indicated that the cleavage facet size corresponded to the dimensions of the individual martensite packets, i.e. 5 to 10 μm. This indicates that when cleavage fracture initiates within an individual packet it must reinitiate at each high angle packet boundary in order to continue propagation through the CGHAZ. The finer the packet size, the higher the resistance to cleavage fracture.

The CGHAZ of the HSLA-80 steel contained entirely coarse bainite after simulation at 4 kJ/mm. This bainitic structure, shown in Figure 6b, consisted of parallel bainitic ferrite laths separated by elongated martensite-austenite (MA) regions. The misorientations between adjacent laths are assumed to be low angle boundaries from fractographic evidence and electron diffraction results in other investigations of bainitic microstructure (9). For the bainitic microstructure, a packet consists of a group of parallel laths. As shown in Figure 6b a prior austenite grain may contain one or as many as three packets. The large cleavage facets (40 to 80 μm) observed by SEM examination of the fractured surfaces of the 4 kJ/mm simulated specimens (Fig. 7b) related to the bainitic packet size shown in Figure 6b.

The prior austenite grain sizes of the 2 and 4 kJ/mm HAZ simulation are similar since they were both heated to the same peak temperature of 1350°C. It should be made clear that the packets in the low C martensite are elongated, narrow (5 to 10 μm wide) and are separated by high angled boundaries. TEM would likely reveal a parallel lath structure within each packet where the subboundaries are low angle. On the other hand, the packets i.e. a unit separated by high angle boundaries, in the coarse bainite are large by comparison being of the order 40 to 80 μm. Within each packet are parallel laths with widths of the order of 3 to 8 μm. The lath boundaries are low angle and contain M-A structure. The M-A constitute may contribute to the lower toughness observed at 4 kJ/mm but the most important factor is that cleavage cracks can easily propagate across a bainite packet with little resistance encountered at the low angle lath boundaries within a packet. The SEM observations of small cleavage facets at 2 kJ/mm and large cleavage facets at 4 kJ/mm support this fracture hypothesis.

Comparison of HAZ Structure and Toughness of HSLA-80, HY-80, HY-100 and Q1N Steels

The notch toughness properties of the Gleeble simulated CGHAZ for the HY-80, HY-100 and Q1N steels are shown in Figures 8 and 9. There was no change in notch toughness properties with energy input for both the HY-80 and HY-100 steels. The CGHAZ toughness for the Q1N steel also did not vary significantly with energy input but the overall level of toughness was higher than either the HY-80 or HY-100 steels. Using the results for the HSLA-80 steel C for comparison purposes, the CGHAZ of HSLA-80 steel simulated at the energy input of 2 kJ/mm has superior toughness to that of the HY series and Q1N steels over the full test temperature range (Fig. 5). However, for specimens of HSLA-80 subjected to the 4 kJ/mm simulation, a low value of notch toughness was observed at temperatures < 10°C which was comparable to that found in the HY-series steels.

A totally martensitic microstructure in the CGHAZ was observed at both the 2 and 4 kJ/mm energy input simulations for the HY-80, HY-100 and Q1N steels. This structure is shown typically in Fig. 10. The microhardness readings for these simulated HAZ specimens were all in excess of 400 DPH as indicated in Table 3. SEM observations of fractured surfaces (Fig. 11) revealed the presence of intercrystalline facets in the simulated HY-80, HY-100 and Q1N specimens. However Q1N contained a significantly lower proportion of intercrystalline discontinuities compared to HY-80 and HY-100.

Table 3. Microhardness of as-received and simulated CGHAZ in the high strength steels

Base Metal		Microhardness, DPH		
			Simulated CGHAZ	
Code	Type	As-received	2 kJ/mm	4 kJ/mm
D	HY-80	217	410	409
E	Q1N	241	429	419
F	HY-100	287	452	450
C	HSLA-80	246	306	242
J	HSLA-100	323	328	---

The notch toughness properties of the CGHAZ of the HY series and Q1N steels can be explained on the basis of these microstructural characteristics. For the high carbon, hardenable quenched and tempered HY series steels, it was felt from previous work (10,11) that the liquation of sulphide inclusions in the CGHAZ during welding was a significant factor in lowering the notch toughness properties of this region. The sulphides which melt in the HAZ near the fusion line during welding penetrate and weaken austenitic grain boundaries by forming sulphide films during cooling (12,13). The tensile stresses generated in the welding cycle can produce discontinuities or microcracking along grain boundaries in the CGHAZ. Liquation embrittlement in the HAZ of HY-80 weldments has been shown to increase with increasing carbon content and to decrease with increasing Mn/S ratio (14). The study of Q1N steel with a lower sulphur (0.005%) and higher Mn/S ratio (58/1) was intended to reduce liquation embrittlement in the CGHAZ and hence to significantly improve the notch toughness properties. The notch toughness results in Figure 8 indicate an improvement in toughness in the Q1N Gleeble simulated specimens compared to the HY-80 steel. Thus, despite the reduction in liquation embrittlement, the presence of high carbon untempered martensite in the CGHAZ of the Q1N steel as well as the HY-80 and HY-100 steels makes the most significant contribution to the overall low toughness of the CGHAZ. As discussed previously, the excellent low temperature notch toughness properties of the CGHAZ of the HSLA-80 steel C, subjected to Gleeble simulation of 2 kJ/mm, was attributed to the characteristics of the low carbon martensite microstructure.

Experimental HSLA Steel (HSLA-100 grade)

The detailed chemical analysis of the composition is shown in Table 1 (steel No. 7). The material was cast as 54 kg ingot of dimensions 127 mm x 140 mm x 305 mm. Ingot was then cut to nominal dimension of 127 mm high x 140 mm wide and 204 mm long. The slab was soaked at 1250°C for 2½ hours and then controlled rolled in 15 passes to a final thickness of 12.75 mm as per

147

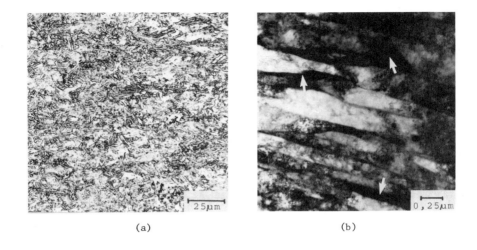

(a)   (b)

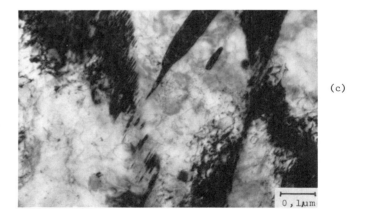

(c)

Figure 12 - Microstructure of experimental HSLA steel
(Steel J). a) optical, b) TEM showing acicular ferrite
laths and M-A (see arrows), c) TEM showing twins and
dislocation networks.

148

Table 4.
Processing Schedule for the Experimental HSLA Steel (steel J)

Soaking temperature = 1100°C
Slab thickness = 125 mm
Soaking time = 2.5 hr

PASS SCHEDULE

Pass No.	Mill Setting		Temperature
	(In)	(mm)	(°C)
1	4.600	116.80	1100
2	4.200	106.70	1080
3	3.800	96.50	1060
4	3.400	86.40	1040
5	3.030	75.80	1020
6	2.700	68.60	1000
7	2.400	61.00	985
8	2.100	53.30	970
9	1.820	45.50	950
10	1.550	39.40	930
11	1.300	32.50	910
12	1.070	26.75	890
13	0.860	21.50	870
14	0.650	16.50	850
15	0.500	12.70	830

Note : After rolling plates were water-quenched

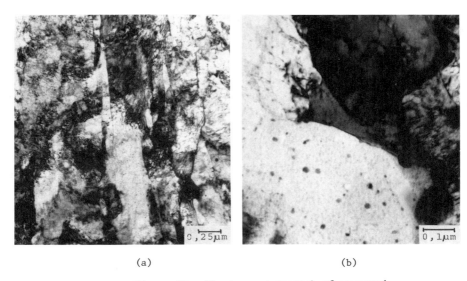

(a)   (b)

Figure 13 - Electron micrograph of tempered
experimental HSLA steel showing a) highly
dislocated lath structure and
b) copper precipitation.

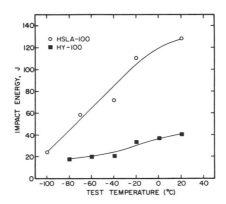

Figure 14 - Charpy transition curves for simulated
CGHAZ in tempered experimental HSLA steel and
HY-100 steel.

Table 5

Hardness and Estimated Yield and Tensile Strength
Values for Experimental HSLA-Steel (tempered)

Hardness		YS		UTS	
Base Metal	HAZ	MPa	Ksi	MPa	Ksi
323	328	853	124	948	137

Note: For copper containing steels, $\dfrac{UTS}{VPHN} = 0.19$
Where UTS is in tons per sq.in. (15)

Table 6

Weldability Indicators

No.	Code	Type	P_{cm}	Ceq	HAZ Hardness
6	F	HY-100	0.316	0.803	452
7	J	HSLA-100	0.240	0.680	328

Note: $P_{cm} = C + \dfrac{Mn+Cu+Cr}{20} + \dfrac{Si}{30} + \dfrac{Ni}{60} + \dfrac{Mo}{15} + \dfrac{V}{10} + 5B$

$C_{eq} \quad C + \dfrac{Mn}{6} + \dfrac{Ni+Cu}{15} + \dfrac{Cr+Mo+V}{5}$

Figure 15 - HAZ microstructure of experimental
HSLA steel (2 kJ/mm).

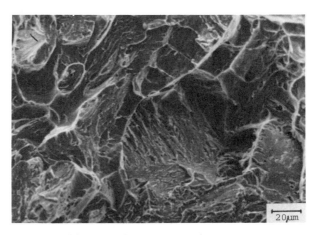

Figure 16 - SEM fractograph of experimental
HSLA steel (tempered sample, Gleeble simulated
and broken at -100°C).

the schedule shown in Table 4. The plate was water-quenched after the final pass. Blanks machined from the plates have been tempered for 1 hour at 600°C in neutral salt furnaces.

Figure 12 shows the microstructure of the as-rolled and quenched plate. Microstructure consists essentially of laths of acicular ferrite interspersed with islands of M-A. Traces of prior austenite elongated in the direction of rolling are also discernable. Extremely fine grained ferrite or acicular ferrite often decorate prior austenite boundaries. Acicular ferrite laths and M-A constituent may be seen in Figure 12(b) whereas Figure 12(c) shows twins and fine dislocation network. On tempering, there were no significant changes in the optical microstructure with the exception that elongated prior austenite boundaries were more clearly revealed. Transmission electron microscopy revealed that the microstructure was still highly dislocated and that a uniform precipitation of copper occured (see Fig. 13). Some carbonitride precipitation was also observed.

Compared to the essentially ferritic matrix of the HSLA-80 steels, the experimental steel, because of the greater hardenability provided by the higher manganese levels, has essentially an acicular ferrite matrix (for example, compare Fig. 2 with Fig. 12). This highly dislocated matrix with the added precipitation strengthening provided essentially by copper and partly by niobium carbonitride, can thus, achieve strength levels beyond that obtainable via HSLA-80 compositions.

HAZ Microstructure and Toughness. Microstructural studies, hardness measurements and impact tests have been carried out on tempered samples after they were subjected Gleeble simulation (2 kJ/mm).

Results of the impact tests are shown in Figure 14. For the sake of comparison, impact test results of Gleeble simulated samples of HY-100 are also shown. Hardness data from the base metal and the HAZ region are shown in Table 5. Table 5 also shows the yield and tensile strength values estimated from the ratio of UTS to VHN (vickers hardness number) which is usually around 0.19 for copper containing steels where UTS is expressed in terms of tons per square inch (15,16). For tempered copper steel the ratio of yield to ultimate is around 0.9 (17).

Hardness readings indicate that yield strength levels of over 690 MPa (100 Ksi) are easily obtained in this experimental composition, along with good HAZ toughness. For example, an upper shelf impact energy over 100 J with a 55 J transition temperature of -70°C or lower are realized in the Gleeble simulated samples (see Fig. 14).

The toughness could, of course, be further improved by sulphide shape control, as studies with HSLA-80 steels have indicated.

The HAZ microstructure (see Fig. 15) was similar to that of HY-80 and HY-100 steel but less hard (328 VHN vs 420 to 450 VPHN). Also fractographs of Charpy specimens broken at low temperatures showed finer fracture facets with considerable tearing and dimples (for example see Fig. 16). There was no evidence of intergranular fracture or grain boundary embrittlement as was the case with HY-80 steels (for example, see Fig. 11).

The experimental steel also has lower P_{cm} and C_{eq} values than HY-100 and HY-80 steels (see Table 6) indicating its superior intrinsic weldability compared to the conventional high strength steels. Further work is in progress to determine its HAZ notch toughness properties over a range of energy inputs.

153

Conclusions

a) In HSLA-80 steels, sulphide shape control, low levels of carbon and proper heat treatment to obtain uniform precipitation of copper are necessary to obtain a combination of high upper shelf and low transition temperatures in the base plate and HAZ. Thus more stringent control of chemistry and heat treatment may be needed for critical applications than called for in the current specifications.

b) For weld simulation of normal energy input conditions (2 kJ/mm), HSLA-80 steels had superior HAZ toughness than HY-grade steels. The low carbon martensite microstructure in the CGHAZ of the HSLA-80 steel conferred effective resistance to crack propagation.

c) Under high energy input conditions (4 kJ/mm) both HSLA-80 and the HY-80 and HY-100 steels exhibited poor HAZ toughness. The coarse upper bainite microstructure in the CGHAZ of HSLA-80 and the hard, high carbon martensite of the HY-steels were responsible for the low toughness.

d) By alloying modifications to the basic Cu-HSLA steel A710 (i.e. HSLA-80) composition and controlled processing, HSLA-100 grades of steels with superior HAZ toughness than HY-100 could be developed.

Acknowledgements

Authors are grateful to the National Research and Engineering Council of Canada and the Department of Energy, Mines and Resources for the financial support received. They also wish to express their thanks to M. Letts, J. Gianetto and R. Orr (CANMET) for their technical assistance in all experimental phases of the program.

References

1) M.R. Krishnadev, "Development and Characterization of a New Family of HSLA Steels", HSLA Steels, Technology and Applications (Metals Park, OH, American Society for Metals, 1984) p. 129-148.

2) M.R. Krishnadev et al., "An Evaluation of The Effect of Low Temperatures on Mechanical Properties of a Commercial Steel to be used in the Arctic", Journal of Testing and Evaluation, 8(1) (1980), 42-47.

3) M.R. Krishnadev and L.R. Cutler, "Strong Tough Steels with Intrinsic Atmospheric Corrosion Resistance for Structural Applications in the Arctic: The Effects of Controlled Rolling and Aging", Metals Technology, 8 (1981), 142-149.

4) M.R. Krishnadev et al., "Fracture - Microstructure - Mechanical Property Relations in an Advanced HSLA Steel", Welding, Failure Analysis, and Metallography, Microstructural Science, vol. 14, Eds. M.R. Louthan, I. LeMay and G.F. Vander Voort, (Metals Park, OH: American Society for Metals, 1984), 189-203.

5) M.R. Krishnadev et al., "Effect of Heat Treatment on the Microstructure and Fracture Toughness of a Precipitation Strengthened HSLA Steel", Microstructural Science, vol. 11, Eds. R.T. Dehoff, J.D. Braun, and J.L. McCall, (Elsevier, New York, 1983), 63-78.

6) L.R. Cutler and M.R. Krishnadev, "Microstructure - Fracture Toughness Relations in Copper Strengthened HSLA Steels", Microstructural Science, vol. 10, Eds. W.E. White, J.H. Richardson and J.L. McCall, (Elsevier,

New York, 1983), 63-78.

7) K. Torronen: <u>Microstructural Parameters and Yielding in Quenched and Tempered Cr-Mo-V Pressure Vessel Steel</u>, Research Centre of Finland, 1979, Publ. 22.

8) M.J. Roberts: "Effect of Transformation Substructure on the Strength and Toughness of Fe-Mn Alloys", <u>Met. Trans.</u>, 1 (1970), 3287-3294.

9) F.B. Pickering: "The Structure and Properties of Bainite in Steels", <u>Transformation and Hardenability in Steels</u>, Climax Molybdenum Co. of Michigan Symp., 1967, 109-129.

10) K. Romhanyi, M.R. Krishnadev, R.D. McDonald, J.T. Bowker and J.T. Mc Grath: "Comparison of Simulated HAZ Microstructure and Toughness for two High-Strength Steels", <u>Microstructural Science</u>, 14 (1985) 115-126.

11) J.T. McGrath, J.A. Gianetto, R.F. Orr and M.W. Letts: "Some Factors affecting the Notch Toughness Properties of High Strength HY-80 Weldments", <u>Can. Met. Quarterly</u>, 1987, in the press.

12) R.H. Phillips and M.F. Jordan: "Liquation Cracking in the Weld Heat-affected Zone of High Strength Ferritic Steels", <u>Metals Tech.</u>, 3 (1976), 571-579.

13) R.H. Phillips and M.F. Jordan: "Weld Heat-affected Zone liquation cracking and hot ductility in high-strength ferritic steels", <u>Metals Tech.</u>, 4 (1977), 396-405.

14) C.F. Meitzner and R.D. Stout: "Microcracking and Delayed Cracking in Welden Quenched & Tempered Steels", <u>Weld. J.</u>, 45 (1966), 393.5 - 400.5.

15) E.A. Wilson, "Copper Maraging Steels", Journal of Iron and Steel institute, 206 (1968), 164-169.

16) M.R. Krishnadev, "Development of Strong, Tough Structural Steels Based on Copper-High Silicon Additions", <u>Metals Technology</u>, 7 (1980), pp. 305-306.

17) M.R. Krishnadev and I. LeMay, "Microstructure and Mechanical Properties of a Commercial Low Carbon Copper Bearing Steels", <u>Journal of Iron and Steel Institute</u>, 208, 1970, pp. 458-462.

CHARACTERISTICS OF TMCP STEEL AND ITS SOFTENING

J. G. Youn and H. J. Kim

Welding & Materials Research Institute
Hyundai Heavy Industries, Co., Ltd.
Ulsan, Korea

Abstract

A TMCP steel manufactured by controlled rolling followed by accelerated cooling process was examined to elucidate the change in mechanical properties by softening phenomena occurring after thermal treatments such as welding and /or PWHT. The accelerated cooled steel exhibited excellent mechanical properties in both strength and toughness, which are mainly derived from its fine microstructure and hardened matrix. This steel also preserved high toughness in its HAZ even after the high heat input welding. After thermal treatments , however, both strength and fatigue properties of the accelerated cooled steel deteriorated. Because the thermal cycles can erase the accelerated cooling effect to some extent, the extent of deterioration in strength was largely dependent on the condition of thermal treatment. That is, the tensile strength of the accelerated cooled steel weldment decreased with welding heat input. Further softening in HAZ occurred after PWHT and was so severe as not to satisfy the specified strength in normal PWHT condition. In addition, the fatigue crack growth rate was higher in the softened HAZ than that in base metal. With supplimentary experiments, it could be concluded that the increase in fatigue crack growth rate in HAZ was due to softening phenomenon occurring in the HAZ rather than the other effects such as residual stress or microstructural change.

I.   Introduction

For the construction of large welded structures, welding with high heat input is preferable for better productivity. However the toughness of weldment generally tends to be deteriorated with an increase in welding heat input. Therefore the advent of steel to have sufficient toughness with high heat input welding has been critically required by the construction parties. To meet these requirement, steel manufacturers have eventually developed Thermo-Mechanical Controlled Process (TMCP) at the end of 1970's(1-3). Since then, one of TMCPs, so called accelerated cooled type TMCP, has been highlighted due to its potential capability in improving both the mechanical properties and the weldability of steel plate at the same time through microstructural modification. In early 1980's, this TMCP steel was commercialized and expanded its application into heavy structures such as shipbuilding and offshore structures.

The authors have already reported on the backgrounds of TMCP development and characteristics of TMCP steel including their problems, and briefly on the HAZ toughness of TMCP steels after high heat input welding (4-5). From these studies, it has been confirmed that the accelerated cooled TMCP steel preserves good toughness in its HAZ even after high heat input welding. However, after welding or thermal treatment such as PWHT, the strength of the accelerated cooled steel is deteriorated to some extent, so called softening phenomenon, because such thermal operations erase the accelerated cooling effects imparted on the strength of base steel plate. For the sake of the safety of weldments, such softening phenomena have to be clarified in more detail

In this study, the softening phenomena of accelerated cooled steel were investigated with respect to hardness and strength variations depending on welding heat input and/or PWHT. With information on softening phenomena, the behavior of fatigue crack growth was also studied.

II.   Characteristics of TMCP steel

From early 1970's extensive research and development have been conducted into TMCP technology to cope with the low temperature toughness problems under severe service condition like Arctic region (1-4). In parallel with the developments of TMCP, the researches on the preventing of HAZ embrittlement after high heat input welding have also been accomplished (6-7). As a results of these studies, the steel manufactured by TMCP comes to have the good combination of mechanical properties in HAZ even after high heat input welding. Comparing with conventional steels, the TMCP steel have some features in steel making process to get good mechanical properties both in strength and in toughness; 1) controlled rolling process, which endows fine microstructure 2) accelerated cooling process, which gives hardening of matrix and 3) reduction of carbon equivalent, which assures good weldability.

In the previous investigations (4-5), the characteristics of TMCP steel (hereafter called ACC steel) manufactured by controlled rolling and accelerated cooling process were evaluated by comparing with conventional steel (hereafter called N steel) manufactured by rolling and normalizing treatment. The manufacturing processes of each steel are illustrated in Figure 1. The materials used in the previous investigations were the 50Kg/mm^2 Grade high tensile strength steel plates (30mm thickness) for ship building, classified as ABS EH36. Their chemical compositions and mechanical properties are listed in Table I and Table II, respectively. These steels would be also used for the present investigation too. As shown in Table I, to achieve the specified strength with normalizing treatment, the N steel was formulated to have its carbon equivalent as high as 0.4. On the other hand, ACC steel has significantly lower carbon equivalent (Ceq. = 0.33) but was manufactured by combined techniques of controlled rolling and accelerated cooling processes (Figure 1)

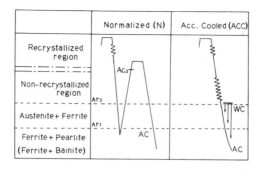

Figure 1   Schematic drawing of manufacturing processes : normalizing treatment v.s. accelerated cooled TMCP.

Table I.   Chemical compositions (wt%) of steels used.

	C	Si	Mn	P	S	Cu	Ni	Cr	Nb	Sol.Al	Ceq.*
N Steel	0.12	0.39	1.45	0.009	0.001	0.28	0.18	0.03	0.024	0.044	0.40
ACC Steel	0.06	0.14	1.33	0.010	0.001	0.31	0.31	0.05	0.015	0.034	0.33

* Ceq. = C + Mn/6 + (Cr + Mo + V)/5 + (Cu + Ni)/15

Table II.   Mechanical properties of steels used.

	Yield stress (Kg/mm^2)	Tensile stress (Kg/mm^2)	Elongation (%)
N Steel	41.2	55.5	33.3
ACC Steel	48.6	60.0	27.0

It is noticed in Table II that in spite of low carbon equivalent (Ceq.= 0.33) the ACC steel has higher yield and tensile strengthes than N steel. Such an advantage in strength with lower carbon equivalent is accounted for the refinement of ferrite grains, the formation of fine dispersed low temperature products and the hardening of matrix acquired from the manufacturing process (5).   Owing to its lower carbon equivalent, moreover, the susceptibility against cold cracking becomes very low, which contributes to the outstanding improvement in weldability.

The microstructures of each steel are shown in Figure 2.   N steel is composed of relatively uniform and equiaxed ferrite grains, which form banded structure with pearlite.   On the contrary, the microstructure of ACC steel consists of very fine ferrite grains with second phase but without banded structure.   More detailed microstructural study was performed under scanning electron microscope (SEM) to identify the second phase of each steel.   As shown in Figure 3, the second phase found in N steel shows typical lamellar structure of pearlite but that in ACC steel does not indicating this phase is more likely bainite.

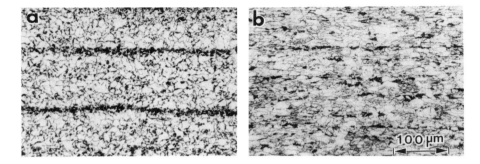

Figure 2 Microstructures of steels used : (a) normalized steel, N, and
(b) accelerated cooled steel, ACC.

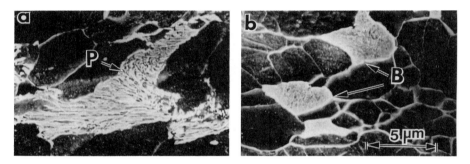

Figure 3 Comparison of second phases shown in Figure 2 : (a) pearlite
in N steel, and (b) bainite in ACC steel.

Because of the difference in microstructure and composition between ACC
and N steel, the weld thermal cycles applied in HAZ would result in different
hardness profiles for each other. Those difference in hardness is shown in
Figure 4. The HAZ of N steel weldment was hardened after welding, whereas t-
hat of ACC steel weldment was softened. If PWHT is carried out on the ACC w-
eldment, further softening in ACC steel weldment is anticipated.

This paper will describe the extent of softening occurred under various
thermal conditions in detail and its effect on fatigue properties.

Ⅲ. Experimental

The materials used in this study were normalized (N steel) and accelerated cooled steel plates (ACC steel) already shown in Table I and Ⅱ. To study the softening phenomena occurred in ACC steel, various thermal conditions
were made ; (1) heat treatments at two different conditions, 580°C for 30 mi-

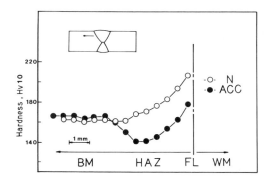

Figure 4   The hardness profile in the HAZs of N and ACC steel
welded with 70KJ/cm.

nutes and 625°C for 75 minutes, (2) welding with various heat inputs between
71 and 76 KJ/cm using SMAW and SAW processes.   The weldment made with 40 KJ/
cm was post weld heat treated at 580°C for 30 minutes and compared with N steel weldment.

Hardness was measured on these weldments using Vicker's hardness (Hv10)
for the qualitative evaluation of softening phenomena.   For evaluating the e-
ngineering importance of softening phenomena, tensile tests were performed w-
ith transverse tensile specimens.   The shape and dimension of tensile specimens taken are shown in Figure 5 (a).

To evaluate the effect of softening phenomena on fatigue properties, fatigue crack growth test was also conducted using compact tension specimens.
For making fatigue specimens, single bevelled groove was prepared and welded
with heat input of 40 KJ/cm and then compact tension specimens were extracted
in two different directions as shown in Figure 5 (b).   Depending on the direction of specimens, the machined notches were made either at 2mm apart from
fusion line parallel or perpendicular to fusion line as shown in Figure 5 (b)
.   Fatigue tests were carried out under constant load or constant stress intensity factor condition with load ratio ($P_{min.} / P_{max.}$) of 0.05.   The fatigue
crack length was measured using compliance method (8) adjusted by microscopic
measurement (X50).

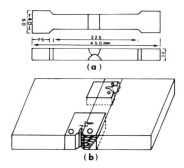

Figure 5   The location and dimension of test specimens ; (a) tensile specimen, and (b) compact tension specimen for fatigue
test

161

IV. Results and Discussion

A. Softening Phenomena of ACC Steel

The strength of ACC steel is mainly derived from accelerated cooling process. Upon thermal cycles such as welding or post heat treatment, the microstructure of ACC steel comes to be stabilized to some extent and thus a part of strength increment derived from accelerated cooling effect is relieved. To confirm these facts quantitatively, the tensile tests were performed after heat treating at 625°C for 75 minutes and after at 580°C for 30 minutes. After the heat treatment at 625°C commonly used for PWHT, the tensile strength of ACC steel was reduced remarkably by the 12.7 Kg/mm^2, resulting in 47.3 Kg/mm^2 which was lower than the specified requirement (minimum tensile strength = 50 Kg/mm^2). The tensile strength of ACC steel after thermal treatment for 30 minutes at 580°C was measured to be about 53.3 Kg/mm^2, which was reduced by the 6.7 Kg/mm^2. From these results, it can be confirmed that the thermal treatment lowers the strength of ACC steel to a large extent and its condition is a critical factor controlling the amount of strength reduction. Furthermore, the normal condition for PWHT would not be applicable in ACC steel weldment.

As mentioned earlier, the welding thermal cycles also can erase the accelerated cooling effect (Figure 4) and thus can cause reduction in tensile strength of weldment. To evaluate the effect of welding heat input on softening phenomenon, hardness profiles were established in the welded joints prepared by different heat inputs. The typical hardness profiles in HAZs with three different heat inputs is shown in Figure 6. The minimum hardness values of HAZs acquired from Figure 6 and from other causes are replotted as a function of welding heat input in Figure 7. These results demonstrate that all the HAZs were softened except the one welded with lowest heat input of 17 KJ/cm and the amount of softening increased with increasing heat input. These results indicate that the softening phenomenon of ACC steel weldment is occurred only when the cooling rate upon welding is slower than critical value, perhaps that applied for accelerated cooled steel plates. It is also worth to note in Figure 6 that minimum hardness value of HAZ decreased sharply and then slowly with increasing heat input. These fact suggests that the minimum hardness occurred in HAZ in accordance with heat input seem to converge a certain limiting value, which is determined by both carbon equivalent and grain size of ACC steel.

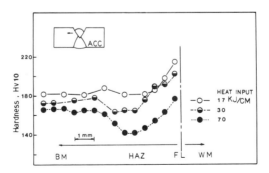

Figure 6   The variation of hardness profiles in HAZ of ACC steel welded with heat inputs of 17, 30 and 70 KJ/cm.

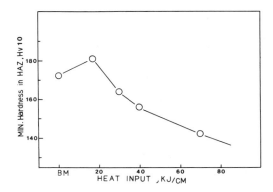

Figure 7　The variation of minimum hardness in HAZ of ACC steel
as a function of welding heat input.

One thing to note in Figure 6 is that the location of minimum hardness value in
HAZ was gradually shifted toward base metal with increasing welding heat input.　Furthermore, as shown in Figure 8, the increase in heat input expands t-
he width of softened region which is denoted Ls and defined as the region having lower hardness than the base metal.　From these results, it could be concluded that the extent of softening phenomenon depends largely on the welding heat input and increases with increasing heat input.

　Figure 9 shows the result of transverse tensile tests.　As expected from
hardness results (Figure 7), the tensile strength of ACC steel weldment decreases significantly with welding heat input.　Most importantly, all the tensile specimens were fractured near HAZ as shown in inserted photo in Figure 9.
This fact demonstrates the critical role of softening in determining the tensile strength of welded joint.　However the actual reduction of tensile strength of ACC steel weldment was smaller than the value anticipated from minimum hardness values.　This difference is attributed to the existence of hard
regions of weld metal and base metal at both sides of softened HAZ and thus t-
o the constraint on the necking of softened HAZ.

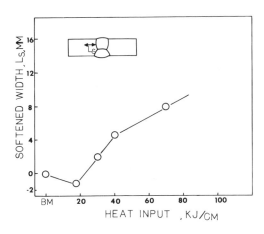

Figure 8　The width of softened HAZ varied with
welding heat input
in ACC steel weld-
ment

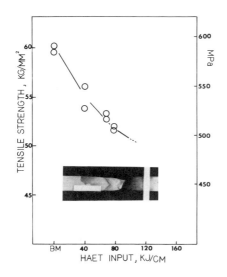

Figure 9   The tensile strength of ACC steel welded joint made with various welding heat input

If PWHT is required on ACC steel weldment, further softening is anticipated. Figure10 shows the results of hardness measurement before and after PWHT at 580°C for 30 minutes. This result clearly shows the secondary softening occurred in HAZ after PWHT as well as the initial softening in base metal.

The variation in tensile property of ACC steel weldment was compared with that of N steel weldment at the same condition. The results are shown in Figure 11. In case of N steel, little change in tensile strength is observed after welding or PWHT. However the tensile strength of ACC steel was lowered successively by applying welding and PWHT, resulting in less than 50 Kg/mm^2 after PWHT. Accordingly the application of PWHT on the ACC steel weldment should be quite limited as mentioned above.

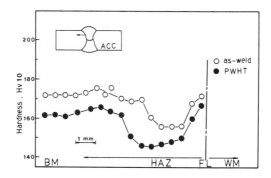

Figure 10   The effect of PWHT on softening of ACC steel weldment.

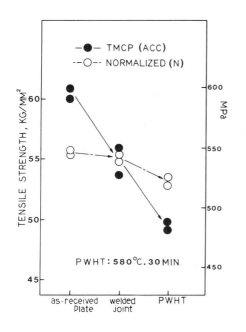

Figure 11   The effect of fabrication
conditions on the tensile
strength of ACC steel
comparing with normalized
(N) steel.

B.   Fatigue Properties of ACC Steel

To evaluate the effect of softening phenomena on the fatigue crack growth rate three sets of fatigue tests were carried out; the first was under constant stress intensity factor (constant ΔK) and the others were under constant load condition.

(1) Under constant ΔK of 30 MPa $M^{1/2}$, the fatigue crack initiated in t-
he weld metal and propagated across the softened HAZ and then to base metal,
i.e. fatigue crack propagated in direction perpendicular to fusion line as s-
hown in inserted figure in Figure 12.   The result obtained is shown in Figure
12 along with the hardness profile of corresponding region.   As shown in this
result, the rate of fatigue crack growth in HAZ was higher than that in base
metal indicating the presence of softening effect on fatigue properties.   The
rate of fatigue crack growth in HAZ was about two times higher than that in
base metal.   However, it is quite difficult to conclude that the softening e-
ffect is a dominant factor in controlling fatigue crack growth rate in HAZ s-
ince the HAZ of ACC steel weldment also includes several other factors such
as microstructural change and residual stresses.

(2)   In order to prove the effect of softening on fatigue crack growth
rate, second sets of fatigue test were carried out under constant load condition.   The first test of the second set was performed with the compact tension specimen having precrack parallel but 2mm away from fusion line (Figure 5
(b)).   In this case, fatigue crack propagated along the softened HAZ.   Figure
13 shows fatigue crack growth rate of HAZ comparing with that of base metal.
As shown in this figure, the fatigue crack growth rate in HAZ is slightly higher than that of base metal in all ΔK range.   This fact conforms the deterioration of fatigue property in the softened HAZ of ACC steel weldment.   The
second test was performed on base metal after heat treatment at 615°C for 1

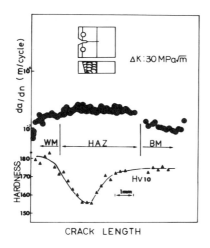

Figure 12 The change in fatigue crack growth rate in ACC
steel weldment under
constant ΔK condition.

hour. The result of tempered ACC steel is shown in Figure 14 along with that
of as-received material. It shows that the fatigue crack growth rate increases after post heat treatment than as-received condition and clearly demonstrates the softening effect on fatigue properties.

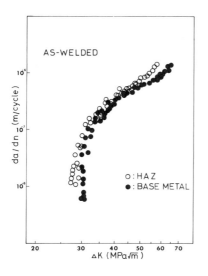

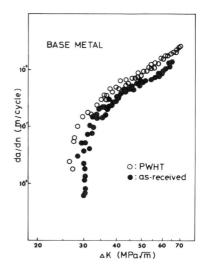

Figure 13 The result of fatigue crack
growth rate tested in HAZ
and in base metal of ACC
steel weldment under constant load condition.

Figure 14 The effect of heat treatment on the fatigue crack
growth rate of ACC steel.

166

(3)   To relieve residual stress, PWHT at 615°C for 1 hour was performed on the ACC steel weldment, which provided further softening in HAZ (Figure 10).   With this specimen, third set of fatigue test was conducted along HAZ and the results was compared in Figure 15 with that obtained in as-welded condition.   After PWHT, the fatigue crack growth rate increased in all ΔK range even though the residual stress activating the fatigue crack growth rate was considered to be fully relieved.

From the above serial tests, it can be concluded that the softening phenomena plays a dominant role in increasing the fatigue crack growth rate and its role becomes greater with increasing the extent of softening.

V. Conclusion

A TMCP steel, described as a ACC steel in this study and manufactured by controlled rolling followed by accelerated cooling processes, was studied to emphasize the softening phenomena occurred in HAZ and after PWHT and its effect on mechanical properties.   The results obtained in this study are summarized as follows;
1. The tensile strength as well as the fatigue properties of ACC steel were deteriorated by welding and/or PWHT.
2. The extent of deterioration in strength was largely dependent on the welding heat input and PWHT condition.
3. The application of PWHT on the ACC steel weldment caused further softening in HAZ and thus fail to meet the specified strength in some cases.
4. The main factor for the increase in fatigue crack growth rate in HAZ of ACC steel weldment was turned out to be the softening phenomena rather than the other factors such as residual stress and microstructural change.

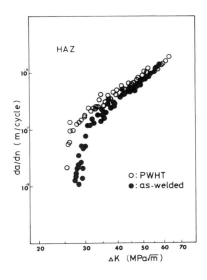

Figure 15   The effect of PWHT on the fatigue crack growth rate in HAZ of ACC steel.

167

Acknowledgements

The authors would like to thank Dr. D. H. Park at the Hyundai Welding & Materials Research Institute for his advise and support.

Ⅵ.　References

1. K. Bessyo, Y. Iida, N. Nakano, I. Seta and Y. Kamata ; the 5th Int. Sym. on OFFSHORE MECHANICS AND ARCTIC ENGINEERING. Apr., 1986 in Tokyo
2. C. Shiga, K. Amano, T. Enami, M. Tanaka, R. Tarui and Y. Kusuhara ; Int. Conf. on the Technology and Applications of HSLA Steels, Oct., 1983 in P-hiladelphia PA
3. K. Tsukada, T. Ohkita, C. Ouchi, T. Nagamine, K. Hirabe and K. Yako ; Nippon Kokan Technical Report (Overseas), 1982, No.35, pp 1-23
4. H. J. Kim ; J. of Korean Welding Society, 1986. Vol.4, No.2, pp 1-11 (in Korean)
5. M. T. Shin, J. G. Youn and H. J. Kim ; J. of Korean Welding Society, 1986 , Vol.4, No.3, pp 43-49 (in Korean)
6. I. Watanabe, M. Suzuki, H. Tagawa, Y. Kunisada, Y. Yamazaki and N. Iwasa-ki ; Nippon Kokan Technical Report (Overseas), 1986, No.47, pp 52-57
7. T. Kitada, K. Fukuda, N. Fukushige, K. Okamoto, K. Mukai and M. Watanabe ; Nippon Kokan Technical Report (Overseas), 1986, No.47, pp 58-64
8. A. Saxena and S. J. Hudax, Jr ; Int. J. of Fracture, 1978, Vol.14, pp453 -468

THE INFLUENCE OF ALLOY CONTENT AND CLEANLINESS ON THE HEAT

AFFECTED ZONE MICROSTRUCTURE OF WELDS ON CAST LOW CARBON STEEL

*Muth, T.R., Frost, R.H., Yorgason, R.R., and *Liby, A.L.

Colorado School of Mines, Golden, Colorado, 80401

*Manufacturing Sciences Corp., Oak Ridge, Tenn.

ABSTRACT

Thermocouple instrumented submerged arc welds were made on ASTM A-216 grade WCA carbon steel castings to evaluate the effect of chemical variations within this specification on heat affected zone microstructures and properties. Steel castings with compositions within this specification were produced by air induction melting with Mn/O ratios ranging from 30 to 188, and sulfur levels from 0.01 to 0.045 wt.%. These castings were heat treated to provide grain sizes and micro- structures typical of that obtained for 152 mm thick plate. Multi pass submerged arc welds were produced at a constant heat input level of 2.45 MJ/m over a range of base metal compositions. These welds showed 3 mm wide heat-affected zones with significant microstructural changes in a one mm wide region next to the fusion line.

Measurements of the thermal cycle in this 1 mm zone showed peak temperatures in excess of 1200°C. The microstructures observed in this region ranged from blocky ferrite and pearlite in high inclusion content-low manganese base metal to bainite in low inclusion content-high manganese base metal. An increase in the number of nucleation sites for austenite decomposition reactions, and hardenability variations caused by changes in the amount of manganese in solid solution were determined to be the reasons for the observed microstructural variations.

INTRODUCTION

The objective of this research was to study the influence of the composition and microstructure of cast low carbon steel base metal on the structure and properties of submerged arc welds. This information is needed for the design and evaluation of candidate materials for nuclear waste overpack containers for the repository burial of high level nuclear waste. A typical burial cask consists of a cast low carbon steel cylinder with a wall thickness of approximately 4 inches, an outside diameter of 20 to 30 inches and a length in excess of 10 feet. Cast end pieces will be welded in place to close the ends of the cylinder. The current design specifies the use of ASTM A-216 Grade WCA low carbon cast steel (1). One proposed production technique would include: centrifugal casting (2,3) to provide a tube with open ends, welding of the bottom closure, remote handing to fill the cask, and remote welding to close the other end.

Steel conforming to the specification ASTM A-216 Grade WCA is a cast low carbon steel with a nominal composition of 0.25% C (max), 0.25% Si (max), and 0.70% Mn (max).(1) This alloy was chosen for ifs castability, weldability, and resistance to corrosion. The design of the experiment was based on the premise that microstructure is the lowest common denominator relating composition, casting, and welding process variables to mechanical properties and corrosion performance of the overpack container. The research focus was on the influence of primary fabrication (casting and heat treating) and secondary fabrication (welding) process variables on overpack container microstructure and service performance.

Chemical variations within the A-216 specification as well as the manufacturing process variables control the volume fraction and morphology of the primary microconstituents as well as the nature and distribution of non-metallic inclusions. Three compositions within the specification were chosen for the investigation. Alloy A was designed to have a low manganese content, a high oxygen content, and a low Mn/O ratio. The low hardenability and the increased number of inclusions promote the nucleation of acicular ferrite (4). Alloy B was designed to have a high manganese content, a low oxygen content, and a high Mn/O ratio. The greater hardenability and smaller inclusion count was designed to minimize ferrite nucleation. Alloy C was designed to have a high sulfur content to analyze the influence of MnS inclusions on ferrite nucleation and hardenability. These composition variations were designed to vary the heat affected zone microstructure through composition variations within the A-216 Specification. The specification allows for some residual elements, but these were kept to trace amounts to eliminate any secondary effects that may be caused by residual impurities.

Submerged arc welding was chosen to evaluate the influence of welding variables on the microstructure of the overpack containers (5). Easterling (6) divides the heat affected zone structure in a 0.15% carbon steel weldment into four sections. The grain growth region is closest to the fusion line, and it sees temperatures in excess of 1100ºC. The grain refined or recrystallized zone sees temperatures in the austenite phase region between 900ºC and 1100ºC. The partially transformed zone sees temperatures in the intercritical region between 750 and 900ºC, and the tempered zone in which carbide spheroidization occurs sees temperatures below the eutectoid in the 700 to 750ºC range.

Extensive grain growth near the fusion line is detrimental to impact toughness and tensile properties. Thus, the grain growth region is considered to be the weak point of the overpack container. The large grains have increased hardenability, and they can transform to lower temperature austenite decomposition products (bainite or martensite) which lower the impact toughness (7).

Since the heat affected zone is too small for sectioning to obtain mechanical test specimens, a technique is required to generate the observed heat affected zone microstructures in macroscopic specimens. The thermal cycle in the heat affected zone near the fusion line can be recorded using a thermocouple, and used with a Gleeble weld simulator to generate macroscopic specimens. These specimens can be used for mechanical or corrosion testing to evaluate the influence of base metal composition variations on the heat affected zone microstructures.

EXPERIMENTAL PROCEDURE

Three steel compositions within the ASTM A-216 specification (1) were selected to demonstrate the influence of base metal composition, cleanliness, and microstructure on the structure and properties of the heat affected zone. The three target compositions (A, B, and C) and the compositions of four experimental heats are given in Table I:

TABLE I – Target Alloy Compositions

Heat No.	%C	%Mn	%Si	%S	%P	%Al	%O	Mn/O
Alloy A	0.18	0.50	0.10	0.01	0.02	0.05	0.020	25
NRC 5	0.17	0.51	0.10	0.01	0.008	0.043	0.011	45
Alloy B	0.18	1.00	0.25	0.01	0.02	0.05	0.007	145
NRC 12	0.17	1.06	0.32	0.01	0.013	0.050	0.017	64
NRC 15	0.21	1.13	0.20	0.01	0.014	0.030	0.006	188
Alloy C	0.18	1.00	0.25	0.045	0.02	0.05	0.007	145
NRC 14	0.15	1.07	0.21	0.045	0.012	0.054	0.022	49

Four heats of steel with target chemical compositions corresponding to those shown in Table I were produced by air induction melting. Four castings were produced from each heat. These included: three 1 x 3.5 x 11 inch (25.4 x 89 x 279 mm) plates and one 1.5 x 6 x 12 inch (38 x 152 x 305 mm) long plate for the production of welds. Vacuum emission spectrometry was used to determine the alloy content of the steel, and the carbon, oxygen, and sulfur were determined by LECO interstitial analysis. Optical microscopy was used to identify the primary microconstituents and scanning electron microscopy was used to determine the identity and volume fraction of non-metallic inclusions.

The ASTM A-216 specification requires a normalizing heat treatment. The heat treatment consisted of austenization for 2 hours at 1050°C followed by slow cooling in a reflective chamber designed to increase the room temperature ferrite grain size to that of a 6 inch (152 mm) thick casting typical of the actual overpack container thickness. Specimens of the normalized plate were evaluated by optical microscopy to identify the primary microconstituents and by scanning electron microscopy to investigate the non-metallic inclusions.

The 1.5 inch (38.1 mm) thick plates were normalized, and one of the plates was machined to provide a one sided, 37.5° edge preparation. This edge preparation was used so that one of the fusion lines would be normal to the surface of the plate for more precise location of subsurface thermocouples. The plates were instrumented with thermocouples placed just below the surface of the 0° edge preparation to monitor the heat affected zone thermal cycle during welding. Two holes were drilled in each plate to a depth expected to place the thermocouple near the fusion line. A 0.005 inch (0.127 mm) diameter bare platinum wire was welded to a 0.005 inch (0.127 mm) bare platinum 13% rhodium wire using a capacitive discharge 150 volt welder and a graphite block. The bare thermocouple was welded to the bottom of the pre drilled holes using the same capacitor discharge welder, and an Al_2O_3 two hole protector tube was used to prevent short circuiting.

A Nicolet 4094 digital recording oscilloscope was used to gather the voltage-time data from the thermocouples. A groundless isolation amplifier was placed in the circuitry to amplify the thermocouple signal and to filter any induced eddy currents or stray voltages caused by the welding process (26). The thermometry circuit also included a cold junction compensator.

Submerged arc welding was carried out on 1.5 inch (38.1 mm) weld plate from heats NRC-5, 12, 14, and 15 using a Hobart RC 500 power supply in the reverse polarity mode. The welds were produced at a constant theoretical energy input of 2.45 MJ/m using a Tibor 22, 3/32 inch (2.38 mm) diameter wire in combination with Oerlikon OP121TT flux. A Hewlett Packard two pen analog chart recorder was used to monitor the current and potential during welding. A flux depth of 1.25 inches (32 mm) was maintained for each pass, and the fused flux was removed between passes.

The welds were sectioned transverse to the welding direction close to the thermocouples, polished through 240 grit, and macroetched with two percent nital to reveal the weld macrostructure. Photomicrographs were taken for documentation and for later comparison with weld simulated specimens. Figure 1 shows macrographs of thermocouple placement with respect to the fusion line.

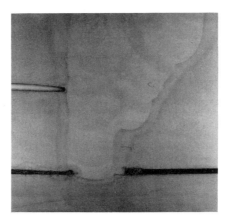

Figure 1 - Macrograph of NRC-9 showing thermocouple placement
placement for thermal cycle measurment. (5.9x).

For microstructural analysis the welds were further sectioned to isolate the thermocouple hot junction, polished to 0.05 micron finish, etched with a boiling saturated picric acid in methanol with one percent Aerosol OT solution for two minutes to reveal the prior austenite grain size next to the fusion lines. Photomicrographs were taken to record the prior austenite grain size. The specimens were repolished to remove the picric etch and etched a second time with two percent nital to reveal the structure directly beneath the thermocouple.

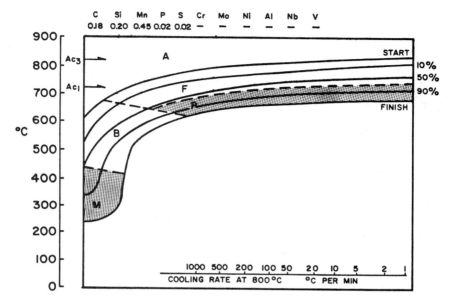

C	Si	Mn	P	S	Cr	Mo	Ni	Al	Nb	V
0.18	0.20	0.45	0.02	0.02	—	—	—	—	—	—

Figure 2 — Continuous cooling transformation diagram for AISI 1018 steel (9).

RESULTS AND DISCUSSION

Variations in the Mn/O ratio and the sulfur content within the ASTM A-216 specification are expected to shift the continuous cooling trans- formation curves and cause changes in the microstructure and properties of the heat affected zone in multi pass submerged arc welds. These influences on austenite decomposition can be illustrated using the continuous cooling transformation curve for AISI 1018 steel as shown in Figure 2 (9). An increase in hardenability causes a shift of the ferrite and pearlite curves on the CCT diagram to longer times. Oxide or sulfide inclusions in the 0.3 to 1.0 micron size range promote ferrite nucleation, and cause a shift of the ferrite and pearlite curves on the CCT diagram to shorter times.

The post weld microstructure in the heat affected zone has an important influence on weld toughness. Large fractions of blocky grain boundary ferrite, bainite, or martensite yield poor toughness (6,7). The best toughness is provided by acicular ferrite which nucleates in the interior of the grain as well as at the grain boundary at an undercooling below the temperature for the formation of blocky ferrite. Acicular ferrite is promoted by the presence of oxide inclusions in the correct size range to act as nucleation sites, and by reduced levels of hardenability agents which promote the formation of bainite.

The four cast base metal compositions are compared with the target compositions in Table I. Figure 3 shows a comparison of the cleanliness in

173

inclusion distributions of the four experimental cast steels using photomicrographs of the polished and unetched structures. Alloy A was designed to have a low Mn/O ratio to provide a low hardenability and a high concentration of oxide inclusions to act as nucleation sites for acicular ferrite. The composition of Heat NRC 5 (top left in Figure 3) is close to the target Alloy A with the exception that the oxygen level is slightly smaller (110 ppm vs. 200 ppm.) and the manganese/oxygen ratio is slightly higher than the target (35 vs. 25). The large inclusions in the micrograph are sulfides. The oxide inclusions are much smaller than the sulfides and they are present in much larger numbers.

Heat NRC 12 (top right in Figure 3) is close to target Alloy B, except for a high oxygen composition which yields a lower Mn/O ratio. Since both the oxygen and manganese values are higher than those in Heat NRC 5, the population of oxide inclusions is greater, and the hardenability is lower. Heat NRC 14 (bottom left of Figure 3) has a very high oxygen concentration,

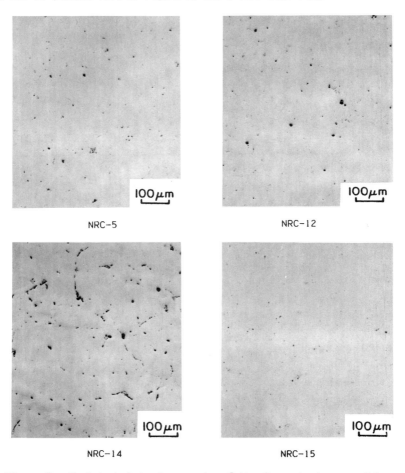

NRC-5   NRC-12

NRC-14   NRC-15

Figure 3 - Unetched photomicrographs of the four steel compositions showing the nonmetallic inclusions. (100x).

a high manganese concentration, a low Mn/O ratio, and a very high sulfur concentration compared to the other heats used in the study. The microstructure shows a substantial concentration of large sulfide inclusions as well as a large number of small oxide inclusions. Heat No. 14 is expected to have the lowest hardenability of the four alloys studied.

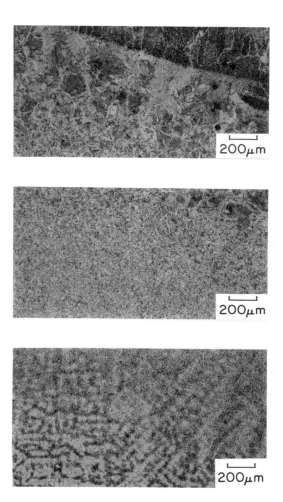

Figure 4 – Photomicrographs of the heat affected zone structure in a 0.18% carbon steel. The top micrograph shows the fusion line and the grain growth region. The center micrograph shows the recrystallized zone, and the bottom micrograph shows the partially transformed and tempered zone. (2% Nital, 50x)

Heat NRC 15 (bottom right hand in Figure 3) is close to target composition B with a low oxygen content, a high manganese content, a high Mn/O ratio, and a low sulfur content. This alloy is expected to have an increased hardenability, and a small population of oxide inclusions to act as acicular ferrite nucleation sites. The microstructure of NRC 15 shows a much smaller concentration of sulfide and oxide inclusions than that found in the other heats, and it was expected that the hardenability would be increased and the development of acicular ferrite suppressed.

Multi-pass submerged arc welds were made on each of the four alloys. The thermal cycle associated with each weld pass was measured by thermocouples placed near the fusion line, and the thermal histories were used in conjunction with heat transport modeling based on the Rosenthal Equation to predict the thermal experience of a one millimeter wide region adjacent to the fusion line. Based on this prediction a Gleeble Weld Simulator was used to produce macroscopic specimens with microstructures typical of this region.

The heat affected zone shows a wide variety of microstructures depending on the heat input, the base plate composition, the composition and distribution of nonmetallic inclusions, and the preweld and postweld heat treatments. The welds and Gleeble specimens were used to correlate the heat affected zone microstructures obtained for the four alloys with the alloy composition, cleanliness, and the thermal cycles experienced.

Figure 4 shows micrographs of the same HAZ regions in a multi pass A-216 steel weld from the present study. The top micrograph shows the fusion line and the grain growth region, the center micrograph shows the recrystallized zone, and the bottom micrograph shows the partially transformed and tempered regions.

Figures 5, 6 and 7 show thermal traces and associated microstructures for multi pass welds produced on Heats 15, 14, and 12. The thermocouples were located near the fusion line, and the thermal traces are shown for the weld passes which most closely approached the thermocouple location and produced the highest peak temperatures.

Figure 5 shows the thermal traces for three of the weld passes on Heat 15 which had a high Mn/O ratio and a low sulfur content. The peak temperatures for weld passes 7 and 11 are in the range of 1000 to 1060°C where recrystallization would be expected to produce a fine grained ferrite-pearlite microstructure. This is confirmed by the micrograph of the region near the thermocouple. Figures 6 and 7 show similar thermal traces and associated microstructures for welds on Heat No. 14, with a low Mn/O ratio and a high sulfur content, and Heat No. 12 with a low Mn/O ratio and low sulfur. A comparison of the microstructures for the three compositions shows that the composition (Mn/O ratio and the sulfur content) and cleanliness have little influence on the heat affected zone microstructure where the peak temperature does not exceed about 1100°C and growth of the austenite grains is not expected.

Figure 8 shows a comparison of the heat affected zone microstructures obtained for the four steel compositions for the case in which the peak temperature was above 1100°C, and Figure 9 is a comparison of the same four microstructures at a higher magnification. In the temperature range between 1100 and 1520°C found near the fusion line the transformation to austenite is complete, grain growth and particle dissolution significantly change the structure, and base metal composition and cleanliness become important variables. Heats 5 and 12 have low sulfur contents and low and medium Mn/O ratios respectively. The heat affected zone microstructures for these compositions show allotriomorphic grain boundary ferrite, widmanstatten ferrite, and some acicular ferrite in the central portion of the grain. The concentrations of oxygen and sulfur, and the volume fractions of sulfide and oxide inclusions were greater in Heat 14 than those in Heats 5 and 12. This caused an increase in the volume fraction of acicular ferrite. Heat 15 had

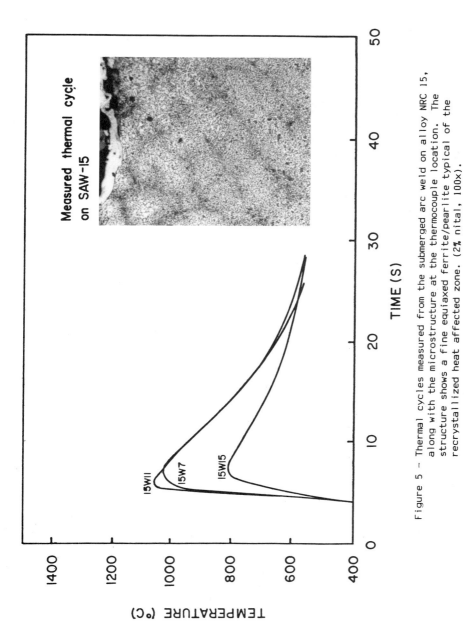

Figure 5 – Thermal cycles measured from the submerged arc weld on alloy NRC 15, along with the microstructure at the thermocouple location. The structure shows a fine equiaxed ferrite/pearlite typical of the recrystallized heat affected zone. (2% nital, 100x).

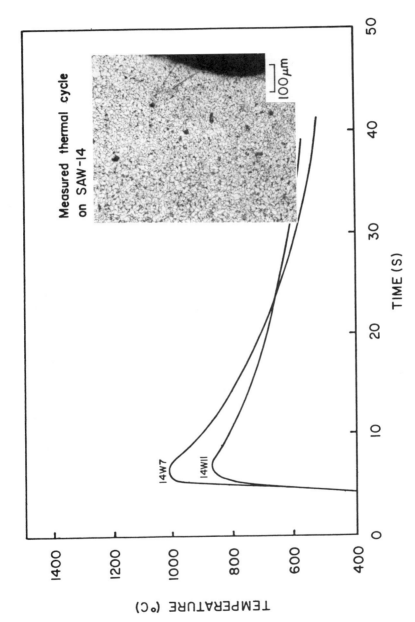

Figure 6 -- Thermal cycles measured from the submerged arc weld on alloy NRC 14, along with the microstructure at the thermocouple location. The structure shows refined equiaxed ferrite and pearlite. (2% nital, 100x).

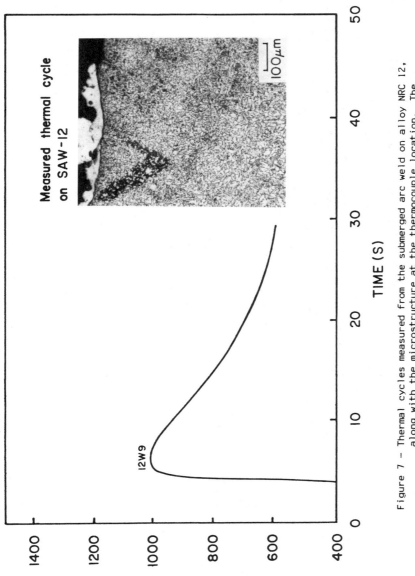

Figure 7 – Thermal cycles measured from the submerged arc weld on alloy NRC 12, along with the microstructure at the thermocouple location. The structure shows refined equiaxed ferrite and pearlite. (2% nital, 100x).

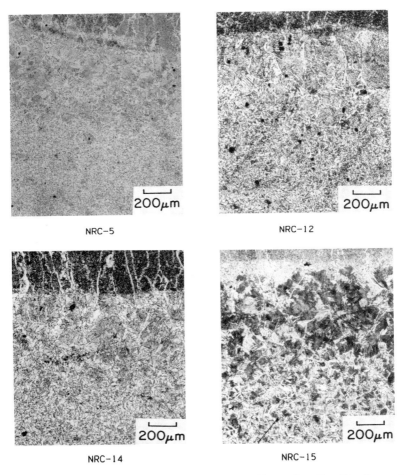

NRC-5

NRC-12

NRC-14

NRC-15

Figure 8 - The fusion zone (top) and grain growth region for the four alloys studied showing the microstructural differences next to the fusion line as a function of composition and cleanliness. (Picral, 50x).

low oxygen and sulfur contents and a manganese content about the same as those of Heats 12, and 14. The high Mn/O ratio and low sulfur resulted in increased hardenability and a lower content of inclusions available for ferrite nucleation. This increased hardenability resulted in the formation of a bainitic microstructure with some widmanstatten ferrite along the grain boundaries. This structure would be expected to provide lower toughness than that provided by the structures which possess large fractions of acicular ferrite.

Figures 10, 11, and 12 show a comparison of the influence of composition and cleanliness on HAZ microstructures for three peak temperatures of about 940, 1250, and 1450°C. The structures were generated using the Gleeble weld simulator with programmed heating and cooling curves based on those observed for the actual welds. The structures in Figure 10 represent a peak

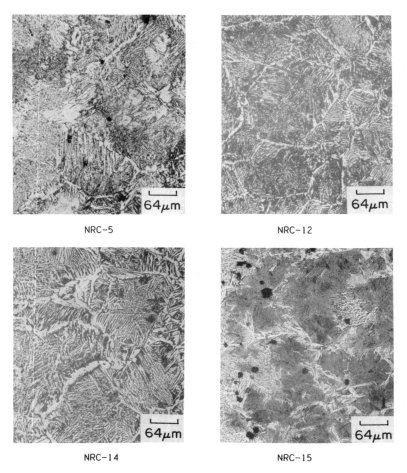

NRC-5   NRC-12

NRC-14   NRC-15

Figure 9 - The fusion zone (top) and grain growth region for the four alloys
studied showing the microstructural differences next to the
fusion line as a function of composition and cleanliness.
(Picral, 156x).

temperature of about 940ºC. The microstructures show some grain refinement,
but all four alloy compositions have approximately the same microstructure.

Figure 11 compares the structures for a peak temperature of around
1250ºC. Here the structure has transformed to austenite, and the effects of
composition and cleanliness can be observed. Alloys 5 and 12, with low and
medium Mn/O ratios show equiaxed ferrite/pearlite microstructures. Alloy 14
with a high sulfur level shows acicular and Widmanstatten ferrite; and Alloy
15, which has a high Mn/O ratio shows a bainitic structure.

Figure 12 compares the microstructures for a peak temperature between
1400ºC and the liquidus, which would be observed in the grain growth region
near the fusion line. Alloy 5 shows large prior austenite grains with
allotriomorphic ferrite and Widmanstatten ferrite decorating the boundaries.

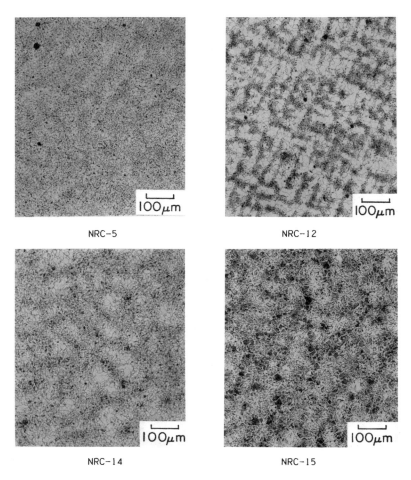

NRC-5   NRC-12

NRC-14   NRC-15

Figure 10 - HAZ microstructures generated for the four steel alloys by the
           Gleeble weld simulator using a peak temperature of about 940°C.
           (2% Nital, 100x).

There is upper bainite in the grain interiors with some acicular ferrite.
Alloy 12 shows more ferrite laths throughout the structure than Alloy 5, and
the upper bainite present is coarse. Alloy 14 shows smaller prior austenite
grains than either Alloy 12 or Alloy 5 with a much coarser ferrite lath
structure than either. Alloy 15 shows predominantly fine upper bainite with
little or no allotriomorphic ferrite.

     The microstructural differences illustrated by Figures 5 through 12
indicate that composition variations within the A-216 specification can have
a significant influence on the heat affected zone microstructure where the
peak temperature exceeds 1100°C and reaustenization occurs. Composition and
cleanliness influence the structure of the reaustenitized region of the heat
affected zone by influencing hardenability and by providing heterogeneous
nucleation sites for austenite decomposition reactions.

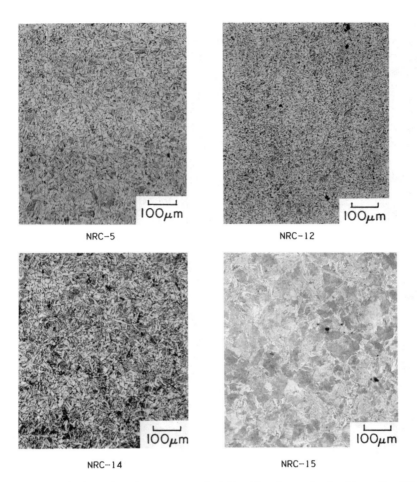

NRC-5

NRC-12

NRC-14

NRC-15

Figure 11 - HAZ microstructures generated for the four steel alloys by the Gleeble weld simulator using a peak temperature of about 1250°C. (2% Nital, 100x).

The heterogeneous nucleation sites provided by oxide and sulfide inclusions allow the formation of a higher ductility acicular ferrite structure at a smaller undercooling below the eutectoid temperature than that necessary for the formation of the lower ductility Widmanstatten ferrite or bainite microstructures. Hardenability is influenced by the presence of austenite stabilizing alloy elements such as manganese and by the action of second phase particles to limit austenite grain growth. A low Mn/O reduces hardenability by both mechanisms and avoids the transformation to lower temperature bainitic decomposition products.

The microstructural similarities observed for Alloys 5, 12, and 14 indicate that both sulfide and oxide inclusions provide nucleation sites for the formation of acicular ferrite. The small austenite grain size observed in Alloy 14 indicates that both oxide and sulfide inclusions play a significant role in limiting austenite grain growth near the fusion line.

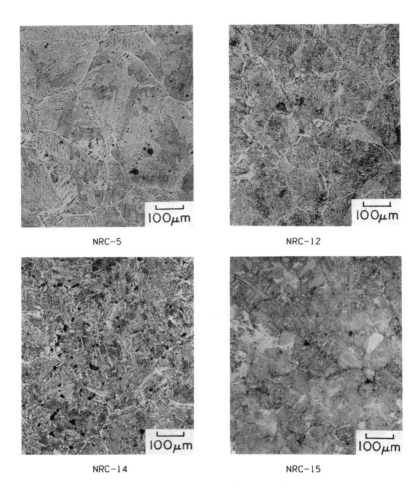

NRC-5   NRC-12

NRC-14   NRC-15

Figure 12 – HAZ microstructures generated for the four steel alloys by the Gleeble weld simulator using a peak temperature between 1400°C and the liquidus. (2% Nital, 100x).

The bainitic structure observed near the fusion zone for Alloy 14 illustrates that a moderate manganese concentration coupled with low oxygen and sulfur contents (greater cleanliness) allow for grain growth in the reaustenitized region near the weld fusion line and the formation of undesirable low temperature austenite decomposition products such as bainite. Thus, cleanliness and alloy content have been shown to be significant variables influencing the structure and properties of welds in low carbon steel castings.

CONCLUSIONS

1. The alloy composition and cleanliness of low carbon steel castings such as those specified by ASTM A-216 have a significant influence on the structure and properties of weld heat affected zones. The significant

composition variables include the concentrations of austenite stabilizing elements such as manganese and the concentration of nonmetallic interstitial elements such as oxygen and sulfur which increase the number of nonmetallic inclusions. The Mn/O ratio is a useful measure of oxygen and manganese contents.

2. The composition and cleanliness of the cast steel base metal influence the heat affected zone microstructure and properties only in the reaustenitized region near the fusion line where the peak temperatures exceed 1100°C. High manganese concentrations stabilize the austenite and increase hardenability, and oxide and sulfide inclusions act to decrease hardenability by limiting austenite grain growth and by providing heterogeneous nucleation sites for the decomposition of austenite to form an acicular ferrite structure rather than a lower temperature bainitic structure with reduced ductility.

3. Composition and cleanliness variations within the ASTM A-216 specification do not influence the microstructures of the castings in the as-cast or normalized condition, or the heat affected zone microstructures in low temperature regions where reaustenization does not occur.

4. In the reaustenitized region of the heat affected zone near the fusion line a high Mn/O ratio yields a bainitic microstructure, a low Mn/O ratio and/or a high sulfur level favor the formation of an acicular ferrite/pearlite microstructure with a minimum bainite content.

ACKNOWLEDGEMENTS

The authors wish to express their appreciation the U.S. Nuclear Regulatory Commission and to the Manufacturing Sciences Corporation for support of this research.

REFERENCES

1. American Society for Testing and Materials, 1977, A-216 Grade WCA.

2. P. R. Beeley, Foundry Technology, Butterworths, London, 1972.

3. H. F. Taylor, M. C. Flemings, and J. Wulff, Foundry Engineering, John Wiley and Sons, Inc., New York, 1959.

4. United States Steel, The Making Shaping and Treating of Steel, Herbick and Held, Pittsburgh, 1971.

5. American Welding Society, Welding Handbook, Seventh Edition, Volume 2, 1978.

6. K. E. Easterling, Introduction to the Physical Metallurgy of Welding, Butterworth and Co., London, 1983.

7. W. C. Leslie, The Physical Metallurgy of Steels, McGraw Hill, 1981.

8. D. Rosenthal, "Mathematical Theory of Heat Distribution During Welding and Cutting," Welding Research Supplement, May 1941.

9. G. Krauss, Principles of Heat Treatment of Steel, ASM, Metals Park, Ohio, 1980.

Weld Metal

CONTROL OF MICROSTRUCTURE AND MECHANICAL PROPERTIES

IN SA AND GMA WELD METALS

B.M. Patchett[*], J.T. McGrath[**], R.F. Orr[**], J.G. Gianetto[**], and A.C. Bicknell[*]

[*] Department of Mining, Metallurgical and Petroleum Engineering
University of Alberta, Edmonton, Alberta, Canada, T6G 2G6

[**] CANMET, Physical Metallurgy Research Laboratory
Ottawa, Ontario, Canada, K1A 0G1

Abstract

The effects of residual oxygen and inclusion population on the microstructure of 2.25%Cr-1%Mo weld metals have been investigated and compared for the SAW process (using fluxes of varying basicity) and the GMAW process (using shielding gases of varying oxidation potential). Some mechanical properties of the deposits, mainly strength and Charpy v-notch properties, have been related to the residual element content.

The results show that the residual oxygen level can be varied in a range of 200 to 900 ppm with the SAW process and 150 to 400 ppm with the GMAW process. Flux basicity determines the residual oxygen level in the SAW deposits. The strength (both yield and ultimate tensile) tends to be significantly higher in the GMAW deposits because the alloy level, particularly for the deoxidants Mn and Si, increases rapidly as the shielding gas oxidation potential decreases.

The sources of the residual oxygen-alloy element balance are discussed as are the resulting structure-mechanical property relationships. Implications for consumable design for both the SAW and GMAW processes are outlined.

Reactor vessels used for hydrotreating of crude oil are often made of Cr-Mo steels in the thickness range of 100-300 mm. Welding of such heavy sections involves large amounts of weld metal if conventional joint designs and procedures are specified, leading to low productivity and possible problems with distortion. Narrow-gap welding processes reduce the volume of weld metal to minimize both problems. "Narrow-gap" welding cannot easily be defined, but one possible way is to specify virtually parallel joint sides (within 5°) and an aspect ratio of material thickness to joint width of 5 or more, for example a 20 mm weld width in material 100 mm thick. The objective of the present work was to identify some of the factors which control microstructures and resulting mechanical properties in SAW and GMAW narrow gap weld metals. Typical mechanical property requirements for weld metals in a Cr-Mo reactor vessel are as follows (1):

Yield strength	420 MPa min.
Tensile strength	515-790 MPa
Charpy v-notch	54 J @ -40°C after PWHT

Experimental Procedure

The SAW process welds were made in two series. In series A, 38 mm thick plates of 2.25%Cr-1%Mo material were welded using a range of commercial consumables, Table 1. Welding procedure parameters are given in Table 2. The joint design and dipass (two passes per layer) technique are shown in Figure 1. In series B, bead-on-plate (BOP) deposits were produced to assess the effects of energy input and weld chemistry on as-deposited weld metal microstructure. These runs were made at the conditions given in Table 1, using only the Oerlikon SD2 wire with both an acid (Linde 124) and a basic (OP76) flux. After assessing the series A and B results, a weld using the consumables giving the optimum microstructure and properties was made in 150 mm Cr-Mo plate, weld CRM-1.

Table I - SAW and GMAW Consumables

Series	Weld No.	Heat Input	Electrode Wire	Flux	Basicity Index
SAW					
	W1	2	Oerlikon	OP121TT	3.00
	W2	2	SD2-Cr-Mo	Linde 124	1.00
	W3	2	(AWS EB-3)	OP76	2.70
A					
	W4	2	Linde	Linde 124	1.00
	W5	2	U 521	OP121TT	3.00
	W6	2	(AWS EB-3)	OP76	2.70
	W7	2	Oerlikon	OP76	2.70
	W8	4	SD2-Cr-Mo	OP76	2.70
	W9	6	(AWS EB-3)	OP76	2.70
B					
	W10	2	Oerlikon	Linde 124	1.00
	W11	4	SD2-Cr-Mo	Linde 124	1.00
	W12	6	(AWS EB-3)	Linde 124	1.00
Thick Section	CRM-1	2	Oerlikon SD2-Cr-Mo (AWS EB-3)	OP76	2.70
GMAW				Shielding Gas Ar Carrier plus Percent Carbon Dioxide	
	G1	3.6	KOBE 2 CM	20	
	G2	3.6	KOBE 2 CM	15	
	G3	3.6	KOBE 2 CM	10	
	G4	3.6	KOBE 2 CM	5	
	G5	3.6	KOBE 2 CM	2	
	G6	3.5	KOBE 2 CM	2 + 0.5% hydrogen	

Table II - SAW Welding Procedures

Weld Series	Heat Input kJ/mm	Current AC Square Wave amps	Voltage volts	Travel Speed mm/sec	Stickout mm	Wire dia mm
A	2	450-500	30-34	6.7-8.3	32	3.2
B	2	550	32	8.5	32	3.2
	4	550	32	4.2	32	3.2
	6	550	32	2.8	32	3.2
Thick Section CRM - 1	2	550	32	8.5	32	3.2
Preheat = 200° C Interpass = 250° C max.						

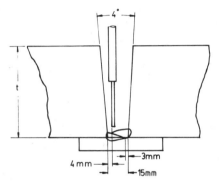

Figure 1. SAW Joint Design.

The GMAW process welds were made with the KOBE "Twistwire" procedure using shielding gases of varying oxidation potential, Table 1. It was necessary to substitute 1.25%Cr-0.5%Mo base plate for this series, due to a shortage of the plate used for the SAW welds. As will be seen in the results, this did not have an unacceptable effect on final deposit chemistry. The welding procedure, Table 3, requires only one pass per layer (monopass) and therefore uses a joint design with parallel sides and a gap of only 13 mm, and produces a low dilution deposit. All welds were made with one reel of wire and one set of 63 mm base metal plates. All SAW and GMAW deposits were heat treated after welding at 690°C for 10 hours. Mechanical testing determined yield and ultimate tensile strength as well as Charpy v-notch impact energy at -40°C and +25°C. The Charpy samples were taken at the weld mid-thickness, with the v-notch oriented perpendicular to the weld surface. The notch was located at the 1/4W position (middle of a bead) in the SAW procedures and at the 1/2W position of the GMAW procedures in order to sample equivalent microstructures. Microstructures were characterized with the light microscope and selective use of a Phillips 400 TEM/STEM for structure and particle analysis.

Table III - GMAW Welding Procedure

Weld Series	Heat Input kJ/mm	Current DC amps Elec +ve	Voltage volts	Travel Speed mm/sec	Stickout mm	Wire dia mm
G1 - G6	3.5 - 3.6	480 - 520	27 - 29	4	35	2 x 2.0
Preheat = 200° C Interpass = 250° C max.						

Results

SAW Deposit Chemistry

The chemical compositions of the SAW welds are given in Table 4. This shows that weld metal composition is dominated by electrode wire chemistry and flux basicity. C, Mn, Si and O levels are attributed to basicity, acid fluxes giving lower C and higher Mn, Si and O levels, e.g. welds W2 and W4. Higher S levels in welds W4-6 are attributed to electrode wire composition. The results also show that increasing the energy input tends to reduce the O content, while flux basicity accounts for the observed differences in C, Mn, Si and O at a given energy input. Series B welds (W7-12) were analysed for residual N, giving a virtually constant 100 ppm, showing that atmospheric contamination did not contribute to O or N content of the welds.

Table IV - Chemical Composition of SAW Consumables and Weld Metals

Designation	% C	% Mn	% Si	% S	% P	% Cr	% Mo	% Ni	% Cu	O ppm	N ppm
21/4 Cr-1Mo Plate	0.10	0.49	0.20	0.028	0.008	2.27	1.03	0.33	0.24	-	-
SD2 Wire	0.12	0.48	0.14	0.007	0.007	2.83	0.82	0.19	0.34	-	-
U521	0.10	0.59	0.20	0.019	0.007	2.36	0.78	0.08	0.24	-	-
W1	0.10	0.55	0.22	0.009	0.013	2.68	1.05	0.21	0.40	260	-
W2	0.06	0.67	0.49	0.011	0.009	2.63	1.06	0.21	0.38	850	-
W3	0.10	0.53	0.19	0.007	0.007	2.70	1.03	0.20	0.40	220	-
W4	0.06	0.66	0.43	0.022	0.008	2.24	1.06	0.09	0.33	770	-
W5	0.10	0.54	0.17	0.016	0.010	2.32	0.98	0.08	0.30	330	-
W6	0.10	0.54	0.16	0.016	0.007	2.32	1.00	0.09	0.30	270	-
W7	0.11	0.56	0.15	0.006	0.008	2.71	1.04	0.19	0.36	216	103
W8	0.10	0.59	0.16	0.005	0.009	2.75	1.01	0.19	0.42	165	107
W9	0.11	0.60	0.19	0.005	0.009	2.75	1.01	0.19	0.39	172	109
W10	0.06	0.63	0.39	0.007	0.007	2.47	0.97	0.20	0.33	835	105
W11	0.06	0.63	0.38	0.007	0.008	2.45	0.98	0.20	0.40	715	105
W12	0.06	0.63	0.38	0.007	0.008	2.45	0.98	0.20	0.40	616	99
CRM-1	0.09	0.55	0.19	0.005	0.006	2.76	0.99	0.19	0.32	190	140
Base Plate	0.16	0.54	0.14	0.002	0.003	2.45	0.98	0.02	-	-	-

GMAW Deposit Chemistry

The chemical composition of the GMAW deposits are given in Table 5. These results show that residual O content decreased as shielding gas oxidation potential decreased and that Mn and Si levels simultaneously increased. These results are shown graphically in Figures 2 and 3. One anomaly can be seen in the result for the shielding gas containing $2\%CO_2$-05%H_2 - the residual O level is similar to the $2\%CO_2$ gas, but the Mn and Si levels are higher, suggesting that the former gas does indeed have a lower oxidation potential, but that the final O level is also influenced by other factors. The Cr and Mo levels are slightly under the specified level due to the use of the 1.25%Cr-0.5%Mo base plate. Nitrogen levels were nearly constant at 180 ppm, again showing that atmospheric contamination was not involved in the residual O or N levels.

192

Table V - Chemical Composition of GMAW Consumables Weld Metals

Designation	% C	% Mn	% Si	% S	% P	% Cr	% Mo	% Ni	% Cu	O ppm	N ppm
1 1/4 Cr-1/2 Mo Plate	0.13	0.51	0.52	0.014	0.006	1.30	0.44	0.21	0.18	-	-
Twistwire 2-CM	0.08	0.96	0.33	0.009	0.004	2.33	1.08	0.15	-	-	-
G1	0.11	0.68	0.25	0.01	0.006	2.11	0.98	-	0.13	401	185
G2	0.11	0.69	0.28	0.01	0.006	2.07	0.95	-	0.13	370	179
G3	0.11	0.72	0.31	0.01	0.006	2.09	0.92	-	0.13	302	196
G4	0.11	0.81	0.32	0.01	0.006	2.15	0.99	-	0.13	215	184
G5	0.11	0.83	0.35	0.01	0.005	2.14	0.98	-	0.13	149	182
G6	0.10	0.86	0.35	0.01	0.005	2.19	1.00	-	0.13	144	187

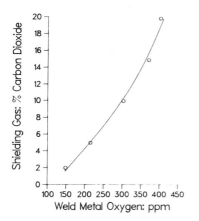

Figure 2. Variation of GMAW Residual Weld Metal Oxygen with Shielding Gas Oxidation Potential

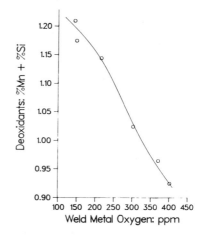

Figure 3. Mn + Si Content of GMAW Deposits vs. Residual Oxygen Level

Microstructures

The SAW weld metals contained a mixture of as-deposited and reheated regions typical of multi-pass welds. Typical distribution of these regions is shown in Figure 4. All areas were bainite. Under the terminology suggested by Torronen (2) and Kotilainen (3), the microstructure was formed of indivdual bainitic packets arranged in bundles. The individual packets were elongated and 3-6 μm wide. High angle grain boundaries surround the packets (and bundles), while the packets are further subdivided into parallel laths by low angle boundaries (misorientation of 1-2°). Packet width and shape varied in these welds. In basic flux deposits, e.g W1 and W3 they were elongated, while in acid flux deposits, e.g. W2 and W4, they were much wider and more irregular, Figure 5. Varying energy input over the range of 2-6 kJ/mm had little effect on packet size, Figure 6. The lower C content and other chemistry differences of the welds made with the acid flux resulted in more irregular, coarser packets than those found in the basic flux welds.

193

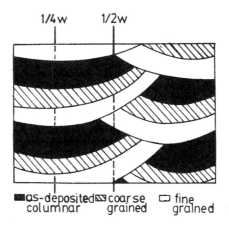

1/4 w   1/2 w

■as-deposited☒coarse   ▢ fine
columnar   grained   grained

Figure 4. Distribution of As-Deposited and Reheated Microstructures in SAW Joints.

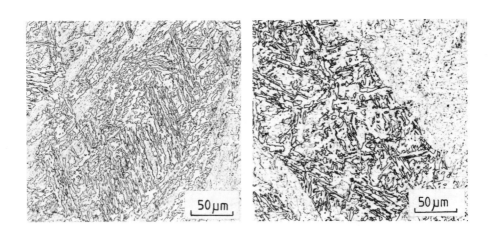

Figure 5. Microstructural Variations in SAW Deposits due to Flux Basicity.

a. Weld W3 - 290 ppm O + S   b. Weld W2 - 960 ppm O + S

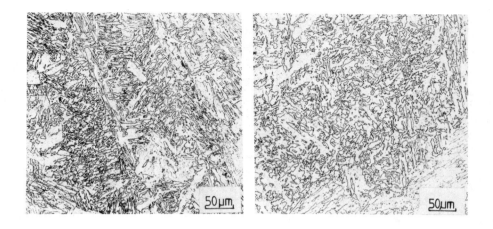

Figure 6. Effect of Heat Input Variations on SAW Microstructures

a. 2 kJ/mm Weld W8 - 215 ppm O + S   b. 4 kJ/mm Weld W11 - 785 ppm O + S

The carbide precipitates found in the weld metals were analysed using welds W7 and W9 to assess the extremes of energy input. After PWHT at 690°C for 10 hours, two stage carbon extraction replicas were taken from the as-deposited areas. The low energy input weld, W7, contained globular precipitates (about 0.1-0.4 μm), rod-shaped particles (0.4-0.6μm in length) and some needle-shaped precipitates (0.1-0.2 μm in length). W9 contained fewer globular particles and more rod and needle-shaped particles. The particles in W9 are shown in Figure 7. The globular and rod-shaped particles were rich in Cr and Fe, while the needles were rich in Mo. Previous studies of Cr-Mo weld metals (4,5) suggest that the former are stable M_7C_3, $M_{23}C_6$ and M_6C carbides and the latter M_2C carbides. The spherical inclusions were not analysed, but were assumed to be mainly oxysulphides and silicates, the former primarily from wire and base plate residuals and the latter from deoxidation processes. The total inclusion content would thus be proportional to %O + %S.

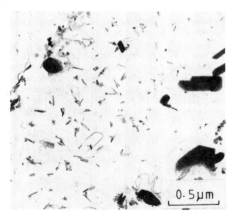

Figure 7. Carbide and Inclusion Particles in SAW Deposit W7.

The GMAW microstructures are very similar to the basic flux SAW microstructures, as shown in Figure 8. The GMAW runs all had virtually identical C levels of 0.10%, similar to the basic flux runs with the SAW process, and significantly higher than the acid flux deposits. There was some tendency for the lower O welds (higher Mn and Si) to have coarser microstructures, e.g. Figure 8b vs. 8a. This occurs despite the virtually constant C level.

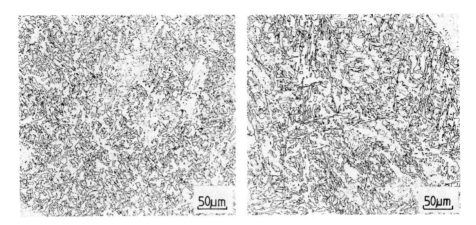

Figure 8. GMAW Microstructures.

a. Weld G1 - 501 ppm O + S   b. Weld G6 - 244 ppm O + S

Mechanical Properties

All series "A" SAW welds achieved the required strength, but only the welds containing under 400 ppm O+S achieved the Charpy requirement (W1 and W3).

The results suggest that the optimum mechanical properties would be produced in a fine bainitic microstructure with a low residual O + S level, and that energy input in the 2-4 kJ/mm range would be best. The thick section weld thus used the SD2/OP76 consumables at an energy input of 2 kJ/mm. The resulting O + S level was 240 ppm and the mechanical properties, including Charpy requirements, were well above the necessary minim.

The GMAW welds all achieved the minimum requirements, with the exception of the Charpy energy for G5. This particular weld was also a bit low in tensile elongation and reduction in area, with no obvious cause. The results in Table 6 show that there is a tendency for the -40°C Charpy results to decrease as the O + S level decreases, which is not as would be predicted. However, the results also show that the Mn + Si levels increase rapidly as the shielding gas oxidation potential decreases, resulting in very high yield and tensile strengths, which would lead to lower Charpy energies.

Table VI - Weld Metal Mechanical Properties

Process	Series	No.	Yield Strength MPa	Tensile Strength MPa	Charpy v-notch Energy - J		O + S ppm
					- 40° C	+ 25° C	
SAW	A	W1	492	610	73	213	350
		W2	468	592	20	98	960
		W3	455	595	140	258	290
		W4	436	565	8	62	990
		W5	476	588	16	142	490
		W6	450	575	18	172	430
SAW	Thick Section	CRM-1	433	562	152	>300	240
GMAW		G1	489	571	78	177	501
		G2	501	576	70	164	470
		G3	503	577	73	194	402
		G4	535	600	86	184	315
		G5	574	634	38	171	249
		G6	585	647	59	202	244

Discussion

The SAW results confirm that the shape and size of the bainitic packets were primarily dependent on the transformation temperature of the weld deposit, in particular the C content (4). The observations also show that fine packet size is also associated with low inclusion (O+S) levels, but this is essentially due to C - O interaction in slag-metal reactions, leading to higher C levels at low O levels. Low inclusion levels reduce the transformation temperature in ferritic weld metals deposited with the SAW process, promoting the formation of acicular ferrite, which suggests that inclusions may play a similar role in bainitic steels. However, work by Ito et al (6) and Chandel et al (7) showed that even at low O+S levels of less than 400 ppm, C has the dominant effect on packet size. There is a suggestion that this is not strictly true in the GMAW deposits, and further work is necessary on the characteristics of the GMAW welds.

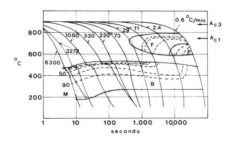

Figure 9. CCT Diagram for 2.25%Cr-1%Mo Steel (8).

The packet size did not alter significantly with energy input due to the relatively constant transformation temperature at a given chemistry for a wide range of cooling rates (8), as shown in Figure 9. Torronen (2) also showed that packet size was independent of prior austenitizing temperature and cooling rate for an alloy of 2.8%Cr-0.6%Mo-0.3%V. All Cr-Mo alloys of 2-3%Cr and 0.5-1.5%Mo have a very wide range of cooling rates which produce bainite. The slight drop in residual O in higher heat input welds is probably associated with a longer solidification time, which allows for inclusion agglomeration and flotation to occur on a wider scale. The greater O level changes in the high O welds (acid flux) supports this view. The lack of an effect of the reducing gas (H_2) in the GMAW shielding gas on residual O level is probably due to the base metal dilution.

197

maintaining a minimum level. Gas-metal reactions are much more intense at the electrode tip than within the weld pool, and O from the base metal would not react rapidly with the arc atmosphere.

The low temperature resistance to cleavage fracture in bainitic steels is inversely related to packet size (3), which is confirmed by the present work. Cleavage facets and packets have similar dimensions, 3-6 μm, which suggests that cleavage cracks reinitiate each time a boundary is crossed, absorbing more energy. Resistance to ductile fracture at higher temperatures decreases as the inclusion content, characterized by the O+S content, increases. This not only reduces the upper shelf Charpy energy, but also shifts the transition region to higher temperatures. This is the cause of the poor performance of SAW deposits W5 and W6, which had good microstructures due to an optimum C level, but higher O+S levels.

Conclusions

1. SAW and GMAW deposits can be produced with a large range of residual O levels by varying flux basicity and shielding gas oxidation potential respectively. Total O+S contents determine inclusion population, for which high levels brings a reduction in Charpy toughness at both low (-40°C) and higher (+25°C) temperatures.
2. Bainite packet size has a strong effect on fracture toughness. A small packet size is promoted by C levels of about 0.10% in both SAW and GMAW deposits and is not significantly affected by other alloying elements or by inclusion content.
3. Reducing the oxidation potential of GWAW shielding gases reduces residual O levels substantially, but also causes large increases in Mn and Si levels, raising the yield and ultimate tensile strengths to levels which adversely affect fracture toughness.
4. SAW consumables should be designed to provide a residual O + S level of under 400 ppm for the optimum strength-toughness balance in Cr-Mo weld metals of about 2.25%Cr-1%Mo composition. This can be achieved with commercially available consumables.
5. GMAW consumables can achieve a wide range of residual O + S levels by controlling the shielding gas oxidation potential. Very low O + S levels produce improved upper shelf Charpy v-notch energies. Low temperature Charpy properties are slightly inferior with low O + S levels due to very high yield and ultimate tensile strengths caused by Mn and Si levels designed for highly oxidizing shielding gases. Modified electrode wire chemistry could exploit the virtues of low residual contents at both high and low temperatures by lowering Mn and Si levels while using low oxidation potential shielding gases.

Acknowledgements

B.M. Patchett and A.C. Bicknell would like to thank CANMET of Energy, Mines and Resources Canada for their financial support of this project. All authors wish to thank Mr. D. Dolan for weld preparation and Ms. J. Ng Yelim for electron microscope analysis of precipitates.

References

1. R.S. Chandel, R.F. Orr, J.A. Gianetto, J.T.McGrath and R.F. Knight, "Mechanical Properties of 2.25 %Cr-1%Mo Welds Deposited by SAW-Narrow Gap Process", Journal of Materials for Energy Systems, 1985, vol. 7, pp137-146.

2. K. Torronen, "Microstructural Parameters and Yielding in Quenched and Tempered Cr-Mo-V Pressure Vessel Steels", (Publication 22, Technical Research Centre of Finland, 1979).

3. H. Kotilainen, "The Micromechanisms of Cleavage Fracture and Their Relationship to Fracture Toughness in a Bainitic Low Alloy Steel", (Publication 23, Technical Research Centre of Finland, 1980).

4. C.D. Lundin, S.C. Kelley, R. Menon and B.J. Cruse, "Stress Rupture Behaviour of PWHT 2.25Cr -1Mo Weld Metal", (Welding Research and Engineering Group Report, Materials Science and Engineering Department, University of Tennessee, 1986).

5. R.A. Swift and H.C. Rogers, "Embrittlement of 2.25Cr-1Mo Steel Weld Metal by Postweld Heat Treatment", Welding Journal, vol. 52, 1973, pp 145s-153s.

6. Y. Ito, M. Nakanishi, N. Katsumoto and H. Tsumura, "Improvement of Toughness and Reduction of Temper Embrittlement in 2.25Cr-1Mo Weld Metal", The Sumitomo Search, no. 30, 1985, pp 41-52.

7. R.S. Chandel, R.F. Orr, J.A. Gianetto, J.T. McGrath and B. Wright, "Some Factors Affecting the Mechanical Properties of SAW Narrow Gap Welds", Proceedings International Conference on Welding for Challenging Environments, Pergamon Press, Toronto, 1985, pp 97-101.

8. T. Wada and G.T. Ellis, "Transformation Characteristics of 2.25%Cr-1%Mo Steel", STP 755, ASTM, 1982, pp343-362.

DISLOCATION SUBSTRUCTURE AND MECHANICAL PROPERTIES

OF LOW-ALLOY FERRITIC WELD METAL

S. Mandziej[*] and A.W. Sleeswyk

Lab. of General Physics, Material Science Centre, University of
Groningen, Westersingel 34, 9718 CM Groningen, The Netherlands.

Summary

Mechanical properties of multipass low-alloy ferritic weld joints are
discussed as regards differences in dislocation configurations of the
ferrite of the inner weld metal layers, heat-affected by subsequent weld
passes. Interpretation of the properties is given based on a recently
presented model of accommodation of minor amounts of second phase
components in the ferritic matrix of the weld. Some dislocation configurations, created during the accommodation, which exhibit an extraordinary
stability to subsequent heat inputs, can play an important role in further
plastic deformation and can influence fracture toughness of the weld metal.

Introduction

Dislocation density in ferritic low-alloy weld metal is one of
characteristics considered to be important to the weld metal properties,
especially as regards ductility of the welds. Generally the substructure in
the welds is described as composed of high or medium density dislocation
tangles or cellular configurations. These configurations are frequently
postulated to influence adversely ductility and fracture toughness of the
welds. In investigations of low-alloy ferritic weld metals of different
multipass weld joints, we observed certain types of dislocation
configurations and proposed that their generation was necessary for
accommodation of martensitic or bainitic second phase islands in the
ferritic matrix. Moreover, presence of these dislocations was always
connected with high Charpy V impact strength and CTOD fracture toughness
values.
In the present paper we intend to explain how different dislocation
configurations influence the fracture toughness of steels containing minor
amounts of hard second phase in a relatively soft ferrite matrix. It was
found that weld metals of almost identical chemical composition, produced
by the same welding procedure and consumables, differed significantly in
CTOD values. No consistent conclusions regarding toughness could be reached
by comparison of either the amount and size of second phase, or content and
size of inclusions, or the amount and grain size of acicular and polygonal

* Now with Foundation for Advanced Metals Science, P.O. Box 8039,
   7550 KA Hengelo, The Netherlands

ferrite. In an attempt to explain differences in ductility, the dislocation substructure of the welds was investigated in detail by TEM. Image quantification of many structural features was further used to compare different welds.

Materials

The observed phenomena are characterised by detailed description of two of the specimens. To illustrate the problem other samples will be also discussed.
The specimens marked B and D (Brittle and Ductile), submerged arc welded with Philips Ph56S (C-Mn) electrode, had chemical composition and properties shown in Table I. The HSLA steel plates of Fe510+Nb type were 25 mm thick and V-joints were made of six passes. Chemical composition of the plates is also given in Table I.

Table I. Chemical compositions and mechanical properties

weld metal B	0.062%C, 0.612%Si, 0.014%P, 0.004%S, 1.19%Mn, 0.036%Ni, 0.042%Cu, 0.002%Al, 0.006%Nb, 0.023%Ti, 237-242 ppm O, 84-88 ppm N
weld metal D	0.055%C, 0.591%Si, 0.007%S, 0.015%P, 1.08%Mn, 0.076%Ni, 0.049%Cu, 0.001%Al, 0.004%Nb, 0.022%Ti, 117-172 ppm O, 61-64 ppm N
steel plate B	0.11%C, 1.50%Mn, 0.019%P, 0.007%S, 0.48%Si, 0.020%Cu, 0.027%Ni, 0.030%Cr, 0.002%Mo, 0.060%Al, 0.031%Nb, Ca-treated
steel plate D	0.08%C, 1.38%Mn, 0.012%P, 0.001%S, 0.044%Si, 0.298%Cu, 0.417%Ni, 0.101%Cr, 0.021%Mo, 0.033%Al, 0.013%Nb, Ca-treated

	yield strength (MPa)	tensile strength (MPa)	area reduc. (%)	elongation (%)	Charpy V-notch (J/cm^2)			
					+20°C	-20°C	-40°C	-50°C
weld metal B	441	528	76	31	178	103	15	11
weld metal D	420	575	72	32	170	139	89	14

Crack Tip Opening Displacement (CTOD) values, determined according to BSI-5762-1979, were 0.168 and 1.235 mm for samples B and D, respectively. For all the CTOD specimens, irrespective of their toughness, the macrocrack re-initiation sites were always on arrays of columnar crystals, close to the V-joint center.

Results

Metallography

On the macro-photos of fractures of the CTOD specimens the difference in ductility may be recognized at once by the shape deformation of the formerly rectangular cross sections (Figs. 1a and 1b).
In these figures the crack initiation sites at the weld mid-thickness and

layers of columnar crystals are also visible. Optical and scanning electron microscopy established slightly different amounts of spheroidal, slag-type, inhomogeneously distributed inclusions in the welds. In the regions of the highest concentration of these inclusions, their average density and size were larger in weld D (see Figs. 2a and 2b), as measured by the SEM-IPS image analyzer.

Fig. 1 – Macrofractographs of CTOD specimen failure surfaces with δ_c of (a) 0.168 mm, weld B, and (b) 1.235 mm, weld D, respectively. Note the columnar crystal arrays in both (a) and (b).

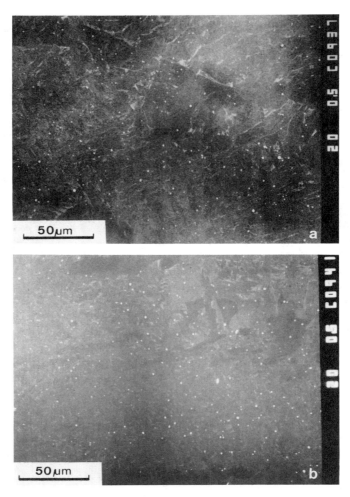

Fig.2 – Scanning electron micrographs of the semi-acicular and polygonal ferrite regions of welds B (a) and D (b), respectively, showing the larger density and size of inclusions in weld D.

Most of the structure of the fine-grained regions which was visible on the fracture surfaces of the CTOD specimens consisted of transformed and/or recrystallized forms of ferrite with minor amounts of second phase. In the as-welded state, i.e. in the last pass, the structure consisted mainly of acicular ferrite in both welds. Quantitative data for the last passes of the welds are given in Table II. It is noteworthy that the coarse-grained zones containing the columnar crystals of both samples B and D contained very similar volume percentages of grain boundary and Widmanstätten ferrite.

Table II.   Structure of as-welded metals

	Phase content (vol.%)			Grain size (μm)			
	acicular ferrite	gr.-bound. ferrite	second phase	second phase	gr.-bound. ferrite	acic.-ferrite Dmax	Dmin
weld metal B	84.7	3.8	11.5	4.27	6.44	6.51	2.51
weld metal D	82.4	8.3	9.3	2.76	8.67	8.33	3.43

Considering further the tertiary and secondary structure of the welds, except for the columnar crystals: the inner weld passes of the weld B contain fine-grained, semi-acicular, ferrite mixed with coarse, equiaxed and elongated ferrite grains. In the weld D the acicular ferrite is seldom present, the structure being mainly medium-sized ferrite grains of irregular or polygonal shape. The quantitative data for ferrite and second phase in the heat-affected inner weld passes are assembled in Table III.

Table III. Heat-affected structure of welds (without columnar crystals)

	ferrite						second phase	
	semi-acicular		polygonal		coarse			
	amount (%)	size (μm)	amount (%)	size (μm)	amount (%)	size (μm)	amount (%)	size (μm)
weld metal B	46.17	7.41	23.51	12.27	24.11	44.83	6.21	3.40
weld metal D	11.28	9.67	71.85	13.90	12.59	27.46	4.28	3.67

The structure of the heat-affected, semi-acicular and polygonal ferrite grains is more pronounced in weld D than in weld B (Figs. 3a and 3b). The microstructure of second phase in weld B is completely pearlitic, whilst in weld D pearlite is a minor component and most of the second phase consists of tempered martensite and bainite (Figs. 4a and 4b). It should be noted that there was very little tendency of pearlite in weld B to coagulate, while in weld D spheroidized second phase islands were sometimes observed (Fig. 5).

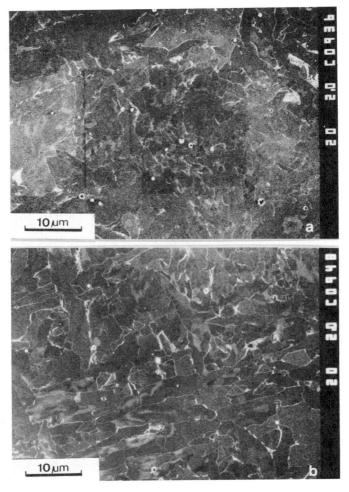

Fig. 3 - Higher magnification, compared to Fig. 2, scanning electron micrographs of the structure of welds B (a) and D (b). Note the more pronounced structure of the semi-acicular ferrite grains in (b).

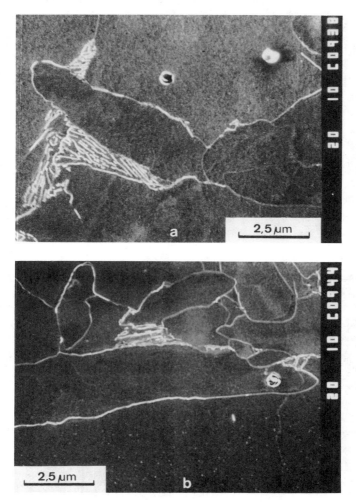

Fig. 4 – High magnification scanning electron
micrographs of the microstructure of welds B (a)
and D (b) showing the characteristic size and
morphology of the second phase, pearlite in (a) and
martensite in (b).

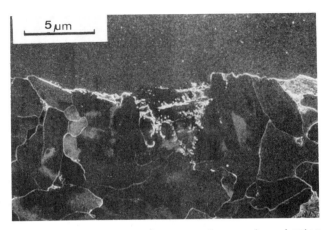

Fig. 5 - Scanning electron micrograph, showing
spheroidized second phase in weld D.

Fractography

The CTOD fractures differ appreciably, even in their macroscopic features. In weld D fracture begins with a large ductile zone, while weld B
fractured nearly entirely by cleavage. Observations in SEM showed ductile
portions in weld B with characteristic dimples only at the origin of the
macrocrack, the remainder of the fracture was brittle (Fig. 6). Fracture of
weld D contains a long, ductile initial zone, and in the cleavage portion
the fracture facets show evidence of plastic shear deformation
(Figs. 7a and 7b).

Fig. 6 - Failure (re)initiation region from the fatigue crack (near bottom of micrograph) in weld B,
showing first a narrow zone of charateristic dimples
and then cleavage, by which final failure took place.

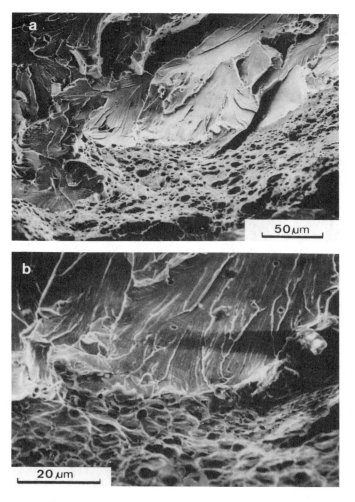

Fig. 7 – Initial crack propagation (from the fatigue crack some 2.5 mm below the region at the bottom of the micrograph) in weld D, (a) showing the characteristic zone of dimples, preceding cleavage with associated plasticity, visible at a higher magnification in (b).

Second phase and dislocation substructure

In order to interpret differences in plastic behaviour between the samples, TEM investigations concentrated on evaluation of the factors influencing the decrease of ductility of semi-acicular and polygonal ferrite in weld B. In this specimen ferrite substructure consists mainly of small subgrains, containing cellular dislocation configurations (Fig. 8). The second phase is pearlitic (Fig. 9). In the semi-acicular and polygonal ferrite of weld D the dislocations are arranged in regular arrays of non-uniform density (Fig. 10) and the second phase consists mainly of tempered martensite and bainite (Figs. 11 and 12). Fine precipitates in the ferrite are present extensively only in weld B. They interact with the cellular network of dislocations in weld B and such interactions in weld D are seldom observed (Figs. 13 and 14).

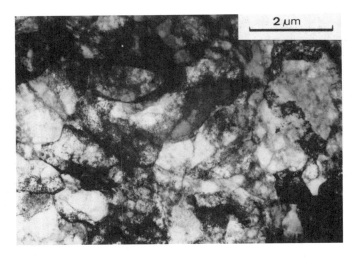

Fig. 8 - TEM micrograph of ferrite subgrains frequently present in weld B.

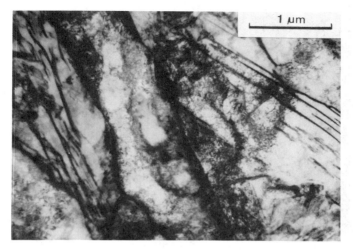

Fig. 9 - TEM micrograph of irregular pearlite colonies and high density cellular dislocation substructure of ferrite in weld B.

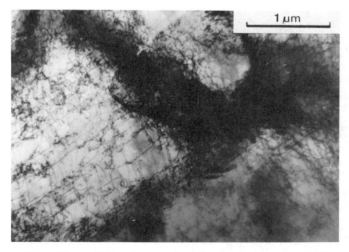

Fig. 10 - TEM micrograph of regular arrays of dislocations dominating in ferrite of weld D.

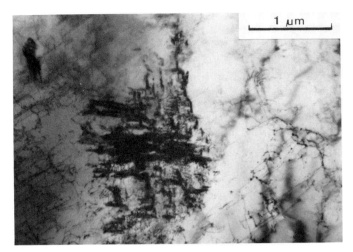

Fig. 11 – TEM micrograph of high-tempered
martensitic (or bainitic) island in weld D,
retaining Bagaryatskii cementite-to-ferrite
orientation-relationship. Note characteristic
dislocation arrays at the island.

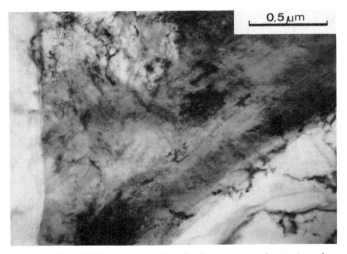

Fig. 12- TEM micrograph of low-tempered, twinned
martensite island in weld D.

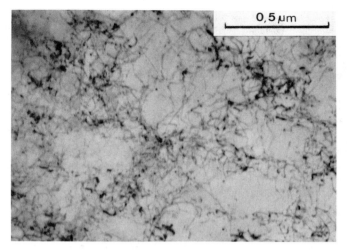

Fig. 13 – TEM micrograph showing interactions between fine precipitates and cellular dislocation network in ferrite of weld B.

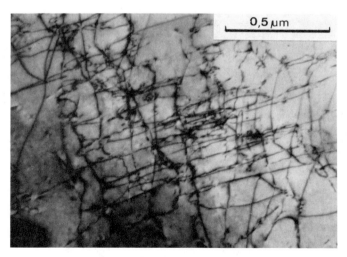

Fig. 14 – TEM micrograph showing characteristic criss-cross of ⟨111⟩ oriented screw dislocations in ferrite of weld D.

Discussion

The secondary weld metal substructure of the ductile specimen D in fact corresponds well to the dislocation configurations characteristic of dual-phase steels. Configurations of this kind were correlated by Korzekwa et al. [1] with an imposed plastic strain of 1%. In our studies similar microstructures were obtained in some grains without any imposed deformation. These configurations are postulated to result from the operation of the 'Catherine wheel' dislocation generation mechanism [2].

One of the most important characteristics of the dual-phase structure is its relatively high ductility. Recently dual-phase steels have been shown to display excellent resistance to crack propagation, which is correlated with the meandering of the crack path [3, 4]. In our research project GNS 44.0704, by optimizing the dual-phase structure, we achieved Upper Shelf Energy values in Charpy V tests as high as 340-420 J/cm^2, and concurrently transition temperatures in the range of -90 to -60°C. For the low-alloy ferritic weld metal, however, any presence of lower-bainitic or martensitic second phase has been associated in the literature with low fracture toughness, especially in the heat-affected zone (HAZ). Recently, evidence was presented that nucleation of cracks occurs on M-A components in intercritically reheated fragments of HAZ and this results in low CTOD values of the whole joint [5, 6]. These results stand in contradiction to the well-known properties of intercritically annealed dual-phase steels. However, any consideration of ductility should not omit a description of dislocation substructure, and this was not presented in the above-mentioned papers.

In our previous work we studied a possible mechanism of accommodation of martensite and bainite islands in dual-phase steels; a model was proposed to explain the formation of the observed dislocation configurations [2]. According to the model, the characteristic criss-cross configuration of screw dislocations, dominating in dual-phase steels of optimum properties, is formed by antisymmetrical accommodation of expanding martensitic islands during their formation from austenite at relatively low temperatures, in which generation of screw dislocations is favourable [7]. These screw dislocations form low-energy configurations which are relatively stable under subsequent heat treatments. Their stability was confirmed by tempering experiments on bulk specimens and also by in-situ heat treatment tests in the TEM. Up to the recrystallization temperature of ferrite, parts of the dislocation half-loops remained virtually unchanged (Figs. 15a and 15b).

High ductility of the dual-phase steels has been studied recently, e.g. by He et al. [8], Cai et al. [9], who described nucleation of cracks by void formation in areas of highest local deformation, e.g. in necked portions of tensile specimens. The voids are explained by differences in deformability between ferrite and second phase. However, crack propagation does not seem to follow the coalescence of such voids, even for tensile specimens. From Charpy V tests and fatigue experiments it is clear that decohesion of the ferrite-second phase interfaces does not contribute significantly to the fracture toughness of the dual-phase structures [10, 3]. In our experiments, sites of macrocrack re-initiation were interfacial regions adjoining columnar crystals nearest to the sample mid-section, regardless of subsequent low or high CTOD values. The differential deformability analogy may perhaps be applicable to the macrocase of two types of ferrite grains: the columnar and the fine. In detailed microstructural investigations of the fracture surfaces using quantitative image analysis, the amount of the second phase on the facets was found to be one-fifth of the amount in the structure (as measured on conventional metallographic sections). Clearly the propagating cracks by-pass the second-phase islands. We conclude that crack propagation by coalescence of microcracks rather than the re-initiation of the macrocrack is influencing fracture toughness.

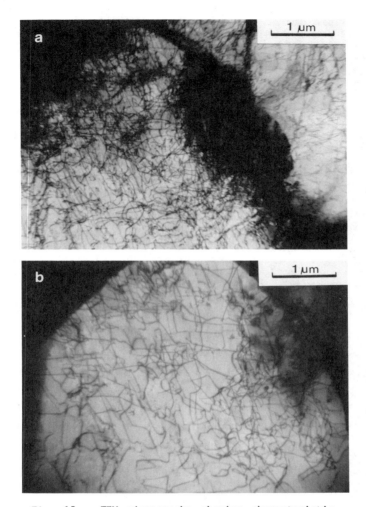

Fig. 15 - TEM micrographs showing characteristic arrays of screw dislocations in as-hot-rolled dual-phase steel after 5% strain (a), and after in-situ tempering for 15 min. at 650°C in TEM hot stage (b).

Indeed, critical values of K_1 and G_1 are the accepted measures of material's resistance to crack propagation, and of CTOD through its relationship with these parameters. It seems that the response of the secondary and/or tertiary structure of the weld metal to the crack propagation is at the origin of the overall fracture toughness of the joint.

Model of crack propagation

Accepting that (fracture) toughness relates to the difficulty of crack propagation, we will now consider cracking with special reference to dual-phase structures. Let us first consider the propagation of a crack simply through a crystalline medium. It is of obvious interest to take into account the possible movement of dislocations around the crack tip, if it propagates not too quickly to preclude that [11]. Calculation of the shear stresses on glide planes in the vicinity of the tip provides some insight in this matter. The glide planes must be not only those radiating from the crack tip; the ones parallel to these that pass it at a certain distance are of more interest.

In a forthcoming paper (Sleeswyk and Mandziej [12]), we present the results of such a calculation on the basis of fracture theory [13, 14] of the stress field around a crack tip in a plane stress field. What follows is a qualitative interpretation of the most important results.

The glide planes considered are inclined at 45° to the plane of the crack. This is the important practical situation of $\{110\}$ glide planes near a $\{100\}$ crack plane in a b.c.c. transition metal. The shear stress on such a plane, $\tau_{45°}$, can be completely described by two parameters. One is the distance d between the glide plane and the crack tip: $\tau_{45°}$ is proportional to d^{-2}. The other parameter is the angle Θ between the plane containing the location on the glide plane and the crack tip, and the plane of the crack. Our calculations show that the shear stress $\tau_{45°}$ on such a glide plane assumes the following extrema; for $\Theta = 0°$, $\tau_{45°} = 0$ and it is at a minimum; for $\Theta = 68°$, $\tau_{45°}$ is at a maximum, for $\Theta = 120°$, $\tau_{45°} = 0$. The functional relationship is illustrated in the accompanying diagram (Fig. 16) (in which $\Theta = 120°$ is not shown).

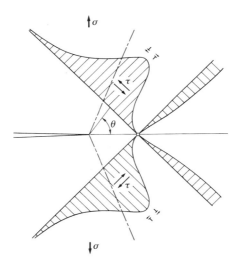

Fig. 16 - Shear stress on glide planes inclined at 45° to the plane of the crack, as a function of the angle Θ between the plane containing the location on the glide plane and the crack tip, and the plane of the crack, in arbitrary units.

For $\Theta < 120°$, the region illustrated in the diagram, the value of $\tau_{45°}$ possesses the same sign everywhere, though $\tau_{45°} = 0$ for $\Theta = 0°$. That means that within this region (the region $\Theta > 0°$ is not important for crack propagation) all mobile dislocations with positive Burgers vectors are forced away from the crack plane for $\Theta > 0°$. For $\Theta < 0°$, the reverse applies, but here the values of $\tau_{45°}$ are generally lower than on the other side of the plane of the crack.

Of course, for dislocations of opposite sign, the opposite holds. The minimum value $\tau_{45°} = 0$ for $\Theta = 0°$ implies that dislocations that have been brought near that position by dislocation interactions, e.g. pile-ups, will have a tendency to linger there. This is of importance if we consider the Cottrell reaction [15] between $\frac{1}{2}\langle 111 \rangle$ dislocations on intersecting $\{110\}$ glide planes in the b.c.c. lattice:

$$\tfrac{1}{2}[111] + \tfrac{1}{2}[1\bar{1}\bar{1}] \rightarrow [100] \tag{1}$$

Agglomerations of [100] dislocations are considered in Cottrell's theory to be crack nuclei on the (100) plane. It is shown in the diagram, presented here as Fig. 17a, that mobile $\frac{1}{2}\langle 111 \rangle$ dislocations on intersecting $\{110\}$ planes can create nuclei in advance of the crack tip. The phenomenon as such is well-known, and accords e.g. with Dugdale's [16] theory of the growing crack.

The existence of the zero shear stress minimum in the crack plane will cause the positions of these nuclei to scatter around the plane of the advancing crack: it will progress on continually changing levels of the crack plane. The phenomenon was observed in polycrystalline steels in 1953 by Boyd [17], but mechanics predicts the same for a single crystal.

More important is that the creation of these crack nuclei is crucially dependent on the symmetry of the flow of $\frac{1}{2}\langle 111 \rangle$ dislocations on the two intersecting glide planes towards the crack plane. What happens if the symmetry is disturbed?

In Fig. 17b the effect of the presence of a 'hard' region on one side of the crack plane is illustrated. The elastic moduli in the 'hard' region are not supposed to be significantly different from the remaining 'soft' material; the region is called 'hard' because it is more difficult to move dislocations already inside it, or those that try to penetrate into it from the 'soft' matrix. As a consequence, the dislocations on only one of the two intersecting glide systems are activated, and these will not react to form $\langle 100 \rangle$ dislocations, but they will tend to pile-up against the interface between 'soft' and 'hard' regions. The properties attributed here to 'hard' and 'soft' regions correspond to those e.g. of martensite islands and the ferrite matrix, respectively, in the dual-phase structure.

In this instance, where the creation of Cottrell cleavage nuclei on $\{100\}$ planes is prevented, the $\{110\}$ are plausible cleavage planes, and the piled-up $\frac{1}{2}\langle 111 \rangle$ dislocations will then act as cleavage nuclei in the manner envisaged by Zener and Stroh [18, 19]. Being in advance of the growing crack, they will deviate its course onto $\{110\}$ planes in the ferrite matrix.

The conclusion is that if the Zener-Stroh mechanism is effective in the case of asymmetry caused by 'hard' regions embedded in a 'soft' matrix, the crack will tend to propagate in the 'soft' material around the 'hard' regions. If the elastic moduli between 'soft' and 'hard' regions were appreciably different, the interfacial tension would very probably cause the crack to be deflected along this interface, especially so if the coherence across the interface were weak.

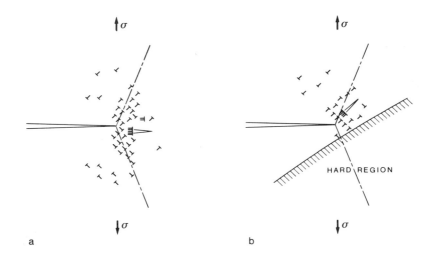

Fig. 17- (a) Formation of crack nuclei in front of advancing crack tip by Cottrell's mechanism. (b) Formation of crack nuclei in 'soft' matrix by pile-ups of dislocations (Stroh's mechanism) in the vicinity of 'hard' regions, and resulting deflection of the propagating crack through 'soft' material.

Conclusions

1. Presence of martensite and bainite in weld metal does not need to be detrimental for fracture toughness of the welds; an optimum distribution of such components resembling a dual-phase structure increases the toughness.
2. The fracture toughness in dual-phase structure, controlled by propagation of cracks, is affected by presence of 'hard' phase, causing deviation of the propagating crack tip.
3. Differences in dislocation substructure, in particular availability of slip dislocations of the 'Catherine wheel' type in the dual-phase structure and lack of glissile dislocations in the cellular substructure, are responsible for different fracture behaviour of the considered weld metals.
4. Various contributions of different dislocation configurations to fracture toughness of the investigated welds can be adequately interpreted by the operation of the presented model of crack propagation in dual-phase structures.
5. It seems plausible that fine precipitates, observed in large numbers in weld metal with high density cellular dislocation substructure, are responsible for the acceleration of the formation of pearlite in the brittle weld B, thus precluding formation of martensite and bainite there.

218

Acknowledgements

The authors would like to acknowledge their gratitude to Dr. A.S. Wronski and Dr. B.H. Kolster for many helpful discussions and contributions to this paper. Additionally, the authors wish to mention that this work developed from cooperation with the Dutch Welding Institute, while investigations on ductility of dual-phase steels were sponsored by The Dutch Foundation for Technical Sciences in project GNS 44.0704.

References

1. D.A. Korzekwa, D.K. Matlock, G. Krauss; Met. Trans. A, 15A, June 1984, p.12.
2. S. Mandziej, A.W. Sleeswyk, B.H. Kolster and J. Beyer; Proc. 7th Int. Conf. on Strength of Metals and Alloys, (ed. H.J. McQueen et al.), Montreal, Canada, August 1985, p. 545.
3. J.L. Tzou and R.O. Ritchie; Scripta Met., 19, No. 6, 1985, p. 751.
4. J.A. Wasynczuk, R.O. Ritchie and G. Thomas, Mat. Sci. Eng., 62, 1984, p. 79.
5. T. Haze and S. Aihara; 'Metallurgical factors controlling HAZ toughness in HT50 steels', IIW Doc. IX-1423-86, R&D Labs II, NSC, Tokyo, Japan, May 1986.
6. J.H. Chen, Y. Kikuta, T. Araki, M. Yoneda and Y. Matsuda; Acta Met., 32, No. 10, 1984, p. 1779.
7. S. Mandziej, A.W. Sleeswyk and B.H. Kolster; 'Effect of the morphology of martensite on its accommodation in dual-phase steels', Proc. of Int. Conf. on Martensitic Transf., The Japan Institue of Metal, 1986, p. 459.
8. X.J. He, N. Terao and A. Berghezan; Metal Sci., 18, July 1984, p. 367.
9. X.-L. Cai, J. Feng and W.S. Owen; Met. Trans. A, 16A, August 1985, p. 1405.
10. H. Suzuki and A.J. McEvily; Met. Trans. A, 10A, 1979, p. 475.
11. M.F. Ashby and J.D. Embury; Scripta Met., 19, No. 6, 1985, p. 557.
12. A.W. Sleeswyk and S. Mandziej; 'On crack propagation in dual-phase steels', to be published, 1987.
13. H.M. Westergaard; Trans. ASME J. Appl. Mech., 61, 1939, p. 49.
14. G.R. Irwin; Fracturing of Metals, ASM, Cleveland, Ohio, 1949, p. 147.
15. A.H. Cottrell; Trans. AIME, 212, 1958, p. 192.
16. D.S. Dugdale; J. Mech. Phys. Solids, 8, 1960, p. 100.
17. M.C. Boyd; Engineering, 175, 1953, p. 65.
18. C. Zener; Fracturing of Metals, ASM, Cleveland, Ohio, 1948, p. 3.
19. A.N. Stroh; Adv. Phys., 6, 1957, p. 418.

DIRECT OBSERVATION OF HYDROGEN INDUCED CRACK PROPAGATION

IN WELDS OF HT-STEEL

K. Hoshino*, R. Yamashita* and T. Shinoda**

*) Daido Institute of Technology
2-21 Daido-cho, Minami-ku, Nagoya, 457, JAPAN
**) Nagoya University, Faculty of Engineering
Furo-cho, Chikusa-ku, Nagoya, 464, JAPAN

Abstract

It is an important trial to observe hydrogen induced crack initiation and propagation during or after welding. This work was done to establish direct observation method of HIC propagation under tensile loading which was simulated in real structures. Specimens of HT-80 were welded with low hydrogen type electrodes under bead on plate condition. Specimens were cut and polished in dry ice temperature to prevent hydrogen evolution from welds. Welded specimens were loaded with hydraulic jack to a certain load under optical microscope observation. Results were analyzed through photos. HIC was initiated at the center of weld bead and propagated into weld metal and heat affected zone. HIC propagation rate was rather high in weld metal but not in HAZ. It seemed to propagate intermittently through HAZ with severe plastic deformation. SEM observation on the fractured surface revealed typical HIC appearances with quasi-cleavage fracture.

221

1. Introduction

It has been clear understanding the reason of the hydrogen induced crack, HIC in high strength steel welds by hundreds of works. However it is still a fundamental problem in actual structure fabrication by lack of knowledge on mutual relationship between metallurgical factors and stress conditions.

It is an important trial to observe the HIC initiation and propagation during or after welding, but there is a few works to observe them. Savage et al.(1,2) were carried out observation using bending type crack test under the optical microscope and found that HIC propagated with severe plastic deformation and hydrogen evolution at the crack front. Matsuda et al.(3) also found the plastic deformation at the crack tip in the scanning electron microscope, SEM.

Under these circumstances, it is necessary to clarify the influence of factors on the HIC initiation and propagation to get more practical knowledge. This work has been done to establish direct observation of the HIC propagation under tensile loading which simulated in real structures, that is a different loading system of above mentioned works.

Specimens of HT-80 were welded with low hydrogen type MMA electrodes under bead on plate condition. Welded specimens were loaded with a hydraulic jack to the certain load under the optical microscope. Crack propagation rates were determined through photos. Specimens after microscopic observation were examined using SEM at the crack tip and the fractured surface.

2.1 Experimental procedure
2.1.1 Newly developed testing machine

A new type of loading equipment was developed shown in Fig. 1. The tensile stress was loaded by the hydraulic jack (2) to the certain stress level and specimen (1) was fixed using a loading bolt (3) to keep the constant strain. Four strain gauges were attached on the loading bar (4) and stress changes were monitored by a digital strain meter (5) and a multi-pen recorder during observation of the hydrogen induced crack propagation. The optical microscope (6) with camera (7) was set above the test specimen.

2.1.2 Specimen and its chemical compositions

Specimens were taken from HT-80 high tensile strength steel plate with 32 mm in thickness. Mechanical properties of this plate were as follows, the tensile strength was 864.7 N/mm^2 and the 0.2% proof stress was 805.6 N/mm^2. Table I indicates the chemical compositions of base metal and deposited metal.

Table I. Chemical Compositions of Base Metal and Weld Metal

	Chemical compositions (wt.%)										
	C	Si	Mn	P	S	V	Cu	Ni	Cr	Mo	Al
Base metal	0.13	0.31	1.38	0.015	0.005	0.01	0.04	0.03	0.03	0.62	0.010
Weld metal	0.08	0.34	1.29	0.014	0.004	<0.01	0.02	1.50	0.12	0.59	0.009

Two kinds of rolling directions were selected to prepare specimens, those were taken from to the transverse and the perpendicular to the rolling direction. The later specimen was assembled using a high power electron beam welding shown in Fig. 2-B. Specimens to be welded were prepared as shown in Fig. 3 and both plate surfaces were polished with normal procedures.

222

Figure 1   Equipment for HIC Observation

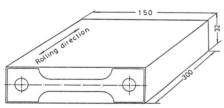

A) Transverse to rolling direction

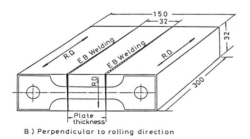

B) Perpendicular to rolling direction

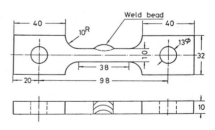

Figure 2   Orientations of HIC Test
Specimens (Unit : mm)

Figure 3 Test Specimen for HIC
Observation (Unit : mm)

223

2.1.3 Welding procedures and HIC observation

A starting and a run-off tab plates were fixed to both sides of the specimen and bead on plate welding was carried out on the center line of the specimen shown in Fig. 3 with 13 kJ/cm of a nominal heat input. A single low hydrogen type electrode with equivalence to E-11016 of the AWS standard, was used for each specimen. The electrode was used in fully moistened and as received conditions, those diffusible hydrogen levels were almost 10 ml and 2.5 ml to 100 grams of the deposited metal by glycerin test according to the Japan Industrial Standard.

The specimen after bead on plate welding was immediately quenched into water bath and kept in dry ice saturated alcohol bath to prevent hydrogen evolution from the weld metal. Tab plates were broken off in this alcohol bath. The fractured weld metal surface to be observed was polished using a normal polishing buff as possible as keeping the specimen in low temperature. Specimen preparation time after welding to observation was kept within 45 minutes. Observation works were carried out under the constant strain condition for 24 hours at room temperature. After loading stress, specimen was observed under the low magnification optical microscope and 35 mm still pictures at the crack tip of HIC were documented in every 5 or 10 minutes at magnifications of 50 x or more.

3 Results and Discussion
3.1 Observation of the HIC propagation
3.1.1 Deposited with fully moistened electrode

Fig. 4 shows examples of series photographs of the HIC propagation in the coarse grained HAZ using a fully moistened electrode. The crack initiated at the center of weld metal and propagated to the transverse direction against the loading axis.

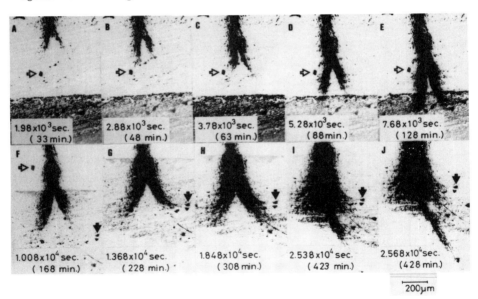

Figure 4   Propagation of HIC in a Transverse Specimen
(Initial Stress : 320 N/mm^2)

224

This crack initiation and propagation in the weld metal were rather quick that was almost within 30 minutes after loading. This quick crack initiation and propagation in weld metal were associated with a sudden load decrement caused stress relaxation by opening crack. Photographs of A-E in Fig. 4 showed that HIC propagated from the deposited metal which located at the top of photographs. Arrow marks in photographs mean the landmarks on the same scratch to assist easier observation of crack propagation. Numbers in photographs indicate lapse times after loading in seconds.

It is clearly seen that the HIC associated with the severe plastic deformation which seemed with a "Y" letter at the crack tip and the width of crack grew wider during propagation. In Fig. 4-I and 4-J, the new sharp crack appeared to propagate.

3.1.2 HIC morphology

Fig. 5 shows that macrophotographs of HIC after release from the testing equipment. Two cracks were observed in Fig. 5-A which ran into the perpendicular direction to the stress axis and turned in 45 degree which was the direction of the maximum shear stress. Cracks initiated at the maximum throat thickness position where was the position of the maximum hydrogen content. It is not clear understood that the two parallel cracks were occurred in case of high stress level, where each crack made stress relaxation at crack tips. However, the lower stress condition made one straight crack initiation and propagation shown in Fig. 5-B. This would be caused by stress levels influenced on the diffusion pass length of hydrogen towards crack tips. Those cracks stopped at the outer of HAZ boundaries.

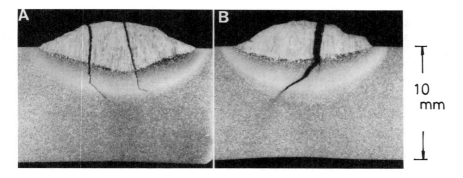

10
mm

Figure 5   Appearances of HIC specimens
A) Initial Stress : 410 N/mm^2
B) Initial Stress : 245 N/mm^2

3.1.3 Crack propagation rate

Crack propagation behavior in the coarse grained HAZ was recorded through a series of 35 mm still films. The total crack length and crack propagation rate were calculated using those films. Fig. 6 shows a typical result of these experiments deposited fully moistened electrode. The initial setting of tensile force was 320 N/mm^2 which was rapidly decreased shown in the upper part of figure. This corresponded to the rapid crack propagation in deposited metal shown in the lower part of figure. The crack propagation chart indicates that crack propagated repeating "go" and "stop" motions in the coarse grained HAZ.

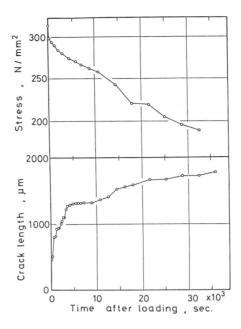

Figure 6   Total Length of HIC from Fusion Boundary
and Stress Changes in a Transverse Specimen

These were caused by hydrogen diffusion at the crack tip. The intermittent crack growth can be explained by Troiano model (4). Savage et al.,(5)
also suggested that the intermittent crack growth based on the hydrogen
diffusion at newly formed crack tips in the tri-axial stress region. This
diffusion process would control the intermittent crack propagation. The crack
propagation rate gradually decreased after 10 hours and stopped after 15
hours in this case shown in Fig. 6.

3.2   Influence of rolling direction
3.2.1 HIC observation in perpendicular specimen to R.D.

A series of experiments were carried out to clarify the effect of
rolling direction on the HIC propagation characteristics. Specimens were
machined according to procedures shown in Fig. 2 and 3.

A fully moistened electrode was used in this experiment. The crack
propagation was recorded shown in Fig. 7. It is clearly observed that a non-
metallic inclusion, existing apart from crack tip, opens without deformation
and that the crack propagated through those inclusions. When crack reached
to an opened crack, newly plastic deformation was formed around the crack
tip. As the results, sharp brittle crack opening and ductile plastic
deformation were repeated. This phenomenon was not observed in the transverse
rolling direction specimen described above.
Crack propagation rate in the coarse grained HAZ was estimated through a
series of photographs. The result is shown in Fig. 8. The HIC propagated in
step wise characteristics and this seems to be 20 or 30 % faster than that of
transverse direction case under the same initial stress condition . This
result is in accord with the results by Kikuta et al., (6) and Savage et
al.,(2).

226

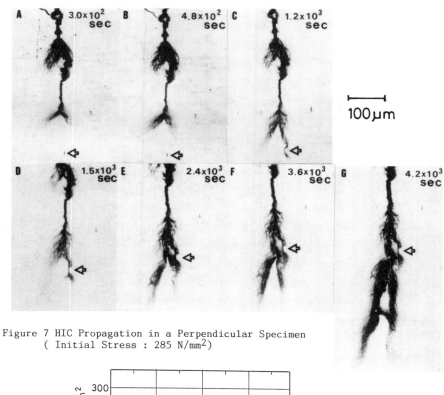

Figure 7 HIC Propagation in a Perpendicular Specimen
(Initial Stress : 285 N/mm^2)

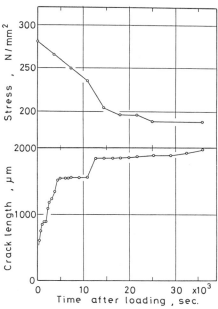

Figure 8   Total Length of HIC from Fusion Boundary and
Stress Changes in a Perpendicular Specimen

227

3.2.2 Micro-structure and crack pass

The mid-thickness specimens were examined under the optical microscope. Fig. 9 shows the HIC pass at the fusion boundary area in case of the high stress and high hydrogen experiment. The main crack ran perpendicular to the stress direction and a secondary crack branched off at the fusion boundary.

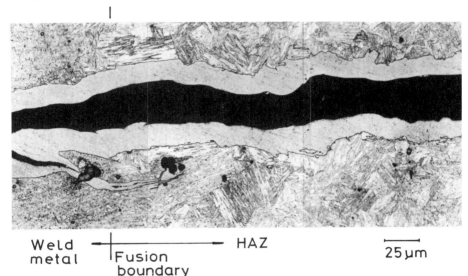

Weld metal ── Fusion boundary ── HAZ   25 μm

Figure 9   HIC passes in the Coarse Grained HAZ
(Initial Stress : 325 N/mm²)

The main crack propagated along or across the martensite laths in the coarse grained HAZ. However it is obvious that crack runs mainly transgranular mode in this region, and made gradually with small branches or concave and convex passes in the medium grained HAZ , shown in the right hand side of this photograph.

3.3   Results of SEM observation
3.3.1 SEM observation at crack tip

After observation of crack propagation under the optical microscope, specimens released from the loading device were observed in the SEM. Fig. 10 shows the SEM photographs at crack tips with different rolling directions. Crack pass was hardly changed in case of the transverse direction with stress and the rolling directions (Fig. 10-A), but was remarkably influenced in case of perpendicular with stress and the rolling directions (Fig. 10-B). The later one indicates step wise propagation mode shown in Fig. 7.

Specimens were fractured in liquid nitrogen and fractured surfaces were examined in the SEM. Fig.11 shows three types of the hydrogen induced crack: those are a quasi-cleavage fracture surface in Fig. 11-A, secondary micro-cracks on the martensite laths in Fig. 11-B, and intergranular fracture in Fig. 11-C (8). Dimple rupture was also found in the case of low hydrogen contents using as-received electrode and under low stress condition or at the end part of HIC propagation.

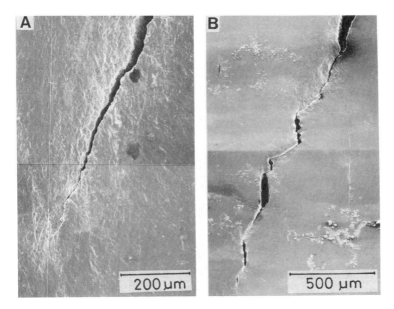

Figure 10   SEM Photographs at Crack Tips
A) Transverse to Rolling Direction
B) Perpendicular to Rolling Direction

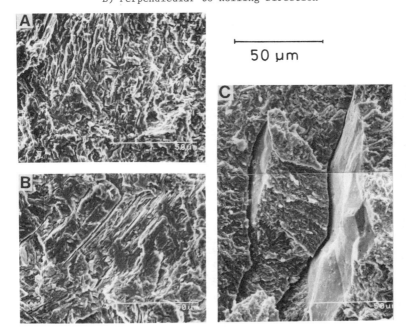

Figure 11 SEM Photograph on the Fracture Surface
of HIC Specimens

229

The differences between rolling direction on fracture appearance were found that the transverse ones showed terrace-like appearance along elongated non-metallic inclusions, but perpendicular one showed no terrace-type appearance. This may be caused by the extremely low sulphur contents, 0.004%, of this steel. Such low sulphur steel is anticipated to have less homogeneity against HIC propagation (7, 8) and increment the maximum HAZ hardness (9). This is now still investigating by authors.

4. Conclusions

The newly developed equipment was used to the direct microscopic observation of the hydrogen induced crack propagation of steel welds. Results were summarized as follows;

1) The hydrogen induced crack was always initiated at the center of weld bead and propagated into weld metal and heat affected zone which run through the maximum throat thickness region. It was not found that toe crack and neither under bead crack.

2) The hydrogen induced crack propagation rate was rather high in the weld metal but not in HAZ which was explained the differences of hydrogen contents in both regions. HIC seemed to propagated intermittently through HAZ with plastic deformation.

3) Observation in SEM on the fractured surface revealed typical HIC appearances with quasi-cleavage fracture and intergranular fracture.

Acknowledgments

The authors are grateful to Prof. Dr. R. Kohno of Daido Institute of Technology for his valuable discussion and helpful comments. The authors would like to thank Mitsubishi Electric Corp., for assistance of assembling test pieces by high power EB welding equipment. Grateful acknowledgment is made to students, in particular Mr. H. Katoh and Mr. M. Kitoh, for their careful experimental works.

5. References

1 W. F. Savage et al., "Hydrogen Induced Cracking in HY-80 Steel Weldments," Welding Journal, 55(1976) 368s-376s.

2 W. F. Savage et al., "Hydrogen Induced Cracking in HY-130 Steel Weldments," Welding Journal, 57(1978) 118s-126s.

3 F. Matsuda et al., "SEM Dynamic Observation of Hydrogen-Induced Cold Cracking in Weld Metal of 80 kg/mm^2 Class Steel," Transactions of JWRI, 10(1981) 81-87.

4 A. R. Troiano, "The Roll of Hydrogen and other Interstitials in the Mechanical Behavior of Materials," Trans. ASM, 52(1960), 54-80.

5 W. F. Savage et al., "Hydrogen Induced Cold Cracking in a Low Alloy Steel," Welding Journal, 55(1976), 276s-283s.

6 Y. Kikuta et al., "Effect of Non-Metallic Inclusions on Hydrogen Induced Crack on High Strength Steel Welds," WM-650-77(1977), Welding Metallurgy Committee of Japan Welding Society

7 Japan Welding Society, "Fractographic Atlas of Steel Weldments," Kuroki Publishing Inc., Tokyo, (1982), 71-72.

8 Y. Kikuta et al., "Fractographic Analysis on Lamellar Tearing Examined under Implant Testing," Journal of the Japan Welding Society, 46(1977), 76-81.

9 P. H. M. Hart, "Resistance to Hydrogen Cracking in Steel Weld Metals," Welding Journal, 65(1986), 14s-22s.

IMPROVEMENT IN MICROSTRUCTURE AND PROPERTIES OF STEEL ELECTROSLAG

WELDMENTS USING LOW ALLOY TUBULAR FILLER METAL

D. W. Yu, H. S. Ann, J. H. Devletian and W. E. Wood

Department of Materials Science and Engineering
Oregon Graduate Center
Beaverton, Oregon

Abstract

The low alloy (2.4%Ni-1.2%Mn-0.5%Mo) steel tubular filler metal deposition improved the fracture toughness of consumable guide electroslag weldments of 50mm and 76mm thick A36 and A588 steels. This alloying promoted equiaxed dendritic growth at the weld center and produced weld metal structures containing approximately 90% acicular ferrite. By utilizing a tubular electrode and "wing" and "web" guides to effectively improve the form factor and decrease the base metal dilution, it was possible to increase welding speed and decrease heat input without the risk of hot cracking and lack of fusion. As a result, the CVN and K_{Ic} toughness was increased at least to that of the base metal. In addition, it is interesting to note that the HAZ size and its behavior were controlled by not only the total heat input but also the geometry of the weld pool which affected the heat transfer towards the HAZ.

Introduction

The attainment of improved toughness in electroslag weldments has been the objective of numerous recent research programs. Excessive heat input and slow welding speeds in electroslag welding (ESW) result in both super-heated weld metal and substantial grain growth in the heat affected zone (HAZ). In general, the morphology of mild steel weld metal basically shows two zones of grain structures: thin columnar grains (TCG) around the weld center, and coarse columnar grains (CCG) near the weld interface. The former is composed largely of grain boundary ferrite and side-plate ferrite, which cannot withstand cleavage crack propagation, and the latter contains a predominantly acicular ferrite substructure which is resistant to cleavage.

Decreasing heat input is one approach to improve the toughness of electroslag weldments (1,2) by increasing the amount of acicular ferrite in the weld metal. Although the adjustment of welding parameters (current, voltage, joint gap and guide plate/tube design) could directly vary the value of total heat input (3), the degree of variation is limited by the susceptibility to hot cracking and lack of fusion (4). Other methods have been reported to decrease heat input, among them are using additional fill-er wires, metallic powders, chopped wires and lump filler metals into the weld pool (5-7). However, problems such as macro-chemical heterogeneity and incomplete fusion limit these approaches. All the methods mentioned above only reduced the proportion of thin columnar grains, but did not substantially change its substructure (4,7).

Aimed at overcoming the heterogeneity in microstructure and proper-ties, alloying measures were also used to develop a fully acicular ferrite weld structure (8). Such efforts were made not only in electroslag weld-ing, but other welding techniques. A number of alloy additions (Mn, Ni, Cr, Mo, Ti, Si, etc.) and their combinations have been reported to enhance

234

a certain amount of acicular ferrite at the expense of grain boundary and side-plate ferrite (9).

Sponsored by the U.S. Department of Transportation/Federal Highway Administration, a low alloy (Ni-Mn-Mo) steel tubular electrode (TW8544) was designed and applied to the narrow gap consumable ESW process at the Oregon Graduate Center. The objective of this investigation was to study the effect of alloyed tubular wire deposition on weld pool geometry, heat input, microstructure and CVN and K_{Ic} toughness. Meanwhile, the influence of alloying elements on the ferrite transformation is discussed. In addition, the effects of heat input on the HAZ size was investigated.

Experimental Procedure

Materials

The welding studies were conducted on ASTM A36 and A588 steel plates with 50mm (2 inch) and 75mm (3 inch) thicknesses. Various filler metals, including solid and tubular filler metal, were supplied in the form of 2.4mm (3/32 inch) diameter wires. Rectangular consumable guide plates were made out of 1010 mild steel. Table I summarizes the chemical compositions of the above materials. All welds were made with a neutral Hobart 201 flux (basicity indices 0.9).

Table I. The Chemical Compositions of Materials (wt.%)

Material	C	Mn	Si	Cr	Ni	Mo	Cu	S	P	Ti	Al
A36	0.17	0.92	0.20	0.02	0.10	--	0.20	0.02	0.01	--	0.024
A588	0.18	1.20	0.37	0.56	0.16	--	0.32	0.02	0.01	--	0.009
Hobart-25p	0.11	1.12	0.50	--	--	--	0.32	0.02	0.02	--	0.004
Linde-WS	0.09	0.50	0.30	0.55	0.50	--	0.30	0.03	0.02	--	0.011
Airco-AX90	0.08	1.40	0.46	0.06	2.10	0.40	--	0.02	0.02	0.007	0.003
TW8544*	0.03	1.20	0.45	--	2.40	0.45	--	0.02	0.02	0.011	0.007
1010	0.10	0.45	--	--	--	--	--	0.05	0.04	--	0.030
(Balance Fe)								(max)	(max)		

*Designed by authors and manufactured by Stoody Company

Welding Process

ESW was performed with DC reverse polarity using weld joint gaps maintained at 19mm (3/4 inch). Reference ES welds were made with standard 31mm (1.25 inch) gap. All welds were deposited vertically with run-in and run-out blocks.

Evaluation

Solidification substructures were revealed by Stead's reagent, while solid phase transformation structures were revealed by 1% Nital solution. Both optical microscopy and SEM analyses were performed.

Standard ASTM E-23 Charpy impact specimens had their notches perpendicular to the welding direction and were located in 1) mid-thickness/weld center, 2) quarter-thickness/weld center, and 3) coarse heat affected zone HAZ-1. Impact tests were conducted in the temperature range of -73° to +65°C (-100° to +150°F). Compact tension ASTM E-399 K_{Ic} specimens were 76mm (3 inch) thick (full weld thickness) and were tested at -18°C (0°F). Fatigue precracks were parallel to the welding direction and were located in the 1) weld centerline, and 2) HAZ-1, respectively.

Results

Alloying and Weld Microstructure

Chemical analyses of several typical narrow gap weld deposits are presented in Table II. Small amounts of Ni and Mo additions dramatically influenced the microstructures of A36 and A588 weld metals. In alloyed welds produced under conventional current and voltage settings, strong constitutional supercooling resulted in equiaxed dendritic growth during solidification near the weld centerline (Figure 1). More microstructural changes were found in weld metal containing more than 0.6%Ni, 0.1%Mo and 1.0%Mn, because the volume fraction of high temperature transformation products (blocky, grain boundary and side-plate ferrite) sharply decreased.

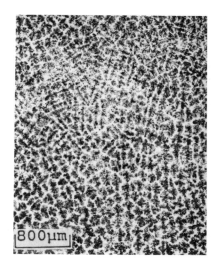

Figure 1 - The solidification substructure in the weld center of 50mm
thick A36 electroslag welds using Ni-Mo filler metal (TW8544).

Table II. Compositions of As-Deposited Weld Metal (wt.%) by ESW
(Balance: Fe)

Plate/ Thickness	Filler Metal	Gap (mm)	Heat Input (KJ/mm)	C	Mn	Si	Ni	Mo	0 (ppm)
A36/50mm	25p	19	63	0.18	1.10	0.40	--	--	240
	AX90+25p	19	61	0.16	1.10	0.36	0.50	0.10	227
	AX90	19	60	0.17	1.10	0.36	0.91	0.15	230
	TW8544	19	36	0.11	1.24	0.30	1.20	0.22	286
A36/76mm	25p	19	92	0.19	1.09	0.39	--	--	233
	TW8544	19	71	0.16	1.01	0.27	0.90	0.14	292
A588/50mm	WS	19	63	0.18	1.05	0.35	0.36	--	230
	AX90	19	62	0.13	1.20	0.36	0.94	0.16	260
	TW8544	19	37	0.11	1.21	0.35	1.47	0.25	289
A588/76mm	WS	19	93	0.19	1.11	0.37	0.32	--	221
	TW8544	19	74	0.11	1.15	0.41	1.25	0.15	294

With the increase of Ni and Mo contents, the prior austenite grain boundaries were no longer decorated by ferrite. In addition, the volume fraction of acicular ferrite steadily increased and reached 90% or more when the weld metal alloy contained 1.0%Ni, 0.2%Mo, and 1.0%Mn. The acicular

237

ferrite plate sizes became finer with increasing alloying. Using TW8544
wire, the highest Ni, Mo, and Mn contents in diluted weld metal could reach
1.5%, 0.13% and 1.2%, respectively, with no bainitic ferrite or martensite
present. In Ni-Mo-Mn alloyed welds (using TW8544 filler metal), acicular
ferrite became the major constituent throughout the weld metal regardless
of the difference in the prior austenite grain zones. Figure 2 presents
typical microstructures of A36 weld metals deposited with conventional mild
steel (25p) vs. alloyed (TW8544) filler metals.

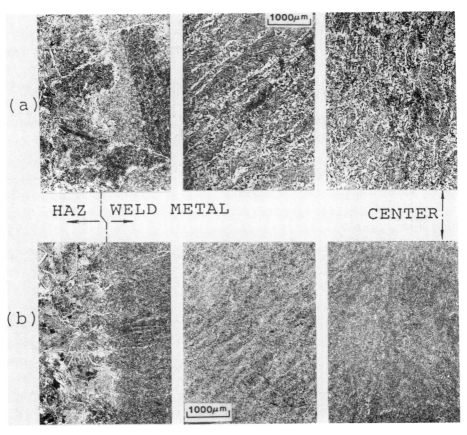

Figure 2 - The typical microstructure of electroslag weldments using
(a) mild steel wire deposited on 50mm A36 weld, and
(b) Ni-Mo-Mn alloyed wire deposited on 50mm A36 weld
(Etchant: 1% Nital).

Tubular vs. Solid Filler Metal

Metal powder-cored tubular filler metals demonstrated several advantages in this investigation. These are as follows:

(1) At a given current, voltage and guide-plate configuration, welds deposited with tubular filler metal exhibited significantly decreased heat input as seen in Figure 3. Using solid wires, the minimum heat input values to make sound narrow-gap electroslag weldments were about 60 KJ/mm for 50mm thick welds and 90 KJ/mm for 76mm thick welds. This represented a substantial reduction in heat input in comparison with conventional gap welds, 80 KJ/mm and 120KJ/mm. However, the heat input of tubular wire depositions could be easily reduced to about 40 KJ/mm for 50mm thick welds and 70 KJ/mm for 76mm thick welds.

(2) Tubular filler metals increased the deposition rate of consumable guide ESW. For example, 1200mm long and 50mm thick steel plates required 23 minutes with tubular wire deposition while a solid wire deposition required at least 30 minutes.

(3) With identical welding parameter settings, the tubular filler metal weld resulted in less base metal dilution and a higher form factor (Figure 4). The former ensured higher weld metal alloy content and the latter allowed increased welding speed with decreased risk of hot cracking. Hence, in narrow-gap ESW of A36 and A588 steels using metal powder-cored tubular wire, the recommended optimum welding current and voltage are 1100A/35V for 50mm welds and 1300A/35V for 76mm welds (instead of 1000A/38V and 1100A/40V in solid wire narrow-gap ESW).

Notch Impact Toughness

Results clearly showed that welds deposited with the Ni-Mn-Mo alloyed tubular wire (TW8544) exhibited a substantially improved weld metal impact toughness by virtue of acicular ferrite-predominant microstructures.

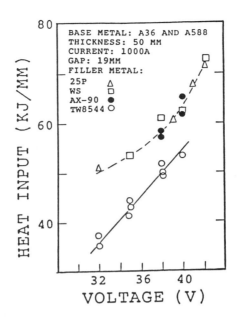

Figure 3 - The effect of tubular (metal powder-cored) filler metal
on heat input of 50mm thick A36 and A588 electroslag welds.

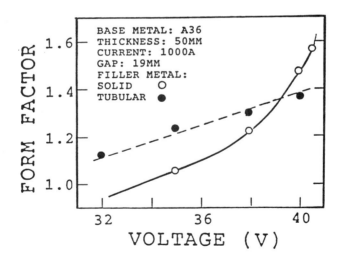

Figure 4 - The effect of tubular filler metal on the form factor
of 50mm thick A36 Electroslag welds.

Figure 5 displays the improvement of CVN toughness corresponding to the increase in weld metal alloy content of A36 welds.

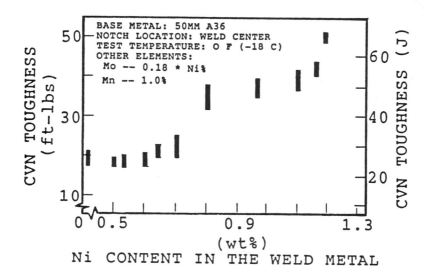

Figure 5 - The CVN impact toughness of A36 weld metals (50mm thick) as a function of Ni content.

Similar trends were displayed by 50mm A588 welds and 76mm A36 and A588 welds. AWS Code D1.1 set a minimum CVN toughness for electroslag weldments: 20J at -18°C (15 ft-lb at 0°F). Usually, with the narrow gap technique, 50mm thick mild steel welds just meet this qualification level while 76mm mild steel welds fail. But, the corresponding impact toughness for Ni-Mn-Mo alloyed (using TW8544 filler) weld metal ranged between 30 to 50J, which was superior to that of both A36 and A588 base metals. Upon comparing the fractographs and the CVN transition curves, the Ni-Mo-Mn alloyed weld metals not only showed good resistance to low temperature cleavage failure, but also exhibited acceptable upper shelf impact energy, Figures 6-8. (In those plots and later, the representations of weld conditions were simplified as follows: SG--31mm joint gap, NG--19mm joint gap, 25p--deposition with Hobart 25p wire, WS--deposition with Linde WA wire,

241

and ST--deposition with TW8544 wire.) Considering the reliability of elec-troslag weldments in different geographical climate zones, Figure 9 dis-plays the test temperature range at which the welds failed to be qualified to 20J.

CVN toughness data from HAZ-1, while improved by reduced heat input was not equivalent to the weld metal. This phenomenon was related to the thermal transfer pattern in electroslag weldments, which will be discussed later. Generally speaking, the reduction in welding heat input is brought about by the tubular wire deposition and the HAZ toughness.

K_{Ic} Fracture Toughness

In this study, most tests of A36 and A588 specimens failed to meet the criteria of E-399 due to their low yield stress, limited thickness and high ductility. For A588 material, the K_{Ic} toughness values of conventional

(a)   **(b)**

Figure 6 - CVN fracture surfaces of 50mm thick A588 electroslag weld
          metals deposited with
          (a) Ni-Mn-Mo alloyed (TW8544) filler metal, and
          (b) mild steel   [Tested at -18°C (0°F)].

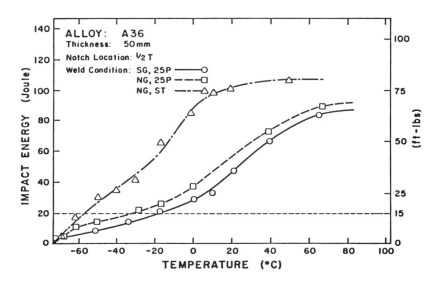

Figure 7 - CVN transition curves for 50mm thick A36 electroslag
weld metals using different weld conditions.

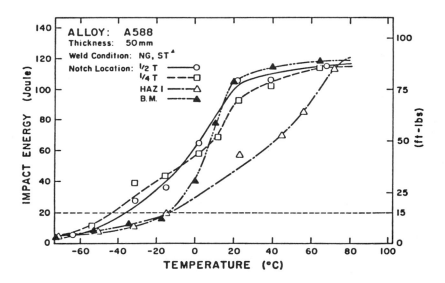

Figure 8 - CVN transition curves for different notch locations
in 50mm thick A588 electroslag welds.

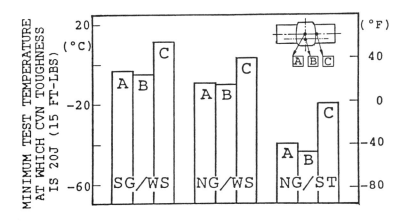

Figure 9 - Minimum test temperatures at which the CVN toughness of
50mm A588 electroslag weldments is 20J (15 ft-lb).

mild steel weld metals were only 55 to 60 MPa-m$^{1/2}$, which was far below
that of the base metal (80 MPa-m$^{1/2}$). The K_{Ic} toughness of Ni-Mn-Mo alloyed weld metals (with TW8544 filler) reached 87 MPa-m$^{1/2}$, Figure 10.

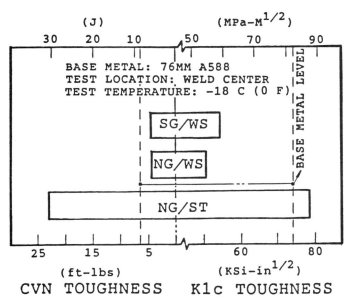

Figure 10 - The CVN and K_{Ic} toughness levels of 76mm A588 weld metals at
-18°C (0°F).

244

Discussion

Alloy Design of Steel Filler Metal

To develop an acicular ferrite weld metal microstructure requires proper alloying elements. Acicular ferrite is an intermediate temperature product of low-carbon austenite decomposition. In order to form it, high temperature products (blocky, grain boundary and side-plate ferrite) must be suppressed while not excessively impeding austenite decomposition, otherwise low temperature transformation products (bainitic ferrite and martensite) will be produced. Most alloying elements affect austenite decompositions. Austenite-stabilizing elements usually depress the transformation temperature from austenite to ferrite and, thus, inhibit nucleation and growth of high temperature products. In this program, Ni and Mn, two important austenite stabilizers, were utilized in this alloying design. In Fe-C-X alloys, there are two different proeutectoid ferrite growth modes in undercooled austenite: 1) growth with partition of the alloying element X between austenite and ferrite under local equilibrium conditions, and 2) growth without partition of X between austenite and ferrite. In the first mode, ferrite grows at a slow rate controlled by the diffusivity of X in austenite. In the second mode, the ferrite growth rate is relatively high because it is controlled by the diffusivity of carbon which is several orders higher than that of metallic elements. As reported (10), the strong austenite stabilizers Ni, Mn and Pt are the only elements to show partition to the austenite phase. Ni has a lower diffusivity in austenite and partitions at the higher austenite decomposition temperatures. These features make Ni very effective in suppressing the growth of high temperature ferrite transformation products.

Chemical analyses revealed that all conventional mild steel weld metals already contained 1.0 to 1.2%Mn. This implied that the Mn level was not sufficient to produce a substantial change in austenite decomposition due to the very slow cooling rate in electroslag weld metals. Though Mn

245

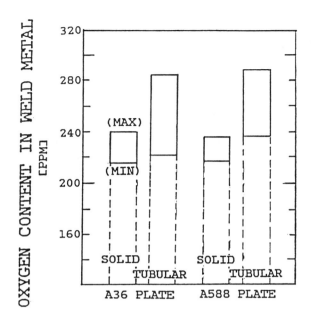

Figure 11 - The oxygen levels in A36 and A588 electroslag weld metals.

seemed cost attractive, further increasing the Mn content is not beneficial because of hardenability, the Mn/Si ratio and deoxidation. In comparison with Mn, Ni is more favorable for the following reasons: 1) its mild strengthening effect (60% less relative to the strengthening effect of Mn), 2) its low affinity for oxygen, and 3) its full solubility in both austenite and ferrite.

Mo, a carbide former, promotes acicular ferrite in weld metals deposited by other welding techniques (9). If the weld metal carbon content could be controlled at a low level, the potential for carbide formation is reduced and a small amount of Mo could effectively reduce the diffusibility of carbon and assist in suppressing high temperature austenite decomposition transformation products.

Based on the above considerations, a Ni-Mn-Mo alloying methodology was established for A36 and A588 electroslag weld metals. Commercially available Airco AX-90 solid wire contains a Ni-Mn-Mo composition (Table I) and

246

was first used to examine the alloying effect on weld metal toughness. This alloying approach proved reliable and successful. Later, a tubular filler metal was developed as a modified formulation and denoted TW8544. The composition of commercial AX-90 wire did not fully conform to the requirement of this electroslag welding study. The major chemical differences between TW8544 and AX-90 include: 1) reduced carbon content (0.3% vs. 0.8%), 2) increased Ni content (2.3% vs. 2.1%), 3) slightly increased Al as a killing agent, and 4) slightly increased Ti. Throughout the whole investigation, TW8544 alloying wire was excellent for both A36 and A588 electroslag weld metals because the resulting strength level was only slightly higher than the base alloys, while the toughness was improved to the level of the base alloy. It appears that, in general, the Ni-Mn-Mo alloying principle could spread to the ESW of other low carbon steels as well as other high heat input welding processes.

Benefits of Tubular Filler Metal

The advantages of metal powder-cored tubular wire was attributed to several unique characteristics during ESW. First, the tubular wire contained metal powder which showed higher electrical resistance than the solid metal wire. At a given voltage and current setting, the main electrical resistance heating in the slag pool (represented by the distance between the electrode tip and the molten metal) was reduced resulting in reduced ohmic heating. Thus, the depth of the weld pool decreased, the weld penetration decreased and the form factor increased. Secondly, since powder metal required less energy to melt, it became more difficult to obtain long extensions into the molten slag than with the solid electrodes. In order to maintain an ohmic relationship for the power supply, an increased filler metal feed rate was needed to attain the specific current, thus accounting for the observed increased deposition rate and decreased heat input. It should be emphasized that the superiority of tubular wire

247

can be maximized by increasing the welding current and decreasing the operation voltage.

One early concern regarding the use of tubular wire was whether it would result in excessive oxidation products in the weld pool. Chemical analysis of weld metal, as sketched in Figure 11, revealed that while the oxygen content was slightly increased, the resulting 200 to 300 ppm is the optimum oxygen level to promote the formation of acicular ferrite. Judging from this, the combination of Ni-Mn-Mo alloying and tubular wire (TW8544) can be expected to consistently improve mild steel electroslag weld metal performance.

Heat Input Effects on HAZ Size

It is generally accepted that reducing the total heat input is the primary means of reducing the size of the HAZ. However, the HAZ size data in this investigation, shown in Figure 12, did not follow this trend. When the weld heat input was changed, the HAZ size of electroslag welds varied considerably. To explain this phenomenon, the thermal transfer pattern of electroslag welds must be taken into account. During ESW, heat is generated from the resistance heating of the slag pool and transferred towards the base metal from the sidewall of the weld pool. Thermal transfer associated with the HAZ can be separated into two parts, as shown in Figure 13a. Variation in the weld pool geometry alters the magnitude of both parts. As the weld pool moves up, some of the base metal associated with part 1 will melt. This implies that the thermal effect related to part 1 is partially offset due to this melting. As weld penetration varies, the degree of offsetting changes. Thus, the total heat built up in the HAZ is not a simple summation of these two parts of thermal transfer. It also extensively depends upon the geometry of the weld pool. To express the variation of the weld pool, it is beneficial to simplify the pool shape into a rectangle and a half ellipse, as shown in Figure 13b, where G is the

248

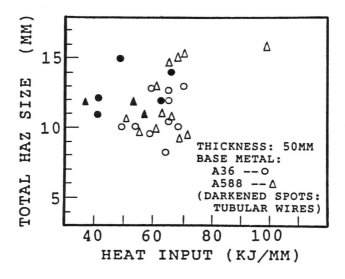

Figure 12 - The HAZ size of various 50mm A36 and A588 electroslag
welds made with different heat input.

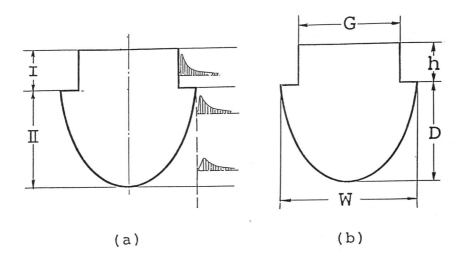

(a)   (b)

Figure 13 - The analysis of thermal transfer in the molten pool of
electroslag welds
(a) two parts of weld delivering heat into the HAZ,
(b) simplified molten pool and its dimensions.

joint gap, W is the weld pool width, h is the preheating height, and D is the weld pool depth. The volume of the molten pool V could be expressed as:

$$V = [hG + (\pi WD/4)] T$$

where T is the thickness. Table III indicates the variation of these factors and their effect on the HAZ size.

Table III. Variation of Weld Characteristics

Characteristic	Variation	Effect	HAZ Size
Heat input	↑	More heat	↑
Form factor (W/D)	↓	Greater part 2	↑
Dilution (W/G)	↓	Less offsetting	↑
h	↑	Greater part 1	↑
W	↓	Less offsetting	↑
D	↑	Greater part 2	↑
G	↑	Less offsetting	↑

It is now clear that, besides heat input, the geometry of the molten pool also influences the final accumulated heat in the HAZ and its size, too. The relative levels of the above factors in various ESW processes are summarized in Table IV.

Table IV. Weld Characteristics of Various ESW Practices

Weld Process	Heat Input	Form Factor	Dilution	G	h
Standard gap	high	high	high	high	high
Narrow gap	low	low	low	low	low
Tubular wire	low	high	low	low	*

* In tubular wire welds, if the same volume of flux is added as in solid wire welds, the value of "h" is high.

The information in Tables III and IV emphasizes the complexity of the heat transfer in ESW. To improve the HAZ microstructure as well as its properties, all the factors mentioned above must be carefully considered and balanced. For instance, in the narrow-gap/tubular wire ESW, since weld penetration is reduced, the offsetting effect of heat transfer of part 2 towards part 1 becomes less, resulting in more heat in the HAZ. However, the combination of reduced penetration and weld pool depth results in a smaller bottom part of the weld pool (half ellipse), resulting in reduced heat transfer to the HAZ by part 2. Furthermore, if the total volume of flux does not decrease, the magnitude of "h" has to increase, which leads to increased heat transfer to the HAZ by part 1. Therefore, in tubular wire depositions, to reduce the total heat transfer from the weld pool to the HAZ, at least two measures should be taken: 1) reduction in gross heat input, and 2) using a minimum volume of welding flux which is sufficient to avoid arcing and lack of fusion.

Conclusions

In attempting to improve the microstructure and toughness of electro-slag welds deposited on A36 and A588 steels, the following conclusions were reached:

1. Alloying with Ni, Mn (austenite-stabilizing elements) plus a small amount of Mo is an effective method to promote acicular ferrite formation in A36 and A588 electro-slag weld metals.

2. Weld depositions using metal powder-cored tubular filler metal exhibited several advantages over the solid filler metal such as lower heat input, higher deposition rate, higher form factor, and higher resistance to hot cracking.

3. Both 50mm and 76mm thick A36 and A588 electroslag welds deposited by Ni-Mn-Mo alloyed tubular filler metal (TW8544) exhibited a great improvement in CVN and K_{Ic} fracture toughness.

4. The HAZ size in electroslag weldments was controlled by not only the total heat input but also the geometry of the molten weld pool.

References

1. B. Paton, Electroslag Welding, (AWS, 1962).

2. M. V. Nolan and R. L. Apps, Weld. & Met. Fab., (November 1980), 464.

3. H. S. Ann, D. Yu, J. H. Devletian and W. E. Wood, "Narrow-Gap High Deposition Rate ESW" (Paper presented at 66th Annual AWS Meeting, Las Vegas, Nevada, March 1985).

4. D. Yu, H. S. Ann, J. H. Devletian and W. E. Wood, Welding: the State of the Art (E. Nippes, ed., ASM, 1986), 28.

5. F. Eichhorn, P. Hirsch, et al., (IIW Document KII-7-76-80).

6. V. P. Andreev, et al., Auto. Weld., (No. 10, 1981), 46.

7. B. I. Medovar, et al., Auto. Weld., (No. 2, 1984), 70.

8. D. Yu, H. S. Ann, J. H. Devletian, and W. E. Wood, "Advances in Welding Science and Technology", (S. A. David, ed., ASM, 1986), 267.

9. D. J. Abson and R. J. Pargeter, Int. Met. Rev., 31 (4)(1986).

10. R. W. K. Honeycombe, "Steels: Microstructure and Properties", (ASM, 1982) 62.

Toughness Evaluation

A USER'S PERSPECTIVE ON HEAT-AFFECTED-ZONE TOUGHNESS

C. P. Royer

Exxon Production Research Company
P. O. Box 2189
Houston, Texas 77001

Abstract

Fracture-mechanics evaluations of weldments focus attention on the zone with the lowest fracture toughness. For modern structural steels and welding consumables, the lowest toughness usually occurs in the heat-affected zone (HAZ). With the evolution of HAZ testing techniques over the past decade, the apparent lower bound of HAZ toughness has decreased significantly, but the frequency and severity of the low-toughness results depend on the test method used. The structural significance of these test results is still unresolved. This paper identifies key research issues to resolve the relationship between HAZ fracture test methods and structural significance. Methods to improve lower-bound HAZ toughness are also reviewed. The user's perspective is emphasized throughout.

Introduction

End users of large welded structures employ fracture-mechanics evaluations of weldments variously for detail design, toughness criteria, inspection criteria, and defect assessment. Weldment integrity is a crucial element of fracture control because of the frequent concurrence of stress concentrations, material property variations, and potential crack sites. Attention is frequently focused on the weldment zone with the lowest fracture toughness. For modern structural steels and welding consumables, the lowest fracture toughness often occurs in the heat-affected zone (HAZ), particularly in the coarse-grain region immediately adjacent to the fusion line.

Over the past decade, HAZ toughness testing methods have evolved considerably. Simultaneously, the apparent lower bound of HAZ toughness for medium-strength structural steels has decreased significantly, to the extent that fractures have occurred under linear elastic loading conditions in some fracture tests. It is not known whether the "true" HAZ toughness has deteriorated because of recent changes in steelmaking or whether the lower results are simply artifacts of more conservative testing practices. Consequently, the relevance of prior service experience with older steels is in question, and the structural significance of the recent test data is not known. This paper highlights recent developments in HAZ toughness testing research and identifies key research issues to resolve the relationship between HAZ fracture test methods and structural significance.

Oil companies are among the largest users of welded structures made from medium-strength structural steel. From the oil company researcher's perspective, the structural significance of low HAZ toughness results should be resolved. But our "customers," the construction and operations managers, want to know how to deal with uncertainties concerning fracture control for present and near-term activities. Methods to improve lower-bound HAZ toughness are also reviewed as well as techniques to reduce reliance on HAZ toughness as a key element of fracture control.

Evolution of HAZ Toughness Testing

Extensive fracture toughness testing of heat-affected zones is a recent development. The vast majority of historical HAZ toughness data consists of Charpy impact tests. Most of these data showed adequate to excellent HAZ toughness. Occasional low values were observed in HAZ tests on some steels, but even these tests did not show very low tough-ness, and they were usually reconciled by conventional Charpy re-testing provisions. Consequently, early HAZ toughness investigations based on Charpy impact testing usually concluded that HAZ toughness was acceptable.

In the latter 1970's, some users began asking for HAZ toughness evaluation by crack-tip opening displacement (CTOD) tests. Early research programs [1,2] found very few low CTOD test results, but a host of testing complexities. Among the issues that arose were specimen size and orientation (surface notch vs. through-thickness), single-side versus two-sided welding, use of single-bevel or K-shaped weld bevels to obtain a straight HAZ perpendicular to the plate surface, welding variables, plate restraint during welding, fusion-line straightness requirements, use of precompression to improve fatigue-crack straightness, and the need for post-test metallurgical evaluation to determine the material sampled by the fatigue-crack front. Investigators, perhaps with a wary eye toward testing complexity, concluded from the early studies that "there would be no need for toughness testing of the HAZ" [3]. Nevertheless, some users continued their requirements for HAZ toughness testing during weld procedure qualification.

Soon thereafter, the offshore oil industry encountered some very low HAZ CTOD values (less than 0.1 mm) during weld procedure qualification. Only a modest fraction of the results were very low, but these could not be resolved by conventional re-testing. Comprehensive metallurgical evaluations of broken CTOD specimens revealed that the low-energy fractures were originating in a coarse-grained bainitic microstructure immediately adjacent to the fusion line. The bainitic coarse-grain regions were discontinuous because of multipass welding, with a tough, fine-grained ferrite microstructure occurring in those portions of the coarse-grain region that were refined by subsequent welding passes. The term "local brittle zone" (LBZ) was chosen to describe a discontinuous, coarse-grain HAZ region that exhibits low fracture toughness. Further study revealed that the low results could not be resolved by justifiable adjustments to testing procedures, by modifying the welding heat input (within the range normally used for platform construction), or by post-weld heat treatment. The propensity for very low CTOD test results was found to be highly correlated with the cumulative amount of LBZ sampled by the fatigue-crack front.

Subsequent research addressed the extent of the LBZ problem. Joint or independent oil-industry research generally revealed that LBZ's (and low CTOD's) can occur in a majority of the 50-ksi-minimum-yield structural steels that were tested. The degree of susceptibility varied significantly, and some steels showed no low CTOD values for the conditions tested. Similar HAZ toughness problems have been encountered in research on line-pipe and pressure-vessel steels. When LBZ's occur, the lower bound of HAZ toughness is characteristically below 0.1 millimeter (0.004 in.), occasionally as low as 0.01 millimeter (0.0004 in.).

In the past few years, yet another test method has been employed, spurred on largely by those who believe the CTOD test methodology for HAZ's is overly rigorous and pessimistic [4]. The surface-notched wide-plate (SNWP) test employs a part-through surface crack placed in or near a heat-affected zone of a weld that lies transversely across the plate width, which is roughly one meter. The plate is generally loaded in uniform axial tension remote from the crack. The technical adequacy, experimental basis, and structural significance of SNWP tests have not been established. Nonetheless, the sheer size of the test specimen implies greater capacity for structural relevance than either Charpy or CTOD tests. Several users have been sponsoring SNWP tests to judge the suitability for use of steels that failed HAZ CTOD testing. By and large, the SNWP tests have "exonerated" the steels in question.

Recent joint industry research, however, has produced linear elastic fractures in SNWP tests on a steel susceptible to LBZ formation. The SNWP results agreed well with CTOD tests, but disagreed markedly with HAZ Charpy tests. When published, this research will underscore the limitations of Charpy testing of heat-affected zones and demonstrate conclusively that the LBZ phenomenon is not simply an artifact of the CTOD test procedure. As for the SNWP test itself, further research will be needed to judge its structural significance, but the testing methodology must be improved and standardized if the test is to reliably measure LBZ susceptibility.

Basis for Research

To place the LBZ issue in perspective, no offshore structural failures have been attributed to LBZ's. A few HAZ fractures have occurred in other structures [5], but it is not known whether any were LBZ-related. It is possible that prior unawareness of the LBZ phenomenon caused mis-diagnosis of past weldment failures, but the overall rate of weldment failures is perceived as low, so even at worst, LBZ's are a minute source of past structural failures. The principal user's motivation to pursue research on the structural significance of LBZ's is to learn why LBZ's are not causing structural failures. We could then ensure that these safeguards remain in place as fracture-control defenses for future construction. Alternatively, the research may identify even more effective means of fracture control in terms of either risk or cost reduction.

257

Until the structural significance issue is resolved, many users will select a cautious course of action for current and near-term fabrication work. This strategy is prudent regardless of how significant LBZ's are ultimately found to be. It is economically justifiable for several reasons:

• Some solutions, such as use of LBZ-resistant steels, are cost-competitive and offer minimum disruption of current practices.

• A less conservative posture risks project delays, as the various interested parties, including certifiers and regulators, must become mutually satisfied of structural integrity.

• The cost of upgrading the integrity of structures after placement in service may prove prohibitive, notably for fixed offshore structures, so preventive efforts are warranted even if the risk is small.

These considerations have already led to user actions in terms of company specifications and practices as well as industry standards and regulatory requirements. One such example is the recently published American Petroleum Institute recommended practice API RP 2Z, for Preproduction Qualification for Steel Plates for Offshore Structures (May 1987), which can be used to identify the LBZ susceptibility of steels through a relatively complex CTOD testing program. But LBZ-resistant steels are not available for all projects because of lack of test data in some cases and restrictions on potential suppliers in other cases. In any case, further research is justified to define and measure LBZ susceptibility in order to streamline future steel selection criteria.

Finally, other means to limit the occurrence or importance of LBZ's have been attempted or considered. Further research to identify such means and to evaluate their feasibility and effectiveness is recommended so that suitable means of fracture control will be available under any eventuality.

Research Issues - LBZ Susceptibility

The overall objectives of this research path are to define LBZ susceptibility and severity. Because the severity and even the existence of the local brittle zone phenomenon is inextricably entwined with the fracture test methods used to evaluate heat-affected zones, many of the research issues concern test method development and analysis of test results.

Charpy Impact Test. HAZ test results generally show that the Charpy test is not very sensitive to LBZ's. The coarse-grain HAZ region usually has the lowest toughness (not always), but values below common acceptance limits are infrequent and often treated as flukes and redeemed by retesting. The insensitivity of the Charpy test is often attributed to the finite notch radius, and possibly to a specimen size effect. Nevertheless, in some HAZ Charpy test programs, a sufficient number of low values occur to suggest occasional sampling of LBZ's. It would be useful to conduct a coarse-grain HAZ Charpy test program on a known LBZ-susceptible steel, with detailed post-test sectioning of specimens, to clarify the degree to which low test results are attributable to LBZ's. The principal value of such a program would be to permit re-analysis of past data sets to identify potentially LBZ-susceptible steels, and thus to determine the relative susceptibility of older steels, on which most of our service experience is based.

CTOD Test. A considerable list of HAZ CTOD testing complexities has already been given. It would be useful to have a large CTOD data base on an LBZ-susceptible steel in which testing parameters are varied systematically to identify individual and synergistic effects on test performance. This would facilitate standardization of HAZ CTOD test procedures and allow adjustment of test parameters to better correlate with large-scale structural tests when the latter become available.

258

Surface-Notched Tension Tests. The surface-notch variation of the wide-plate test needs further development and standardization in order to become a reliable tool for HAZ toughness evaluation. Present indications of interlaboratory variability must be resolved. Surface-notch severity must be calibrated, preferably in a way that allows interpolation for crack shapes not tested. The effect of notching procedure (e.g., straight vs. zig-zag) needs further study, particularly for LBZ-susceptible steels. Large-scale residual stresses should be modeled in some tests, perhaps with axial crossing welds as in classical wide-plate tests; this is particulary important for studies of transverse-weld mismatch. The statistical signifi-cance of SNWP tests must be studied to determine the number of tests re-quired for a given study. The relative absence of constraint common to all surface-notch tension tests must be rationalized in comparison with more constrained structural configurations. Post-test sectioning to determine which zones were sampled also should be developed further. If these ques-tions are resolved, the SNWP test could be a valuable tool for studying geometrical effects of LBZ severity. But it is doubtful that the SNWP test will ever reach the status of a routine HAZ evaluation test, primarily because of testing cost.

Precracked Charpy Test. For LBZ evaluation purposes, the superior notch acuity of the precracked Charpy specimen seems preferable to the standard Charpy. Specimen size and impact loading effects must be con-sidered, but recent research suggests this can be done with sufficient testing and statistical interpretation. A nice research feature is the ability to remove specimens normal to the fusion face in thick weldments without special bevel preparations for toughness-testing purposes. Instru-mented, precracked Charpy (IPC) tests allow approximate separation of initiation and propagation energies. With post-test sectioning and statis-tical analysis, IPC tests could be useful in quantifying the fracture toughness of LBZ's in the absence of tough surrounding microstructures, thus contributing to efforts to define LBZ severity.

Some research laboratories are exploring weld thermal-cycle simulation in Charpy-size specimens machined from parent steel in order to study microstructural and toughness variations as a function of welding variables such as heat input. IPC testing of simulated HAZ regions could prove to be a rapid and relatively inexpensive method of judging LBZ sensitivity of various steels. The accuracy of HAZ simulation must be demonstrated and maintained for this approach to gain broad acceptance.

Research Issues - LBZ Significance

The overall objective is to understand the structural significance of low HAZ toughness results from smaller-scale fracture tests. Besides the HAZ test-method developments described above, at least four technologies can contribute to an understanding of the many factors involved.

1. Welding Engineering. There is still considerable uncertainty about the frequency and size of LBZ-susceptible microstructural zones in both structural weldments and fracture-test idealizations of same. Extensive weldment sectioning and characterization, including transmission electron microscope work, will be needed to generate the data. Then the data must be statistically analyzed and correlated to welding variables and differences in steel composition. Finally, we need to learn how to produce LBZ's of various desired sizes and shapes for experimental testing programs.

2. Elastic-Plastic Fracture Mechanics (EPFM). EPFM is a developing technology itself, and it has remained generally within the bounds of continuum mechanics, upon which fracture mechanics is founded. This status is inconsistent with the needs of HAZ toughness evaluation, for which large variations in microstructure and properties occur on the order of grain spacings. Consequently, there are several opportunities for fracture-mechanics research. We need more understanding of crack-tip plastic-zone

259

development in heterogeneous weldments. Shallow-crack effects need better characterization, particularly how they interact with LBZ's. The effect of weldment strength overmatching has been studied, but only on simplified configurations, which may lead to overly optimistic conclusions. Residual-stress contributions need further clarification. Crack-tip constraint comparisons are needed between fracture tests and structural configurations, including non-homogeneity effects such as overmatching welds. We need to resolve whether constraint differences or weak-link considerations are responsible for size effects observed in fracture testing and to adjust our testing strategies accordingly. Developments in elastic-plastic finite-element analysis interpretation, three-dimensional modeling capability, and photoelasticity evaluations make this an opportune time to address these issues. These findings, in particular, will have widespread applicability throughout fracture mechanics, not just for LBZ evaluations.

3. <u>Probabilistic Applications of Fracture Data</u>. It seems very likely that much of the difference between fracture test expectations and actual structural performance is due to simplified deterministic applications of fracture test data. A working, comprehensive probabilistic model of even the fracture initiation event--only the first step toward structural collapse--seems unattainable at this time. But some effort seems essential if we are ever to truly understand LBZ significance. Even modest success would permit us to prioritize variables that must be characterized and, thus, would motivate and guide future research. The ultimate reward would be new, more reliable fracture control practices that would be more cost-effective and less contentious among interested parties.

A probabilistic model could also answer many questions about the fundamental statistical aspects of LBZ influence. A detailed hindcast study of past structural performance would be needed to calibrate the model.

4. <u>Large-Scale Structural Tests</u>. Eventually, large-scale tests such as tests on tubular joints containing precracks in well designed and placed LBZ's will be required to confirm fracture test predictions, but such tests are presently premature. Many side issues must be clarified in advance to limit impractical expansion of the test matrix. The large-scale test matrix should be carefully designed to address central issues of competing prediction methodologies and to establish statistical significance of the results. Each interested industry will probably customize the large-scale testing to its individual needs.

<u>Research Issues - Reduction of LBZ Significance</u>

By this point, it is probably clear how little we know about projecting the structural significance of low HAZ toughness. Nevertheless, we can already identify potential ways to <u>reduce</u> the significance of LBZ's, whatever the absolute magnitude. It might seem that such research is premature until the need for reduction is established. In fact, this research is the more urgent and pragmatic because it addresses questions that users must confront in the interim before the larger issues are resolved. I have sorted my suggestions into two categories: (1) those that improve structural integrity and (2) those that improve confidence (reduce uncertainty) in structural integrity. In user's terms, the first category lists active precautions that a construction manager might consider, whereas the second deals primarily with whether and where to look for potential damage in existing structures.

1. <u>Integrity Improvements</u>. Where circumstances permit, users can and frequently will seek out and select LBZ-resistant steels. This comes close to an ideal solution except that LBZ susceptibility is somewhat arbitrarily defined, and some important steel sources may not be able to provide LBZ-resistant steels. But other potential solutions also exist. The occurrence of LBZ-susceptible microstructures is dependent upon rate of cooling from welding. Identification of the relationships between HAZ microstructure and

260

welding variables such as heat input can permit control of LBZ production during welding, although with a potential loss of productivity. LBZ's can be reduced in size and number by special controls on weld-bead placement and shape, but we need to establish practical and relevant control parameters. Temper-bead welding is a possibility for capping beads that otherwise would not obtain grain refinement. Weld bevels could be adjusted so that the maximum stress is less normal to the fusion line, but work is needed to determine how much bevel is sufficient. Fabrication inspection practices could be enhanced to reduce initial flaw sizes, but we need to know which nondestructive test methods to use and how much the effort will help. We can also grind weld contours to reduce stress concentration and improve fatigue initiation life. Lastly, post-weld heat treatment (PWHT) can be performed to reduce residual stresses and, possibly, to improve LBZ toughness. However, the effect of PWHT on LBZ toughness is not yet resolved; some studies have shown a benefit, while others have shown no improvement or some degradation. The reasons for these apparently contradictory conclusions are unknown, but worth attention.

2. Confidence Improvements. The most fertile territory is in structural redundancy, at least for structures like tubular space frames. Modern analytical procedures can be employed to determine whether the loss of any single weldment jeopardizes structural integrity. (Those same procedures can also be used in design to eliminate fracture-critical members wherever feasible.) Fracture-mechanics analyses can be refined to reduce conservatism in HAZ toughness criteria, but only to a limit that is frequently not sufficient in the face of very low test data. Finally, in-service inspection can be increased, at least temporarily, to monitor for evidence of cracking or other damage attributable to LBZ's, but this measure can be very expensive for some applications such as fixed offshore structures.

Concluding Remarks

The emergence of the local-brittle-zone phenomenon in fracture testing has increased uncertainty regarding the adequacy of fracture initiation resistance of heat-affected zones of weldments in some common structural steels. I have outlined the development of the phenomenon and described a number of potential research tasks to help reduce the uncertainty to a more comfortable level. I have no illusions that all this research will be performed; nor is that necessary, because a sufficient number of satisfactory solutions will emerge for effective fracture control to be implemented without a complete scientific understanding. But enough tasks must be successfully pursued to provide practical, low-cost solutions for any application. Because the LBZ phenomenon and related research apply to several industries and elements within industries, and because the eventual consensus resolution of the issue is years away, many sources can contribute relevant research. This is fortunate for us because recent economic reverses have made it impractical for the oil industry to support the research single-handedly. There are numerous opportunities for joint industry programs and standing committee research as well as independent contributions. Some of these are already underway, and a few will be reported in other papers at this symposium. This paper describes one perspective on how these efforts contribute to resolution of the local brittle-zone issue.

References

1. H. G. Pisarski and J. D. Harrison, "Fracture Resistance of Materials Used in Offshore Structures" (Paper 2, Select Seminar: European Offshore Steels Research, Cambridge, England, November 1978).

2. G. H. Eide, "Fracture Mechanics for Assessing the Safety Against Brittle Fractures in Offshore Installations" (Welding Institute Seminar - COD Fact or Fiction, Sheffield, England, April 1980).

3. J. D. Harrison, "Material Selection Considerations for Offshore Steel Structures" (Proceedings, International Conference - Steel in Marine Structures, Paris, October 1981), 148-193.

4. R. Denys, "Advances in Fracture Toughness Testing," ASME Paper No. OMAE 87-659.

5. J. D. Harrison, "Why Does Low Toughness in the HAZ Matter?" The Welding Institute Seminar, Coventry, England, June 1983.

EXPLORATORY STUDIES ON THE FRACTURE TOUGHNESS OF MULTIPASS WELDS

WITH LOCALLY EMBRITTLED REGIONS

H.G. Pisarski* and J. Kudoh**

* Fracture Department, The Welding Institute,
Abington Hall, Abington, Cambridge,
CB1 6AL, England.

** Kawasaki Steel Corporation, Mizushima,
Kurashiki, Japan.

Abstract

This paper examines the effect of the size of locally embrittled regions, in 25mm thick multipass submerged-arc welds made in a C-Mn steel, on Charpy V and CTOD fracture toughness. Local embrittlement was effected by modifying the aluminium content of weld beads through aluminium powder additions to the flux. The results for Charpy V specimens demonstrate that if the region of embrittlement occupies a length of more than 30% along the notch tip then toughness is controlled by the lowest toughness region. However, for CTOD specimens, notched through the plate thickness, the presence of embrittled weld metal occupying a length of only 10% of the crack tip and located in the central part of the crack significantly reduces fracture toughness. Evidence is also presented to show that surface notched CTOD specimens can exhibit lower toughness than through thickness notched specimens.

Introduction

Multipass welds in C-Mn steels are generally metallurgically heterogeneous and consequently the mechanical properties can vary from region to
region. When it is necessary to measure the fracture toughness of welds,
especially where this information is used to select a weld metal or welding
procedure, it is usual practice to employ rectangular section specimens
with through-thickness notches (1). In such specimens the line of the crack
tip samples a variety of microstructures in various proportions. However,
at the present time there is no clear understanding on how the size of any
low toughness region that is sampled influences the fracture behaviour of
the specimen. During production welding the sizes of potentially low
toughness regions may vary. Consequently, for procedure qualification, it
would be helpful if the minimum size of such a region could be specified so
that the fracture toughness measured, in the specimen, is independent of
the size of any local low toughness region larger than the specified minimum.

This paper describes some exploratory studies on weld metal to examine
how the size of artificially created local regions of low toughness influence overall fracture toughness as measured using Charpy V and CTOD specimens. Various local toughness distributions were produced in multipass
submerged arc welds, deposited in a C-Mn steel plate, by varying the
aluminium content of individual weld runs through the addition of aluminium
powder to the flux. Previous studies showed that weld metal toughness would
deteriorate in proportion to the aluminium content in the weld metal (2).

Experimental Details

Welding

Weld panels were prepared using 25mm thick, normalised, C-Mn steel
plates to BS 4360 Grade 50D. The welds were deposited by a single wire
submerged-arc process which employed 3.25mm diameter SD3 wire (to AWS
A5.17-80 EM12K mod.) and Oerlikon OP121TT flux. Most of the welds were made
using the joint preparation and weld deposition sequence shown in Fig. 1.
Some of the welds made at an early stage in the programme employed a
slightly different joint preparation which had a more open joint angle on
the first side. This joint preparation required 11 passes to fill it
instead of 7, however, both welds contained about 6 layers. All welding was
carried out at a heat input of 2.2kJ/mm and the preheat/interpass
temperatures were 50°C to approximately 100°C. Aluminium powder, with a
particle size of between 64μm and 178μm, was mixed with the flux to produce
low toughness weld metal. The amount of Al powder mixed was varied in the
weld passes according to the plan shown in Fig. 2. Welds W6, W12 and W13
can be considered as reference welds because they contained no deliberate
Al powder additions to the flux. On the other hand, welds W7, W4 and W5
contained deliberate additions of 1.2, 1.9 and 4.0 weight percent Al powder
to all weld passes. In welds W9, W10, W11 and W16 Al powder additions were
made to specific weld passes according to the plan shown in Fig. 2.

Testing Procedure

The chemical compositions of each weld were analysed with an optical
emission spectrometer, however, in this paper only the results for Al are
given. The analyses were carried out on planes parallel to the original
plate surface, after the welds had been sliced along these planes. The

264

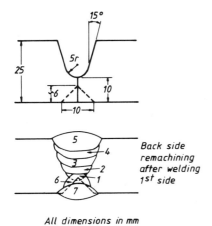

15°
5r
25
6 10
10

5
4
3
2
6 1
7

Back side
remachining
after welding
1st side

All dimensions in mm

Weld panel	1st side							Pass number	2nd side		
	8	7	6	5	4	3	2	1	9	10	11
W4					1·9%						
W5					4%						

Weld panel	1st side				Pass number	2nd side	
	5	4	3	2	1	6	7
W6			0%			+	+
W7			1·2%				
W9	0%	19%		0%		+	+
W10	0%		1·9%				0%
W11	0%	2·5%		0%			
W12			0%				
W13	0·6%		0%			0·6%	
W16			1·9%				

+No welding was carried out

Figure 1 - Weld joint preparation and deposition
sequence.

Figure 2 - Weld procedure plan
(percentages refer to Al powder
additions to the flux).

slices were taken at various positions through the thickness of the weld so
that the distribution of Al could be established. This involved determining
chemical composition at about 7 different locations in each weld.

Charpy and crack tip opening displacement (CTOD) specimens, with
through thickness notches (T-T-N), were machined from various locations in
the welds, after taking into account the analysed Al distributions through
the thickness of the welds. Standard 10 x 10mm Charpy V notch specimens
were employed. The CTOD specimens were generally rectangular in section and
either 15mm or 25mm thick (the latter equal to the full weld thickness)
depending on the desired Al distribution. Some welds which were made with
the same Al powder additions to each pass were found, after analysis, not
to have uniform Al contents through the thickness. In particular, Al
content was found to vary through the thickness of welds W4 and W5. Both
the Charpy V and CTOD specimens (in this case 15mm thick) were removed from
positions through the thickness of the weld (25mm thick) which contained
the desired Al distribution.

Prior to fatigue pre-cracking the CTOD specimens were locally
compressed according to recommended procedures (1). CTOD values were
calculated in accordance with BS 5762 (3) using yield strengths derived
from weld metal hardness (appropriate to the region from which fracture
occurred) following recommended procedures (1). Some square section CTOD
specimens were also tested. These were notched from the plate surface (S-N)
and were not locally compressed prior to fatigue pre-cracking.

265

Results

Weld Metal Composition and Hardness

As expected, increasing additions of Al powder to the submerged arc flux resulted in a proportional increase in the Al content of the weld metal. However, as Fig. 3 shows, there is a large amount of scatter. This is thought to be caused by uneven mixing of the Al powder in the flux and variations in Al dilution due to subsequent weld passes. The actual variations in Al content through the thickness of the weld are described together with the Charpy V and CTOD test results below.

Increases in weld metal Al content were accompanied by a decrease in oxygen content but an increase in carbon, manganese, silicon and titanium contents. In total, these changes are expected to result in a significant decrease in weld metal toughness.

The relationship between weld metal hardness and Al content is shown in Fig. 4. Hardness rises with increasing in Al content but becomes constant above about 0.1 wt. % Al.

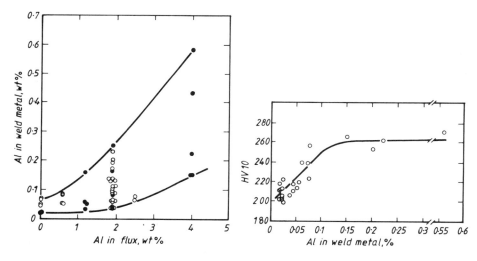

Figure 3 – Relationship between Al content in weld metal and amount of Al powder mixed with flux. (Solid points refer to analyses where the whole weld was made using flux containing Al powder. Open points refer to analyses on particular weld beads where Al powder was added.)

Figure 4 – Relationship between hardness and Al content in weld metal.

Charpy V Tests

As might be expected from the relationship between Al content and hardness, Charpy V toughness decreases as the percentage of Al increases. Fig. 5 shows that in specimens taken from weld regions containing rela-

266

tively uniform levels of Al, the temperature for 50J increases rapidly from the reference Al content (0.02 wt. %) and then starts to stabilise at Al contents of 0.13 wt. % and above.

10% of the notch line in the Charpy specimens, taken from a particular position in weld W10, sample weld metal with an Al content of 0.12 wt. %. The remainder of the notch samples 0.07 wt. % Al, which is similar to the uniform Al distribution sampled by the specimens taken from weld W7. However, despite these differences both welds have similar toughness; for example, the 50J temperature is about −15°C. If the local region of high Al content in the specimens taken from weld W10 were to control fracture behaviour, a transition curve similar to weld W4 would be expected and then the 50J temperature would be approximately +30°C. Clearly, this did not occur and the result suggests that Charpy toughness is unaffected if the most embrittled region is narrow, occupying a length of approximately 10% of the notch tip line, and is located at the specimen edge.

Fig. 6 shows that specimens taken from welds W9 and W11 sample weld metal with Al contents which vary from 0.02 to 0.07 wt. % but the shape of the Al distribution is different in each weld. The central 30% of the notch line in specimens from weld W11 samples weld metal with the higher Al content, whilst for specimens taken from weld W9 the figure is 70%. The toughness of the weld metal flanking the central region, in both specimens, is known to be high, (see results from W13). Nevertheless, the actual toughness measured in both welds is relatively poor (in comparison with W13) and similar to that measured in W10 where 90% of the notch line sampled weld metal with 0.07 wt. % Al.

These results show that if the central 30% of the notch line samples low toughness material, this is sufficient to control the toughness of the Charpy specimen irrespective of the toughness of the adjacent material.

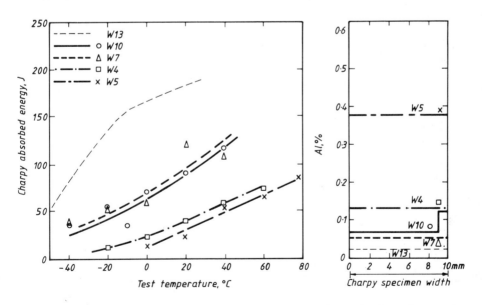

Figure 5 - Charpy energy transition curves and Al distributions in specimens, from welds W4, W7, W10, W13 and W5.

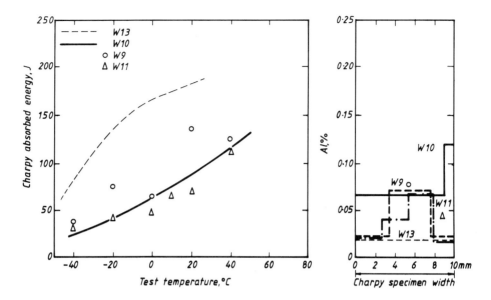

Figure 6 - Charpy energy transition curves and Al distributions in specimens, from welds W9, W10, W11 and W13.

CTOD Test Results

Welds W12 and W7 contain uniform but different levels of Al. However, as shown in Fig. 7 an increase in the aluminium content from 0.02 to 0.05 wt. % reduces CTOD toughness significantly. Within the central 60% of the notch tip line, the specimens taken from weld W4 contain a similar level of Al (0.06 wt. %) as W7 but the CTOD transition curve is shifted to significantly higher temperatures than that for W7.

The 15mm thick specimens from weld W5 (originally 25mm thick) sample weld metal with higher Al contents than weld W4 (originally 25mm thick), especially at the edges of the crack, but at the positions from which the specimens were taken the Al distribution profiles are similar. The toughness of weld W5 is low in comparison with weld W4, and this is attributable to the low toughness material sampled by the edges of the crack, (occupying a total length of approximately 30% of the crack tip line).

These results show that the fracture toughness of the CTOD specimens is controlled by the region of low toughness material located at the edge of the specimen when this material occupies a length of approximately 30% of the crack tip line.

In specimens from weld W10 the outer edges of the crack tip line samples weld metal with a relatively low Al content which is similar to the uniform level measured in weld W7. However, the central 60% of the crack tip line samples weld metal with a higher Al content which is almost half way between the peak values measured in W7 and W4. As shown in Fig. 8, the CTOD transition curve for specimens from this weld (W10) lies between those of W7 and W4. Clearly, when the central 60% of the crack tip line samples low toughness material the fracture behaviour of the specimens is controlled by this region. However, as shown below, specimen behaviour can be influenced by smaller regions of local low toughness.

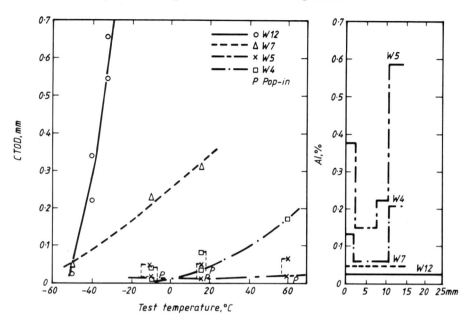

Figure 7 – Crack tip opening displacement transition curves and Al distributions for welds W12, W7, W4 and W5.

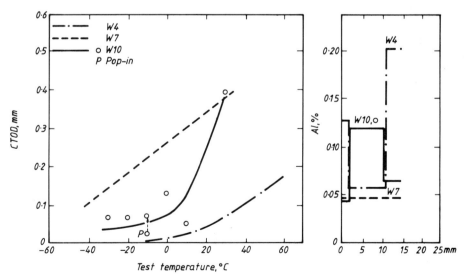

Figure 8 – Crack tip opening displacement transition curves and Al distributions for welds W4, W7 and W10.

Weld W16 contains double Al peaks, as shown in Fig. 9. The lower of the two is similar to the peak Al content in W10, whilst the higher peak is similar to W4. This latter peak occupies a length of approximately 15% of the crack tip line. Nevertheless, the CTOD transition curve for specimens from this weld is similar to that for W4, where the specimens sample 30% low toughness material at the edge of the crack. The CTOD results from W16 show that the fracture behaviour of the specimens is controlled by as little as 15% low toughness material when this is located towards the centre of the crack tip line.

The CTOD results from weld W11, see Fig. 10, show similar characteristics to weld W16 (Fig. 9) in terms of a small local region dominating the fracture behaviour of the specimen. In comparison with weld W12, weld W11 contains a region of relatively high Al content (0.067 wt. %) which occupies a length of about 10% of the crack tip at the centre of the specimen. Even so, three CTOD values out of the four from weld W11 are considerably lower than those from W12, which had a low but uniform distribution of Al (~ 0.02 wt. %), but similar to those obtained from weld W7 which contains a higher, uniform distribution of Al (~ 0.05 wt %). The CTOD values measured in specimens from weld W11 are significantly reduced by the presence of a narrow locally embrittled zone which occupies about 10% of the crack tip line in the central part of the specimen.

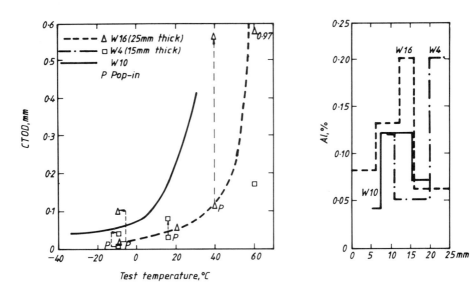

Figure 9 — Crack tip opening displacement transition curves and Al distributions for welds W16, W4 and W10.

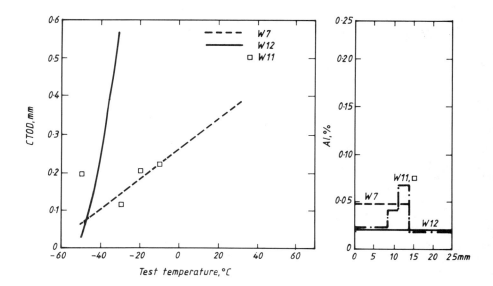

Figure 10 - Crack tip opening displacement transition curves and Al distributions for welds W7, W12 and W11.

Effect of Notch Orientation

The CTOD transition curves obtained using rectangular section through-thickness notched (T-T-N) and square section surface notched (S-N) specimens are shown in Figs 11 and 12. The results in Fig. 11 were from specimens extracted from welds W12 and W13, both of which were welded with no Al powder additions to the flux. In these welds Al is low (0.020 wt. %) and is distributed uniformly throughout the weld thickness. Although in the transition regime the two transition curves are similar in shape, the CTOD values for the T-T-N specimens are greater than those for the S-N specimens despite the higher constraint associated with the rectangular specimen geometry. Similar differences in specimen behaviour can be observed in Fig. 12, which presents the test results on the welds with locally high Al contents. In the S-N specimens the crack tips were completely located in the high Al region of the weld. Although the maximum levels of Al are similar in both welds, the T-T-N specimens give higher CTOD values than the S-N specimens. The T-T-N specimens sampled a length of approximately 10% low toughness material toward the centre of the crack tip line. The apparent effect of notch orientation on the CTOD value may be due to the differences in the proportion of the various microstructural constituents sampled by the crack tip in the S-N and T-T-N CTOD specimens.

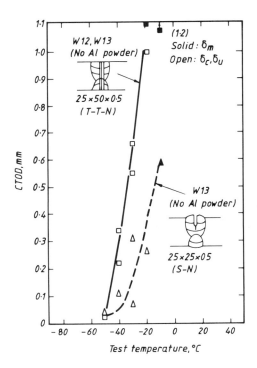

Figure 11 — Crack tip opening displacement transition curves for T-T-N and S-N specimens taken from welds W12 and W13 which contained no deliberate Al powder additions.

Discussion

Shibasaki (4) showed that Charpy transition temperature rises almost linearly with increasing proportion of low toughness material at the notch tip. However, the present work does not confirm this. Instead, it was found that the Charpy transition temperature of the composite weld is equal to that of the lowest toughness material when it occupies more than 30% of the notch tip line. However, when 10% of the notch tip line samples low toughness material (at the edge) Charpy toughness is unaffected and is the same as that of the higher toughness material sampled.

The CTOD results show that with rectangular section T-T-N specimens, toughness is dominated by the presence of the lower toughness constituent, even when it only occupies a length of approximately 10% of the central region of the crack tip, or, a total of approximately 30% at the edges. These observations agree with those of Kaihara (5) and Satoh (6). However, since embrittled weld beads occupying lengths of less than 10% of the crack tip were not studied, it is not possible to confirm Shibasaki's (4) results which show an increase in CTOD toughness if the proportion of low toughness material occupies less than approximately 10% of the crack line. Therefore provided that the region of suspect low toughness is large enough (i.e. occupying 10% or more of the crack line), it would appear that T-T-N specimens can be employed to characterise weld toughness.

272

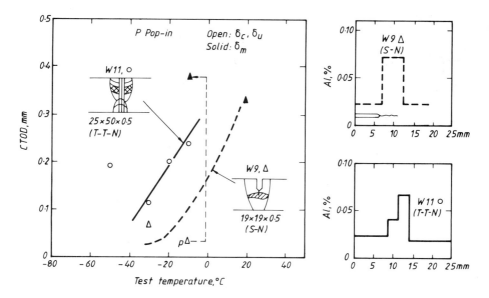

Figure 12 — Crack tip opening displacement transition curves for T-T-N and S-N specimens taken from welds W9 and W11 which contained deliberate Al additions to one pass.

However, this conclusion needs to be considered in the light of the results from the square section S-N specimens (welds W12, W11 and W9, Figs 11 and 12). The surface notched specimens appear to show that the true toughness of the locally embrittled region is overestimated by the T-T-N specimen despite the increased constraint associated with this specimen design. Further test work is necessary to confirm this. Nevertheless, these observations are consistent with those made by Bokalrud and Karlsen (7) and Dawes et al (8). The former examined the CTOD fracture toughness of thick section multipass MMA welds using square section S-N and T-T-N specimens. Although the constraint provided by specimens was the same, lower CTOD values were obtained with S-N specimens.

Dawes et al (8) examined the behaviour of a composite weld in which some beads were deposited with a MMA electrode giving poor toughness (occupying approximately 20% of the weld thickness) relative to the electrode used to make the majority of the weld. Both square section S-N and T-T-N CTOD specimens were employed but the lowest toughness values were measured in the S-N specimens when the crack tip completely sampled the low toughness weld metal.

In carrying out defect assessment analyses it is usual practice to use the lowest toughness value (from a small number of tests) appropriate to the region in which the defect is expected to lie. As discussed above, the standard T-T-N specimen may provide an overestimate of toughness for the local region and, therefore on the basis of the work presented here, it would be more appropriate to use a S-N specimen. However, it is not always possible to know in advance the region of lowest toughness. Therefore, where avoidance of brittle fracture is of utmost importance, it may be necessary to employ T-T-N specimens initially. Subsequently S-N specimens could be used to measure the lower bound toughness of the welded joint with crack tips placed in the regions from which fractures were identified to have initiated in the T-T-N specimens.

Nevertheless, despite the reservations concerning CTOD testing procedures, this work has shown that where a crack tip samples a variety of microstructures fracture will be influenced by quite small regions of low fracture toughness material. The smallest region of low toughness material examined in this programme occupied a length of about 10% of the crack tip, but this was sufficient to affect CTOD toughness. The effect on fracture behaviour of a crack tip occupying less than 10% low toughness material is a subject worthy of further investigation.

Conclusions

CTOD and Charpy tests have been carried out on multipass submerged-arc welds, in 25mm thick plate to BS 4360 Grade 50D, exhibiting various toughness distributions through the thickness of the weld. The toughness distributions were produced by modifying the aluminium content of the weld metal through aluminium powder additions to the flux. The main conclusions drawn from the tests carried out are as follows:

1. The Charpy absorbed energy measured in heterogeneous multipass welds is dominated by the lowest toughness region when it occupies a length of more than 30% along the notch tip. However, Charpy energy is unaffected if the low toughness region occupies a length of less than 10% of the notch tip at the edge of the specimen.

2. The CTOD values measured by the T-T-N specimens are significantly reduced by the presence of a locally embrittled region, which occupies a length of about 10% of the crack tip in the central part of the specimen, or, approximately 30% if located at the edges.

3. Square section surface notched (S-N) CTOD specimens may exhibit lower fracture toughness than through-thickness notched (T-T-N) rectangular section geometry specimens.

Acknowledgements

The authors wish to acknowledge the financial support of the Research Members of The Welding Institute and of the Materials, Chemicals and Vehicles Requirements Board of the Department of Trade and Industry. Thanks are also due to the Kawasaki Steel Corporation for providing financial support to one of the authors (J. Kudoh) during his period of secondment to The Welding Institute.

References

1. S.J. Squirrell, H.G. Pisarski and M.G. Dawes, "K_{Ic}, CTOD and J tests on weldments" (Paper presented at the Third ASTM International Symposium on Non-Linear Fracture Mechanics, Knoxville, October 1986.)

2. H. Terashima and P.H.M. Hart,"Effect of flux TiO_2 and wire Ti content on tolerance to high Al content of submerged-arc welds made with basic fluxes" (Paper presented at The Welding Institute International Conference, 15-17 Nov. 1983, London), 27.

3. "Methods for crack opening displacement (COD) testing," The British Standards Institution, BS 5762, 1979.

4. M. Shibasaki, "Fracture toughness of layered steel with different toughness level," J. of the Iron and Steel Inst. of Japan, vol. 68, No. 5, (1982), S630.

274

5. S. Kaihara et al., "Study on variation of critical COD value for welded joints of low temperature service steels," Quarterly J. of the Japan Welding Society, vol. 2, No. 1, 1984, 104.

6. K. Satoh, M. Toyoda and F. Minami, "A probabilistic approach to evaluation of fracture toughness of welds with heterogeneity," Quality and Reliability in Welding: Int. Conf. The Welding Institution of the Chinese Mechanical Engineering Society, vol. 3, Session C, Hanzhou, China, (1984)

7. T. Bokalrud, and A. Karlsen, "Fracture toughness and significance of defects in heavy offshore structures" (Paper presented at the Select Seminar: European Offshore Steels Research, 27-29 November 1978, Cambridge), 37.

8. M.G. Dawes et al., "Fracture mechanics measurement of toughness in welded joints" (Paper presented at the Int. Conf. 'Fracture toughness testing - methods, interpretation and application', The Welding Institute, June 1983, London), 33.

FRACTURE MECHANICS TESTING OF WELD METAL FOR

LOW CARBON MICROALLOYED STEEL

C. Thaulow, A.J. Paauw, G. Rørvik

SINTEF, Division of Materials and Processes
7034 Trondheim - NTH
Norway

Abstract

The fracture toughness of the base material and HAZ of a low carbon
microalloyed steel has been examined. The main variables were the thickness
(30, 70, 100, 130 and 160 mm), welding technique (SMAW, SAW), heat
treatment (AW, PWHT), rate of loading during testing (static, dynamic) and
the tempera- ture (in the range -60 to +20° C.

During the testing it was realized that the toughness of the SMAW weld
metal was poor, and the majority of the brittle fractures from these weldments initiated from the primary weld metal.

The paper discuss the effect of weld metal in CTOD testing. Relationships
between $CTOD_c$ and the number, or total size, of brittle zones hit by the
crack tip are presented.

It is concluded that the most important factor in CTOD testing of weldments
is the positioning of the fatigue precrack.

Introduction

Fracture mechanics analysis has been applied during the planning and fabrication of recent offshore structures in Norway (1, 2). Major savings in cost have been reported by avoiding unecessary PWHT, accepting burried defects and root defects in one side weldments. In one case, the cost savings of not performing PWHT have been calculated to about 9 mill. $, and about 100 "deviations" reported during the fabrication of a platform could be accepted (3).

A supposition for the application of fracture mechanics in optimum selection of steel and welding consumables/procedures and in defect evaluation, is that relevant fracture toughness data are available. During the last years, it has been widely accepted that the present fracture mechanics testing procedures, originally developed for the testing of plain material, must be revised when applied in weldment testing (4-6).

During the last 5 - 7 years the steel composition of structural steels for offshore application, has changed. The major changes are the reduction of the carbon content to about 0.10 %, a very low sulphur content and the introduction of new thermo mechanical control processes (TMCP). This should primarily serve an increased weldability.

However, during welding procedure testing of the first platform to be built of this low carbon steel in Norway, very low CTOD values were found in the HAZ. This was of great surprise, since the HAZ toughness was considered to be of no problem. As a consequence, the HAZ toughness of the low carbon microalloyed steels has been examined in detail (7-8).

Major oil/gas reservoirs in deep water areas (300-400 m) will be developed in the near future, and the application of heavy steel structures, with thicknesses up to 150 - 200 mm, has been in question. But since it is generally accepted that the fracture toughness of structural steel is reduced with thickness, there has been a demand to quantify this effect on the base material and HAZ of low carbon structural steels. A research project has been carried out on this subject and the major findings will be presented elsewhere (9). However, during the progress of the project it was experienced that brittle fracture often was related to the weld metal. The present paper will concentrate on the experience with the fracture toughness of the weld metal.

Experimental background

HAZ Toughness Evaluation

Weldments.   The investigated low carbon micro alloyed steel is a a typical representative for the modern offshore steel, table 1.   The base material and the heat affected zone (HAZ) of weldments were investigated with respect to the transition curve for CTOD. The main variables were: thickness (30, 70, 100, 130 and 160 mm), welding technique (SMAW and SAW), heat treatment (AW and PWHT) and loading rate during testing (static and dynamic), table 2. All welding was performed at an offshore shipyard with manual welding (SMAW) using OK 48.00 (ESAB) and Kryo 1. mod (Smitweld), for the main and supplementary program respectively, and submerged arc welding (SAW) with the consumables Autrod 12.32/Flusc 10.71 (ESAB) and SW60/P240 (Smitweld) respectively.

Experimental procedure.   Three point bending CTOD specimens
were extracted from the weldments with the preferred B x 2B geometry with a
through thickness notch, according to BS 5762:1979. The notch was
positioned in the HAZ in order to hit a maximum fraction of the coarse
grained zone i.e. close the fusion line, fig. 1. After testing, the
specimen broken halves were sectioned in order to evaluate the notch
location by identifying the microstructures along the fatigue precrack and
the areas of unstable crack initiation, fig. 2 and 3. Several specimens
were examined in detail in a scanning electron microscope with further
sectioning through the crack initiation area. Special data sheets were
worked out to summarize the results from the sectioning procedures.   An
example is presented in fig. 4 for the 160 mm SMAW-weldment in the PWHT
condition. The toal fraction of weld metal and coarse grained HAZ was
determined.

Test results evaluation.   The CTOD transition curves are presented in fig. 5 for the SMAW weldments in the AW and PWHT condition. The
data are plotted as a function of the specimen thickness in fig. 6. The degree of scatter is high, and post test sectioning identified that the majority of the specimens had fracture initiation in the primary weld metal.
The point of crack initiation in the through thickness position is presented in fig. 7. Since the area of fracture initiation was related to the weld
metal, the $CTOD_c$ values have been plottet as a function of the total
fraction of primary weld metal hit by the fatigue crack, fig. 8.   The criterium "total fraction of primary weld metal" has been further developed.
It is suggested that a better criterium should be based on a "weakest link"
mechanism of crack initiation e.a. the number of brittle zones or the total length of the brittle zones, fig. 9. Only specimens with crack initiation in the primary weld metal have been included in this figure.

Weld Metal Toughness Evaluation.

Weldments.   Charpy V-notch tests clearly indicated that the
SMAW weld metal of the main program was poor, fig. 10. On this background a
weld metal fracture toughness evaluation programme was carried out on 30 mm
plates, table 3, in order to select the welding consumables for an additional programme, table 2. It was also examined if it was possible to produce
a more straight fusion line by introducing a 1/2 V groove instead of the K
groove. As expected a relatively straight fusion line was obtained with
the 1/2 V groove, fig. 11, while no significant effect on the weld metal
toughness could be observed.

Test results.   The data for the CTOD and Charpy V-notch tests
are presented in table 4 and 5 respectively, and the chemical composition
in table 6. The specimen broken halves were sectioned for the evaluation
of the brittle fracture initiation areas.

Comparison.   The best results for the SMAW weldments were obtained with Kryo 1-mod., with the highest level for both CTOD and Charpy
V-notch toughness. With respect to brittle fracture initiation, the
$CTOD_c$ was always related to the primary weld metal. The primary weld
metal of Kryo 1 and Kryo 3 was dominated by a side plate structure, Fig.
12a, b, while an increasing amount of accicular ferrite and proeutectoid
ferrite was found in Tenacito 38 R, Fig. 12c. However, Kryo 1 mod. had
no side plate structure, Fig. 12d, and differs considerably from for
instance Kryo 1. Considering the chemical composition of the weldments,
Kryo 1, Kryo 1-mod. and Tenacito 38 R have about the same content of
nickel table 6. The main difference between Kryo 1 and Kryo 1-mod. is the

279

somewhat higher content of manganese and lower content of oxygen in Kryo
1-mod. Kryo 3 deviates with a high nickel content and a reduced manganese
content.

The Charpy V-notch and CTOD results were more scattered for the SAW
condition. All welds have both high and low Charpy V-notch values, while
SW 60/P240 with a 1/2 V groove reached the best CTOD result.

The brittle fracture initiation occured in the primary weld metal for
SW 60/P240, while Fluxocord 35.21/OP121TT also showed two cases of initiation in the transformed weld metal. The primary weld metal for SW 60/P240
is characterized as sideplate ferrite, Fig. 13a, while Fluxocord reveals a
complete accicular ferrite microstructure, Fig. 13b.

SW60/P240 has about the same chemical composition as Tenacito 38R, and
the coarser microstructure resulted from the higher heat input. The Fluxocord weld metal deviates in chemical composition with addition of Mo and
Ti-B, but without Ni. The fine grained accicular ferrite microstructure is
probably due to the effect of Ti-B (14).

Further Experience With Weld Metal Fracture Toughness.   Low
CTOD values in the weld metal, in experiments where the main purpose was
HAZ testing, have recently been experienced in the testing of offshore
steel. And, again, a limited welding procedure testing programme was performed. Four steels were tested, one traditional CMn-steel (0.18 % C) and
three low carbon microalloyed steels. Two SMAW consumables were tested,
table 7. With Kryo 1, the three low carbon steels weld metals had brittle
fractures at -10° C, with CTOD values in the range 0.05 - 0.25 mm,
while the CMn steel had values in the range 1.17 - 0.44 mm. With the Ti-B
electrode L55SN (14) all the weld metals had usually $CTOD_m$ values
($CTOD_m$ >2 mm). The results underline the importance of using welding
consumables developed for the class of steel in question.

Discussion

The Thickness Effect.   Theoretical analysis and practical experiments have shown that the fracture toughness of structural steels and
their weldments, decreases with increased specimen thickness (10-13). The
effect is explained on the basis of the degree of triaxiality at the crack
tip (constraint effect) and the probability of hitting an area (eg. a local
brittle zone, LBZ) where fracture initiation can take place (statistical
effect). When the specimen thickness is increased, both the degree of
plane strain and probability of hitting a LBZ will increase, hence the
fracture toughness will be reduced.

The statistical effect can simplified be expressed as (13):

$$CTOD(B_1) = CTOD(B_2) \ (\frac{B_2}{B_1})^{\frac{1}{2}} \qquad (1)$$

B_1, B_2: specimen thicknesses

$CTOD(B)$: The $CTOD_c$ value for thickness B

The relationship can schematically be presented as shown in Fig. 14. Systematic deviations with higher $CTOD_c$ values than expected for small specimens can be explained on the basis of reduced degree of triaxiality. This "weakest link" model was originally developed for plain material, but progress has been made in order to include weldments (10).

The results of CTOD testing of a 200 mm thick narrow gap weldment, with the fatigue precrack positioned at the fusion line, Fig. 15, indicated that the CTOD values for the 200 mm thick specimen could be calculated on the basis of the results from thinner specimens extracted from the same weld (10). The calculations were based on expressions of the form (simplified):

$$\frac{CTOD(B_2)}{CTOD(B_1)} = \left[\frac{\ln [1-(1-0.5^{1/r})/p]^{1/\alpha}}{\ln (1-0.5/p)} \right] \tag{2}$$

B_1, B_2: refers to the "effective" thickness, eg. the thickness except for the surface region dominated by plane stress.
r: B_2/B_1
α: Weibull shape parameter
p: Probability of fracture

With an estimate of $\alpha = 2$ (13), and P = 1, the following results are obtained for eq 1 and 2 with B_2 = 30 mm and B_1 = 160 mm:

eq (1): CTOD (160) = CTOD (30) 0.43

eq (2): CTOD (160) = CTOD (30) 0.43

Hence, the same results are obtained with the two equations.

The $CTOD_c$ results from the SMAW-PWHT series, with fracture initiation primary weld metal, are summarized in table 8.

According to eq (1), the expected middle values at 160 mm thickness, at -20° C, are:

$CTOD_c$ (160) = $CTOD_c$ (30) 0.43 = 0.06 mm
$CTOD_c$ (160) = $CTOD_c$ (70) 0.66 = 0.17 mm
$CTOD_c$ (160) = $CTOD_c$ (100) 0.79 = 0.06 mm
$CTOD_c$ (160) = $CTOD_c$ (130) 0.90 = 0.14 mm

The result from 30 mm is indicated with a broken line on Fig. 16. It must be underlined that the number of parallels are far too low to allow for a valid statistical analysis, and the present results must only be regarded as indications. The result of the calculations shows that extrapolation based on the 70 and 130 mm specimens will give too optimistic $CTOD_c$ values at 160 mm, while the results from 30 and 100 mm would fit the measured values at 160 mm.

The largest scatter is found in the 70 and 130 mm series, and in Fig. 17 the two parallels at 70 and 130 mm are indicated with 7 and 13.

281

The 130 mm specimen with the highest CTOD$_c$ value has hit significantly less primary weld metal than the other. The highest CTOD$_c$ value at 70 mm may also be explained on the basis on the relative low number of hits, but the extremely low value with only 3 hits in primary weld metal, can not be explained directly.

The results show that when the CTOD values from weldments are evaluated, the effect of notch positioning must be included in addition to the degree of constraint and the statistical effect. The statistical effects in the weakest link model are based on a homogeneous distribution of LBZ, with increased probability of hitting the zones with increased specimen thickness. The LBZ in a weldment is distributed at preferential positions, and the fracture toughness depends very much on the positioning procedure.

The results show a gradual decrease in the mean CTOD$_c$ with increased number of LBZ hit, Fig. 9, but at the same time low values can be found even at low hit values. There will always be a chance for low CTOD values, even when the number of LBZ is low, but the probability will decrease and the scatter in CTOD values increase as the size and number of LBZ at the fatigue precrack front decreases.

Most weldments will have some LBZ, and low CTOD values are to be expected from time to time. It will therefore be important to know the significance of low CTOD values from LBZ. A possible philosophy is to tolerate low CTOD values from LBZ if it can be documented that the surrounding material has sufficient crack arrest properties. Recent results indicates that even relatively large pop-ins from LBZ's in CTOD testing can be regarded as uncritical (15), and crack arrest testing of low carbon microalloyed steels has indicated that the crack arrest properties are better than the higher carbon steels. The crack arrest properties in the weldments of modern low carbon microalloyed steels will now be further examined.

ACKNOWLEDGEMENT. The financial support from Statoil, and the discussions with representatives from Statoil are highly acknowledged. Especially thanks to B. Lian, Å. Gunleiksrud and T. Simonsen.

CONCLUSION

- The welding consumables for low carbon microalloyed steels should be carefully selected. The existing consumables for BS 4360 50D CMn steels should be modified before they are used on low carbon steels.

- The best relationship between CTOD$_c$ and specimens thickness was established by correlating CTOD$_c$ with the number, or total size, of brittle primary weld metal hit by the notch tip.

- The most important factor in CTOD testing of weldments is the positioning of the fatigue precrack with respect to the local brittle zones (LBZ).

- It is proposed that the next step in the evaluation of LBZ, could be to permit low CTOD values, initiating in LBZ, and then rely on crack arrest in the nearby surrounding material.

REFERENCES

1. Braathen, C.M.: "Collection and application of fracture mechanics data by design, fabrication and service of offshore structures". (In Norwegian) Norwegian Society of civil Engineers. January 1987.

2. Valland, G.: "Experience with the application of fracture mechanics on offshore projects". Ibid.

3. Braathen, C.M., Aker Engineering. Private communication.

4. Squirell, S.J., Pisarski, H.G., Dawes, M.G.: "K_{IC}, CTOD and J tests on weldments". Paper presented at Third ASTM Int. Symp. on Nonlinear Fracture Mechanics. Knoxville, October 1986.

5. Satoh, K., Toyoda, M.: "Guidelines for Fracture Mechanics Testing of WM/HAZ". IIW Doc. X-1113-86.

6. Andersson, H. et al: "Evaluation of standards for fracture mechanics testing". SINTEF report STF16 A86138, Trondheim, Norway.

7. Thaulow, C. et al: "Heat affected zone toughness of a low carbon micro- alloyed steel". Met. Constr. Vol 17 no. 2, 1985, pp. 94R - 99R.

8. Thaulow, C, Paauw, A.J., Guttormsen, K.: "The heat affected zone toughness of low carbon microalloyed steel". To be published in Welding journal, spring 87.

9, To be presented at SIMS (Steel in Marine Structures), Int. Offshore Conference, Delft, 15-18 June 1987.

10. Toyoda, M.: "Fracture toughness evaluation of steel welds (review)". Osaka University, July 1986.

11. Pisarski, H.G.: "Influence of thickness on critical crack opening displacement (COD) and J-values". Int. Journal of Fracture, 17 (1981), pp. 427 - 440.

12. Karlsen, A.: "Fracture toughness of heavy welded joints". Det norske Veritas report no. 79-0448. 1979.

13. Wallin, K.: "The size effect in K_{IC} results". Eng. Frac. Mech. Vol 22, no. 1, 1985, pp. 149 - 163.

14. Horii, Y. et al: "Development of welding materials for low temperature service". Presented at the International Trends in Welding Research, ASM, Tennessee, May 1986.

15. Arimochi, K., Isaka, K.: "A study on pop-in phenomenon in CTOD test for weldment and proposal of assessment method for significance of pop-in". IIW Doc. X-1118-86. July 1986

	Plate thickness		(mm)		
	30	70	100	130	160
Chemical composition in wt%					
C	0.09	0.09	0.09	0.09	0.09
Si	0.38	0.35	0.35	0.34	0.38
Mn	1.46	1.45	1.45	1.50	1.46
P	0.008	0.006	0.006	0.006	0.009
S	0.002	0.001	0.001	0.001	0.001
Cr	0.02	0.02	0.02	0.01	0.01
Ni	0.25	0.35	0.35	0.35	0.35
Cu	0.24	0.24	0.24	0.25	0.25
Mo	0.01	0.01	0.01	0.01	0.01
Nb	0.025	0.023	0.023	0.026	0.024
V	0.03	0.025	0.025	0.025	0.025
Ti	0.011	0.008	0.008	0.006	0.009
Al	0.029	0.034	0.034	0.040	0.040
N	0.0039	0.0033	0.0035	0.0035	0.0035
Mechanical properties					
R_{eH} (MPa)	373	339	338	333	296
R_m (MPa)	494	487	491	479	471
A (%)	32	36	36	38	37
$KV_{-40°C}$ aver. (J)	341	348	319	364	289

Table 1 : The chemical composition and mechanical properties of the investigated low carbon micro alloyed steel.

Table 2 — Test programme (rotated table, Plate thickness in mm).

Base material	CTOD		30	70	100	130	160
		Static	X	X	X	X	X
		Dynamic	X	X	X	X	X

Buttweldments			30				70			100			130		160
			SMAW		SAW		SMAW		SAW	SMAW		SAW	SMAW	SAW	SAW
			AW	PWHT	AW	PWHT	AW	PWHT	PWHT	AW	PWHT	PWHT	PWHT	PWHT	PWHT
Main program	Charpy V-notch			X				X			X			X	X
	CTOD	Static	X	X	X	X	X	X	X	X	X	X	X	X	X
		Dynamic	X	X				X			X				
Additional program	CTOD	Static	X	X	X	X		X	X		X				
		Dynamic	X	X											

Table 2 : The test programme for the evaluation of the
heat affected zone and base material toughness.

Welding method	Heat input (MJ/m)	Groove	Welding consumables	
			Name	Company
SMAW	2		Kryo 1 Kryo 1-Mod. Kryo 3	Norweld
			Tenacito 38R	Oerlikon
SAW	5		Wire : SW60 Flux : P240	Norweld
			Wire : Fluxocord 35.21 Flux : OP121TT	Oerlikon
			Wire : SW60 Flux : P240	Norweld
			Wire : Fluxocord 35.21 Flux : OP121TT	Oerlikon

Table 3 : The weldments used for the evaluation of the weld metal toughness level.

Weldment (indicated with welding consumable)	CTOD at -20°C	
	mm	Remark
SMAW – weldments		
Kryo 1	0.04	c
	0.08	c
	0.37	c
	0.45	u
	0.65	u
Kryo 1-Mod.	0.32	u
	2.06	m
	2.21	m
Kryo 3	0.26	c
	0.49	u
	0.60	u
Tenacito 38R	0.23	c
	0.31	u
	0.41	u
SAW – weldments		
SW60/P240 K-groove	0.18	c
	0.20	c
	0.35	c
	0.58	u
	3.56	m
SW60/P240 ½V-groove	0.32	c/u
	1.87	m
	2.27	m
Fluxocord 35.21/ OP121TT K-groove	0.28	c
	0.36	c
	0.59	c
Fluxocord 35.21/ OP121TT ½V-groove	0.20	c
	0.21	c
	0.22	ˑ

Table 4 :

The CTOD results of the Bx2B specimens with the notch positioned in the weld metal of the weldments presented in table 3.

Weldment (indicated with welding consumable)	Surface				Root	
	KV at -40°C (J)	aver.	KV at -20°C (J)	aver.	KV at -40°C (J)	aver.
SMAW - weldments						
Kryo 1	131 / 126 / 145	134	184 / 172 / 177	171	75 / 50 / 93	73
Kryo 1-Mod.	122 / 113 / 152	129			98 / 114 / 159	124
Kryo 3	116 / 61 / 82	86			55 / 37 / 91	61
Tenacito 38R	92 / 39 / 104	78			49 / 81 / 107	77
SAW - weldments						
SW60/P240 K-groove	27 / 45 / 58	43	142 / 144 / 100	129	136 / 122 / 133	180
SW60/P240 ½V-groove	119 / 141 / 77	112			58 / 102 / 26	62
Fluxocord 35.21/ OP121TT K-groove	76 / 102 / 46	75			96 / 63 / 70	76
Fluxocord 35.21/ OP121TT ½V-groove	112 / 68 / 98	93			39 / 33 / 59	44

Table 5 : The Charpy V-notch results of the weld metal of the weldments presented in table 3.

	SMAW – weldments								SAW – weldments							
	Kryo 1		Kryo 1-Mod.		Kryo 3		Tenacito 38R		SW60/P240				Fluxocord 35.21/OP121TT			
	Surface	Root	Surface	Root	Surface	Root	Surface	Root	Surface	Root	Surface	Root	Surface	Root	Surface	Root
C	0,06	0,06	0,06	0,07	0,08	0,07	0,07	0,06	0,09	0,09	0,07	0,08	0,09	0,06	0,07	0,06
Mn	1,32	1,45	1,58	1,68	0,88	0,75	1,14	1,17	1,11	1,17	1,13	1,19	1,53	1,51	1,50	1,47
Si	0,29	0,29	0,29	0,31	0,41	0,34	0,37	0,31	0,18	0,21	0,18	0,23	0,23	0,22	0,24	0,28
P	0,008	0,009	0,010	0,009	0,007	0,007	0,009	0,007	0,013	0,011	0,014	0,010	0,015	0,018	0,014	0,012
S	0,005	0,005	0,003	0,004	0,003	0,004	0,007	0,006	0,004	0,004	0,004	0,003	0,011	0,011	0,010	0,009
Cr	0,04	0,03	0,03	0,03	0,03	0,03	0,03	0,03	0,06	0,05	0,05	0,0;	0,04	0,04	0,04	0,04
Ni	0,91	0,79	0,93	0,82	2,6	2,4	1,1	0,85	1,0	0,87	0,88	0,77	0,11	0,06	0,12	0,29
Cu	0,04	0,08	0,03	0,09	0,04	0,06	0,03	0,09	0,15	0,17	0,16	0,17	0,13	0,11	0,13	0,14
Mo	0,01	0,01	0,01	0,01	0,01	0,01	0,01	0,01	0,01	0,01	0,01	0,01	0,24	0,30	0,24	0,14
V	0,01	0,01	0,01	0,01	0,01	0,01	0,02	0,01	0,01	0,01	0,01	0,02	0,02	0,02	0,02	0,01
Nb	0,009	0,010	0,008	0,012	0,010	0,010	0,007	0,011	0,009	0,013	0,016	0,067	0,015	0,012	0,014	0,020
Ti	0,010	0,007	0,009	0,007	0,017	0,007	0,005	0,005	0,003	0,003	0,005	0,035	0,015	0,014	0,014	0,014
B	-	-	-	-	-	-	-	-	-	-	-	-	0,0019	0,0015	0,0015	0,0015
Al	0,006	0,005	0,007	0,005	0,009	0,004	0,003	0,004	0,012	0,013	0,013	0,016	0,015	0,014	0,014	0,017
N	0,008	0,008	0,007	0,005	0,007	0,010	0,008	0,009	0,006	0,006	0,007	0,0;5	0,009	0,010	0,009	0,005
O	0,041	0,013	0,029	0,006	0,038	0,038	0,044	0,038	0,030	0,026	0,005	0,0;4	0,027	0,028	0,027	0,005

Table 6 : The chemical composition of the weld metal at the surface and the root of the weldments presented in table 3.

Welding consumable	Groove	Type of current	Current A	Voltage V	Speed mm/s	Heat input MJ/m
Kryo 1		AC	240	24	4.2	1.4
L-55SN		DC+	250	27	3.3	2.0

Table 7 : Welding characteristics of two SMAW electrodes.

Testing temp. °C	$CTOD_c$ (mm)				
	Plate thickness (mm)				
	30	70	100	130	160
0	0.20 0.66	0.30 0.59	0.12 0.20 0.29 0.50 0.55	0.04 1.56	0.12 0.13 0.25 0.91
-20	0.12 0.17	0.04 0.49	0.05 0.08 0.12	0.08 0.24	0.04 0.05

Table 8 : $CTOD_c$ values referring to crack initiation in the primary weld metal for SMAW-PWHT test series with the fatique crack positioned along the fusion line.

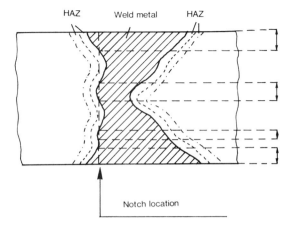

Fig. 1 : The position of the through thickness notch in the
Bx2B CTOD specimens along the fusion line of the
weldment.

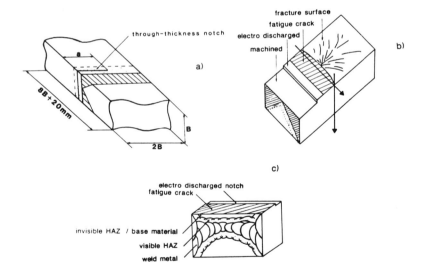

Fig. 2 : The sectioning procedure of the fracture surface of
the CTOD specimens: a) The CTOD specimen;
b) The broken half after testing; c) The specimen
section for the identification of the microstructures
along the crack front.

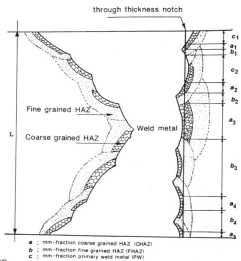

through thickness notch

Fine grained HAZ

Coarse grained HAZ

Weld metal

L

a : mm–fraction coarse grained HAZ (CHAZ)
b : mm–fraction fine grained HAZ (FHAZ)
c : mm–fraction primary weld metal (PW)

Fig. 3 : A schematical presentation of typical microstructural zones along the crack tip.

$$\% \, CHAZ = \frac{a_1 + a_2 + a_3 + a_4 + a_5}{L} \cdot 100$$

Test temp.	Spec. no	CTOD (mm)	Notch location & position of initiation	Fraction PW (%)	Fraction CHAZ (%)	Initiation	Comments
+20°C	16EK 4	0.38 c		22.2	19.3	weld defect	macro cavity
	9	10.60 u		3.2	11.9	–	
0°C	16EK11	0.12 c		14.6	6.4	primary weld metal	
	13	0.13 c		15.7	7.0	— " —	
	1	0.25 c		40.9	4.5	— " —	SEM & fracture profile
	6	0.91 c		9.9	7.5	— " —	
	16 EC	2.17 c		0.5	0.7	deep in weld	SEM; inclusions
	16EK10	5.81 u		0	4.4	–	
-20°C	16EK12	0.04 c		14.3	6.3	primary weld metal	
	3	0.05 c		32.0	14.0	— " —	SEM
	8	2.12 c		2.2	12.3	partially transf. HAZ	SEM & fracture profile
	16BK 2	2.89 c		0	0	— " —	— " —
	16EK 5	3.10 u		1.1	7.5	–	

CTOD – data : Spec. no, CTOD (mm)
Data from fracture surface examination

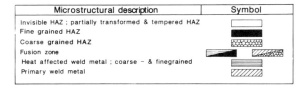

Microstructural description	Symbol
Invisible HAZ ; partially transformed & tempered HAZ	
Fine grained HAZ	
Coarse grained HAZ	
Fusion zone	
Heat affected weld metal ; coarse – & finegrained	
Primary weld metal	

Fig. 4 : A data sheet summarizing the information from the metallographic examination: Example for a 160 mm thick weldment (SMAW-PWHT).

291

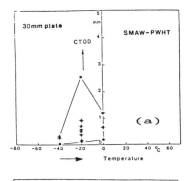

Fig. 5 : The CTOD transition
curves for the SMAW-
PWHT weldments with the fatigue
crack positioned along the
fusion line.

    (a)   30 mm plate thickness
    (b)   70 mm plate thickness
    (c) 100 mm plate thickness
    (d) 130 mm plate thickness
    (e) 160 mm plate thickness

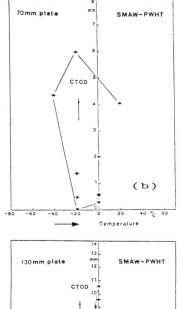

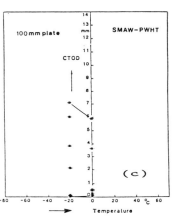

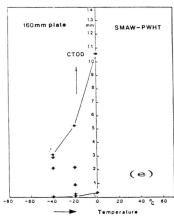

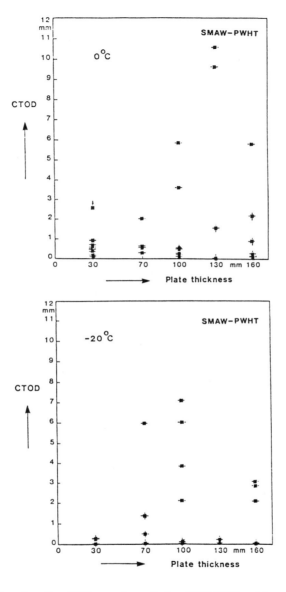

Fig. 6 : The CTOD results of the SMAW-PWHT weldments
with the fatigue crack positioned along the
fusion line as a function of the plate thickness for two testing temperatures.

(a) 0° C ; (b) -20° C

293

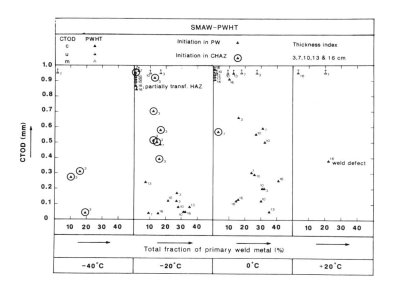

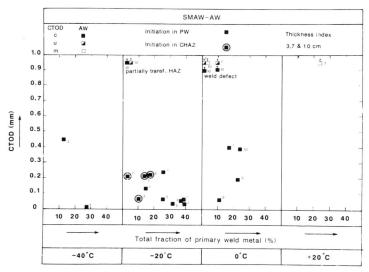

Fig. 8 : The CTOD results as a function of the total
fraction of primary weld metal (%) sectioned
by the crack tip.

(a)   SMAW-PWHT weldments
(b)   SMAW-AW   weldments

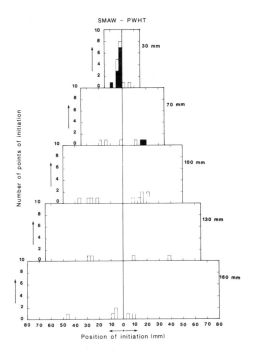

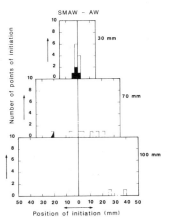

Fig. 7 : The position of the brittle fracture initiation area along the crack front as a function of the specimen thickness.

    (a)   SMAW-PWHT weldments
    (b)   SMAW-AW   weldments

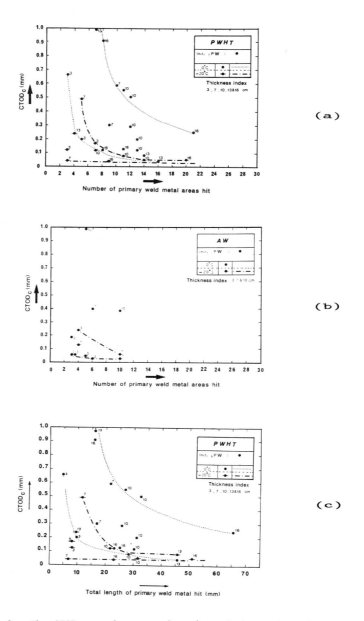

(a)

(b)

(c)

Fig. 9 : The CTOD results as a function of the number of
areas ((a),(b)) and the total lenght ((c)) of
primary weld metal sectioned by the crack front.

(a)   SMAW–PWHT weldments ; and (c)
(b)   SMAW–AW   weldments

296

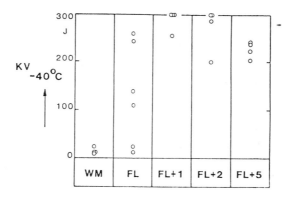

Surface

Fig. 10 : Results from the Charpy V-notch tests of the
100 mm SMAW-PWHT buttweldment.

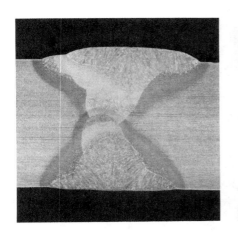

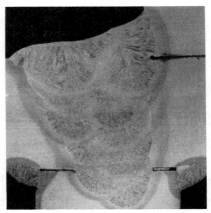

(a) K-groove   (b) ½V-groove

Fig. 11 : Macro sections of the SAW weldments with
different grooves. Welding consumable:
SW60/P240 (see table 3).

297

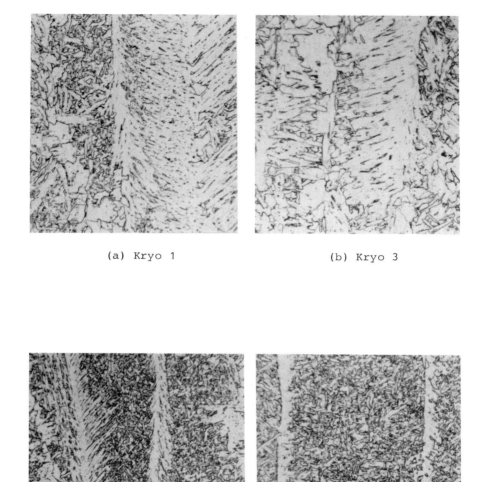

(a) Kryo 1   (b) Kryo 3

(c) Tenacito 38R   (d) Kryo 1-Mod.

0.01mm

Fig. 12 : The microstructures of the primary weld metal
of the SMAW weldments presented in table 3.
Etched in 2% Nital.

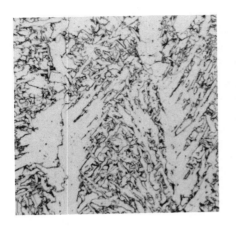 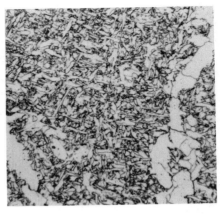

(a) SW60/P240   (b) Fluxocord 35.21/OP121TT

0.01mm

Fig. 13 : The microstructures of the primary weld metal
of the SAW weldments presented in table 3.
Etched in 2% Nital.

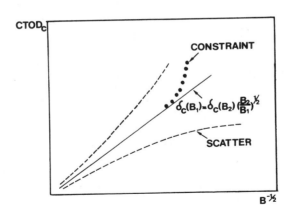

Fig. 14 : Schematical presentation of the expected
relationship between CTOD and the specimen
thickness from a statistical point of view.

299

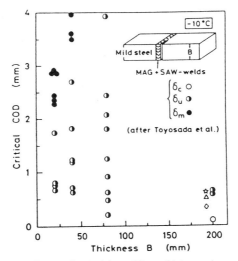

◇ : Median estimated from 20mm thick specimens
☆ : Median estimated from 40mm thick specimens
△ : Median estimated from 80mm thick specimens

Fig. 15 : The thickness dependence of the critical CTOD
of an originally 200 mm thick weldment (10).

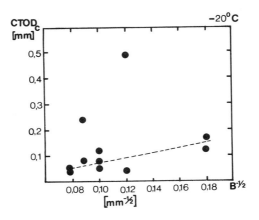

Fig. 16 : The CTOD results of the SMAW-PWHT test series,
see table 8: $CTOD_c$ at $-20^{\circ}C$ as a function of
the specimen thickness ($B^{-\frac{1}{2}}$).

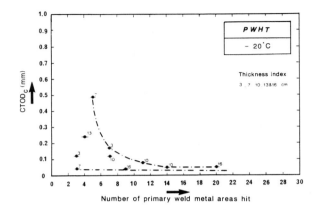

<u>Fig. 17</u> : The CTOD results at -20° C of the SMAW-PWHT
weldments as a function of the number of areas
with primary weld metal sectioned by the crack
tip.

LOCAL BRITTLE ZONES IN STRUCTURAL WELDS

D. P. Fairchild

Exxon Production Research Company
Houston, Texas 77001

Abstract

In the past four years, there has been some concern about the heat-affected-zone (HAZ) toughness of offshore platform welds in thick plate (>50 mm). When the crack tip opening displacement (CTOD) test is used, HAZ toughness values as low as 0.02 mm can be measured, and this represents linear elastic fracture. These values are generally associated with small areas of low toughness within the coarse-grain HAZ (CGHAZ) called "local brittle zones" (LBZ's). In a multi-pass weld, CGHAZ areas can be reheated by subsequent weld passes, and the location of LBZ's becomes complex. The appearance of LBZ's within a multi-pass weld is explained. When the CTOD test is used to assess LBZ behavior, post-test sectioning techniques are required to determine the HAZ areas actually sampled by the fatigue crack. These techniques are also discussed. Potential solutions to the LBZ issue are given.

Introduction

Low-carbon, microalloyed structural steels have been in use for about 30 years, and as early as 1966, it was known that these materials may develop low HAZ toughness associated with the coarse-grain region.[1] As HAZ studies of these materials continued in the '60's and '70's,[2-8] it was found that the heterogeneous nature of the HAZ, particularly in multipass welds, creates a large amount of toughness data scatter, and unacceptable results can occur periodically. Engineers have learned to deal with sporadic low toughness results in various ways. Criteria exist that permit one low result to be replaced by two acceptable results. This method, however, may simply take advantage of the natural data scatter where a high probability exists of passing the two additional tests. Another technique for dealing with low results is to place heavy emphasis on the mean toughness value instead of the low bound. A CTOD data set for the coarse-grain HAZ of a 50-mm-thick, C-Mn-Nb-V offshore platform steel is shown in Figure 1. While the mean CTOD toughness is 0.62 mm (0.02 in.), which is quite acceptable, values as low as 0.016 mm (6 x 10^{-4} in.) were measured, and this represents linear elastic behavior.* It is clear from the differently shaded regions in Figure 1 that two distinct populations of test results exist and that one population lies significantly below the mean. This type of data scatter is not uncommon, and this paper will focus on the causes of the low-toughness population.

During the early 1980's, the North Sea offshore industry noted a change in the HAZ toughness results reported for some low-carbon, microalloyed platform steels. The frequency of low HAZ CTOD results escalated to a higher-than-normal level. Initially, the authenticity of these data was questioned; however, after some research,[9-14] their existence is now accepted. These low CTOD results are caused by local brittle zones (LBZ's), which are small areas of limited cleavage resistance within the coarse-grain HAZ. Major research emphasis has changed from proving authenticity to studying the metallurgical causes and the structural significance of LBZ's.

This paper discusses LBZ formation in a multipass weld and explains CTOD techniques in relation to LBZ's. Artist drawings will be used to avoid visual clarity problems that can occur (particularly during reproduction) when using micrographs and fractographs to define specific metallurgical features. Some of the factors that contributed to the early 1980's transition and made the low CTOD results difficult to accept will also be described.

HAZ and LBZ Formation

The various regions of a single-pass weld HAZ are defined in Figure 2. This information can be reduced to the simpler picture shown in Figure 3. Although the SCHAZ is often referred to as the region heated to a temperature below Ac_1, Figures 2 and 3 define a low-temperature boundary for the SCHAZ: 500~600°C. This boundary was chosen based on the assumption that no significant metallurgical changes occur below 500~600°C for the holding times related to a typical weld HAZ. In a two-pass weld (Fig. 4a), some regions of the first pass are eliminated, while

---

*Linear elastic validity was conservatively estimated using E399 criteria:

$$B \geq 2.5 \left(\frac{K_{IC}}{\sigma_{ys}}\right)^2, \quad K^2 = mE\sigma_{ys}\delta_c, \quad E = 207\text{GPa}, \quad \sigma_{ys} = 345\text{MPa}, \quad m = 2$$

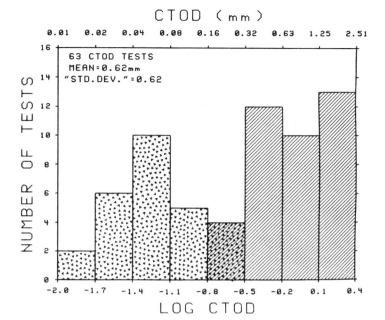

Figure 1 - CGHAZ CTOD (TxT-notched) data for a 50-mm-thick,
C-Mn-Nb-V offshore platform steel (Y.S. 345 MPa). A K-bevel
weld was used with heat inputs of 1.0-3.5 kJ/mm. Test temperature -10°C.

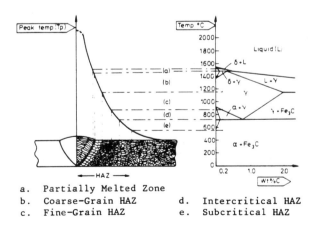

a. Partially Melted Zone
b. Coarse-Grain HAZ   d. Intercritical HAZ
c. Fine-Grain HAZ   e. Subcritical HAZ

Figure 2 - A single-pass weld with respect to the
iron-carbon phase diagram [14].

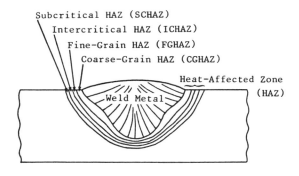

Figure 3 - The various regions of a bead-on-plate weld.

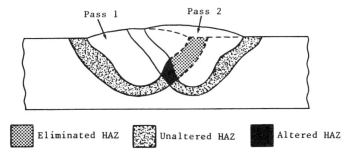

Figure 4 (a) - The eliminated, altered, and unaltered regions of a two-pass, bead-on-plate weld.

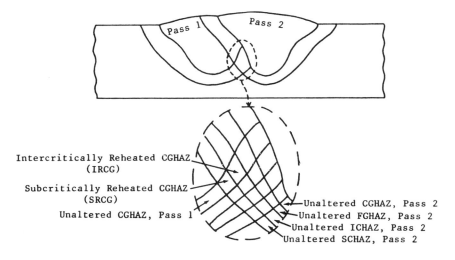

Figure 4 (b) - Identification of some altered and unaltered regions.

others are significantly altered. Figure 4b identifies some of these altered and unaltered regions. In a single-bevel, multipass weld (often used for platform steel qualification), the overlapping HAZ's that penetrate the unbeveled edge appear as shown in Figure 5a. Redrawing Figure 5a so that only the unaltered CGHAZ regions are depicted results in Figure 5b. Notice that the unaltered CGHAZ is directly adjacent to the fusion line, where the columnar weld metal contacts the base metal. If the intercritically reheated CGHAZ (IRCG) and the subcritically reheated CGHAZ (SRCG) regions are added to Figure 5b, then Figure 5c results.

The CGHAZ, IRCG, and SRCG will be referred to as the coarse-grain (CG) regions. LBZ's are comprised of low-toughness CG regions. The following metallurgical factors are believed to be the main contributors to the limited toughness of the CG regions.

<u>Upper Bainite Formation</u>.[1,5,10,15] Under certain combinations of HAZ thermal cycle (related mainly to heat input) and steel chemistry, upper bainite can form in the CGHAZ. This structure can have very low toughness, and because of its high transformation temperature (thermal stability), it is resistant to decomposing during subsequent HAZ thermal cycles and/or PWHT. It can exist in all CG regions.

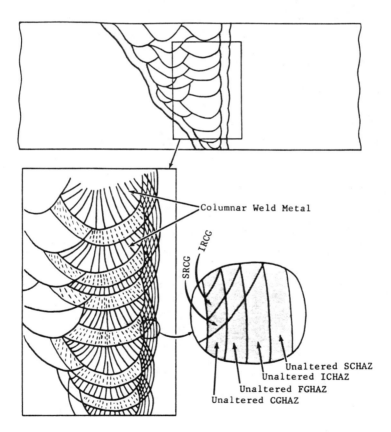

Figure 5 (a) - The HAZ regions in a single-bevel, multipass weld.

307

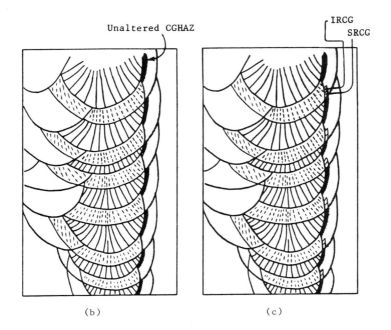

Unaltered CGHAZ

IRCG
SRCG

(b)   (c)

Figure 5 - The HAZ regions in a single-bevel, multipass weld:
(b) the unaltered coarse-grain HAZ (CGHAZ); (c) the intercritically and subcritically reheated CGHAZ regions (IRCG and
SRCG).

Precipitation.[1-7,10,14,15]   When microalloying elements such as
niobium and vanadium are present, precipitates or pre-precipitation
clusters can form and reduce toughness. In the CG regions, most precipitates near the fusion line will dissolve during the initial weld thermal
cycle. Upon reheating by subsequent weld passes or PWHT, irregularly
shaped particles may coalesce that strain the surrounding matrix. In the
as-welded condition, the IRCG and SRCG may be affected, but during PWHT,
all CG regions may experience precipitation.

Grain Size.[16]   Although specific effects on HAZ fracture toughness
are unclear, it is generally believed that larger grains reduce toughness. As their name indicates, the CG regions have a large prior-austen-
ite grain size, typically 75-150 microns (ASTM 5-2). This affects all CG
regions.

Martensite Islands.[13,15]   During HAZ thermal cycles between Ac_1
and Ac_3 (i.e., the intercritical region, 750~850°C), the austenite
becomes enriched with carbon, while the ferrite is depleted. Upon cooling, the austenite transforms to martensite, but the ferrite remains
unchanged. This final structure is sometimes referred to as dual phase
or M-A constituent. These islands of high-carbon martensite may have
very low toughness. The phenomenon typically affects the IRCG region in
the as-welded condition. Subsequent weld thermal cycles or PWHT can
restore toughness by decomposing (tempering) the martensite.

308

It should be mentioned that not all steels will produce CG regions of unacceptable toughness. The previously mentioned factors will occur to varying degrees depending on steel chemistry, steel manufacturing route, and weld thermal cycle, so some materials may be more resistant to LBZ formation than others. Because the above factors are thermally activated phenomena, it is even possible that a particular steel may generate low-toughness CG areas at certain high heat inputs, but not at low heat inputs. The term "local brittle zone" refers to coarse-grain regions that produce low toughness. Although an exact definition of low toughness will depend on application, a general rule of thumb might be a CTOD less than 0.10 mm.

LBZ Size Effects

The extent to which LBZ's affect the fracture toughness of a welded joint will depend on the size, number, and distribution (location) of the individual CG regions. It seems logical that a weld with many large LBZ's will be more likely to incur a low-toughness fracture event than one with just a few small LBZ's. Several researchers have investigated this subject.[12,17] Figure 6 shows that CTOD is reduced by the occurrence of large individual CG regions.

In an actual weld, it is possible to eliminate or at least reduce the size of the CG regions depending on the amount of HAZ overlap. A weld fusion line with a high degree of HAZ overlap and one with a small amount of overlap are shown in Figure 7. A comparison of these welds shows that CG regions are eliminated or reduced in size if a larger number of weld beads penetrate the plate edge. This suggests that certain welding techniques involving bead placement could be used to prevent LBZ formation.[14] This will be discussed later.

CTOD Testing

In-house research has shown that when LBZ's are present, the lower-bound CTOD magnitude is the same for through-thickness and surface-notch specimens. Therefore, this paper will not address the debate concerning which notch orientation is best, but will discuss only the through-thickness technique.

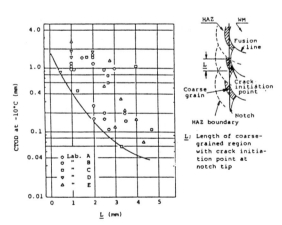

Figure 6 - The relationship between CTOD and the length (L) of the coarse-grain region at the point of crack initiation [12].

309

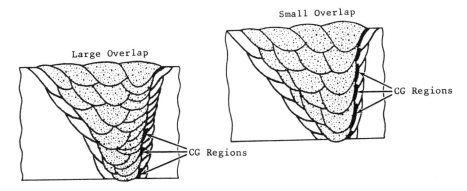

Figure 7 - The relationship between weld-bead overlap and CG region size.

During material or weld procedure qualification, the CTOD test can be used to judge LBZ susceptibility. To accomplish this, it is necessary that the CTOD fatigue precrack sample the appropriate CG regions. Fatigue-crack placement can be controlled during in-house testing by carefully positioning a sharp machine notch (before fatiguing) as follows. Once the rectangular CTOD blank is removed from the welded plate, but before notch machining, one side of the specimen is etched with Nital to reveal the HAZ geometry. A fine surface grind or a light polish helps provide a better etch. Using a straight-edge and a metal scribe, a line is marked on the specimen to represent the intended fatigue-crack location. This scribe line is drawn so that it intersects the maximum number of CG regions. The position of the scribe line must be referenced to the specimen boundaries because the line itself will be lost in subsequent machining operations. A fairly wide notch (possibly several millimeters) is cut to begin the precrack. At the bottom of this notch, a thin slit approximately 0.15 mm (0.006 in.) wide is cut using a spinning carbide disk. Because this slit is flat-bottomed and the corners are even sharper than the 0.15-mm width, fatigue-crack initiation is more uniform than when a blunter, machine notch is used. The location of the slit should coincide with that of the scribe line. The depth of the slit is maximized, while the length of the fatigue crack is minimized (without violating BS 5762 precrack geometry criteria) to reduce fatigue-crack wandering problems.

Because even the best fatigue-precrack placement efforts may fail and the HAZ regions of interest may not be sampled, each half of a broken specimen is cut open to determine actual fatigue-crack location (Figs. 8 and 9). The saw cut into the fracture surface should be placed so that the resulting cross section exposes the area sampled by the front portion of the fatigue crack. The fatigue crack may tunnel (propagate further into the central area of the specimen), and the saw cut may cross the fatigue-crack/fracture-face boundary (Fig. 8). This crossover is limited to the specimen edges. The two cross sections are polished, etched, placed together (Fig. 9), and photographed at two to six times magnification. To enable placement of the two sections close together, it is important that the pieces contain the machine notch and not the fracture face. If they contain the fracture face, the torn and distorted material on the face will not allow the pieces to be closely mated. The percentage of CG regions that were sampled is calculated as shown in Figure 9. This can be done by examining the specimen and the photograph together and/or by examining the specimen at higher magnification (100-200X) under a microscope. The percent CG is used as a measure of how well the fatigue crack sampled the suspected LBZ's.

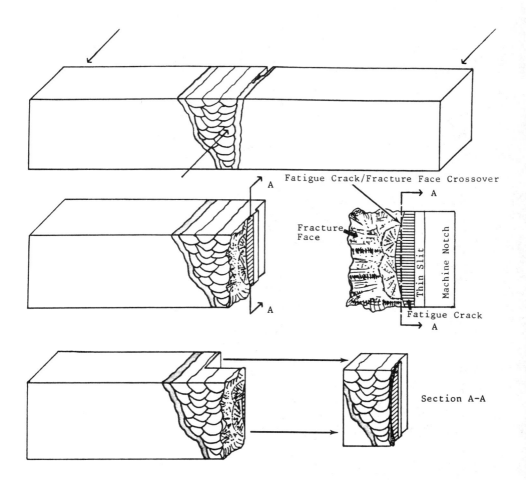

Figure 8 - Sectioning a HAZ CTOD specimen.

When calculating percent CG, it is necessary to define the width of the CGHAZ (i.e., the CGHAZ/FGHAZ boundary) in order to identify the individual ligaments, L_i (Fig. 9). The specific manner in which this is done depends on how the CTOD cross sections are examined, with a microscope or with enlarged photographs. If a microscope is used, the CGHAZ/FGHAZ boundary is based on prior austenite grain size, usually about 70 microns. The L_i ligaments are the areas where the fatigue crack samples CG regions of greater than, say, 70-micron grain size. When enlarged photographs are examined, a distance method is used in which CGHAZ width is predetermined (for the heat inputs of interest) by observing just a few specimens under the microscope at 100-200X. It is important to note that once these widths are recorded, they can be used for hundreds of percent CG calculations. Defining the L_i's then becomes a two-step process. First, the points at which columnar weld metal contacts the base metal are identified. These are the points where the CG regions exist. Second, each of these points is inspected to determine if the fatigue crack has come within the required distance from the fusion line. If it has come close enough to the fusion line, the length (L_i) over which it has done so is measured.

311

If the CTOD value does not correlate with the percent CG sampled by the fatigue crack (e.g., a low CTOD occurs, but the fatigue crack is shown to sample nearly all FGHAZ), then further examination may be required to understand the test result. Sometimes a small amount of stable tearing and a clear initiation point are evident, as shown in Figure 10. Because the initiation point is a small distance from the fatigue-crack cross section, the microstructure at the initiation site may not be the same as that sampled by the fatigue crack. The stable tear may have advanced away from the fatigue-crack plane at a sharp angle, i.e., torn from one HAZ region into another. A cross section can be removed in the same manner as mentioned above to reveal the micro-structure at the initiation site.

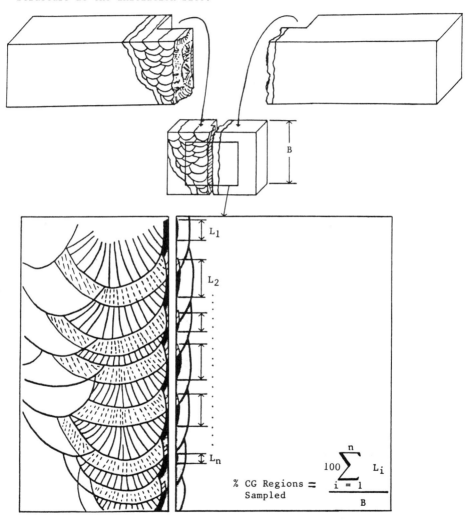

$$\% \text{ CG Regions Sampled} = \frac{100 \sum_{i=1}^{n} L_i}{B}$$

Figure 9 - Calculating the percentage of CG regions sampled by the fatigue crack.

Realization of Low Toughness Values

The critical joints (node joints) in most conventional North Sea platforms have been fabricated using C-Mn (possibly microalloyed) structural steels that are normalized to give a minimum yield strength of approximately 345 MPa (50 ksi). BS 4360 is a very common British Standard designation for this material; some United States equivalents include A537, A633, and API 2H. Because this steel has been used successfully for about a decade, initial reports of low HAZ CTOD toughness were heavily scrutinized. By the early 1980's, a "state of mind" had developed accepting the belief that HAZ toughness was adequate. The following factors will help explain how this state of mind evolved.

Charpy Testing. Although the CTOD test has become quite popular within the offshore industry, Charpy testing has produced most of the toughness data on platform steels. In the author's opinion, the Charpy test may be unable to detect the presence of LBZ's. Because LBZ's are isolated and discontinuous and because the Charpy specimen is relatively small, it is possible to remove a Charpy specimen from a thick-section weldment without sampling any of the suspect microstructure. Even if an LBZ is sampled, the large notch-root radius of the Charpy specimen will allow other microstructures of high toughness (e.g., weld meld or FGHAZ) to influence the energy absorbed. It must also be remembered that Charpy energy is comprised of both initiation and propagation energy; however, LBZ's affect mainly initiation phenomena. Therefore, the Charpy test may not give a realistic low-bound value for LBZ-related fracture initiation. The Charpy data that helped develop the early 1980's "state of mind" may not have accurately identified LBZ behavior.

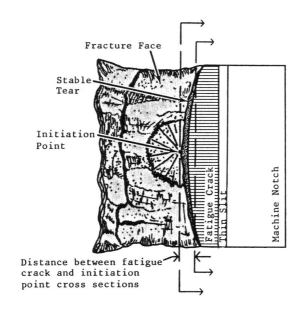

Figure 10 - Sectioning a HAZ CTOD specimen at the point of crack initiation.

CTOD Techniques. When the CTOD test was first developed (1960's), it was used primarily to determine base and weld-metal toughness, but not HAZ toughness. In the late '70's, when a large part of the HAZ CTOD data was generated that influenced the state of mind, the notch placement and sectioning techniques described earlier were rarely (if ever) used. The lack of use of these techniques could have contributed to the absence of low CTOD results. From the standpoint of HAZ testing, the CTOD method was somewhat in its infancy during the late '70's.

Industry Position. At an international conference entitled "Steel in Marine Structures" (Paris, October 1981), HAZ toughness was addressed during a plenary presentation, "Material Selection Considerations for Offshore Steel Structures." Two references were cited[18,19] that reported a total of 413 HAZ CTOD tests conducted on BS 4360 50D and 40E weldments up to 82 mm thick (1-4 kJ/mm, single and multipass, as-welded and PWHT). It was concluded that "...the HAZ toughness of steels used in node manufacture is excellent...there is no fracture problem in the HAZ over the range of variables studied." At the time this conference took place, the British Standard Draft for Development of Fixed Offshore Platforms, DD55, was being re-drafted as a Code of Practice. Based on the above conclusions, the statement proposed for the re-draft was "It was decided that, for heat inputs in the range studied...and provided that the steel was killed and Al treated, there would be no need for toughness testing of the HAZ." This event certainly affected the state of mind of the industry. Why LBZ's were not identified at this time, one can only speculate. It is possible that the welding procedures used produced high HAZ overlap and eliminated LBZ's. Since detailed sectioning techniques were not popular at the time these studies were conducted, the HAZ overlap phenomenon would have gone undetected. It is also possible that the steels used were simply not susceptible to LBZ formation and the CG regions had sufficient toughness.

Service Failures. One of the most important sources of feedback to the engineer in evaluating a successful/unsuccessful design is the absence/presence of service failures. To the knowledge of the author, no failures, catastrophic or otherwise, of North Sea structures have been attributed to LBZ-related brittle fracture. Although this fact does not preclude the existence of LBZ's or prove them insignificant, it does play a role in developing a state of mind concerning HAZ toughness. If brittle fractures do not occur, it is natural to assume that material toughness is adequate.

Steel Manufacture. Throughout the 1970's and early 1980's, steel manufacturers gradually reduced the carbon content of their platform steels in response to fabricator and operator demand for a more weldable steel (i.e., with better HAZ cracking resistance). Microalloying was used to compensate for carbon-related strength loss. A reduction in carbon content, however, may adversely affect HAZ toughness because coarser bainitic microstructures can develop.[5,8] In general, this trend in steelmaking may have promoted bainitic transformation and precipitation behavior, which are two factors previously mentioned as LBZ contributors. Although the change in steelmaking technology did not affect the state of mind, it may have influenced the observation of LBZ's by making their occurrence more frequent.

After reviewing the above information, it is understandable why LBZ's in platform steels went undetected until the early 1980's and why their existence was difficult to accept.

Discussion

Because LBZ's have been responsible for linear elastic CTOD fractures in platform steel qualification welds, their effect on structural integrity must be considered. On one hand, the linear elastic CTOD results can be interpreted to indicate that LBZ's may become critical and could cause failure in some form (e.g., arrested brittle crack, severed non-critical member, collapse, etc.). On the other hand, many platforms are in successful operation in the North Sea with no apparent brittle-fracture problems. This paradox indicates that LBZ-related failure is a statistical phenomenon, and the simultaneous occurrence of certain critical conditions has not yet taken place. The necessary critical conditions will include

- Susceptible steel - LBZ's exist only if the steel produces low-toughness CG regions when welded.

- Welding procedure - HAZ overlap will not be so great as to eliminate (refine) the CG regions.

- Stress concentration - Presence of sharp defect, fatigue crack, HAZ hydrogen crack, etc. located in or near an LBZ.

- Applied stress - Occurrence of necessary wave/wind loading, etc.

- Temperature - Occurrence of a minimum temperature.

In light of good service performance, when a HAZ fracture test (CTOD, wide-plate, etc.) produces an LBZ-related low-toughness result, it is tempting to conclude that the test method is not structurally relevant. It must be remembered, however, that during this fracture test, all of the critical events listed above were _forced_ to occur simultaneously. The test itself may, in fact, be structurally relevant, but only for the situation where all critical events take place at one time. Before a test method is deemed irrelevant, we must gain a better understanding of the probabilistic aspects regarding LBZ-related structural failure. At this time, it is premature to discount any particular test method because of limited correlation with service performance.

The offshore industry will need to find solutions to prevent LBZ-related failures. The absence of platform failures suggests that the statistics are in favor of successful service performance. However, because there is a significant number of unknowns surrounding the LBZ issue, it would seem prudent to pursue a solution. A discussion of potential solutions follows.

Heat Input Control. It was mentioned previously that the metallurgical causes of LBZ's depend on weld thermal cycles (i.e., mainly heat input). It may be possible to avoid LBZ formation during platform fabrication by limiting heat input. The limits could be established for each steel by exploring various heat inputs and conducting CTOD tests during material or weld procedure qualification. Heat input limits, however, may have economic drawbacks due to reduced weld-metal deposition rates and slower fabrication. Also, when manual welding is used during fabrication, quality control may be a problem because the travel speed (an important heat-input variable) will be difficult to monitor.

Bead Placement Control. LBZ's can be eliminated or reduced in size by using certain bead placement techniques. During weld procedure qualification, techniques could be developed to maximize the number of beads that penetrate the plate edge (Fig. 7). Procedure acceptance might be in terms of successful CTOD testing. Potential drawbacks are in the area of quality control and practical enforcement of bead placement procedures during fabrication.

Large-Scale Significance. Although LBZ's have been shown to affect the outcome of CTOD tests, their effect on full-scale performance is not well documented. It is argued that LBZ's will be protected by overmatching weld metal and will not influence large-scale behavior.[20] Proving that LBZ's are insignificant is, of course, a potential solution. The wide-plate test seems to be the leading candidate for such studies. If the wide-plate test is used, certain test details such as notch geometry and placement, residual stress levels, etc. will need to be adequately addressed. Disadvantages will be the expense of large-scale testing and possible controversy surrounding test details. If LBZ's are shown to adversely affect the wide-plate test, then other resolutions to the issue will have to be pursued.

Statistical Analysis. Because LBZ-related failure is a statistical phenomenon, presumably an analysis could be conducted to predict the probability of failure. If the risk of failure were proven to be small, then platform operators might decide to accept the risk. Such an analysis, however, would be tremendously complex, and a large number of unknowns would have to be addressed.

Steel Selection. From the user's standpoint, one of the most efficient solutions may be to use a steel that does not produce low-toughness CG regions. Purchasing such a steel would eliminate the need for fabrication welding controls and the need to research the subject of LBZ significance and/or statistical risk. Also, there would be no risk of costly fabrication delays due to unacceptable toughness results that occur during weld procedure qualification. In fact, if LBZ resistance can be proven during material qualification, then weld procedure qualification HAZ CTOD testing becomes unnecessary. The disadvantage of the steel-selection approach would be the verification of LBZ resistance. The CTOD testing procedure necessary might be difficult to perform.

In closing, it is important to mention that the offshore industry may not be the only group to experience the LBZ phenomenon. It may, however, be the first to deal with the problem. Any industry using relatively thick, high-strength, low-alloy (HSLA) steels at a service temperature near the transition temperature may generate LBZ's. The pipeline and shipbuilding industries fit this definition. The most significant difference between the offshore industry and others in the area under discussion is probably plate thickness. Few structures require HSLA steels in the 75-125-mm (3-5-in.) range.

Conclusions

1. Local brittle zones (LBZ's) can cause linear elastic fracture during a CTOD test.

2. LBZ's are comprised of low-toughness, coarse-grain regions, which include the coarse-grain HAZ (CGHAZ), the intercritically reheated CGHAZ (IRCG), and the subcritically reheated CGHAZ (SRCG).

3. The main contributors to the low toughness of LBZ's are:

    a. Upper bainite.
    b. Precipitation.
    c. Large grain size.
    d. Martensite islands.

4. LBZ-related failure is a statistical phenomenon, and the critical conditions required for failure include:

    a. Using an LBZ-susceptible steel.
    b. Using a welding procedure that produces LBZ's.

c. The presence of a defect located in or near an LBZ.
d. Sufficient applied stress.
e. Occurrence of a low temperature.

5. Potential solutions to the LBZ issue include:

a. Using a bead placement technique or a heat input that precludes the production of LBZ's.
b. Proving that LBZ's do not affect large structures.
c. Proving that the risk of LBZ-related failure is acceptably low.
d. Selecting a steel that is not susceptible to LBZ formation.

References

1. A. H. Aronson, "The Weldability of Columbium-Bearing High Strength Low Alloy Steel," Welding Journal, June 1966, 266-271.

2. N. E. Hannerz and B. M. Johsson-Holmquist, "Influence of Vanadium on the HAZ Properties of Mild Steel," Metal Science, 8, 1974, 228-234.

3. N. E. Hannerz, "Effect of Cb on HAZ Ductility in Constructional HT Steels," Welding Journal, 54, 162-168.

4. J. Malcolm Gray, "Weldability of Niobium Containing High Strength Low Alloy Steel," WRC Bulleton 213, February 1976, 1-19.

5. R. E. Dolby, "The Weldability of Low Carbon Structural Steels," The Welding Institute Research Bulletin, August 1977, 209-216.

6. B. A. Graville and A. B. Rothwell, "Effect of Niobium on HAZ Toughness Using the Instrumented Charpy Test," Metal Construction, October 1977, 455-456.

7. E. G. Signes and J. C. Baker, "Effect of Columbium and Vanadium on the Weldability of HSLA Steels," Welding Journal, June 1979, 179-187.

8. R. E. Dolby, "Factors Controlling HAZ and Weld Metal Toughness in C-Mn Steels," Institution of Metallurgists, South Africa, Johannesburg, 1979, 117-134.

9. J. D. Harrison, "Why Does Low Toughness in the HAZ Matter?" The Welding Institute Seminar, Coventry, England, June 1983.

10. H. G. Pisarski and R. J. Pargeter "Fracture Toughness of Weld HAZ's in Steels Used in Constructing Offshore Platforms," Welding in Energy Related Products, Toronto, Canada, September 1983, 415-428.

11. C. Thaulow, A. J. Paauw, A. Gunleiksrud, and O. J. Naess, "Heat Affected Zone Toughness of a Low Carbon Microalloyed Steel," Metal Construction, February 1985, 94-99.

12. "HAZ Toughness of Offshore Steels," Nippon Steel Corporation, 1985.

13. T. Haze and S. Aihara, "Metallurgical Factors Controlling HAZ Toughness in HT50 Steels," IIW Doc. IX-1423-86, May 1986.

14. O. Grong and O. M. Akselsen, "HAZ Toughness of Microalloyed Steels for Offshore," Metal Construction, September 1986, 557-562.

15. Private communication from Dr. J. Y. Koo, Exxon Research and Engineering Company regarding unpublished research.

16. R. W. Hertzberg, _Deformation and Fracture Mechanics of Engineering Materials_, John Wiley & Sons, New York, 1976, 356-362.

17. J. Kudoh and H. G. Pisarski, "Exploratory Studies on the Fracture Toughness of Multipass Welds with Locally Embrittled Regions," The Welding Institute Research Report, No. 294, January 1986.

18. H. G. Pisarski and J. D. Harrison, "Fracture Resistance of Materials Used in Offshore Structures," Paper 2, Select Seminar: European Offshore Steels Research, November 27-29, 1978, Cambridge, England.

19. G. H. Eide, "Fracture Mechanics for Assessing the Safety Against Brittle Fractures in Offshore Installations," Welding Institute Seminar - COD Fact or Fiction, Sheffield, England, April 1980.

20. I. R. M. Denys, "Advances in Fracture Toughness Testing," presented at the OMAE Symposium, Houston, Texas, March 1-6, 1987, ASME Paper No. OMAE 87-659.

THE EFFECT OF DEFECT SIZE ON WIDE PLATE TEST PERFORMANCE

OF MULTIPASS WELDS WITH LOCAL BRITTLE ZONES.

Rudi M. DENYS

Rijksuniversiteit Gent - Belgium
Sint Pietersnieuwstraat, 41, B 9000 GENT.

Abstract

Nowadays, the surface fatigue precracked wide plate test is often used to assess the engineering significance of low CTOD values for the cases where the tip of the fatigue crack has intercepted the coarse grained HAZ (local brittle zones - LBZ). An important aspect of wide plate testing which often goes unrecognized is the effect of defect size on the ultimate wide plate test deformation behaviour. The purpose of this paper is to comment on the effect of both defect size and the proportion of LBZ's sampled to help to assess the results of surface notched welded wide plate tests. The evaluation is based on the requirement of gross section yielding in order to permit the factors important to wide plate test performance to be treated systematically. It is concluded that wide plate test performance is related to defect size and that for this reason due consideration should be given to this effect to obtain a full understanding of the behaviour of cracks located in regions of low fracture toughness as identified by CTOD testing.

Introduction

Standard CTOD testing of the HAZ fracture toughness in low carbon steel weldments indicates that, occasionally, low level CTOD values can be obtained. The low CTOD values are normally observed when the fatigue crack occupies a certain percentage of the CGHAZ (further denoted as local brittle zones – LBZ's). The engineering significance of such test data as well as the the critical amount of LBZ's or the critical length of an individual LBZ causing low CTOD values it is not precisely known. This problem is currently a subject of active research in Japan, in the USA, in the UK and in Belgium. Mainly on the basis of work under contract, it is attempted to establish the influence of the proportion of CGHAZ sampled by a fatigue crack tip on both CTOD and wide plate performance. Furthermore, it is pertinent to note that the results of wide plate tests are primarily used to assess the engineering significance of LBZ's [1].

Although the detailed results of these research efforts are not yet published, it is known that some wide plate test results indicate rather poor wide plate test behaviour, whereas the results of other wide plate tests indicate adequate wide plate test behaviour i.e. at the end of test (or at fracture), it could be demonstrated that gross section yielding occurred. A straightforward explaination for the difference in wide plate behaviour is not yet available. The difference in wide plate test performance depends certainly on either the proportion of LBZ's sampled or the individual length of each intercepted LBZ. The lowest strain at fracture will be seen when the fatigue crack tip occupies either a high amount of LBZ's or a long single LBZ. In this connection, it should be noted that the crack dimensions (i.e. its length as well as its depth) are important as well. It is believed, therefore, that the ultimate deformation behaviour of the wide plate test specimen is affected by the interaction between the proportion (or individual length) of LBZ's intercepted by the fatigue crack front and defect size.

Demonstration of Adequate Fracture Resistance via Wide Plate Testing.

The criterion to be used in assessing the results of a wide plate test depends on the conditions of service to be evaluated. In order to interpret the wide plate test result for service performance, it was proposed in the 1960's to measure the average (gross) strain over a considerable gauge length. For C-Mn pressure vessel steel grades the required gross strain has been set at a semi-arbitrary value of 0,5 % [2]. This value corresponds to about four times the yield strain[1],[2]. For higher strength steels (σ_Y > 350 N/mm^2), the four times the yield strain criterion implies that the required gross strain would be considerably higher than 0,5 %. To demonstrate adequate fracture resistance for higher strength steel weldments and because of the lack of any other practical criterion for the assessment of both elastic/ plastic and plastic material behaviour, it is proposed to use the Gross Section Yielding (GSY) concept[3]. The requi-

---

[1] – Experimental observations [2] suggested that the plastic strain at a nozzle with a SCF of about 2,5 loaded to 1,3 (hydraulic pressure test) x 2/3 yield pressure test conditions correspond with that value.
[2] – This criterium was also used for selecting materials for pressure vessels and storage tanks via Charpy V notch testing (cfr. BS 5500 App. D).
[3] – Remark that some authorities claim that GSY can be identified with plastic collapse. To the authors opinion, plastic collapse should be referred to when net section yielding occurs.

rement of GSY is satisfied when, before fracture initiation occurs in the wide plate test specimen, the gross stress in the remote (defect free) cross sections reaches the plate materials yield strength[4].

As the concept of GSY is a simple and straightforward approach for the evaluation of wide plate test performance, and because it also satisfies automatically the generally accepted survival criterion of "four times yield strain", it is logic to utilize the concept of GSY for an investigation into the effect of defect size on the wide plate deformation behaviour and thus wide plate test performance.

Effect of Defect Dimensions on Plain Plate Deformation Behaviour.

Before a quantitative description can be given of the effect of defect size on the performance of HAZ surface cracked wide plate test specimens, the basic facts about the effect of defect size on plain plate wide plate specimens behaviour will be discussed first. Ref.[3][5] provides the basis for such a description and the major findings of this data base will be discussed first.

Through thickness defects.

The most convienent way to appreciate the role of defect dimensions on the ultimate deformation behaviour of a plain plate wide plate test specimen is obtained from an analysis of through thickness notched wide plate specimens. With the help of Fig. 1a, which gives a schematic representation of the relationship between the gross stress at fracture versus defect length, it can be observed that for cracks smaller than a_{gy}, the plate material's yield strength will be reached in the plate sections remote from the plane of the crack.

Since the strain hardening characteristics of the crack tip material provoke the occurrence of GSY, the value of a_{gy} decreases for higher strength steels [3]. A simple quantification of the effect of the material tensile properties on a_{gy} can not be given, but is has been observed [3,4] that the ratio of the yield strength to the ultimate tensile strength V (= $_Y/_{UTS}$) is an useful parameter to evaluate this effect. For 30 mm thick plates it has been observed that in case :

(a) $V = 0,669$ ($_Y = 397$ N/mm^2), $a_{gy} = 28,0$ mm (T = $-20°$C)
(b) $V = 0,774$ ($_Y = 269$ N/mm^2), $a_{gy} = 14,5$ mm (T = $-50°$C)
(c) $V = 0,829$ ($_Y = 747$ N/mm^2), $a_{gy} = 9,5$ mm (T = $-55°$C)
(d) $V = 0,888$ ($_Y = 1061$ N/mm^2), $a_{gy} = 5,5$ mm (T = $-20°$C)
(e) $V = 0,898$ ($_Y = 776$ N/mm^2), $a_{gy} = 12,0$ mm (T = $-20°$C)

These experimental data illustrate quite clear that the value of a_{gy} for high strength steels decreases, whereas in turn a_{gy} further decreases as the temperature is lowered.

---

[4] - For situations where high stress concentrations have to be taken into account, the requirement of gross section yieling alone would be unconservative. For this reason, it is suggested in order to establish full confidence in the overall integrity of such structural details which contain a defect with the proposed dimensions, to supplement the GSY requirement with an additional gross requirement of (σ_{design} x SCF)/E.

[5] - The data base consists of some 500 through thickness as well as surface notched wide plate tests on plate material having a yield strength varying between 280 and 1050 N/mm^2.

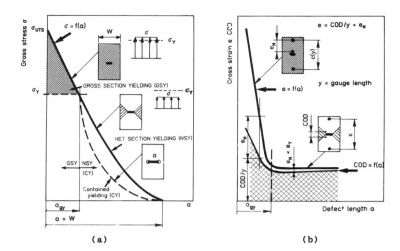

Figure 1 – Gross stress, gross strain and COD dependence of defect length a

It was further observed that the gross strain as well as the COD[6] at fracture are large when GSY occurred (Figs. 1a&b). This may be explained as follows: once GSY occurs, the defect is less constrainted and the conditions for a large crack tip opening are provided. At the same time the plastic deformation of the remote cross sections (defect free part of the test plate) provides a further contribution to the gross strain. The relationship between the (overall) gross strain e, the COD and the remote strain e_N (y = gauge length spanning the defect) can be cast in the following equation:

$$e = \frac{COD}{y} + e_N \qquad (1)$$

With the help of this equation it is easy to establish the link between gross strain and COD. For example, for the situation of long defects, the value of e_N is below the plate material's uniaxial yield strain e_Y, and the gross strain is directly propertional with COD. For short defects, not only the value of e_N is beyond yield but also, because of the reduced constraint at the crack tip, a larger COD will occur, so that the gross strain at fracture will be high.

Surface breaking defects

Surface breaking defects have, owing to their common occurrence in practice, a greater practical interest than through the plate thickness defects. Surface defects are characterized by their length and depth. A dis-

---

[6] – The term COD indicates the crack mouth opening as measured at the plate surface. Although the value of COD is not fully compatible with the commonly encountered definition of crack tip opening displacement, detailed measurement at the actual crack tip revealed that the use of COD does not alter the implications of the general findings reported hereinafter.

tinction can be made between (a) infinitely long and (b) semi-elliptical surface defects which have a finite length.

Infinitely long surface defects. Much of what has been said about through thickness defects also holds for infinite long surface defects (Fig. 2a). When the defect is deep, the plastic deformation will be confined to the ligament between the defect tip and the back plate surface (i.e. NSY or Contained yielding will occur); for shallow defects the situation of GSY can be obtained. The demarcation between both deformation patterns is now characterized by t_{gy}^7. It is obvious that for defect depths smaller than t_{gy} (1 = infinite), large strains and large COD's will be measured (Fig. 2b).

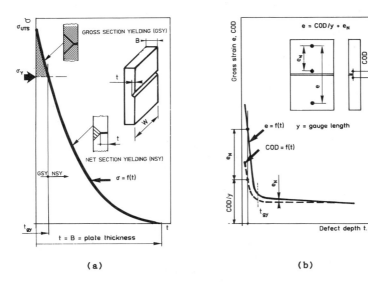

(a)   (b)

Figure 2 - Gross stress, gross strain and COD dependence of defect depth t. (infinite defect length)

Semi-elliptical surface defects. For the present study, the gross stress at fracture is presented as a function of defect depth (a similar presentation as a function of defect length is equally possible). As can be seen in Fig. 3, an infinite number of length/depth combinations can be studied. The position of the dashed lines in Fig. 3 correspond to the situations of an infinitely long (curve CF – see Fig. 2) and a short defect whose length is equal to a_{gy}. For the latter, point D corresponds with the situation of a through thickness defect. The solid curved line BXE represents one of the many parameter curves $\sigma = f(t)$ for a = cte. The point of intersection X indicates the demarction between GSY and NSY. The value of t_{gys} is smaller than that for an infinitely long surface defect. In itself, for each value of t_{gys} the corresponding defect length can be denoted by a_{gys}. Moreover, it has been observed that the principle of area equating applies, i.e. $t_{gys} \times a_{gys} = a_{gy} \times$ plate thickness.

---

7 – a_{gys} is the maximum length of a semi-elliptical surface defect, while t_{gys} is the maximum depth of the same semi-elliptical surface defect giving rise to GSY.

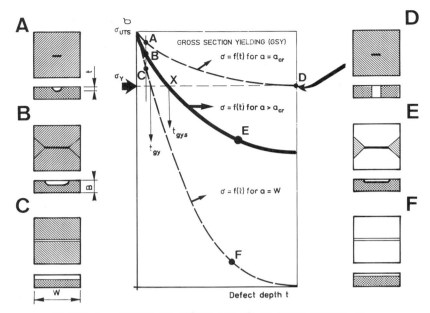

Figure 3 — Effect of defect length on gross stress as
a function of defect depth t.

The corresponding curves of gross strain versus defect depth and COD
versus defect depth are shown in Fig. 4. Two particular cases are
considered :
(a) For 'ductile' material behaviour, it can be seen in Fig. 4a that the
    gross strain as well as the COD for long semi-elliptical surface defect
    will be larger than for a long through thickness defect. The increase
    of gross strain and COD for the former situation is caused by an addi-
    tional contribution of plastic deformation of the ligament between the
    crack tip and the back face of the plate.
(b) For relatively brittle material behaviour (Fig.4b), this ligament does
    not contribute to the plastic deformation process and the position of
    the gross strain and COD curves for long semi-elliptical defects will
    nearly coincide with that of an infinitely long surface defect.

Welded Joints.

In welded joints, the problem is compounded by the non homogeneous
material properties sampled by the tip of the defect. Since it virtually
excluded that the leading edge of the defect samples either 100 % brittle
(e.g. grain coarsened) or 100 % ductile (e.g. fine grained or plate) material [7], the combined effects of ductile and poor material toughness on
wide plate test performance should be considered.

As it is excluded to produce wide plate test data for which the same
proportions of brittle material are sampled consisently and because there
is no precise information available on the amount of LBZ's which causes
poor wide plate behaviour, there is no alternative but modelling this situation in the context of present discussion. On the other hand, it should
also be pointed out that the experimental results presently available con-

324

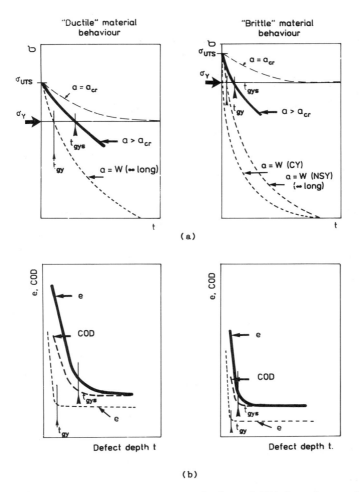

Figure 4 — Gross stress, gross strain and COD dependence of
defect depth for three typical defect lengths [3]

cerning the effect of LBZ's on wide plate test behaviour are highly diverse
(different plate thicknesses, crack length and crack depth, etc...) and not
always published to verify the next theoretical considerations.

Effect of defect depth on wide plate performance of welded plates.

The approach adopted in the following discussion is that it is possible to perform a series of wide plate tests in such a way that the leading
edge of a surface defect samples either a small (denoted as : lower incidence LBZ) or a high (denoted as : higher incidence LBZ) proportion of
LBZ's (Remark that the term LBZ can interchangeably be used with the terms
brittle material or grain coarsened — CGHAZ material). To facilitate the
comparisons, it is further assumed that the length of the defect is the
same for each of the above defined cases.

325

The ultimate tensile strength of the weld HAZ which, as a result of welding contains LBZ's, will not be affected drastically and will be characterised by that of the plate material. On the other hand, the amount of Lüders strain, the notch toughness and the yield strength of the weld HAZ will decrease [8]. The gross stress-crack depth curves for the lower and higher incidence LBZ cases cannot be represented by that of ductile material behaviour, instead, these curves will shift towards to left when compared with to the situation of ductile material behaviour. In Fig 5a, the σ_{uts} of both the lower and higher incidence LBZ cases is characterized by σ_{uts} of ductile material behaviour (point A). The position of the yield strength of both cases is not indicated, since it is only the plate material's yield strength which is of importance to verify whether GSY is obtainable. Furthermore, for the sake of simplification in presentation, the actual curved lines of gross stress versus defect depth are replaced by straight lines. The variation of the gross stress with defect depth is plotted for the following cases: (a) ductile material wide plate behaviour, line AB, (b) lower incidence LBZ wide plate behaviour, line AC, and (c) higher incidence LBZ wide plate behaviour, line AD. The accentuated solid lines AC and AD identify the occurrence of GSY. As indicated in Fig. 5a, the presence of LBZ's along the tip of the crack causes an decrease in t_{gy}, i.e. point B moves towards points C or D. The largest decrease will be observed for the higher incidence LBZ case (point D). Another look at the effect of defect depth effect can be obtained by comparing the deformation modes associated with increasing crack depth.

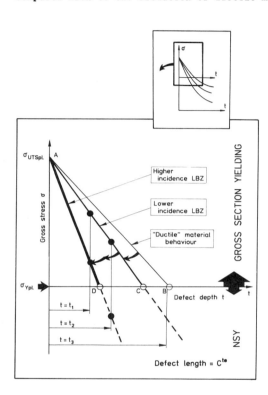

Figure 5a - The effect of LBZ on the relationship between the gross stress versus defect depth.

(a) For $t = t_1$, both the lower and higher incidence LBZ cases give rise to GSY.

(b) For $t = t_2$, the lower incidence LBZ case gives still rise to GSY, whereas the deformation mode for the higher incidence LBZ case will be either net section yielding (NSY) or contained yielding (CY).

(c) For $t = t_3$, GSY is no longer possible. At fracture, either net section (NSY) or contained yielding (CY) will occur.

A clearer picture of the implication associated with the change in deformation mode on wide plate performance can be obtained from the corresponding gross strain and COD curves plotted in Fig. 5b. The curves are

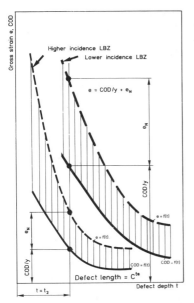

shown for both lower and higher incidence LBZ (remark that the COD curves for ductile and brittle behaviour are not plotted). It is apparent from an examination of the curves in Figure 5b that, depending on the deformation mode at fracture, there will be a large difference in gross strain and COD. For the lower incidence LBZ case, the gross strain will be larger than for the higher incidence LBZ case; a similar difference will be observed in COD. As indicated above, this difference could be explained by the reduced effect of strain hardening for materials with a high stress ratio. In this respect, compare the value of both strain and COD for $t = t_2$.

Figure 5b – The effect of LBZ on the relationship between the gross strain and COD versus versus defect depth

Effect of defect length.

Longer defects will shift the gross stress curves (lines) AC (lower incidence LBZ) and AD (higher incidence LBZ) towards lower defect depths (Fig. 6).

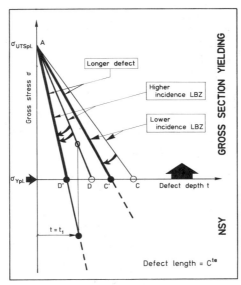

Figure 6 – Effect of defect length on t_{gys} (see also Fig. 5a)

Longer defects causes a change in deformation mode; e.g. for defect depth $t = t_1$, it can be seen in Fig. 6 that for the higher incidence LBZ case, GSY is no longer possible. Accordingly, the values of gross strain and COD will also decrease.

Effect of plate material properties.

Another variable that must be discussed at this point is the effect of the plate material properties. Tests on plate material revealed that an increase in yield strength lowers the defect size for GSY [3]. For welded joints made with high strength steels, the gross stress curves of Fig. 5a will shift to smaller defect depths. In other words, provided that a similar amount of LBZ's is intercepted by the tip of the defect, an increase of the plate material yield strength causes similar effects on wide plate test behaviour as increasing the defect depth (or length) when the yield strength of the parent plate remains unchanged.

Discussion

This semi-empirical study demonstrates/suggests that not only the proportion of brittle material sampled by the leading edge of the defect but that also the defect dimensions play a dominant role in the ultimate deformation of a wide plate test specimen.

The effect of defect dimensions on wide plate test behaviour is of primary importance when a substantial volume of brittle material is intercepted. In other words, the defect size which ascertained GSY of the parent alone seems to be a key factor in the evaluation of the engineering significance of localised areas of brittle material on structural performance.

Where it is wished to obtain GSY, it is expected that crack depths exceeding 20% of the plate thickness ("deep cracks") with an aspect ratio of approximately 0,10 to 0,15 will not produce satisfactory wide plate test behaviour. It might further be tentatively suggested that, provided the total length of all single LBZ regions does not occupies more than 15 to 20% of the fatigue crack front, the following defect sizes may yield successful wide plate test results:

Plate thickness 25 mm.
    Defect dimensions (depth x length) - 5 x 100 mm^2 (500 mm^2)
Plate thickness 50 mm.
    Defect dimensions (depth x length) - 10 x 200 mm^2 (2000 mm^2)
Plate thickness 100 mm.
    Defect dimensions (depth x length) - 20 x 250 mm^2 (5000 mm^2)

The suggested defect sizes are based on the consideration that they will give rise to GSY in a surface notched plain material wide plate test. They were arrived at by taking into account that (a) the maximum length of a through thickness defect for which GSY can be still achieved (a_{gy}) is about 25, 40 and 50/60 mm for 25 mm, 50 mm and 100 mm thick plates respectively and (b) by using the simple defect conversion based on area equation (i.e. plate thickness x a_{gy} for a through thickness defect = a_{gys} x t_{gys} for a surface defect). Furthermore, the estimate of the above defect dimensions are partly supported by the experience accumulated on offshore steel grades with a yield strength of about 350 N/mm^2.

The suggested percentages of intercepted LBZ's are believed to represent a realistic estimate when this percentages is composed of scattered and localised brittle material pockets separated by notch tough material.

Whether this observation still holds when one single long portion of brittle material is sampled is not known (effect of spacial variation). In the latter case, it is most likely that fracture will occur at lower strains, since the beneficial effects of relief of constraint caused by the adjacent ductile material is no longer warranted[8].

Accepting that the above indicated defect sizes yield satisfactory results, shallower defects will equally give adequate wide plate test results, because shallower defects provide less constraint and thus more favourable condition for plastic deformation. Moreover, in such cases, larger proportions of brittle material could be intercepted without initiating a low-strain brittle fracture. One particular wide plate test result obtained on a 70 mm thick plate reported in [5] tends to support this observation: an infinitely long surface defect (length 797 mm) having a depth of 5 mm (equivalent crack = 797 mm x 5 mm = 3985 mm^2 = e.g. 14 mm x 285 mm) and whose tip was located in the CGHAZ of the cap bead over nearly its whole length, showed a pop-in at a gross strain level of 2,47 % (crack tip opening was 1,0 mm). The unstable cracking extended over a length of 600 mm to a total depth of 10 mm. Notwithstanding this, the wide plate panel survived a gross strain of 3,13 % at the end of testing (no fracture).

For deep cracks, i.e. for the cases where yielding of the plain plate material alone can no longer be ascertained in the areas remote from the plane of the crack, local brittle zones will most likely cause premature fracture and thus result in poor wide plate performance.

At this stage, it is to be re-emphasized that to the author's experience defect dimensions which not exceeds the proposed ones, and provided the total amount of intercepted LBZ's is less than some 20% of the crack length, will yield a satisfactory wide plate test behaviour. However, only time and research will confirm whether this point of view is valid. It must further be pointed out that the aforementioned defect sizes reflect the behaviour of welded steel grades with a yield strength of about 350 N/mm^2. For higer strength steel weldments, the above defect size limitations are no longer applicable. Since the plain material value of a_{gy} has a fairly large influence on the potential criticality of local brittle zones, it is a small step to appreciate that in case of higher strength steels, not only the value of a_{gy} (and its derivates for surface defects obtained via area equating) will be lower, but that also the percentage of intercepted LBZ's should be kept small because of the reduced strain hardening characteristics of the plain plate material. This means also that, when both the defect size and the amount of brittle material sampled by the defect tip remains unaltered, wide plate behaviour in the case of medium strength steels (characterised by a pronounced yield plateau and a high strain hardening ratio) might be adequate, whereas for the case of high strength steels (characterized by a rather smooth stress-strain curve and a low strain hardening ratio) poor wide plate test performance might be seen.

The effects of defect size and the proportion brittle material sampled by the defect tip described above are assumed to be relevant for perfectly

---

[8] — Metallographic examination of the various microstructures sampled by the crack tip revealed that the crack tip opening of ductile material is mainly characterised by blunting, whereas the crack tip opening of the material identified as LBZ is associated with a deformation mechanism of slip towards the softer material side of the HAZ [7]. The latter crack opening mode is considered to be caused by the "constraining" effects of the adjacent ductile material, in which crack tip opening occurs more easily.

straight crack fronts. This assumptions is not in line with the experience. In reference [7], it has been observed that the likelihood of sampling a substantial volume of LBZ's is small. Because of the occurrence of weld bead waving along the length of the weld and because of the tendency of the fatigue crack path to deviate away from the brittle regions into softer material during propagation, the chance of sampling long portions of brittle material are considered to be small. Moreover, service cracks are often irregularly shaped so that the LBZ's will not be sampled at the deepest points along the leading edge of crack. In other words, it could be so that the plate material's maximum defect size which ascertain the occurrence of GSY is the main parameter in assessing the engineering significance of localised areas of brittle material on structural performance.

Last but not the least, a further factor affecting wide plate test performance is the possible benefit arising from overmatching weld metal yield strength relative to parent plate. Defects entirely contained within high strength weld metal (or the HAZ) will initially be protected from the effect of the remote stress. This protection will be less of the crack tip samples along its length a mixture of weld metal, HAZ and plate material. Also, other benefits accrue from this overmatching effect. Firstly, because of the overmatch, GSY will be more easily obtained. Secondly, relief of constraint[9] by the onset of gross section yielding will occur because the plastic deformation at the crack tip can extend more readily in material that is already stresses to the yield point. Provided that the size of the defect is such that all regions of the weld are not forced to deform equally, it is thus apparent that the effects of weld metal overmatching on wide plate performance consists of two different mechanisms :
   (a) because of the protection, the crack tip opening during the initial loading stage i.e. up to the plate material's yield strength will be small [8] and,
   (b) because of the occurrence of GSY, a subsequent increase in load will be associated with a large crack tip opening.

Obviously, this deformation mechanism will not take place if the total amount of intercepted LBZ's along the length of the crack is too large or when the LBZ's are sufficiently brittle to lead to fracture under elastic conditions. Anyhow, it is believed that the protection associated with the use of overmatched weld metal overmatching may positively contribute to the observed interaction between LBZ's and defect size.

Concluding Remarks

1. Recognising that the wide plate test is the best possible laboratory test configuration to assess to engineering significance of local brittle zones in welded joints implies that a reliable measure of the available fracture toughness is required to produce evidence of adequate wide plate test performance. The gross section yielding concept is considered to provide such a measure to assess the engineering significance for local brittle zones on weld joint performance.

2. In order to establish the significance of local brittle zones (LBZ) on weld joint performance, the results of the present semi-theorical study

---

[9] — Constraint depends on both the crack depth and the extend of plastic deformation (flow lines) to the front face (GSY) and below the crack (NSY). Obviously, this process can occur more readily in low strain hardening material and in thinner material sections.

illustrate that it is necessary to know (a) the yield properties in which the defect lies predominantly, (b) the strain hardening characteristics of the plate material, (c) the size as well as the shape of the defect, to allow a proper evaluation of the proportion of LBZ's which cause poor weld joint deformation behaviour.

3. If the results of a wide plate test are used to provide information on the significance of low CTOD values, it is necessary to stipulate both the defect dimension and the amount of LBZ's intercepted by the crack tip for which they have been obtained.

4. For practical purposes it would be more convienent to test fixed defect sizes (length x depth), for which it should be attempted to derive the maximum allowable percentage of brittle material giving satisfactory wide plate behaviour.

5. Provided that the defect sizes in testing covers the worst expected service defect, the findings of the present analysis lend further support to the view that the results of standard small scale fracture mechanics tests give a large margin of conservatism when account is taken of the many variables affecting structural performance. Beyond that, it is rather doubtful whether all these factors will to occur at the same time, so as to create the conditions for poor material behaviour.

6. There is clearly a need for engineering judgement when it is the aim to relate material toughness, applied stress and defect sizes. This expertise is apparently not always available, and this is most probably the main reason why the subject of local brittle zones is currently (over) emphasised.

Acknwoledgement.

The author wishes to thank A. Vinckier, professor and director of the Laboratory Soete for Strength of Materials and Welding Technology, Gent University − Belgium, for permission to publish this paper. The financial support of IWONL and NFWO is also acknowledged.

References

1. H.G. Pisarski and E.F. Walker, "Wide plate testing as a back up to the CTOD approach," TWI Report for the Department of Energy 3915/4/86.
2. F.M. Burdekin and M.G. Dawes, "Practical use of linear elastic and plastic fracture mechanics with particular reference to pressure vessels", Conf. on Practical Application of Fracture Mechanics to Pressure Vessel Technology, Inst. of Mech. Eng. Paper 5, May 1971.
3. R.M. Denys, "Plastische Breukmechanica", Doctors Thesis, Gent 1976 .
4. R.M. Denys, "The wide plate test and its application to acceptable defect", Conf. Fracture Toughness Testing − Methods, Interpretation and Application, TWI London, 1982.
5. R.M. Denys, and B. Musgen, "Gross Section Yielding crack tolerance of StE 36 and StE 70 steels, ibid ref. 4
6. R.M. Denys, A. Dhooge, A. Lefevre, "HAZ fatigue precracking of welded plate specimens", Conf. Proc. The state of the art in materials testing. Gent, Nov. 1986.
7. H.T. Hahn, W.S. Owen, B.L. Averbach and M. Cohen, "Micromechanism of Brittle Fracture in a Low −Carbon Steel", Welding Research Supplement, Sept. 59, No.9, pp. 367s−376s.
8. R.M. Denys, "The relevance of CTOD in cross welded joints with weld metal overmatching in strength", Conf. Proc. Welding for Challenging environments, (Pergamon Press ISBN 0-08-031866-5), Toronto, 1985.

Factors Controlling Properties of Weldments

THE EFFECTS OF Mn AND Si ON THE MICROSTRUCTURE AND PROPERTIES OF SMA STEEL WELD METAL

Stephen A. Court* and Geoffrey Pollard**

* Department of Materials Science and Engineering, University of Illinois, 1304 W. Green St., Urbana, IL 61801, U.S.A.

** Department of Metallurgy, University of Leeds, Leeds, LS2 9JT, West Yorkshire, ENGLAND.

Abstract

The microstructures within a range of multi-pass, shielded metal arc (SMA) weldments in C-Mn steel have been characterized and quantified using optical microscopy, and transmission electron microscopy (TEM) and associated microanalytical techniques. The samples were in two series, the first showing a variation in manganese of ≈0.9-1.8 wt.% at a silicon level of ≈0.7 wt.%; the second consisting of two deposits containing ≈1.4 wt.% Mn and ≈1.8 wt.% Mn, respectively, at a silicon level of ≈0.3 wt.%. Both elements were found to cause refinement of the microstructure and increased the proportion of acicular ferrite, consistent with their effect on increasing hardenability.

The yield stress, tensile strength and hardness of the deposits increased with increasing manganese and silicon, whereas the impact properties indicated that a correct balance of both elements is required to achieve high levels of notch toughness.

Microanalysis (EDX) revealed that the non-metallic inclusion composition could be related to that of the overall weld deposit. Specifically, the Ti content and the Mn/Si ratio of the inclusions could be related directly to that of the overall weld composition, as could the inclusion Mn content. Furthermore, varying the basicity index (B.I.) of the electrode coating was found to have little effect on inclusion composition.

Introduction

The factors which affect the strength, microstructure and toughness of C-Mn weld metal have been the subject of a recent review [1]. Weld metal strength is controlled principally by the degree of alloying, i.e. the level of elements such as Mn, Si, Ni and Mo, which exert a solid solution hardening influence, and the level of elements such as Nb, V, Ti and Mo, which in combination with C and/or N may exert a precipitation hardening effect. However, important contributions to strength are also made by the grain size and the dislocation substructure, although in weld metals a fine grain structure is required primarily to improve toughness rather than for high strength. A reduction in grain size is generally regarded as being beneficial to weld metal toughness, as the grain boundary area and the number of grain boundaries are increased, which serves to hinder the propagation of cleavage cracks. In this respect, it is generally accepted that to obtain a strong, tough welded joint, the deposit should contain a high proportion of fine-grained acicular ferrite [1-4]. High proportions of both grain boundary ferrite allotriomorphs and Widmanstatten ferrite sideplates are considered to be detrimental to toughness [1-3], as is the presence of bainite or martensite [3].

It has been suggested that the major alloying elements (e.g. C, Mn, Si and Ni) could influence toughness by affecting the deposit microstructure, solid solution hardening, and by modifying the deposit oxygen content by taking part in the deoxidation reactions [5]. In the present study, the effects of Mn and Si on the microstructure and properties of a series of shielded metal arc (SMA) deposits have been investigated.

Experimental

The welds investigated form part of a Welding Institute programme initiated to examine the effects of manganese on the microstructure and mechanical properties of vertical-up, SMA welds in thick section structural steel. All welds were produced with experimental, low hydrogen electrodes of the AWS E7016 type. Details of the welding procedure are given elsewhere [6,7]. The chemical analyses of the deposits, and the base plate composition (38mm thick BS 4360 50D, microalloyed with niobium (0.025wt.%Nb)) are given in Table 1. The welds had similar compositions other than for their Mn and Si levels. Welds W8,W5,W9,W6 and W7 show increasing Mn level at a constant, high, Si level of $\approx$0.7wt.%. Welds W9, W7, W15 and W20 are used to show the effect of Si at two Mn levels ($\approx$1.4wt.% and $\approx$1.8wt.%). Finally, welds W18, W19 and W20 were made with varying electrode coating basicity. In the first, high Si, series of welds the arc energy employed was $\approx$1kJ/mm, with a typical range of 0.8-1.2kJ/mm, whereas the other welds were deposited at a slightly higher arc energy, with a typical range of 1.1-1.4 kJ/mm.

Metallographic examination, both optical microscopy and TEM, utilised transverse sections and concentrated mainly on the as-deposited capping passes. The proportions of the various microstructural constituents in the as-deposited microstructure of each weld were determined at The Welding Institute (U.K.) using an optical microscopy/point counting method [6,7]. Mean linear intercept measurements were made to characterize the size of the prior austenite grains and the grain-refined, reheated regions. A similar technique utilising carbon extraction replica specimens was used to determine the grain size of the acicular ferrite. Thin foil and carbon extraction replica specimens were prepared using standard procedures, and were examined in a JEOL 200CX scanning transmission electron microscope (STEM), operating at 200kV and interfaced with a Link energy dispersive x-ray analysis (EDX) system. A STEM/EDX technique (described elsewhere [8]) was used to determine inclusion compositions of between 20 and 30 inclusions for each deposit.

The hardness (HV30) of each deposit was determined, as was the microhardness (using a 200g load) of the acicular ferrite in welds W8, W9 and W7. Details of the generation of the tensile data and the Charpy toughness evaluation are given elsewhere [6,7].

336

(A) Composition

From the chemical analyses (Table 1), it can be seen that other than the intentional variations in composition (i.e. in the levels of Mn and Si) there is little variation in the other elements. Specifically, there is no systemmatic reduction in oxygen content with increasing Mn, but the lower oxygen levels of the higher Si deposits confirm the deoxidation potential of Si. It can be seen, however, that a reduction in the weld metal Si has resulted in a reduction in the level of Ti from ≈0.01-0.02 wt.% at the high Si level (≈0.7%) to ≈0.005-0.01 wt.% at the lower level of weld metal Si (≈0.3%). This was unintentional, and was probably a consequence of the lower level of Si effecting a higher degree of deoxidation by Ti in the weld pool [7]. Furthermore, comparison of welds W18 (B.I.=1.8), W19 (3.5) and W20 (3.0) shows that no significant variations in deposit composition are observed for changes in the electrode coating basicity index (B.I.) [8]. Of particular interest in this respect is the oxygen content, which ranges from ≈300-330 p.p.m., and suggests that for the values of basicity index employed in this investigation, oxygen content is independant of electrode coating B.I..

Table 1.   Weld metal and base plate chemical analyses (wt.%).

Weld	Composition (wt.%)*								p.p.m	
	C	S	P	Si	Mn	Ti	Al	Cu	N	O
W8**	0.06	0.011	0.018	0.74	0.92	0.021	<0.005	0.03	88	272
W5	0.06	0.011	0.019	0.72	1.15	0.020	<0.005	0.03	110	250
W9	0.07	0.009	0.017	0.66	1.40	0.016	<0.005	0.03	112	295
W6	0.06	0.009	0.017	0.76	1.70	0.021	<0.005	0.03	90	230
W7	0.06	0.009	0.018	0.71	1.81	0.017	<0.005	0.03	89	271
W15	0.06	0.009	0.016	0.32	1.38	0.008	<0.003	0.02	110	360
W20	0.06	0.010	0.017	0.35	1.81	0.010	<0.003	0.03	130	330
W18	0.06	0.009	0.017	0.34	1.75	0.006	<0.003	0.04	120	300
W19	0.05	0.009	0.016	0.32	1.69	0.005	<0.003	0.03	110	320
Base Plate	0.14	<0.005	0.016	0.39	1.40	0.006	0.025	0.23	N.D.+	N.D.

*   Ni ≤0.04, Cr ≤0.06, Mo,V ≤0.01, Nb <0.005 (0.025wt.%Nb in base plate).
** B.I. = 3.0, except W18 (1.8) and W19 (3.5).
\+   N.D. = Not Determined.

(B) Metallography

The results of the metallographic analysis of welds from the first, high Si, series are shown in Figs.1-3. The quantitative data for the proportions of the individual microconstituents in the as-deposited microstructures are represented graphically in Fig.4, together with the results for the low Si deposits (welds W15 and W20) described below. These results show that the proportion of acicular ferrite increases with increasing deposit Mn content at the expense of the grain boundary nucleated ferrite morphologies, i.e. grain boundary allotriomorphs and Widmanstatten ferrite sideplates. It should be noted that under recent guidelines for microstructural characterization, these microconstituents would be termed grain boundary nucleated primary ferrite (PF(G)) and ferrite with aligned second phase (FS(A)), respectively [9]. The grain size of the columnar grains (prior austenite) increases with increasing Mn (see Fig.1). Examination of the acicular ferrite (Fig.2) shows a reduction in the size of the laths and an increase in its aspect ratio with increasing Mn. The reduction in the grain size of the fine-grained, reheated (grain-refined) regions as Mn is increased is shown in Fig.3. Quantitative data for the prior austenite grain size, the acicular ferrite grain size and the grain-refined regions are presented in Table 2, together with data from welds W15 and W20 at the lower level of weld metal Si.

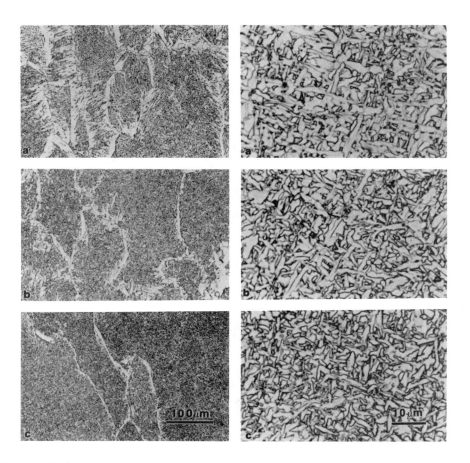

Fig.1. Optical micrographs taken from welds: (a) W8 (0.92% Mn); (b) W9 (1.40% Mn), and (c) W7 (1.81% Mn).

Fig.2. Optical micrographs of acicular ferrite in welds: (a) W8 (0.92% Mn); (b) W9 (1.40% Mn), and (c) W7 (1.81% Mn).

Table 2. Mean linear intercept measurements for the prior austenite (γ) grains, the acicular ferrite and the grain-refined regions.

Weld Identity	Mn wt.%	Si wt.%	Prior γ μm	Acicular Ferrite μm	Grain-Refined μm
W8	0.92	0.74	130	1.92	6.5
W5	1.15	0.72	140	N.D.*	6.5
W9	1.40	0.66	170	1.45	4.3
W6	1.70	0.76	185	N.D.	N.D.
W7	1.81	0.71	190	1.19	3.7
W15	1.38	0.32	N.D.	1.57	6.3
W20	1.81	0.35	N.D.	1.28	N.D.

* N.D. = Not Determined.

338

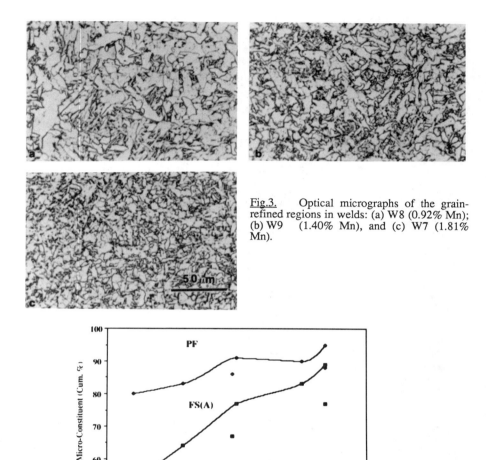

Fig.3. Optical micrographs of the grain-refined regions in welds: (a) W8 (0.92% Mn); (b) W9 (1.40% Mn), and (c) W7 (1.81% Mn).

Fig.4. The effect of manganese (and silicon) on the proportions of the various microstructural constituents in the as-deposited microstructures. The nomenclature used is that recommended in ref.9. The value of PF for all the high Si deposits (curves) includes 1% Intragranular Ferrite (PF(I)).

Transmission electron microscopy was used to further characterize the individual microconstituents. Typical micrographs showing the refinement of the acicular ferrite with increasing Mn are shown in Fig.5. Only at the highest level of Mn (1.81%) does the microstructure appear to be truly acicular in nature. The grain boundary ferrite allotriomorphs were often found to contain retained austenite microphase regions (Fig.6(a)) [10] or interphase cementite precipitation and grain boundary cementite films (Fig.6(b)) at the lower levels of Mn, whereas in the higher Mn deposits, the grain boundary ferrite appeared as thin, generally precipitate-free veins (Fig.7). Areas of Widmanstatten ferrite sideplates were evident at all levels

339

of Mn, and were invariably found to contain regions of retained austenite at inter-lath sites [10] (Fig.8). Examination of the intragranular microphase regions showed them to be predominantly lath martensite at low levels of Mn (Fig.9(a)), with some cementite present in weld W8 (0.92%Mn), whereas twinned martensite regions predominated at the higher levels of Mn (Fig.9(b)). However, various anomalies existed and twinned martensite microphases were observed in all the deposits. Furthermore, it was suspected that the majority of the microphase islands also contained retained austenite. From Fig.9, it can be seen that the microphases at the higher Mn level are more angular in shape and the overall impression was for the acicular ferrite to appear more lath-like.

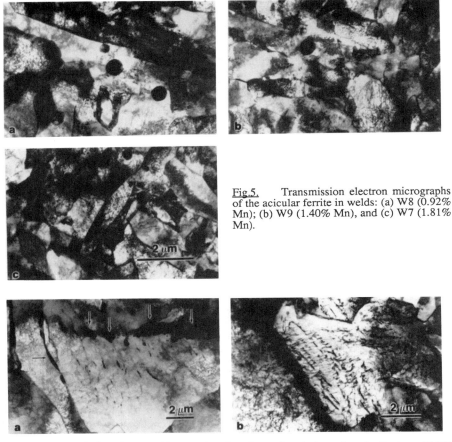

Fig.5. Transmission electron micrographs of the acicular ferrite in welds: (a) W8 (0.92% Mn); (b) W9 (1.40% Mn), and (c) W7 (1.81% Mn).

Fig.6. Transmission electron micrographs of regions of grain boundary ferrite in weld W8 (0.92% Mn) showing: (a) areas of retained austenite (arrowed), and (b) interphase cementite precipitation and grain boundary cementite films.

The results of the metallographic analysis of the second, low Si, deposits are shown in Figs.10-12. As at the higher level of weld metal Si, it is evident from Fig.10 that increasing the Mn content (from ≈1.4% to ≈1.8%) increases the proportion of acicular ferrite in the as-deposited microstructure at the expense of the grain boundary nucleated ferrite morphologies, and reduces its grain size (Figs.11 and 12). The quantitative data for the proportions of the individual microconstituents in the as-deposited microstructures are represented graphically in Fig.4.

340

Fig.7.   Transmission electron micrograph of a region of grain boundary ferrite in weld W7 (1.81% Mn).

Fig.8.   Transmission electron micrograph of a region of Widmanstatten ferrite sideplates in weld W8 (0.92% Mn), showing retained austenite at inter-lath sites.

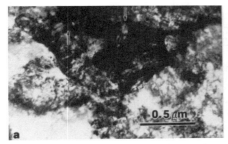 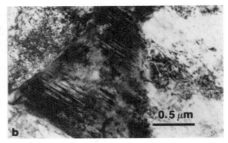

Fig.9.   Transmission electron micrographs of microphase regions in welds: (a) W8 (0.92% Mn) and (b) W9 (1.4% Mn), showing lath and twinned martensite, respectively.

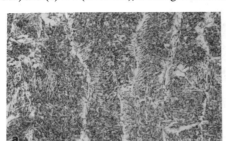 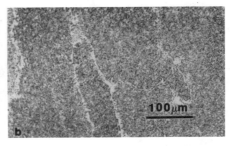

Fig.10.   Optical micrographs taken from welds: (a) W15 (1.38% Mn) and (b) W20 (1.81% Mn).

By comparing welds from the two series, it is possible to assess the effect of Si on microstructure. Firstly, considering the proportions of the microstructural constituents in the as-deposited microstructure, it can be seen from Fig.4 that lowering Si from ≈0.7% to ≈0.3% results in a decrease in the proportion of acicular ferrite by approximately 10% at the two levels of Mn. Comparison of the relevant micrographs in Figs.2 and 11 show that at the lower level of weld metal Mn, increasing Si reduces the acicular ferrite grain size, whereas at the higher level of weld metal Mn, the effect of increasing weld metal Si is less well pronounced. However, the quantitative data presented in Table 2 shows that there is a reduction in the grain size of the acicular ferrite with increasing Si at both levels of weld metal Mn. The effect of Si on the aspect ratio of the acicular ferrite is unclear. At the lower level of weld metal Mn, increasing Si appears to effect an increase in the aspect ratio of the laths, whereas at the higher level of Mn, any effect of increasing Si on acicularity is not immediately obvious. From Table 2 (c.f. welds W15 and W9), it can be seen that increasing Si also reduces the grain size of the grain-refined regions. As with increasing Mn, the nature of the microphase regions changed from lath martensite to twinned martensite as Si was increased from ≈0.3% to ≈0.7%, at the lower level of weld metal Mn.

Electrode coating basicity was found to have little effect on weld metal microstructure and mechanical properties, although it is recognised that these observations for the range of B.I. will be specific to the coating formulations employed. The results from this part of the investigation are discussed in detail elsewhere [7].

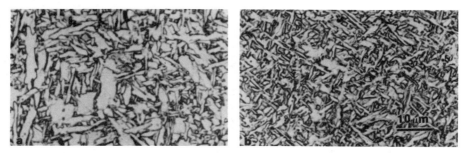

Fig.11. Optical micrographs of acicular ferrite in welds: (a) W15 (1.38% Mn) and (b) W20 (1.81% Mn).

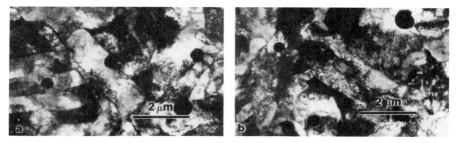

Fig.12. Transmission electron micrographs of the acicular ferrite in welds: (a) W15 (1.38% Mn) and (b) W20 (1.81% Mn).

(C) Mechanical Properties

The tensile and hardness data are summarised in Table 3, and the tensile data are presented graphically in Fig.13. It can be seen that there is a progressive increase in yield stress (Y.S.), tensile strength (T.S.) and hardness with increasing Mn at the expense of only a small reduction in ductility, at both levels of weld metal Si. On assuming the tensile properties to be linearly related to Mn, the following regressions were obtained for the high Si deposits:

Y.S. = 356 + 142 Mn (MPa).
T.S. = 453 + 133 Mn (MPa).

At both levels of weld metal Mn, the yield stress, tensile strength and hardness increased with increasing Si, the increase generally being the greater in the deposits containing ≈1.4% Mn (Fig.13).

The Charpy Impact results are presented graphically in Figs. 14 and 15. Values of absorbed energy read from full Charpy transition curves [6,7], at temperatures of -20°C and -40°C, versus weld Mn content are shown in Fig.14. For the high Si deposits, it can be seen that notch toughness increases with increasing Mn up to a level of ≈1.7%Mn, after which the toughness falls-off. At the lower level of Si, increasing Mn from ≈1.4% to ≈1.8% also leads to an increase in toughness. It can also be seen that increasing Si from ≈0.3% to ≈0.7% leads to an increase in toughness at the lower (≈1.4%) Mn level (although slight at the -40°C criterion), whereas at the higher level of weld metal Mn (≈1.8%), increasing Si is detrimental to toughness. The peak in toughness at ≈1.7% Mn in the high Si deposits is also apparent in Fig.15, which shows values of temperature read from full Charpy transition curves [6,7], for an absorbed energy of 27J and 40J, versus weld metal Mn. There is also an increase in toughness with increasing Mn at the lower level of weld metal Si (although slight at the 27J criterion). From Fig.15, however, it can be seen that the effect of increasing Si is more complex. At the lower level of Mn, increasing Si has little effect on toughness (although slightly detrimental), whereas at the higher level of Mn, increasing Si increases toughness at the 27J criterion and leads to a decrease in toughness at the 40J criterion.

Table 3.   Tensile and hardness data.

Weld Identity	Yield Stress MPa	Tensile Strength MPa	Elongation %	Hardness HV30	Microhardness HV(0.2)*
W8	495	587	30	220	181
W5	512	597	28	236	N.D.**
W9	555	649	27	245	191
W6	593	668	26	252	N.D.
W7	622	699	28	254	196
W15	483	567	31	215	N.D.
W20	582	644	25	228	N.D.

* HV(0.2) = Hardness of the acicular ferrite (200g load).
** N.D. = Not Determined.

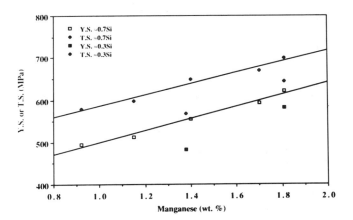

Fig.13.   Tensile data as a function of manganese content.

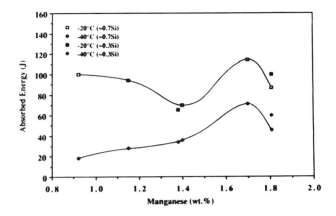

Fig.14. The effect of manganese content on Charpy toughness, showing the absorbed energy at temperatures of -20°C and -40°C.

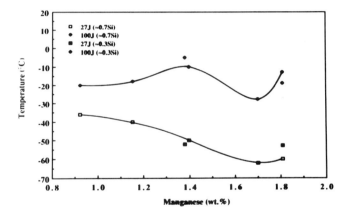

Fig.15. The effect of manganese content on Charpy toughness, showing the temperature for absorbed energies of 27J and 40J.

(D) Inclusion Analysis

During the course of the investigation, many examples of an inclusion/acicular ferrite association were observed (e.g. Fig.16), consistent with the many other previous observations of the inclusion nucleation of acicular ferrite [see ref.1]. Microanalysis of the inclusions within all the deposits generally showed them to be multiphase particles rich in Mn, Si and Ti, often with lower levels of Al, Cu and S, as observed previously [11,12]. The results of the investigation of inclusion compositions within individual welds are presented in Table 4, the values in parentheses represent the percentage element in wt.% in each deposit. The relation between weld metal composition and inclusion composition is most evident for Ti, the Ti level in the inclusions increasing with increasing weld metal Ti. It also appears as though the inclusion Mn level

344

increases with increasing weld metal Mn, for a given level of weld metal Si and electrode coating basicity index. By comparing welds W20, W18 and W19, it is apparent that a variation in electrode coating basicity has had little effect on inclusion composition; the slightly higher Ti level of the inclusions in weld W20 is consistent with the increased deposit Ti. The deposit and inclusion Mn/Si ratios are also presented in Table 4. Considering welds W8, W9 and W7, the inclusion Mn/Si ratio increases as the deposit Mn/Si ratio increases, as is the case for welds W15 and W20 at the reduced level of weld metal Si, all of them being at the same basicity.

Table 4. Inclusion compositions and Mn/Si ratios.

Weld	B.I.*	Apparent Composition (%)				Mn/Si ratio	
		Mn	Si	Ti	Others+	Weld	Inclusions
W8	3.0	24 (0.92)**	27 (0.74)	41 (0.016)	8	1.24	0.89
W9	3.0	48 (1.40)	28 (0.66)	20 (0.011)	4	2.12	1.71
W7	3.0	45 (1.81)	20 (0.71)	27 (0.012)	8	2.45	2.25
W15	3.0	55 (1.38)	29 (0.32)	10 (0.008)	6	4.31	1.90
W20	3.0	59 (1.81)	26 (0.35)	10 (0.010)	5	5.17	2.27
W19	3.5	59 (1.69)	31 (0.32)	5 (0.005)	5	5.28	1.90
W18	1.8	60 (1.75)	32 (0.34)	5 (0.006)	3	5.15	1.87

* B.I. = Basicity Index (after ref.8).
+ Others = Al, Cu, S.
** The values in parentheses represent the wt.% element in the deposit.

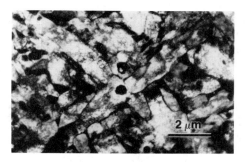

Fig.16. Transmission electron micrograph of a nucleant inclusion from weld W7 (1.81% Mn).

Discussion

The influence of Mn on the as-deposited microstructure was similar to that observed in previous investigations [e.g.5,13,14], in that the proportion of acicular ferrite increased at the expense of the grain boundary nucleated ferrite morphologies, both grain boundary ferrite and ferrite sideplates, with increasing Mn. Furthermore, as deposit Mn was increased there was a refinement of the acicular ferrite grain size and an increase in its acicularity (i.e. there was an increase in the aspect ratio of the laths), and also a reduction in the grain size of the grain-refined regions. The microphase islands were found to be predominantly lath martensite (with some cementite) at the low levels of Mn (say ≤1.2% Mn), whereas at the higher levels of Mn, twinned martensite regions predominated. Qualitatively, it also appeared as though the proportion of microphase increased with increasing Mn. These observations are consistent with the effect of Mn on hardenability, in that at a given cooling rate it promotes the formation of the lower temperature transformation products, i.e. nucleation occurs at larger undercoolings as Mn is increased.

The observed increase in yield stress, tensile strength and hardness with increasing Mn can be accounted for in terms of solid solution hardening and the grain refinement described above. The Charpy test results showed that, in agreement with previous observations [e.g.2-4], high levels of toughness were associated with a high proportion of acicular ferrite in the as-deposited microstructure. In the high Si deposits (≈0.7% Si), maximum toughness was achieved at a Mn content of ≈1.7%. The increasing toughness can be related to the microstructural refinement that occurs with increasing Mn, although it would be expected that the morphology and distribution of the microphase regions, which change from lath to twinned martensite islands, would also contribute towards the improved toughness [3]. Above this Mn content the toughness deteriorated, which can be accounted for in terms of increasing strength with little or no further microstructural refinement [2]. Taylor [14] has also suggested that the formation of small pockets of martensite; excessive solid solution hardening, and increased segregation of Mn to dislocations may reduce the toughness. It seems likely that all these factors would have some effect in the deposits of this investigation. The cementite interphase precipitation, which was most evident at the lower levels of Mn would not be expected to have had an adverse effect on toughness. In the lower Si deposits (≈0.3% Si), Charpy toughness increased as Mn was increased from ≈1.4% to ≈1.8%.

In its role as a deoxidising element Si was seen to effect a decrease in the weld deposit oxygen content of ≈60 p.p.m. as Si was increased from ≈0.3% to ≈0.7%. This is consistent with the findings of previous investigations [e.g.15]. A microstructural refinement was also observed as deposit Si was increased, consistent with previous observations of the beneficial effects of Si on microstructure [e.g.15-17]. Increasing Si from ≈0.3% to ≈0.7% was found to increase the proportion of acicular ferrite in the as-deposited microstructure by ≈10% at the two levels of Mn, at the expense of both the grain boundary ferrite and ferrite sideplate morphologies. Furthermore, a reduction in the grain size of the acicular ferrite was observed. These observations are similar to those of Abson [17], although Evans [15] observed no decrease in the acicular ferrite grain size with increasing Si. Evans did, however, observe an increase in the aspect ratio of the acicular ferrite laths (i.e. increased acicularity) with increasing Si, but in the present investigation, it is not clear whether such an effect exists. The grain size of the grain-refined regions also decreased with increasing Si, which is consistent with the findings of Tuliani et al. [8]. These microstructural differences are attributed to the increased Si, the slightly higher arc energies of the deposits containing ≈0.3% Si probably made only a small contribution. From Table 2, it can be seen that the acicular ferrite grain-refining potential of Mn is between two to three times greater than that of Si. The microphase islands were found to be predominantly lath martensite at the lower level of Si, whereas at the higher level of Si, twinned martensite predominated. Although having the most adverse effect on toughness [3], the former would be expected to constitute a smaller volume fraction of the microstructure, and so their effect on toughness is uncertain. These microstructural observations are again consistent with the effect of Si on hardenability, in that it promotes the lower temperature transformation products. Although it has been suggested that Si is a promoter of sideplate structures [18], this does not appear to be the case in this investigation, possibly because the high hardenability of the ≈0.7% Si deposits outweighs all other factors.

Silicon increased the yield stress, tensile strength and hardness at the two levels of Mn examined. This can again be accounted for in terms of solid solution hardening and grain refinement effects. The Charpy toughness results show that the effect of weld metal Si is complex, and is dependant upon both the level of weld metal Mn, and the criteria by which the toughness is assessed.

Summary and Conclusions

The results pertaining to the effects of manganese and/or silicon on weld metal microstructure and mechanical properties have shown the following to occur with increasing manganese and/or silicon content:

- The proportion of acicular ferrite in the as-deposited regions increased at the expense of the grain boundary nucleated ferrite morphologies.
- The grain size of the acicular ferrite decreased.
- The grain size in the grain-refined regions decreased.
- The morphology of the microphase islands changed from being predominantly lath martensite (with some cementite at the lowest levels of Mn) to twinned martensite.

- The yield stress, tensile strength and hardness increased.

In addition, the following results were obtained which were specific to an increase in either the weld metal manganese or silicon content:

- With increasing Si, the weld metal oxygen content decreased.
- With increasing Mn, the prior austenite grain size increased.
- With increasing Mn, the aspect ratio of the acicular ferrite increased.
- With increasing Mn at the ≈0.7% Si level, the notch toughness increased up to an optimum level of Mn of ≈1.7%, and then deteriorated. Increasing Mn from ≈1.4% to ≈1.8% at the ≈0.3% Si level increased the notch toughness.

The Charpy results generally indicated that a correct balance of both Mn and Si is required to achieve high levels of notch toughness. The highest values of toughness were achieved in weld W6 (1.7%Mn, 0.76%Si), quite independently of the criteria used in assessing the toughness.

The results of the inclusion analyses generally showed them to be rich in Mn, Si and Ti, often with lower levels of Al, Cu and S [10,11], and a good correlation was observed between the deposit and the average inclusion compositions.

Acknowledgements

At the time the work was carried out, both authors were in the Department of Metallurgy at the University of Leeds. One of us (S.A.C.) wishes to thank The Welding Institute, U.K. and the Science and Engineering Research Council (SERC) for financial support, and Professor J. Nutting for provision of the research facilities. The authors are particularly grateful to Dr. D.J. Abson of The Welding Institute for permission to use some of his data.

References

1. D.J. Abson and R.J. Pargeter, Int. Met. Rev., 1986, vol. 31 (4), pp. 141-194.
2. R.E. Dolby, Factors Controlling Weld Toughness - The Present Position. Part II - Weld Metals, The Welding Institute Research Report, 1976, # 14/1976/M.
3. J.G. Garland and P.R. Kirkwood, Metal Construction, 1975, vol. 7 (5), pp. 275-283 and vol. 7 (6), pp. 320-330.
4. Y. Ito, M. Nakanishi and Y. Komizo, The Sumitomo Search, 1979, vol. 21, pp. 52-67.
5. G.S. Barritte and D.V. Edmonds, in Proc. Conf. "Advances in the Physical Metallurgy and Applications of Steels", The Metals Society, London, 1982, pp. 126-135.
6. D.J. Abson, The Influence of Manganese and Preheat on the Microstructure and Mechanical Properties of Vertical-up MMA Welds in 38mm Thick Steel Plate-Part 1, The Welding Institute Research Report, 1982, # 194/1982.
7. D.J. Abson, The Influence of Manganese and Coating Basicity on the Microstructure and Mechanical Properties of Vertical-up MMA Welds in 38mm Thick C-Mn-Nb Steel Plate-Final Report, The Welding Institute Research Report, 1986, # 309/1986.
8. S.S. Tuliani, T. Boniszewski and N.F. Eaton, Welding and Metal Fabrication, 1969, vol. 37 (8), pp. 327-339.
9. "Guidelines for the Classification of Ferritic Weld Metal Microstructural Constituents Using the Light Microscope", Welding in the World, 1986, vol. 24 (7/8), pp. 145-148.
10. S.A. Court, "The Structure of Shielded Metal Arc Steel Weldments", Ph.D. Thesis, University of Leeds (U.K.), 1985.
11. S.A. Court and G. Pollard, to be submitted to "Metallography".
12. S.A. Court and G. Pollard, J. Mater. Sci. Letts., 1985, vol. 4, pp. 427-430.
13. G.M. Evans, Weld. J. Res. Supp., 1980, vol. 59 (3), pp. 67-s-75-s.
14. D.S. Taylor, Welding and Metal Fabrication, 1982, vol. 50 (9), pp. 452-460.
15. G.M. Evans, Metal Construction, 1986, vol. 18 (7), pp. 438-444.
16. S.S. Tuliani and R.A. Farrar, Welding and Metal Fabrication, 1975, vol. 43 (7), pp. 553-558.
17. D.J. Abson, Welding Research International, 1979, vol. 9 (5), pp. 1-23.
18. R.C. Cochrane, Welding in the World, 1983, vol. 21 (1/2), pp. 16-24.

EFFECT OF NITROGEN ON THE TOUGHNESS

OF HSLA WELD DEPOSITS

T.W. Lau*,   M.M. Sadowsky**,   T.H. North***,   G.C. Weatherly****

*   Research Associate, Welding Institute of Canada, Oakville, Ont.
**   Graduate Student, University of Toronto, Toronto, Ont.
***   WIC/NSERC Chair in Welding Engineering, University of Toronto, Ont.
****   Professor, Department of Metallurgy & Materials Science, University of
        Toronto, Ont.

Abstract
The microstructure and toughness of Ti-B containing weld metal can be effectively modelled assuming the sequence of interactions is titanium/oxygen, titanium/nitrogen and then boron/nitrogen. As weld nitrogen content rises the titanium content required to provide optimum toughness values is increased. When the titanium/oxygen balance is such that titanium is not available for interaction with nitrogen, boron controls the mechanical properties and optimum toughness occurs when $B = 0.7(N) + 8$ (ppm).

Nitrogen contents exceeding 70 ppm produce a marked change in acicular ferrite content, and in FATT values.

349

Introduction

The final nitrogen content of submerged arc weld metal depends on the nitrogen content of the plate and filler wire materials, and on nitrogen absorbed from air entrapped in the flux burden. The method of plate manufacture has a major influence on the nitrogen content. Canadian continuously cast steels generally contain around 80/100 ppm nitrogen while special steels for sour gas applications have higher nitrogen levels (up to 140 ppm) as a direct result of the calcium silicide injection treatments involved in sulphide-shape control. Even the highest quality Japanese steels having nitrogen contents in the range 20/40 ppm have increased nitrogen levels as a result of sulphide-shape control treatment.

Not only is deposit nitrogen content affected by plate and filler wire compositions, but also by the choice of welding parameters employed during fabrication. Factors such as flux burden height, flux composition and particle size distribution, and welding speed markedly influence final deposit nitrogen content. Deposit nitrogen content increases at higher welding speeds, as the particle size distribution coarsens, and the burden height is lowered, and when high nitrogen content tack welds are melted during welding operations (1)(2)(3). Changing to high productivity welding techniques can exacerbate nitrogen problems, e.g. 3-electrode tandem welding at 1.45 m/min produces higher weld nitrogen contents than 2-wire tandem welding at 0.8m/min (2).

Titanium-boron containing welding consumables represent the ideal method of optimizing toughness properties in high heat input welding situations. Increasing deposit nitrogen content has a strongly detrimental effect on toughness properties (4)(5)(6)(7). Given this fact, how can weld metal toughness be optimized when the plate material contains 70-90 ppm nitrogen? Furthermore, how should the welding consumables be formulated to provide the ideal balance of titanium, boron, oxygen and nitrogen in weld metal? Also, much recent research has indicated a major influence of plate aluminum on final weld toughness, viz: too high an aluminum level for a given flux oxygen potential produces very deleterious effects (8)(9)(10)(11). As aluminum is a much more effective deoxidant than titanium, variation of plate aluminum content will decrease the titanium content tied up in oxide inclusions, and directly affect toughness since the titanium/oxygen and titanium/boron interactions are modified.

This paper evaluates the effect of systematic changes in nitrogen content on the microstructure and mechanical properties of Ti-B containing deposits with the aim of answering the questions posed above. A prime requirement of this program was the development of a technique capable of varying nitrogen content independently of other important alloying elements.

Background

Much research has shown a clearcut relation between deposit microstructure, mechanical properties and oxygen content. In the case of nitrogen, two viewpoints have been put forward to explain the detrimental influence of deposit nitrogen content on toughness properties, viz: microstructure-independent and microstructure-dependent views. For example, Den Ouden(12) and Horii(13) have shown that the toughness decrement due to increasing nitrogen content is similar in titanium-boron containing and in titanium-boron free welds. In effect, the prime control vis-a-vis weld toughness is soluble nitrogen content, not deposit microstructural changes.

In the microstructure-dependent approach(3)(4) the prime roles of boron are thought to be the control of the austenite/ferrite transformation (minimizing polygonal ferrite content) and BN formation. Titanium protects the boron from oxidation, and BN formation in preference to the more thermodynamically favourable TiN formation is explained by means of an inter-dendritic segregation effect. Both Mori(4) and Kawabata(3) noted nitrogen content had no effect on deposit microstructure (acicular ferrite content) for nitrogen changes from 45 to 93 ppm(4) and 40 ppm to 80 ppm (3). However, Okabe(14) observed a decrease in acicular ferrite content for a nitrogen content variation in the same range (29 to 79 ppm).In welds free of Ti-B additions increasing nitrogen content was associated with higher contents of M-A phase, not with changes in acicular or polygonal ferrite content(15)(16).

This paper evaluates the effect of nitrogen variations on deposit microstructure and mechanical properties when using a single welding consumablecombination. The welding consumable combination employed involves titanium supply from the electrode wire and boron (borate reduction) from the flux. Since the titanium/oxygen interaction is so critical to the performance of Ti-B containing welding consumables, repeat tests were carried out using a filler wire having a very low titanium content. This effectively allowed comparison of two different welding situations, one where boron was active at prior austenite grain boundaries and one where boron was inactive due to boron oxidation during welding.

Experimental

a) Nitrogen Variation

Nitrogen variation was accomplished by:

. making gas-tungsten-arc melt runs on HSLA plates containing known nitrogen contents (plate A containing 70 ppm N, plate B containing 40 ppm N, and plate C containing 30 ppm N). Gas-tungsten-arc shielding gases comprised argon - $2\%CO_2$/nitrogen mixtures containing 50%, 55% and 60% nitrogen by volume. Different combinations of welding speed and/or nitrogen content in the shielding gas allowed close control of the final weld zone nitrogen level.

. remelting of the GTA melt runs using submerged arc welding. A single flux formulation (SAF 72) was used throughout. Two electrode chemistries were employed (electrodes A and B in Table I).

Four test series were examined, viz:

Series A   Bead-on-plate welding on plate A using electrode A at 550 amp, 28 volt, 300 mm/min.

Series T   Bead-in-groove welding on plate B using electrode A at 550 amp, 28 volt, 280 mm/min.

Series N   Bead-in-groove welding on plate B using electrode B at 550 amp, 28 volt, 280 mm/min.

Series S   Bead-in-groove welding on plate C using electrode B at 575 amp, 28 volt, 250 mm/min.

351

The flux formulation employed in all tests had the formulation, 19/24%SiO$_2$, 16/21%Al$_2$O$_3$, 9.5/14.5%TiO$_2$, 17/23%MnO.

Table I: Compositions of Base Plates and Filler Wires

	C	Mn	P	s	Si	V	Nb	ASA	Mo	Ti
Base Plate A	0.07	1.40	0.015	0.005	0.30	0.07	0.02	0.04	–	0.01
Base Plate B	0.13	1.45	0.004	0.005	0.30	0.083	0.030	0.060	–	0.015
Base Plate C	0.08	1.47	0.009	0.002	0.21	–	–	–	–	0.017
E'rode Wire A	0.13	1.13	0.011	0.015	0.44	–	–	0.017	0.49	0.19
E'rode Wire B	0.097	1.02	0.011	.011	0.05	–	–	0.01	0.49	0.02

The technique for nitrogen variation in each test series was extremely successful. In each of the T and A series the weld metal hardenability was constant, if the effect of variability in nitrogen content is discounted. In the N-series both nitrogen and oxygen varied (over a nitrogen range from 37 to 73 ppm). The nitrogen/oxygen relation in the N-series resulted from the technique used to vary nitrogen across a wide range. The GTA technique employed a shielding gas comprising argon-2%CO$_2$/nitrogen and changes in welding speed led to both nitrogen and oxygen content variations(17). In the case of T-series weld deposits, higher titanium and silicon in the electrode filler wire effectively deoxidized the weld pool and counteracted any effect due to GTA melt run oxygen content variations.

This effect of deoxidants in gettering oxygen during welding was used to solve the nitrogen/oxygen variations existing in the N-series deposits. 0.5wt% aluminum powder was added to the flux formulation prior to submerged arc welding using electrode B and plate C (see Table I). Fig. 1 shows that this technique was successful in permitting nitrogen variation without concomitant changes in deposit oxygen content.

Fig. 1. Oxygen/nitrogen relation in S-series weld deposits:
▲ with zero aluminum added to the flux
△ with 0.5wt% aluminum in the flux formulation

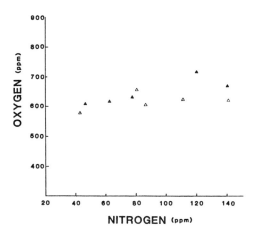

It is important to emphasize that the high titanium (and silicon) contents of electrode A ensure that boron is adequately protected from oxidation during welding. In this case active boron is available at prior austenite grain boundaries during weld solidification. However, in N and S series weld deposits a very low titanium content filler wire was employed (electrode B) and the boron analyzed must be considered as oxidized boron in weld deposits. The importance of this active and inactive boron content in series A and T inactive boron, and in series N and S weld deposits, will be discussed later.

b) Chemical Analysis

Final weld metal chemistries are given in Table II. The test series cover the range from 33 to 192 ppm nitrogen.

Weld oxygen and nitrogen levels were analyzed using the LECO inert gas fusion process. Boron was analyzed using wet chemical analysis techniques. Oxygen and nitrogen values were the average of five analyses on each deposit. The scatter observed in oxygen values was typical of that in submerged arc welds.

Soluble nitrogen content was evaluated using the hot extraction method developed by Kawamura(22) and has been detailed elsewhere (23). This technique provides soluble nitrogen values which must be considered as lower boundary values since the thermal cycle involved might allow pre-precipitation clustering and/or very fine precipitate formation during testing.

Table II - Weld Metal Analyses

Sample	C	Mn	P	S	Si	Cu	Ni	Cr	Mo	V	Nb	ASA	Ti	B	O(ppm)	N(ppm)
NA	0.098	1.27	0.009	0.007	0.23	0.025	0.012	0.008	0.201	0.047	0.013	0.014	0.014	0.0023	626	37
ND	0.092	1.29	0.012	0.006	0.23	0.029	0.013	0.009	0.215	0.045	0.013	0.017	0.013	0.0024	664	44
NE	0.094	1.25	0.012	0.007	0.21	0.027	0.015	0.010	0.207	0.046	0.013	0.015	0.014	0.0022	692	63
NF	0.090	1.27	0.010	0.007	0.23	0.021	0.012	0.009	0.190	0.051	0.014	0.016	0.012	0.0023	705	73
NM	0.091	1.33	0.013	0.006	0.25	0.027	0.012	0.008	0.192	0.050	0.015	0.017	0.014	0.0021	775	102
NK	0.092	1.28	0.011	0.006	0.25	0.021	0.011	0.011	0.180	0.048	0.015	0.015	0.014	0.0019	752	138
NJ	0.099	1.28	0.013	0.007	0.23	0.024	0.013	0.014	0.191	0.049	0.015	0.015	0.013	0.0021	760	156
TA	0.106	1.35	0.007	0.008	0.35	0.039	0.016	0.016	0.206	0.055	0.014	0.015	0.045	0.0022	418	33
TE	0.106	1.39	0.008	0.008	0.35	0.041	0.018	0.015	0.217	0.055	0.013	0.014	0.037	0.0024	410	38
TF	0.105	1.36	0.008	0.007	0.34	0.041	0.018	0.017	0.214	0.054	0.015	0.014	0.044	0.0019	438	60
TG	0.101	1.38	0.009	0.007	0.35	0.039	0.017	0.020	0.211	0.056	0.015	0.013	0.034	0.0024	443	68
TH	0.110	1.35	0.008	0.008	0.34	0.040	0.018	0.021	0.215	0.055	0.014	0.014	0.039	0.0021	410	77
TN	0.109	1.40	0.010	0.008	0.35	0.042	0.016	0.017	0.205	0.057	0.017	0.014	0.031	0.0026	465	132
TK	0.111	1.41	0.011	0.008	0.35	0.042	0.018	0.021	0.207	0.055	0.014	0.014	0.039	0.0025	421	96
TJ	0.110	1.39	0.010	0.008	0.33	0.043	0.018	0.019	0.210	0.055	0.017	0.015	0.039	0.0024	432	154
TB	0.093	1.52	0.011	0.010	0.39	0.027	0.018	0.139	0.214	0.050	0.014	0.017	0.052	0.0020	430	40
TD	0.093	1.46	0.011	0.009	0.39	0.017	0.039	0.071	0.195	0.050	0.016	0.017	0.056	0.0021	435	50
A 1	0.060	1.54	0.014	0.009	0.40	–	0.187	0.210	0.234	0.045	0.011	0.017	0.035	0.0032	553	68
A 2	0.057	1.57	0.014	0.010	0.41	–	0.186	0.211	0.250	0.044	0.010	0.016	0.036	0.0031	517	79
A 3	0.055	1.54	0.015	0.010	0.40	–	0.185	0.210	0.246	0.043	0.011	0.016	0.033	0.0028	540	101
A 4	0.062	1.52	0.015	0.011	0.38	–	0.192	0.215	0.230	0.044	0.010	0.018	0.036	0.0029	525	106
A 5	0.056	1.53	0.015	0.010	0.37	–	0.182	0.202	0.247	0.042	0.009	0.014	0.028	0.0033	566	123
A 6	0.057	1.59	0.018	0.010	0.41	–	0.192	0.219	0.241	0.044	0.010	0.018	0.036	0.0032	533	129
A 7	0.058	1.59	0.018	0.010	0.41	–	0.192	0.219	0.241	0.044	0.011	0.018	0.036	0.0032	533	129
A 8	0.061	1.49	0.015	0.010	0.38	–	0.186	0.215	0.236	0.044	0.011	0.016	0.040	0.0028	586	145
A 9	0.055	1.56	0.015	0.010	0.39	–	0.186	0.207	0.250	0.042	0.009	0.016	0.034	0.0030	590	160
A10	0.054	1.56	0.017	0.010	0.40	–	0.186	0.210	0.250	0.043	0.011	0.016	0.034	0.0031	537	166
A11	0.055	1.58	0.016	0.010	0.39	–	0.193	0.212	0.234	0.043	0.009	0.016	0.033	0.0029	525	171
A12	0.059	1.55	0.16	0.009	0.38	–	0.185	0.208	0.246	0.043	0.010	0.016	0.040	0.0033	612	192
S1	0.075	1.32	0.015	0.008	0.22	0.03	0.025	–	0.210	0.007	0.010	0.024	0.015	0.0015	583	42
S2	0.075	1.31	0.022	0.009	0.21	0.04	0.025	–	0.270	0.009	0.010	0.016	0.008	0.0020	660	80
S3	0.09	1.29	0.017	0.009	0.21	0.03	0.025	–	0.220	0.005	0.006	0.018	0.013	0.0010	605	86
S4	0.09	1.34	0.018	0.008	0.25	0.03	0.025	–	0.220	0.007	0.010	0.015	0.013	0.0015	625	107
S5	0.07	1.30	0.016	0.008	0.21	0.03	0.025	–	0.230	0.006	0.010	0.020	0.013	0.0010	620	139

c) Microscopy

Quantitative analysis of the acicular ferrite and proeuctectoid ferrite contents was carried out using a point counting technique at X500 magnification (using a binocular eyepiece with an 11 X 11 grid). Two cross-sections of each weld were examined and on each cross-section five fields of view were counted.

Carbon extraction replicas were taken from the centreline region of submerged arc deposits. Inclusion and particle compositions were evaluated using a Hitachi H800 STEM/KEVEX 7000 operating at 100kV. The uncertainty in the levels of major elements (Al, Si, Ti, Mn) detected in the inclusions was estimated to be ±15%.

Table III - Average Inclusion Compositions
(wt% of elements excluding oxygen

Code	Al	Si	S	Ti	Mn	Fe
Average for TM	29	5	2	34	21	9
Average for TJ	23	5	5	40	23	4
Average for TA	12	9	3	41	30	5
Average for TH	32	1	5	44	15	3
Average for NA	20	19	2	11	42	6
Average for NF	19	20	1	11	42	7
Average for NM	18	19	2	11	43	7
Average for NJ	22	17	1	14	40	6
Average for A1	19	4	4	46	23	4
Average for A8	25	4	2	48	16	5
Average for A12	25	6	2	43	20	4
Average for S1	13	22	4	10	51	–
Average for S2	16	21	3	9	51	–
Average for S5	21	21	4	10	44	–

d) Mechanical Properties

All weld metal, longitudinal Houndsfield tensile specimens were taken from seven welds in the A-series (A1, A2, A4, A8, A11, A12 and A13). Tensile specimens were 4.5 mm diameter and had 32 mm guage lengths. In the case of the T and N series deposits, transverse tensile specimens were taken (from TA, TH, TK, TM, TJ, NA, NE, NF, NK and NM). In this case diameter sections were removed and extension pieces were friction welded to make sub-standard tensile specimens. The guage length was 8 mm. Longitudinal tensile specimens were also taken from weld deposits TA, TB, TD, TG, TD and TH. In this case, the specimen diameter was 4.5 mm and the gauge length was 32 mm.

In toughness testing 10 mm x 10 mm Charpy specimens were taken from N and T series deposits and the notches were cut parallel to the weld axis. In the case of S-series deposits substandard Charpy specimens having dimensions 10mm x 5mm were machined and notches were again cut parallel to the weld axis.

354

Table IV - <u>Mechanical Properties</u>

Code	N	Transverse Tensiles		Longitudinal Tensiles		VHN$_{500}$	F.A.T.T
.	(ppm)	0.2% Yieldstress (MPa)	U.T.S (MPa)	Yieldstress (MPa)	U.T.S. (MPa)		(50%)
NA	37	554.8	660.8	–	–	237	-78°C
NE	63	512.2	632.5	–	–	255	-76°C
NF	73	562.2	666.9	–	–	242	-72°C
NK	138	531.8	643.3	–	–	236	-62°C
NM	102	557.5	679.8	–	–	246	-48°C
TA	33	609.5	737.9	578.4**	736.5	275	-90°C
TF	60						-112°C
TG	68			621.7(638.6)	743.3		-115°C
TH	77	613.5	733.8	660.8(675.7)	750	279	-110°C
TK	96	585.2	687.9	–	–	276	-85°C
TM	132	588.5	701.4	–	–	279	-72°C
TJ	154	617.6	739.2	–	–	274	-74°C
TB	40	–	–	589.9**	750	272	–
TD	50	–	–	607.5**	743.3	268	–
A1	68	–	–	625 (655.4)*	690.6	242	–
A2	79	–	–	629.1(663.6)	691.9	240	–
A4	106	–	–	627.1(682.5)	692.6	240	–
A8	145	–	–	629.1(689.2)	692.6	238	–
A11	171	–	–	628.4(669.6)	692.6	241	–
A12	192	–	–	629.8(665.6)	700.0	241	–

* Upper yield stress values given in brackets
** 0.2% yield stress values

Results

Increasing weld nitrogen content decreased the acicular ferrite content in three test series (T, N and A)(see Fig. 2). However, in the case of N-series deposits the nitrogen relation was confounded by an oxygen variation from 620 to 760 ppm (see Table II). It follows that the marked change in acicular ferrite content exhibited in N-series deposits may result from a combination of nitrogen and oxygen changes. In A and T series weld deposits the oxygen levels did not change with change in nitrogen content since titanium deoxidation (via the filler wire) was effective in gettering the oxygen during welding.

Fig. 2.   Acicular ferrite/nitrogen relations in series A, T and N-deposits

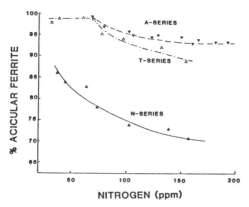

In S-series weld deposits, small additions of aluminum powder in the flux again gettered the weld oxygen content, and nitrogen variation was independent of oxygen content variations (see Table II). In this test series increased nitrogen had no influence on acicular ferrite content (see Fig. 3). Also, there was no clearcut relation between the content and morphology of M-A phases and deposit nitrogen content.

Fig. 3.   Acicular ferrite/nitrogen relation in S-series weld deposits:
▲ no aluminum added to the flux
△ 0.5wt% aluminum added to the flux

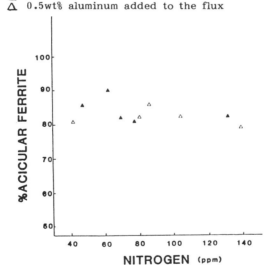

Comparison of results in T and S series illustrates the effect of nitrogen variation in welds where boron was active (T-series) and inactive (S-series) at prior austenite grain boundaries. In the case of T-series weld deposits the acicular ferrite content markedly decreased when the weld nitrogen content exceeded 70 ppm (see Table V).

Table V - Summary of Microstructural Constituents			
Sample	(ppm N)	Proeutectoid/Grain Boundary Ferrite %	Acicular Ferrite %
NA	37	14%	86%
ND	44	16%	84%
NE	63	17%	83%
NF	73	22%	78%
NM	102	26%	74%
NK	138	27%	73%
NJ	156	29%	71%
TA	33	2%	98%
TE	38	1%	99%
TF	60	1%	99%
TG	68	1%	99%
TH	77	5%	99.5%
TK	96	6%	94%
TM	132	8%	92%
TJ	154	11%	89%
A1	68	1.5%	98.5%
A2	79	3%	97%
A3	101	4.5%	95.5%
A4	106	5%	95%
A5	123	5.5%	94.5%
A6	129	5%	95%
A7	129	5%	85%
A8	145	5.5%	94.5%
A9	160	7%	93%
A10	166	6%	94%
A11	171	6.5%	93.5%
A12	192	6.5%	93.5%
S1	42	19%	81%
S2	80	14%	83%
S3	86	12%	86%
S4	107	12%	85%
S5	139	17%	78.5%

Fig. 4 shows FATT values for T and N-series weld deposits. For the T-series weld metal toughness was optimum when the nitrogen content was approximately 70 ppm. At lower nitrogen contents there was a fall off in toughness properties. This fall off in toughness was confirmed in further testing of low nitrogen content welds (TB and TD) (see Fig. 5). In S-series weld deposits toughness values were relatively unaffected by nitrogen content variations from 42 to 139 ppm (see Fig. 6).

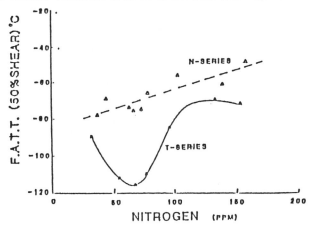

Fig. 5.   Effect of increasing titanium content in deposits containing 40 ppm N(TB) and 50 ppm N(TD)

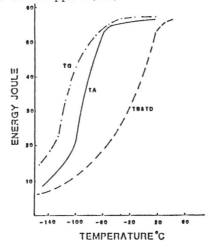

Fig. 6.   Toughness in S-series deposits (where 0.5wt% aluminum was added to the flux)

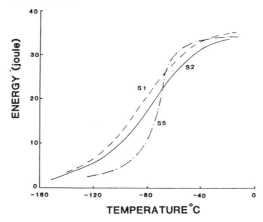

The soluble nitrogen content increased as weld nitrogen content increased in the A-series (see Fig. 7) and both ultimate tensile strength and weld metal hardness values were unaffected by nitrogen variation (see Table III). In A-series deposits the difference between upper and lower yield stress values peaked at around 130 ppm nitrogen and this peak in upper yield stress occurred when the peak free nitrogen content was observed. In the case of T and N-series weld deposits, transverse tensile test specimens did not show upper and lower yield strength values. However, longitudinal testing of samples TH (77 ppmN), TG (68 ppmN), TA (33 ppmN), TB (40 ppmN) and TD (50 ppmN) showed clearcut upper and lower yield values in the case of samples TH and TG. In low nitrogen content welds (TA, TB, TD) no upper and lower yield strength values were apparent.

Fig. 7. Soluble nitrogen content in A-series welds

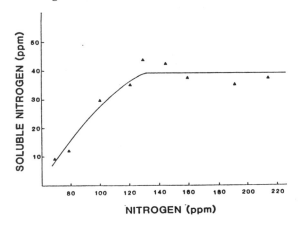

Inclusion Analysis

Inclusion morphologies varied from spheroidal to cuboidal. Average inclusion analyses were evaluated by counting 20 inclusions in random fields. Table IV summarizes the results for T, N, A and S-series weld deposits. The average inclusion compositions in two size ranges, viz. 0-1 um and 1-2 um were evaluated. No inclusion composition differences were apparent between the two size ranges. More detailed results concerning inclusion composition and precipitation in test welds are provided elsewhere(17).

Nitrogen variation had no observable influence on inclusion chemistry in any test series. However, within any test series T and A-series inclusions had higher Ti and lower Mn and Si contents than in N-series weld deposits. In the case of S-series weld deposits aluminum additions in the flux burden promoted inclusion compositions containing lower titanium levels.

Discussion

Final weld metal chemistry is dependent on the complex interplay of titanium, oxygen, boron and nitrogen during welding. Bearing in mind the type of welding consumable combination used in welding, large additions of titanium in the filler wire are available for interaction with oxygen, and with nitrogen since titanium is a strong oxide and nitride former. Since inclusion chemistry is largely unaffected by nitrogen content variations in all test series this suggests that the prime role of titanium is in titanium/oxygen interaction. It has been shown elsewhere (17) that the assumption of titanium being combined with oxygen in its highest oxidation state (TiO_2) provides an effective means of calculating the proportion of titanium tied up in oxide inclusions. In the case of T-series deposits, the total average titanium content is 0.0375% (see Table II). It can be calculated that 0.023% titanium is combined as oxide inclusions, and 0.0155% titanium is available for interaction with nitrogen. In the case of N-series weld deposits all titanium is tied up in oxide inclusions.

Ti-B welding consumables are formulated on the basis that soluble (effective) boron inhibits ferrite formation at prior austenite grain boundaries. In T-series weld deposits the total boron is 0.0023% (see Table II). The boron content effective in increasing the hardenability of wrought low-alloy steels, high heat input gas-metal-arc, and shielded metal arc weld metals had been given as 6 to 10 ppm [7][18][19][20]. The nitrogen content above which the effective boron content at prior austenite grain boundaries will be depleted can be calculated assuming a reaction sequence Ti:O, Ti:N and B:N. For an effective boron content of 8 ppm the nitrogen content tied up as TiN and BN in T-series weld deposits can be calculated as 45 ppm and 20 ppm.

It follows that the effective boron at prior austenite grain boundaries should be affected when deposit nitrogen content exceeds 65 ppm. The results presented in Figs. 2, 4 and 7 support this argument. Significant changes in acicular ferrite content, in FATT values, and in soluble nitrogen content occur when the weld metal nitrogen content exceeds 70 ppm. Also, longitudinal tensile testing of low nitrogen welds TA, TB and TD did not show distinct upper and lower yield point values. In the case of welds TG and TH containing 68 ppm and 77 ppm nitrogen upper and lower yield point values were clearly apparent.

In the case of S-series welds, where boron was inactive at prior austenite grain boundaries, no significant change occurred in acicular ferrite content (Fig. 3) or in weld metal toughness values (Fig. 6). In this connection the variation in acicular ferrite content in N-series weld deposits can be wholly ascribed to oxygen content changes (Fig. 8) Liu[24] and Fleck[25] have recently emphasized a relation between polygonal ferrite content and increased deposit oxygen content. In the case of S-series weld deposits there was no clearcut relation between M-A phase content and nitrogen level. Increasing nitrogen content in Ti-B free weld deposits has previously been associated with higher contents of M-A phase[15][16].

Fig. 8.   Nitrogen/oxygen relation in N-series weld deposits

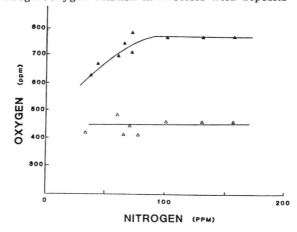

The argument given above essentially depends on sequential TiN and BN formation after completion of the Ti:O interaction, and optimum toughness occurs when the ideal active boron content is present at prior austenite grain boundaries. This argument has important consequences, viz:

a) Fall off in weld toughness properties at low nitrogen contents

Figs. 4 and 5 show a marked decrease in toughness values in low nitrogen content weld deposits. The soluble titanium content in welds TB and TD is much higher than in weld TG (the point at which maximum toughness occurs). However, there were no microstructural differences between welds TG, TB and TD and no change in tensile strength properties (see Table III). It is suggested that TiN formation in preference to BN formation leads to excess boron content, and to borocarbide formation, in weld deposits. It has been well documented that increasing boron content beyond that required to deliver maximum hardenability change leads to a decrement in toughness properties in the case of wrought steels(19), and in weld metal(26).

b) Increased oxygen and nitrogen contents

Since weld titanium and oxygen contents must be in balance, increasing titanium levels are required at higher oxygen contents in order to optimize weld zone toughness values. In this connection, Nakanishi (21) observed that optimum toughness values occurred at an oxygen/titanium ratio around 1.1, for deposit oxygen levels varying from 250 to 650 ppm.

Also, as weld metal (plate) nitrogen content increases, higher titanium contents are needed to maintain optimum toughness values (for any given weld metal boron level). Fig. 9 shows the calculated titanium/oxygen/nitrogen balance for optimum toughness properties for weld metal containing 0.0023%B. In Fig. 9 the titanium content tied up in oxide inclusions is that preserved in T-series weld deposits (see Table IV) and applies directly to a welding situation involving titanium deoxidation via the electrode wire. This argument is supported by the observation that peak toughness values occur at higher titanium levels as deposit nitrogen rises(2).

Fig. 9.   Calculated   titanium/oxygen/nitrogen   balance   for   optimum toughness properties in Ti-B bearing weld metal containing 0.0023% boron

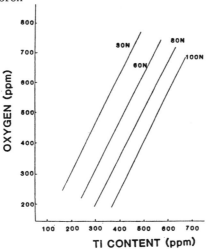

If the titanium/oxygen balance is such that no excess titanium is available for interaction with nitrogen, weld metal toughness essentially depends on BN formation and the availability of active boron at prior austenite grain boundaries. Optimum toughness will occur when:

$$B^* = B_{(total)} - \frac{10.8}{14.1} \; [N_{total}]$$

where B^* = effective boron content at prior austenite grain boundaries.

Assuming an effective boron content of 8 ppm, optimum toughness will occur at:

$$B_{(total)} = 0.7 \; (N_{total}) + 8 \quad (ppm)$$

This relationship is similar to that found by Kawabata(3) when he regressed all his toughness data on Ti-B weld metals. For a boron range of 0-0.0074% and a deposit nitrogen level less than or equal to 80 ppm, the optimum boron level corresponds to:

$$B = 0.7 \; (N) \pm 15 \quad (ppm).$$

c) Plate aluminum content

Much research has emphasized a relation between deposit toughness and plate (weld metal) aluminum content. Grong(11) has shown that optimum toughness occurs when deposit aluminum and oxygen contents are such that the ratio of deposit aluminum content divided by the square of the oxygen content $[Al]/[0]^2$) = 28:1. This effect has been ascribed to changes in inclusion composition, number and diameter i.e. in terms of intragranular acicular ferrite nucleation at inclusions. Horii(27) has recently provided data in Ti-B containing GMA deposits to support Grong's hypothesis(6).

In T-series deposits the $[Al]/[0]^2$ ratio is around 13 and optimum toughness should occur when deposit aluminum content approaches 0.05% (for the same deposit oxygen content in Table II). Since aluminum has little or no tendency to form AlN in submerged arc weld deposits (16) it follows that increasing aluminum content will free up titanium to interact with nitrogen, and BN formation will be limited. At some point increasing aluminum content will lead to a fall-off in deposit toughness properties due to excess boron.

This counterbalance between the beneficial influence of higher aluminum on intragranular acicular ferrite nucleation (on inclusion size and number) and the detrimental effect of increasing boron content (borocarbide formation) is important in terms of deposit nitrogen content. Higher aluminum contents in T-series weld deposits (moving the $[Al]/[0]^2$ ratio from 13 towards 28) will make the weld metal more tolerant to increased nitrogen content. In effect, the point to which the polygonal ferrite content suddenly increases will move to higher weld nitrogen levels (see Fig. 10).

Fig. 10. Effect of increased weld metal (plate) aluminum content on microstructure of Ti-B bearing weld deposits

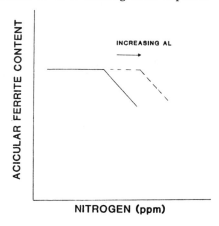

Conclusions

A systematic study on the influence of nitrogen on deposit micro-structure and mechanical properties has shown that:

1) the microstructure and toughness of Ti-B containing weld metal can be effectively modelled assuming the sequence of interactions is titanium/oxygen, titanium/nitrogen and then boron/nitrogen. As weld nitrogen content rises the titanium content required to provide optimum toughness values is increased,

2) when the titanium/oxygen balance is such that titanium is not available for interaction with nitrogen, boron controls the mechanical properties and optimum toughness occurs when B = 0.7(N) + 8 (ppm),

3) nitrogen contents exceeding 70 ppm produce a marked change in acicular ferrite content, and in FATT values. In the case of weld deposits where boron is effectively inactive at prior austenite grain boundaries, increasing nitrogen content has no influence on deposit microstructure (acicular ferrite, polygonal ferrite or M-A phases) and negligible influence on deposit toughness values.

363

References

1. Lau, T., Weatherly, G.C., McLean, A. "Gas/metal/slag reactions in submerged arc welding using $CaO-Al_2O_3$ based fluxes. Welding Journal, 65(2)(1986), 31-38.

2. Akao, K., Ishihara, T., Kitada, T., Nishino, Y., Okudu, N., Wada, T., Nagatani, K. Improvement of notch toughness and soundness of longitudinal seam weld of line pipe. Trans. I.S.I.Japan 26(5)(1986), 379-386.

3. Kawabata, F., Matzuyama, J., Nishiyama, N., Tanaka, T. "Progress in productivity and weld quality in UOE pipes by 4-wire submerged arc welding. Trans. I.S.I. Japan, 26(5)(1986).

4. Mori, N., Homma, H., Okita, S. and Wakabayashi M. "Mechanism of notch toughness improvement in Ti-B bearing weld metals". IIW Doc. IX-1196-81; 1981

5. Mori, N., Homma, H., Okita, S. and Asano K. "The behaviour of B and N in notch toughness improvement of Ti-B bearing weld metals". IIW Doc. IX-1158-80; 1980.

6. Horii, Y., Ohkita, S., Wakabayashi, M., Namura, M. "Development of welding materials for low temperature service". Nippon Steel Technical Report, 1986, No.5.

7. Watanabe, I., Kojima, T. "Study on notch toughness of weld metal in large current, MIG arc welding (reports 1 and 2). (Proceedings Welding Institute Conference "Effect of residual impurity and micro-alloying elements on weldability and weld deposit properties", Nov. 1983, London.)

8. North, T.H., Bell, H.B., Nowicki, A., Craig, I. "Slag/metal interaction, oxygen content and toughness during submerged arc welding". Welding Journal, 57(3)(1978), 63s-75s.

9. Devillers L. "Metallurgie et properties mecaniques du metal fondu en soudage multipasse sous flux d'acier au CMn-micro-allie. IRSID report Oct., 1983, No.10.

10. Terashima, H., Hart, P.H.M. "Effect of flux TiO_2 and wire Ti content on tolerance to high Al content of submerged arc welds made with basic fluxes". (Welding Institute Conference "Effect of residual impurity and microalloying elements on weldability and weld properties", Nov, 1983, London.)

11. Grong O., Matlock, D.K. Microstructural development in mild and low alloy steel weld metals. International Metallurgical Reviews, 31(1)(1986) 27/48.

12. Den Ouden, G., Verhagen, J.G., Tichelaar, G.W. "Influence of chemical composition on mild steel weld metal toughness". Welding Journal, 54(1975), 87s-94s.

13. Horii, Y., Miaki, K. "Development of the high-COD titanium-boron bearing covered electrode". (Conference proceedings, 4th International Symposium of the Japan Welding Society, Osaka, 1982.)

14. Okabe, R., Koshizuka, N., Tanaka, M., Katamine, A. and San-Nomiya, Y. Improvement in toughness of weld joint of steel plate for large heat input welding. Transactions I.S.I Japan 23(10)(1983), B-390.

15. Ito, Y. and Nakanishi, M. "Study on Charpy impact properties of weld metal with submerged arc welding". IIW Doc. XII-113-75; 1975.

16. Oldland, R.B. "The influence of aluminium and of nitrogen on the microstructure and properties of single-pass submerged arc welds". (Ph.D. thesis, Monash University 1985.)

17. Lau, T., Sadowski, M.M., North, T.H., Weatherly, G.C. "Nitrogen and submerged arc weld metal properties" - to be published, Materials Science & Technology, The Metals Society, U.K.

18. Habu, R. "Effect of boron on hardenability of Al-B-N low alloy steels. Trans. ISI. Japan. 18(8)(1978), 492.

19. Kapadia, B.M. "Prediction of the boron hardenability effect in steel - a comprehensive review". (AIME Symposium "Hardenability concepts with applications to steel". 1978).

20. Brownrigg, A., Chong, C-J., Glover, G., Banks, E.E. "HAZ hardenability and boron". (Proceedings Welding Institute Conference "Effect of residual impurity and microalloying elements on weldability and weld properties, Nov. 1983, London), 33.

21. Nakanishi, M., Hashimoto, T., Komizo, Y., Akasaka, M. "Weldability of low CNb TiB steel for linepipe". Trans I.S.I. Japan, 26(5)(1986).

22. Kawamura, K., Oksubu, T., Mori, N. "Nitrogen analysis method using hot-hydrogen extraction". Trans I.S.I. Japan, 14(1974), 374.

23. Lau, T., North, T.H., Weatherly, G.C. "Effects of nitrogen on submerged arc weld metal properties". (CANMET/CSIRA Conference "Inclusions and Residuals in Steels", March 4/5, 1985, Ottawa)

24. Liu, S., Olson, D.L. "The role of inclusions in controlling HSLA steel weld microstructures". Welding Journal. 66(5)(1986), 139s-149s.

25. Fleck, N.A., Grong, O., Edwards, G.R., Matlock, D.K. "The role of filler wire and flux composition in submerged arc weld metal transformation kinetics". Welding Journal, 66(4)(1986), 113s-121s.

26. Delvetian, J.H., Heine, R.W. "Effect of boron content on carbon steel welds". Welding Journal, 55(2)(1975), 44s-53s.

27. Horii, Y., Ohkita, S., Wakabayashi, M., Namura, M. "Development of welding materials for low temperature service". Nippon Steel Technical Report, 1986, No. 5.

EFFECTS OF SULFUR CONTENT AND STEEL COMPOSITION ON CRITICAL

PREHEAT TEMPERATURES IN STRUCTURAL STEELS

P. J. Konkol

Westinghouse Electric Corporation
Machinery Technology Division
Box 18249
Pittsburgh, Pennsylvania 15236

Abstract

Using self-restrained Lehigh and Tekken weld cracking tests, the effects of sulfur content on the critical preheat temperature to prevent weld cracking (T_c) were investigated in A656 Grade 50, HY-100, and A514 Grade Q steels. The results showed that sulfur in the range 0.002 to 0.006 percent had no effect on the T_c of A656 Grade 50 steel, and that sulfur in the range 0.006 to 0.031 percent had no effect on the T_c of A514 Grade Q steel. A514 Grade Q and A514 Grade F steels had similar T_c's although A514 Grade Q had much higher carbon equivalents. The T_c of HY-100 steel with 0.002 percent sulfur was 50°F higher than the T_c of HY-100 steel with 0.005 percent sulfur.

Introduction

With the development of modern steelmaking and ladle metallurgy practices, it is possible to reduce the sulfur content of steels to less than 0.001 percent on production facilities. This development, along with reductions in other impurity elements and nonmetallic inclusions, has resulted in very clean steels. However, there have been conflicting reports (1,2,3) over whether the weldability of these clean steels is adversely affected in terms of increased susceptibility to hydrogen-induced cold or delayed cracking in the weld heat-affected zone (HAZ). Evidence exists which indicates that in some cases ultralow sulfur increases the critical preheat temperature (T_c) necessary to prevent cracking.

Several theories exist to account for the increased crack susceptibility. One theory is that sulfides and other nonmetallic inclusions act as nucleation sites for ferrite transformation during cooling after welding (2). Thus clean steels can have increased hardenability (increased amounts of lower bainite and martensite), which are more susceptible to cold cracking. Another theory is that inclusions act as traps or sinks where atomic hydrogen in steel combines to form molecular hydrogen and thus prevent hydrogen from diffusing to potential crack initiation sites (3). A third theory is that steelmaking practices used to produce low sulfur may result in increased hydrogen absorption in liquid steel, which may remain in the final product and thus contribute to the hydrogen absorbed from the welding process (3). A fourth theory is that the low number of sulfides in low sulfur steels reduces the number of preferred sites for boron segregation and boron nitride formation (4). The residual boron, which is normally combined with nitrogen and oxygen in conventional steels, is "activated" as free boron in clean steels and thus available to increase hardenability.

To obtain additional information on the effects of low sulfur on weldability, a series of weld restraint cracking tests were conducted to determine the T_c of both ferritic and martensitic structural plate steels which differed essentially only in sulfur content. The ferritic steel was a control-rolled ASTM A656 Grade 50 and the martensitic steel was a quenched-and-tempered HY-100. The effects of higher levels of sulfur on weldability were also examined in another quenched-and-tempered martensitic structural steel, A514 Grade Q. The weldability of A514 Grade Q was also compared to that of the leaner-alloy A514 Grade F. The results are described herein.

Materials and Experimental Work

Materials

All steels used in the present program were obtained from production-size heats that were produced on commercial melting, rolling, and heat-treating facilities. The original plate thicknesses and chemical compositions are shown in Table I.

ASTM A656 Grade 50 is a 50 ksi minimum yield-strength low-carbon-Mn-Cb control-rolled (low-finishing temperature) microalloyed ferritic steel. As shown in Table II, the three heats of A656 investigated had similar carbon contents and carbon equivalents as calculated by several different formulae (5,6,7) and differed essentially only in sulfur content.

Table I. Chemical Composition of the Steels Investigated

Code	Plate Thickness, Inches	Chemical Composition, Percent (Check Analyses)																
		C	Mn	P	S	Si	Cu	Ni	Cr	Mo	V	Nb	Ti	Al	B,ppm	N,ppm	O,ppm
A656 GRADE 50 SERIES																	
6-2*	1 1/2	0.088	1.25	0.007	0.002	0.23	0.16	0.08	0.03	0.02	-	0.031	-	0.037	-	12	31
6-4*	1 1/2	0.093	1.21	0.006	0.004	0.24	0.16	0.09	0.05	0.02	-	0.029	-	0.030	-	10	30
6-6*	1 1/2	0.083	1.24	0.006	0.006	0.23	0.14	0.09	0.05	0.02	-	0.029	-	0.030	-	10	19
HY-100 SERIES																	
H-2	2	0.19	0.36	0.010	0.002	0.23	0.10	2.87	1.44	0.42	0.02	-	0.010	0.011	-	9	11
H-5	2	0.19	0.32	0.008	0.005	0.27	0.04	2.98	1.53	0.46	0.02	-	0.007	0.059	-	11	15
A514 GRADE Q SERIES																	
Q-6	3	0.17	1.27	0.008	0.006	0.19	0.06	1.24	1.23	0.46	0.034	-	-	0.033	-	-	-
Q-11	3	0.14	1.24	0.008	0.011	0.24	0.06	1.24	1.28	0.45	0.048	-	-	0.033	-	-	-
Q-15	6	0.16	1.13	0.007	0.015	0.24	0.05	1.26	1.18	0.48	0.036	-	-	0.018	3	6	32
Q-31	6	0.16	1.17	0.016	0.031	0.22	0.04	1.18	1.26	0.43	0.044	-	-	0.021	3	7	19
A514 GRADE F SERIES																	
F-4*	2	0.17	0.97	0.009	0.004	0.25	0.21	0.78	0.58	0.46	0.058	-	0.028	0.014	51	12	32
F-7**	2	0.13	0.74	0.008	0.007	0.22	0.24	0.85	0.53	0.43	0.037	-	0.041	0.044	12	8	22
F-11	2	0.15	0.83	0.004	0.011	0.21	0.25	0.82	0.54	0.42	0.048	-	0.016	0.014	12	6	-
F-13	2	0.14	0.82	0.011	0.013	0.27	0.24	0.80	0.53	0.45	0.038	-	0.052	0.062	19	11	24
F-15	1	0.18	0.80	0.010	0.015	0.25	0.26	0.89	0.56	0.42	0.034	-	0.015	0.039	33	5	41
F-16	2	0.18	0.85	0.006	0.016	0.24	0.17	0.92	0.51	0.50	0.052	-	0.044	0.040	36	10	32

* Calcium treated
**Rare-earth treated

Table II. Carbon Equivalents of the Steels Investigated

Code	C,%	CE*	P_{cm}**	CEN***
A656 GRADE 50 SERIES				
6-2	0.088	0.32	0.17	0.08
6-4	0.093	0.33	0.17	0.09
6-6	0.083	0.32	0.17	0.08
HY-100 SERIES				
H-2	0.19	0.82	0.37	0.77
H-5	0.19	0.85	0.38	0.79
A514 GRADE Q SERIES				
Q-6	0.17	0.81	0.36	0.76
Q-11	0.14	0.79	0.33	0.68
Q-15	0.16	0.77	0.34	0.71
Q-31	0.16	0.78	0.35	0.72
A514 GRADE F SERIES				
F-4	0.17	0.62	0.34	0.61
F-7	0.13	0.53	0.27	0.44
F-11	0.15	0.56	0.29	0.51
F-13	0.14	0.55	0.29	0.49
F-15	0.18	0.59	0.33	0.59
F-16	0.18	0.61	0.34	0.60

$$*CE = C + \frac{Mn}{6} + \frac{Cr + Mo + V}{5} + \frac{Ni + Cu}{15}$$

$$**P_{cm} = C + \frac{Si}{30} + \frac{Mn + Cu + Cr}{20} + \frac{Ni}{60} + \frac{Mo}{15} + \frac{V}{10} + 5B$$

$$***CEN = C + A(C) \times [\frac{Si}{24} + \frac{Mn}{6} + \frac{Cu}{15} + \frac{Ni}{20} + \frac{Cr + Mo + Nb + V}{5} + 5B]$$

Where

$$A(C) = 0.75 + 0.25 \tanh [20 (C-0.12)]$$

HY-100 is a 100 ksi minimum yield-strength quenched-and-tempered martensitic alloy steel used as a hull material for naval combatant vessels. Because of the very high carbon equivalents and similar carbon contents, both of the heats investigated would be expected to have fully hardened HAZ of similar hardness when welded at low energy input.

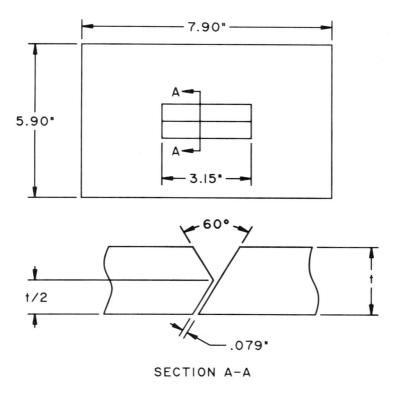

SECTION A-A

Figure 1 - Tekken restraint-cracking test specimen.

ASTM A514 Grade Q (A514-Q) is a 100 ksi minimum yield-strength
quenched-and-tempered martensitic constructional alloy steel. This grade
is a boron-free version of A514 with higher manganese, nickel, and
chromium. The major difference among the heats of A514-Q investigated was
in sulfur content, which varied from 0.006 to 0.031 percent. To compare
the weldability of A514-Q with the leaner-alloy (lower carbon equivalent)
boron-containing A514 Grade F (A514-F), several heats of A514-F were
evaluated. As shown in Tables I and II, these heats had normal variations
in sulfur (0.004 to 0.016%), carbon (0.13 to 0.18%), and carbon equivalents
typical of production practice.

Cracking Test Description

Tekken Test. The Tekken, or Japanese Y-groove restraint-cracking
test, is used to simulate a root pass in a highly restrained butt weld (8),
Figure 1. The test specimen is preheated to one of a series of
temperatures and a single weld bead is deposited in the groove. The
geometry of the groove is such that hydrogen-induced cold cracking is

371

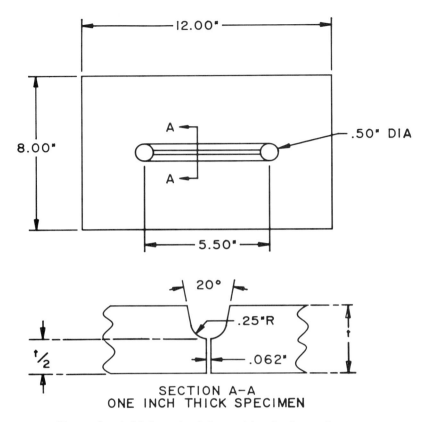

Figure 2 - Lehigh restraint-cracking test specimen.

induced at the weld root, usually in the HAZ. After a delay of at least 48
hours after welding to permit occurrence of cold cracking, if any, the
weldment is sectioned transverse to the welding direction in three
locations and examined metallographically for cracking. The intensity of
restraint, K_s, in the Tekken test is calculated to be 70 kg/mm^2 per mm
of plate thickness (9), whereas the observed K_s of weld grooves in most
actual structures is seldom more than 40 kg/mm^2 per mm of plate thickness
(10). Thus the Tekken test may be useful for comparing the HAZ cracking
susceptibility of various steels but may be overly conservative when used
to establish safe welding procedures.

Lehigh Test. The Lehigh restraint cracking test (11), Figure 2,
is conducted in a manner similar to the Tekken test. However, in the
Lehigh test the groove is longer, resulting in a lower K_s. In addition,
the groove geometry is symmetrical, which may result in cracking in either
the weld metal or HAZ, depending on which region has greater cracking
susceptibility. The measured K_s of 44 kg/mm^2 per mm of plate thickness
(9) is similar to that of a highly restrained weld in an actual structure;

Steel Code	Sulfur Content,%	Electrode*	Preheat, °F	Cracking,%**
6-2	0.002	E7018	70	0
6-4	0.004	E7018	70	0
6-6	0.006	E7018	70	0
6-2	0.002	E7018	36	0
6-4	0.004	E7018	36	1
6-6	0.006	E7018	36	3
6-2	0.002	E11018-M	70	100, 0
6-4	0.004	E11018-M	70	100, 0
6-6	0.006	E11018-M	70	0

*5/32-inch-diameter SMAW electrode deposited at 27.6 kJ/inch
energy input.
**Individual specimens.

thus the Lehigh test is a good test for determining safe welding procedures
in a _weldment_, that is, for a particular combination of plate, consumable,
and procedure.

Experimental Work

A656 Grade 50 Series. Tekken tests were conducted using the
shielded-metal-arc welding (SMAW) process with baked E7018 or E11018-M
electrodes and the conditions shown in Table III. The maximum HAZ hardness
was determined by conducting SMAW bead-on-plate welds at 15 and 30 kJ/inch
energy input and submerged-arc (SAW) bead-on-plate welds at 60 and 120
kJ/inch energy input, all at 70°F preheat. The Vickers diamond-pyramid
hardness (HV10) of the HAZ was measured on polished transverse sections.
The amount of nonmetallic inclusions in each heat of steel was measured by
quantitative television microscopy microcleanliness ratings on polished
longitudinal sections at the quarterthickness and midthickness locations of
the plates.

HY-100 Series. Tekken tests were conducted by SMAW with baked
E11018-M electrodes on the HY-100 heats using the conditions shown in Table
IV. The maximum HAZ hardness was determined on polished transverse
sections through a 30 kJ/inch SMAW bead-on-plate weld and a 60 and 120
kJ/inch SAW bead-on-plate weld, all at 70°F preheat. The microcleanliness
of the plates was measured on longitudinal sections.

A514 Grade Q and A514 Grade F Series. To obtain results that could be
directly applied to fabrication practices, the Lehigh test was used for
determining T_c's of the A514 steels. Because the plates were obtained in
thicknesses from one to six inches, the plates were machined to
one-inch-thick Lehigh test specimens with the top of the specimen as the
original plate surface. This ensured that each specimen had the same
degree of restraint and cooling rate. The Lehigh tests were conducted in a

Table IV. Results of Tekken Tests on 2-Inch-Thick
HY-100 Steels*

Steel Code	Sulfur Content,%	CRACKING,%** Preheat,°F				
		150	175	200	225	250
H-2	0.002	100	-	100,100	20	0
H-5	0.005	100	100	0,0	-	0

*Deposited with 3/16-inch-diameter E11018-M SMAW
electrode at 35 kJ/inch energy input.
**Individual specimens.

manner similar to the Tekken tests over a range of preheat temperatures in
25°F increments and with baked E11018-M electrodes used at 35 kJ/inch
energy input.

Results and Discussion

A656 Grade 50 Series

As shown in Tables I and II, the three heats were very similar in
chemical composition and carbon equivalent, and differed essentially only
in sulfur content. Sulfur in the range 0.002 to 0.006 percent did not
affect the maximum hardness or hardenability of the HAZ of the
bead-on-plate welds as shown in Figure 3. The hardness of the 0.004
percent sulfur heat was slightly greater due to its slightly higher carbon
content. The results of the microcleanliness analysis showed an inclusion
density of 26, 38, and 58 particles/mm^2 for the 0.002, 0.004, and 0.006
percent sulfur heats, respectively; thus inclusion content increased with
sulfur content. No effect of sulfur in the range of 0.002 to 0.006 percent
on T_C was found, as shown in Table III. When welded using the E7018
electrode, which is suitable for steels of this strength level, no cracking
occurred in any of the Tekken test specimens at 70°F preheat. In an effort
to induce cracking, the preheat temperature was lowered to 36°F. At this
temperature, an insignificant amount of HAZ cracking occurred. Thus the
data indicate that sulfur in the range of 0.002 to 0.006 percent has no
effect on the T_C of A656 Grade 50.

In an effort to induce HAZ cracking in A656 Grade 50, the Tekken tests
were repeated with a higher strength E11018-M electrode to impose higher
stress across the HAZ. Initial tests indicated that the 0.002 and 0.004
percent sulfur heats exhibited 100 percent HAZ cracking but the 0.006
percent sulfur heat did not. However, repeat tests on the 0.002 and 0.004
percent sulfur heats did not result in any cracking. Thus, with a
highly-restrained test and higher than normal strength weld metal, the
effect of sulfur in the range 0.002 to 0.006 percent could not be
determined because of scatter in the test data as shown in Table III.

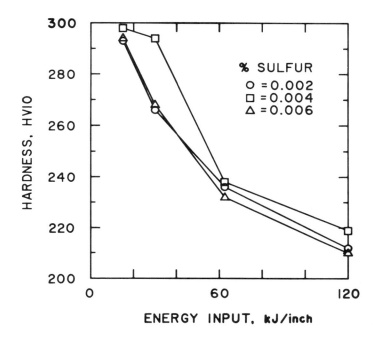

Figure 3 - Effect of sulfur content on maximum HAZ hardness in A656 Grade 50 steels.

HY-100 Series

The major differences in chemical composition of the two heats is that Steel H-2 contained 0.002 percent sulfur and 0.011 percent aluminum and Steel H-5 contained 0.005 percent sulfur and 0.059 percent aluminum, which resulted in Steel H-5 having a higher inclusion content. Microcleanliness ratings showed that Steel H-5 had an inclusion area fraction of 0.056 percent, versus 0.002 percent for Steel H-2. Because HY-100 is an alloy steel with high hardenability, differences in inclusion content would not be expected to affect the hardenability or hardness of the HAZ. The HAZ of both heats were martensitic at the 35 kJ/inch energy input used for the Tekken test. The results of the bead-on-plate hardness tests, Figure 4, show that although the test data exhibited scatter due to the inherent difficulty in measuring maximum HAZ hardness the HAZ hardness of both heats did not significantly change over the energy input range of 30 to 120 kJ/inch, and that the hardness of the 0.002 percent sulfur heat may have been slightly lower than that of the 0.005 percent sulfur heat. The hardnesses of the Steel H-2 and Steel H-5 base metals were 279 HV10 and 282 HV10, respectively.

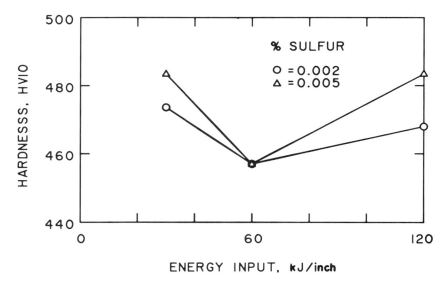

Figure 4 - Effect of sulfur content on maximum HAZ hardness in HY-100 steels.

However, the T_c's of the two heats were significantly different. As shown in Table IV, the 0.002 percent sulfur heat exhibited 100 percent HAZ cracking at 200°F preheat in duplicate Tekken tests whereas the 0.005 percent sulfur heat exhibited no cracking at 200°F in duplicate tests. The 0.002 percent sulfur heat exhibited partial HAZ cracking at 225°F preheat and a 250°F preheat was required to prevent cracking in the Tekken test. Because both heats are fully hardenable, the increase in T_c at low sulfur levels may possibly be attributed to the fewer inclusions present that act as sinks to remove hydrogen from the matrix (3).

Note that the current fabrication specification for HY-80/100 (12), which has been used successfully for years, specifies a 200°F preheat for plate thicknesses greater than 1-1/8 inches. The present results suggest that a 200°F preheat is insufficient for two-inch-thick plates of both heats investigated; however, as mentioned previously, the Tekken test simulates a worse-case condition, i.e., the restraint is higher than found in most actual joints and the weld is deposited in a single pass and allowed to cool to room temperature without maintenance of interpass temperature or subsequent weld passes. Thus the Tekken test a useful laboratory test to determine the effects of sulfur content or other variables, but the results should not be extrapolated to predict actual service performance.

A514 Grade Q and A514 Grade F Series

The A514 heats varied not only in sulfur content but also had normal variations in carbon and other alloying elements, Table I, which resulted in variations in carbon equivalent, Table II. Although the carbon ranges

Table V. Effects of Sulfur Content and Carbon Equivalent on
Critical Preheat Temperatures in One-Inch-Thick
Lehigh Specimens* of A514 Steels

Code	S,%	C,%	CE	Pcm	CEN	Tc,°F
A514 Grade Q Series						
Q-6	0.006	0.17	0.81	0.36	0.76	175
Q-11	0.011	0.14	0.79	0.33	0.68	175
Q-15	0.015	0.16	0.77	0.34	0.71	(175-300)**
Q-31	0.031	0.16	0.78	0.35	0.72	(200-300)**
A514 Grade F Series						
F-4	0.004	0.17	0.62	0.34	0.61	175
F-7	0.007	0.13	0.53	0.27	0.44	175
F-11	0.011	0.15	0.56	0.29	0.51	70
F-13	0.013	0.14	0.55	0.29	0.49	175
F-15	0.015	0.18	0.59	0.33	0.59	200
F-16	0.016	0.18	0.61	0.34	0.60	125

*3/16-inch-diameter E11018-M SMAW electrodes deposited at 35
kJ/inch energy input.
**See Figure 5.

overlapped, the carbon equivalents of the A514-Q heats were all greater
than those of the A514-F heats. The results of the one-inch-thick Lehigh
tests are shown in Table V. Several observations can be made:

1. Sulfur in the range 0.006 to 0.031 percent for A514-Q and 0.004 to
0.016 percent for A514-F had no effect on T_c except that at very high
sulfur levels in A514-Q the results exhibit considerable scatter, and at
very high sulfur levels HAZ cracking may be increased. As shown in Figure
5, scatter in individual test results was observed in the 0.015 and 0.031
percent sulfur A514-Q heats. This suggests that in high-sulfur steels some
HAZ cracking may occur which is possibly due to decohesion at the
metal-sulfide interfaces and linkup of the decohesion cracks among the
elongated sulfide inclusions (13,14). Note that in the 0.031 percent
sulfur heat, a small amount of cracking was still present at preheat
temperatures as high as 300°F.

2. Little difference in T_c was observed between the A514-Q and
A514-F steels in spite of A514-Q having a much higher alloy content and
carbon equivalent. This is in agreement with the theory that for fully
hardened steels (100% martensite in the HAZ) the cracking susceptibility is
a function of the stress, hydrogen level, and hardness of the
microstructure rather than carbon equivalent (5). By increasing the
hardenability until a steel is fully hardened, further increases in
hardness (or cracking susceptibility) will occur only with increasing
carbon content; thus alloy content becomes irrelevant to cracking
susceptibility in fully martensitic microstructures.

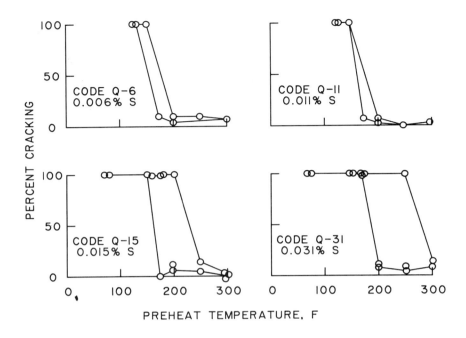

Figure 5 - Results of one-inch-thick Lehigh tests on A514 Grade Q steels.

3.   The T_c of a steel may vary with test location through the plate thickness. For example, Steel F-15, which had the highest T_c of the A514-F heats, was machined from one-inch-thick plate, which placed the weld root at the plate centerline, Figure 2.   In the other heats one-inch-thick Lehigh specimens were machined from one surface of a thicker plate.

Summary

A series of Tekken weld restraint cracking tests was conducted on three heats of ferritic A656 Grade 50 steel and two heats of quenched-and-tempered martensitic HY-100 steel that differed essentially only in sulfur content.   Lehigh weld restraint cracking tests were conducted on four heats of quenched-and-tempered martensitic A514 Grade Q steel with variations in sulfur content at higher sulfur levels, and the results were compared with results of tests on six heats of A514 Grade F steel.   The results of the program are summarized as follows:

1.   Sulfur in the range 0.002 to 0.006 percent did not affect the maximum HAZ hardness, hardenability, or critical preheat temperature (T_c) to prevent weldment cracking in A656 Grade 50 steel.

2. The T_c of the 0.002 percent sulfur HY-100 heat was 50°F higher than the T_c of the 0.005 percent sulfur HY-100 heat.

3. Higher sulfur contents (0.015 and 0.031%) were deleterious to the cracking resistance of A514 Grade Q steel.

4. A514 Grade Q and A514 Grade F steels had similar T_c's although A514 Grade Q had much higher carbon equivalents.

References

1. H. Suzuki, "Weldability of Modern Structural Steels," 1982 Houdremont Lecture, Welding in the World, 20(7/8),(1982), 121-148.

2. P. H. M. Hart, "The Influence of Steel Cleanliness on HAZ Hydrogen Cracking: The Present Position," IIW Document No. IXB-49-82 (1982).

3. Y. Hirai, S. Minakowa, and J. Tsuboi, "Effects of Sulfur on Hydrogen-Assisted HAZ Cracking in Al-Killed Steel Plates," IIW Document No. IX-1160-80 (1980).

4. P. H. M. Hart, "Effects of Steel Inclusions and Residual Elements on Weldability," Metal Construction, 18(10)(1986), 610-616.

5. F. R. Coe, Welding Steels Without Hydrogen Cracking, (The Welding Institute, 1973).

6. K. Satoh, S. Matsui, H. Horikawa, K. Bessyo, and T. Okumura, "JSSC Guidance Report on Determination of Safe Preheating Conditions Without Weld Cracks in Steel Structures," IIW Document No. IX-834-73 (1973).

7. N. Yurioka, H. Suzuki, S. Ohshita, and S. Saito, "Determinations of Necessary Preheating Temperature in Steel Welding," Welding Journal, Research Supplement, 62(6)(1983), 147s-153s.

8. H. Kihara, H. Suzuki, and H. Nakamura, "Weld Cracking Tests of High-Strength Steels and Electrodes," Welding Journal, Research Supplement, 41(1)(1962), 36s-48s.

9. K. Masubuchi and N. T. Ich, "Computer Analysis of Degree of Constraint of Practical Butt Joints," Welding Journal, Research Supplement, 49(4)(1970), 166s-176s.

10. H. Suzuki, "Cold Cracking and Its Prevention in Steel Welding," Transactions of the Japan Welding Society, 9(2)(1978), 140-149.

11. R. D. Stout, S. S. Tor, L. J. McGeady, and G. E. Doan, "Some Additional Tests on the Lehigh Restraint Specimen," Welding Journal, Research Supplement, 26(11)(1947), 673s-682s.

12. MIL-STD-1688(SH), "Military Standard, Fabrication, Welding, and Inspection of HY-80/100 for Submarine Applications," (1981).

13. W. F. Savage, E. F. Nippes, and E. S. Szekeres, "Hydrogen Induced Cold Cracking in a Low Alloy Steel," Welding Journal, Research Supplement, 55(9)(1976), 276s-283s.

14. W. F. Savage, E. F. Nippes, and J. M. Sawhill, "Hydrogen Induced Cracking During Implant Testing of Alloy Steels," Welding Journal, Research Supplement, 55(12)(1976), 400s-407s.

Acknowledgements

The author expresses his appreciation to the USS Division of USX Corporation, where the experimental work was supported and conducted.

EFFECT OF WELDING HEAT INPUT AND FLUX BASICITY ON THE WELD METAL TOUGHNESS

J. M. Lee, H. J. Kim and D. H. Park

Welding & Materials Research Institute
Hyundai Heavy Industries, Co., Ltd.
Ulsan, Korea

Abstract

Three aggromerated-type fluxes having different flux basicities were used to produce bead-on-plate and multipass welds at three different heat inputs - namely, 2.3, 4.0 and 6.5 KJ/mm. Based on the chemical analysis including oxygen content, the chemical behavior of this flux system has been characterized with respect to silicon, manganese and oxygen. The silicon and manganese contents of weld metal were strongly dependent on the amount of corresponding oxides in the flux but was not on the welding heat input. However t-
he oxygen content of weld metal was strongly affected by the change in heat
input but slightly by the variation in flux basicity. Because the oxygen co-
ntnet decreases with increasing heat input regardless of flux basicity, the
preferential effect of high basicity on weld metal toughness was not obtained
in the high heat input welds. From this result, it was discussed that the r-
eappraisal of the relationship between flux basicity and weld metal toughness
has to be made for evaluating consum a bles for the high heat input welding.

I. Introduction

To understand and predict weld metal properties such as strength, toughness, and solidification cracking behavior, it is important to be able to estimate the composition and the microstructure of weld metal. Besides, the inclusions in ferritic weld metal play very important roles not only in controlling the microstructure as nucleation sites of acicular ferrite but also in fracture process by acting as sites for cleavage or void formation (1,2). Consequently, the weld metal composition (including oxygen content) and the welding heat input which determines the cooling rate and thus controls the microstructure are the two main factors that determine the mechanical properties, especially the toughness of the weld.

Alloying elements in the weld metal may come from base metal, welding electrode, and welding flux. Among these, since the chemistries of base metal and welding electrode are generally known the chemical behavior of welding flux has to be understood for estimating the composition of weld deposit. This is why the majority of previous studies on submerged arc welding (SAW) have concentrated either on the effect of flux composition on weld metal chemistry or on the effect of flux basicity on the oxygen content in weld metal. One of the first studies for flux basicity was by Tuliani et al.(3). They found that the weld metal oxygen content drops from about 900 ppm to 300 ppm for a basicity index change from 0.5 to 1.5 and then remains constant with further increasing basicity, and that high basicity improves toughness. Along with these findings and others reported later (4-6) has grown up a quite widely held impression that basicity (or oxidation potential proposed lately (7)) would be a most important factor in determining the oxygen content of weld deposit and that in any welding condition a change to a more basic flux composition will always improve toughness.

However, under the condition that the product of slag-metal reaction such as deoxidation products can be removed or floating into the slag, the slag-metal reaction would only be important in determining the content of alloying elements such as Mn and Si but not in oxygen. Since the oxygen content of weld metal accounts for mostly in the form of oxide inclusions it would be controlled not only by the oxygen content carried into the weld pool but also by the separation of oxide particles as a result of weld pool fluid flow or particle flotation (1). It implies that the removal of oxide inclusions to the slag phase the slag-metal reactions prior to freezing both can play a part in determining the final oxygen content.

In this study, for the reasons outlineal above, the oxygen content was considered to be influenced not only by flux basicity but also by the extent of oxide separation which would depend on welding heat input or on individual welding variables. It is important that the effects of procedual variables are recognized and examined along with the effect of flux basicity since only then can a reliable basis for consumable selection and development be expected to emerge. As a first step to reappraise the influence of flux basicity, therefore, it is proposed to concentrate attention on the effect of welding heat input and flux basicity separately. The information obtained is used to demonstrate the interaction between flux basicity and welding heat input in determining weld metal toughness.

II. Experimental Procedure

A. Material

The chemical compositions of steel plates and welding wire used in this study are given in Table I. The steel plate of EH36 grade (ABS specification

) was used for bead-on-plate welding and that of A36 were for multi-pass weld joint. The welding wire employed throughout this study was commercially available medium Mn type (EM12K by AWS classification) and this was used over a range of heat inputs with three different welding fluxes.

Table I. Chemical compositions of steel plates and welding wire used.

Item	Grade	C	Si	Mn	P	S
Steel Plate	EH36	0.07	0.21	1.32	0.020	0.002
(25mm thick.)	A36	0.15	0.23	0.98	0.014	0.006
Welding Wire	EM12K	0.10	0.18	1.27	0.014	0.007
(4mm in dia.)						

Table II shows the composition and the basicity index of welding fluxes, which are all aggromerated type and alumina-based. The flux basicity index (BI) shown in the last column of Table II were calculated using the widely quoted formula (3).

$$BI = \frac{CaO + MgO + CaF_2 + 0.5 \ (MnO + FeO)}{SiO_2 + 0.5 \ (Al_2O_3 + TiO_2)}$$

Table II . Composition (wt.%) and basicity index (BI) of the fluxes used in this study.

Flux	SiO_2	MnO	Al_2O_3	CaO	CaF_2	MgO	Na_2O	BI
BI (1.11)	22.0	4.0	30.5	1.5	12.5	24.0	1.8	1.11
BI (1.43)	19.2	4.5	27.5	2.5	15.3	25.8	1.7	1.43
BI (1.67)	17.0	5.0	25.8	3.7	16.0	27.0	1.6	1.67

Following their basicity index, they were designated as BI (1.11), BI (1. 43) and BI (1.67). In this table, it is worth to note the variation of SiO_2 and MnO contents with respect to the change in flux basicity since it will affect the Si and Mn contents of weld metal, which will be described later.

B. Experiment

Two sets of experiments were performed ;
(1) Bead-on-plate welds were made on EH36 steel plates with each one of the fluxes at the heat inputs of 2.3, 4.0 and 6.5 KJ/mm. The welding parameters employed for each heat input are given in Table III. Each weld was sectioned in transverse direction to measure the dilu-tion of base metal (expressed as a percentage) and its averaged values are given in last column of Table III. The value of dilution increases slightly with heat input but at a given heat input it was little influenced by the compositional difference of present fluxes.

Chemical analysis including oxygen content was performed on each weld metal. The content of alloying elements was determined by wet analysis while oxygen content was analysed by inert gas fusion method. Based on the result of chemical analysis, the chemical interaction between the molten flux and weld pool was also studied.

Table III. Welding parameters and the value of dilution measured on the bead-on-plate welds.

Heat Input (KJ/mm)	Potential (volts)	Current (amperes)	Travel Speed (mm/min.)	Dilution (%)
2.3	28	550	400	52
4.0	34	700	360	58
6.5	38	800	280	61

(2) Second set of experiment was all weld deposit test, which was performed as recommended in AWS A5.17-80 specification (8). For this test, A36 steel plate as a base metal was prepared for welding as shown in Figure 1. The welding fluxes and conditions employed for this second set of tests were the same as those used for the bead-on-plate welding and are shown in Table II and III respectively. Each weld was sectioned and tested to provide all weld tensile and Charpy impact data as well as the chemical analysis. The locations of mechanical test specimens are also shown in Figure 1.

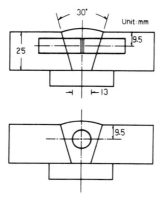

Figure 1   Joint configuration for welding and location of specimens for mechanical test.

III.   Result and Discussion

A.   Weld Metal Composition of bead-on-plate Welds

The final composition of weld metal is made up to contributions from the welding electrode, the base metal, and the elemental transfer from (or to) slag. A convenient way of studying the effect of flux constituents on the elemental transfer is to compare the analysed composition with that expected. The expected composition is based simply on the wire and plate compositions, and the amount of dilution, which is determined by metallographic cross-sections of the bead-on-plate welds.

The difference between analysed and expected values, ΔM defined as following equation, was used in an attempt to rationalize the elemental transfer as a function of flux composition.

384

$$\Delta M = \text{analysed comp.} - \text{expected comp.}$$

A negative value of delta (Δ) means that a given element has been transferred from the weld metal to slag and/or been lost by evaporation; a positive delta indicates the element going from the slag to the weld metal.

The chemical composition and the elemental delta values of the bead-on-plate welds are detailed in Table IV. As noticed in this table, the variation of heat input little affects the weld metal chemistry but that of flux basicity changes it significantly. Such trends are presented graphically in Figure 2. Figure 2(a) showing the relationship between the weld metal chemistry and the heat input for the case of flux BI (1.11) clearly demonstrates that the content of alloying elements such as C, Si and Mn is little influenced by heat input. As shown in Figure 2 (b), however, the variation in flux basicity leads to significant change in Si and Mn contents of weld metal. These trends are the same in the other levels of heat input and flux basicity (Table IV).

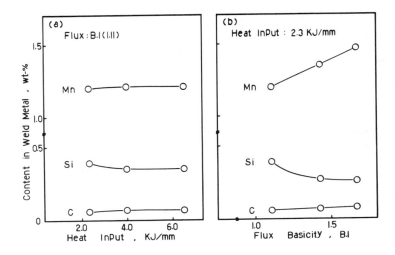

Figure 2 Variation of weld metal chemistry with respect to change in (a) heat input and (b) flux basicity.

Another thing to note in Table IV is the variation of oxygen content. Unlike in the alloying elements the oxygen content varies not only with the flux basicity but also with the heat input, as illustrated in Figure 3. The oxygen content decreases sharply with increasing heat input but slightly with increasing flux basicity within the range studied here. As for the present flux system, therefore, it can be said that welding heat input is a more important factor in controlling the oxygen content than the flux basicity.

Table IV. Chemical composition (wt % except oxygen) of the bead-on-plate welds and the delta value of each element.

Flux	Heat Input (KJ/mm)	C (ΔC)	Si (ΔSi)	Mn (ΔMn)	P (ΔP)	S (ΔS)	O ppm
BI (1.11)	2.3	0.06 (-0.02)	0.39 (+0.19)	1.20 (-0.10)	0.009 (-0.007)	0.009 (+0.006)	400
	4.0	0.07 (-0.01)	0.35 (0.15)	1.21 (-0.09)	0.006 (-0.01)	0.008 (+0.005)	334
	6.5	0.07 (-0.01)	0.27 (0.15)	0.35 (-0.09)	1.21 (-0.012)	0.005 (+0.005)	265
BI (1.43)	2.3	0.07 (-0.02)	0.27 (0.07)	1.35 (0.05)	0.013 (-0.003)	0.012 (+0.006)	380
	4.0	0.08 (0.00)	0.25 (0.05)	1.44 (0.14)	0.010 (-0.006)	0.010 (+0.007)	298
	6.5	0.08 (0.00)	0.26 (0.06)	1.35 (0.05)	0.014 (-0.002)	0.011 (0.008)	256
BI (1.67)	2.3	0.08 (-0.02)	0.26 (0.07)	1.47 (0.17)	0.016 (-0.001)	0.010 (+0.006)	365
	4.0	0.08 (0.00)	0.27 (0.07)	1.54 (+0.24)	0.011 (-0.005)	0.008 (+0.005)	237
	6.5	0.08 (0.00)	1.23 (0.03)	1.49 (0.19)	0.015 (-0.002)	0.008 (+0.005)	237

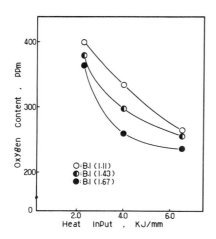

O:BI (1.11)
◑:BI (1.43)
●:BI (1.67)

Figure 3 Variation of oxygen content with respect to the change in heat input and flux basicity

B. Effect of Flux Constituents

In Figure 2, it is apparent that the change in flux basicity results in a significant change in Si and Mn contents of weld metal ; Mn increases but Si decreases with increasing flux basicity. However, since these changes are only possible through slag-metal reaction, they had better be related to SiO_2

386

and MnO contents of a given welding flux than to flux basicity. In this connection, all the delta values of Si and Mn presented in Table IV are plotted
against the SiO_2 and the MnO contents of the fluxes and are shown in Figure 4
and 5 respectively. As expected, the   Si and   Mn increase with the corresponding oxides, SiO_2 and MnO, independently of flux basicity. However, it is
very interesting to note that silicon has positive delta values in all three
fluxes but manganese has both positive and negative values within a narrow
range of MnO, i.e.   Mn for BI (1.11) containing 4.0% MnO is negative but t-
hat of BI (1.43) containing 4.5% MnO is positive.

Since all the fluxes employed in this investigation contains SiO_2 more
than 17%, they would be expected to add silicon to the weld and thus to have
positive delta values. The positive   Si stems from the Si pick-up due to
SiO_2 reduction being greater than the loss in Si as a result of deoxidation,
meanwhile the latter becomes greater with increasing basicity because of the
decrease in SiO_2 content resulting in almost zero   Si with BI (1.11). Similarly, the transition behavior of manganese (loss to gain) indicates that there are competing reactions which involve this element, that is, the loss of
Mn by evapor-ation and/or oxidation and the gain of Mn by decomposition of
MnO. Previously, Chai and Eager (9) studied CaF_2- based fluxes containing v-
arious amount of MnO and found that the pure CaF_2 flux resulted in 25% reduction in manganese from the expected value. Considering such a significant
loss of Mn, they proposed it is due to the evaporation of manganese in the
welding arc. Similar observations were made by Lau et al (10). They found
that, using an MnO-free flux, near 50% of the Mn content of the wire was lost
in the arc column, i.e. during the time interval available from the electrode
melting to the droplet stage. Considering the observed loss by evaporation
they predicted the temperature of droplets as to about $7200^{\circ}C$.

With MnO-bearing fluxes like the present flux system, on the other hand,
the rate of evaporation would be offset to a considerable extent by decomposition of MnO. In fact, it was reported (9) that as little as 5% MnO addition
in CaF_2 caused an increase in manganese content over an expected value resulting in positive   Mn. Based on this and the result shown in   Figure 4, it
can be concluded that approximately 4.25% of MnO is necessary to recover the
evaporation loss of Mn in the present flux system.

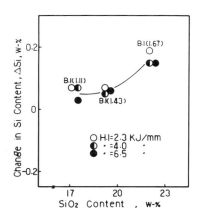

Figure 4   Delta manganese plotted
as a function of SiO_2
content of fluxes.

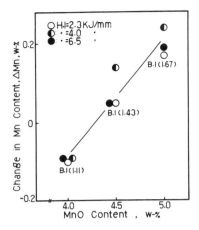

Figure 5   Delta manganese plotted
as a function of MnO
content of fluxes.

C. Effect of Heat Input on Oxygen Content

Allowing the chemical equilibrium, which is debatable but widely accepted to be reached in the molten weld pool despite the short time-scales in welding (11, 12), the final composition of weld metal will not depend on the w-
elding condition ; actually Si and Mn do not as shown in Figure 4 and 5.   Only exception of this is oxygen. As shown in Figure 3, heat input strongly i-
nfluences the oxygen content of weld metal. It implies that the behavior of
oxygen is different from that of other alloying elements and can not be explained by the equilibrium consideration alone. During slag-metal reaction,
the alloying elements are mostly in solid solution but the oxygen is in the
form of oxide particles, which can be removed prior to freezing. Consequently, the equilibrium consideration would only be important in determining the
level of alloying elements such as Si and Mn, however, for determining the o-
xygen content, some other factors have to be considered such as the density
and composition of inclusions, the weld pool motion and the time available
for removal.

As for the similar flux system, the shape or composition of inclusion
would not differ appreciably with the change in heat input or welding parameters. However, the change in welding parameters results in the variation in
arc temperature, penetration, weld contour, weld pool motion and duration time in liquid state. The change in these factors can influence the extent to
which the immersible reaction products formed can seperate into the slag phase prior to freezing. Previously, Iwamoto et al.(13) have studied the effect of welding speed on the oxygen content of weld metal under the fixed current and voltage condition. Their result showed that the oxygen content became larger with increasing welding speed. Moreover, a recent study (14) reported that under the condition of constant heat input and voltage the oxyg-
en content increased with increasing welding current. Since the welds were
made under the constant heat input, and the voltage was kept constant, this
result also can be interpreted in terms of welding speed, i.e. the oxygen content increases with the increase in welding speed. This latter result is

same as that reported by Iwamoto et al.(13).

Accordingly, it would be meaningful to explain the result shown in Figure 3 in terms of welding speed rather than heat input and then same conclusion can be drawn that the oxygen content increases with the increase in welding speed. As a result, it is quite evident that the oxygen content of weld metal depends so strongly on the welding variables employed that the oxygen data can be compared only when they were obtained under the same welding condition. Although it is not quite certain which welding variable is most dominent the previous and present studies suggest that at least welding speed is quite important in determining the oxygen content of weld metal. More systematic study on the effect of an individual welding parameter is undergoing

D. Weld Metal Toughness

Preliminary result on strength, toughness and oxygen content of multipass weld metal are shown in Table V and Figure 6 and 7 respectively. The test specimens were taken at the center of the all weld metal test plate as described earlier. Although the tensile properties of weld metal are rather consistant in all weldments (Table V) the weld metal toughness varies widely not only with flux basicity but also with heat input. Most importantly, a progressive increase in flux basicity improves the weld metal toughness at low heat input levels but does not make any improvement at the highest heat input examined, 6.5 KJ/mm. It is also worth to note that the toughness of BI (1.11) is little affected by the change in heat input even though the microstructure of weld metal would become coarse with increasing heat input.

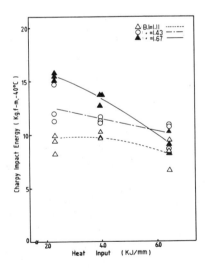

Figure 6   The result of impact test done at -40°C.

Table V. Tension properties of multi-pass weld metal

H.I (KJ/mm)	BI	T.S (Kg/mm^2)	Y.P (Kg/mm^2)	E.L (%)	R.A (%)
2.3	1.43	57.0	48.7	26.4	71.2
4.0	1.11	60.7	51.8	30.2	68.0
	1.43	58.7	48.9	29.2	72.9
	1.67	61.1	52.2	28.0	71.2
6.5	1.43	58.4	48.0	33.4	68.1

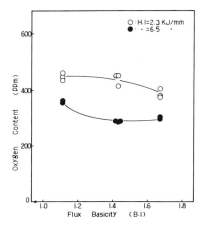

Figure 7   Effect of heat input
and flux basicity on
the oxygen content of
multi-pass weld deposit

Once the absence of basicity effect at high heat input has been pointed out by Garland et al.(15), but no further studies were found. They explained this behavior in terms of the relative effect of the microstructure and the cleanliness of weld metal, which was originally proposed by Tayer and Farrar (16). They seperated the influence on mechanical properties as being either inclusion control or microstructural control depending on the oxygen content. Above 600 ppm oxygen, inclusions were found dominating the mechanical behavior ; and below 300 ppm oxygen, the microstructural effort was considered dominent. This suggestion is very useful in explaining the toughness behavior shown in Figure 6. As the oxygen content in the weld metal or BI (1.67) is relatively low compared with those of lower basicity fluxes (Figure 7), its weld metal toughness will be dominently controlled by its microstructure which becomes coarse as heat input increases. Therefore, the weld metal toughness of BI (1.67) is quite sensitive to heat input and becomes poor as the heat input increases. On the other hand, the weld metal of BI (1.11) contains a large amount of oxygen and maintains this level of oxygen over a range of heat input studied. Consequently, the weld metal toughness becomes less sensitive to microstructural change associated with heat input and thus can be uniform over a range of heat input. Based on these results, it is quite evident that an approach to the selection of submerged arc welding consumables based on the maximization of flux basicity has clear limitations and may give an error in selecting a flux for high heat input welding.

IV. Conclusion

(1) The content of alloying elements in weld metal is mainly determined by slag-metal reaction resulting in a consistant value regardless of heat input while that of oxygen is strongly affected by some other factor(s) controlled by heat input or individual welding variables. Apparently, the lower the welding speed is the lower the oxygen content.

(2) Because of the significant decrease in oxygen content with heat input regardless of flux basicity, the preference of higher basicity flux for the better toughness can not be made in selecting or developing welding consumables for high heat input welding.

Reference

1. O. Grong and D. K. Matlock ; Int. Met. Reviews, 1986, Vol.31, No.1, p.27.
2. D. J. Abson and R. J. Pargeter ; Int. Met. Reviews, 1986, Vol.31, No.4, p.141.
3. S. S. Tuliani, T. Boniszewski and N. F. Eaton ; Weld. Met. Fab., 1969, Vol.37, No.8, p.327.
4. P. Colvin ; Weld. Met. Fab., 1974, Vol.42, No.11, p.362.
5. V. M. Baryshev, A. I. Shebaron, A. M. Birbrover and V. V. Malov ; Weld. Prod., 1978, Vol.25, No.7, p.43.
6. Y. Yoshino and R. D. Stout ; Weld. J., 1979, Vol.58, No.3, p.59-S.
7. T. H. North, H. B. Bell, A. Nowicki and I. Craig ; Welding J., 1978, Vol. 57, No.3, p.63-S.
8. AWS Specification A5.17-80, "Specification for carbon steel electrodes and fluxes for submerged arc welding".
9. C. H. Chai and T. W. Eagar ; Welding J., 1982, Vol.61, No.7, p.229-S.
10. T. Lau, G. C. Weatherly and A. McLean ; Welding J., 1986, Vol.65, No.2, p.31-S.
11. G. R. Belton, T. J. Moore and E. S. Tankins ; Welding J., 1963, Vol.42, p.289-S.
12. M. L. E. Davis and F. R. Coe ; Weld. Inst. Members Report 39/1977/M, 1977 , May.
13. N. Iwamoto, Y. Makino, K. Shimizu and T. Akamatsu ; Trans. JWRI, 1983, Vol.12, No.1, p.143.
14. T. Lau, G. C. Weatherly and A. McLean ; Welding J., 1985, Vol.64, No.12, p.343-S.
15. J. G. Garland and P. R. Kirkwood ; Weld. Met. Fab., 1976, Vol.44, No.5, p.217.
16. L. G. Taylor and R. A. Farrar ; Weld. Met. Fab., 1975, Vol.43, p.305.

INVESTIGATION OF SUBMERGED ARC WELD METAL INCLUSIONS

By:R.J. Pargeter

Materials Department, The Welding Institute
Abington Hall, Abington, Cambridge CB1 6AL.

Abstract

The results of determinations of size distribution and energy dispersive X-ray analyses carried out on submerged arc weld metal inclusions using an automated Particle Analysing Scanning Electron Microscope (PASEM) are reported. The suitability of the technique for the investigation of weld metal inclusions is discussed, and inclusion size distribution and type have been considered in relation to flux, wire, plate and weld metal composition, and also with regard to weld metal microstructure. It is concluded that, although there are limitations to the precision of the X-ray analysis, this technique does have value owing to the large amount of data which can be collected within a reasonable time. It has been possible to show that aluminium, silicon and titanium can all be transferred from the flux to weld metal inclusions. Many inclusions have complex compositions, and the significance of different inclusion types with regards to the austenite ferrite transformation is discussed.

Introduction

It is becoming increasingly widely recognised that non-metallic inclusions play an important part in the formation of ferritic steel weld metal microstructures and hence the mechanical properties attainable. Cochrane and Kirkwood (1) suggested that, in high oxygen deposits (>600ppm), the microstructure is controlled by the nucleation of ferrite on inclusions in the austenite grain boundaries, while Abson et al (2) proposed that, in lower oxygen deposits (~250-600ppm), the formation of acicular ferrite is the result of intragranular nucleation of ferrite on a suitable inclusion distribution. It is recognised in both these papers, however, that any attempt to correlate microstructure and inclusions would have to take into account the chemical type and size distribution of the inclusions, as well as total number and spatial distribution. Few such data exist and the present work was therefore carried out to investigate inclusions in weld metals, using the particle analysing scanning electron microscope (PASEM) at the Swedish Institute for Metals Research, which provides data on the size and chemical composition of particles down to about 0.25 micron diameter.

Specimens were taken almost entirely from a number of submerged-arc welds which had already been examined in previous programmes at The Welding Institute (3,4,5), with the aim of revealing differences in the inclusion distributions due to flux type and plate deoxidation practice. It was also hoped to correlate inclusion distributions and microstructure, and two further pairs of welds which were closely similar in all except microstructure, were included for this purpose. In one pair, the welds were nominally identical in terms of plate, wire, flux batch and welding conditions; in the other pair, the only difference was the plate, and in particular the plate deoxidation practice. Details of all the welds are summarised in Table I.

Table I   Summary of welds submitted to PASEM

Weld number	Flux[a] Type	Flux[a] Code	Wire[b] Type	Wire[b] Code	Plate[c] Type	Plate[c] Code	Thickness, mm	Weld type*
7200/109	CSHS	F15	S4	s	C-Si-Al	J	32	Multipass
7200/11	CS	F1	S3	l	C	B	32	Multipass
7200/9	CSLS	F2	S3	l	C	B	32	Multipass
7200/7	MS	F5	S1	b	C	B	32	Multipass
7200/3	MS	F3	S1	a	C	C	32	Multipass
7200/15	AR	F7	S2	e	C	C	32	Multipass
7200/13	AR	F6	S2	f	C	E	32	Multipass
7200/53	AB	F8	S3	n	C-Si-Al	G	32	Multipass
7200/21	B	F9	S3	k	C	D	32	Multipass
7200/113	B	F18	S3	u	C-Si-Al	J	32	Multipass
7200/5	B	F16	S3	n	C-Si-Al	G	32	Multipass
7200/63	CS	F1	S3	q	C-Si-Al	F	32	Two-pass
7200/29	CSLS	F2	S3	r	C-Si-Al	F	32	Two-pass
7200/27	MS	F5	S1	d	C	F	32	Two-pass
7200/35	AR	F7	S2	g	C-Si-Al	F	32	Two-pass
7200/37	AB	F8	S2	h	C-Si-Al	F	32	Two-pass
7200/41	B	F9	S3	l	C-Si-Al	A	32	Two-pass
7319/3	B	F16	S3	x	C-Si-Al	J	32	Two-pass
20000/1	B	F16	Mo-Ti-B	–	C-Mn	DH34	25	Triple arc, two pass
20000/10	B	F16	Mo-Ti-B	–	C-Mn	DH34	25	Triple arc, two pass
20000/3	B	F17	S3Mo	–	C-Mn	DH34	25	Triple arc, two pass
20000/7	B	F17	S3Mo	–	C-Mn-Si-Al	1336J	25	Triple arc, two pass

a see Table VII   b see Table III   c see Table II

Description of PASEM

The particle analysing scanning electron microscope consists of a Philips PSEM 500 scanning electron microscope with an EDAX X-ray spectrometer coupled to a TRASK P300 mini computer. A full description of the system can be found in Ref.6. The instrument requires a flat polished specimen and is operated in backscattered electron mode. The whole field of view is recorded in a special image memory (COMTAL 5000) and each object in this memory is examined (automatically) in turn. A small frame is drawn around it and a local grey level threshold is determined; the particle is sized by area, perimeter, x and y diameters, and, if the area is less than a predetermined threshold (usually 10 picture points = 0.28 microns diameter for the conditions used in this work), it is discarded. If not, these measurements, along with the co-ordinates of the centre of the particle, are stored in a matrix of data.

The computer then drives the electron beam to each pair of co-ordinates, and the intensities of a predetermined set of peaks in the X-ray spectrum are added to the matrix of data for each particle. It is from this matrix that all other data are derived. In order to classify an inclusion chemically, a threshold level is set for the X-ray intensity to eliminate background counts. The classification is then purely in terms of which elements are found to be present above the thresholds; it is in no way quantitative. It should be noted that only a chosen set of elements (in this case Mg, Al, Si, S, Ca, Ti, Cr and Mn) is considered, and it is possible for an inclusion to contain none of these (e.g. iron oxide). Iron has been omitted because of inevitable interference from the matrix, and light elements (notably boron, nitrogen, oxygen and fluorine) have been omitted because of the limitations of the X-ray energy analysis.

Experimental details

A small transverse section from the same region as used for photomicrography in previous work (3-5) was taken from each weld. This was polished to a 1 micron diamond finish and the area of interest was indicated with a scribe mark, prior to submitting the specimens to the Swedish Institute for Metals Research. Particle analysing scanning electron microscope examination was carried out at a magnification of x 2,500 on a total of 50 fields of view within the area of interest, giving a total area of $0.052mm^2$.

Parent plate compositions for all welds are presented in Table II, although the effects of parent plate composition were minimised in the case of the multipass welds, as the low dilution, upper runs were examined. Wire compositions are presented in Table III, and it should be noted that the wires used for multipass welds 7200/7 and 3 were of <0.01wt.%Si content, and that welds 7200/21 and 113 were made with wires of slightly higher Al content than the rest (0.010wt.% cf <0.007wt.%), one (used in weld 7200/113) also having a very low sulphur content (0.008wt.%). In the single arc two-pass welds, a wire of <0.01wt.%Si content was used for weld 7200/27, and wires containing 0.027 and 0.015wt.%Al were used for welds 7200/29 and 7319/3 respectively. This last wire also contained <0.008wt.% sulphur. The majority of these welds were made on C-Si-Al plate of 0.014-

Table II Chemical compositions of plates

Plate	Element, wt.%													Welds with which used		
Code	C	S	P	Si	Mn	Ni	Cr	Mo	Cu	Nb	Al	O	N	Multipass	Two-pass	Triple arc, two-pass
J	0.16	0.016	0.015	0.20	0.90	0.01	<0.01	<0.01	0.03	<0.005	<0.005	0.034	0.012	7200/109, 113	7319/3	
B	0.25	0.038	0.018	<0.01	1.15	0.01	<0.01	<0.01	0.01	<0.005	<0.005	0.002	0.005	7200/11,7,9		
C	0.21	0.047	0.027	<0.01	0.99	0.02	<0.01	<0.01	0.01	<0.005	<0.005	0.010	0.004	7200/3,15	7200/27	
E	0.23	0.055	0.030	<0.01	1.01	0.02	<0.01	<0.01	0.02	<0.005	<0.005	0.009	0.004	7200/13		
G	0.15	0.019	0.015	0.30	0.83	0.02	<0.01	<0.01	0.05	<0.005	0.037	0.004	0.008	7200/53,55		
D	0.22	0.032	0.014	<0.01	1.11	<0.01	<0.01	<0.01	<0.01	<0.005	<0.005	0.010	0.004	7200/21		
F	0.21	0.025	0.037	0.18	1.00	0.02	<0.01	<0.01	0.05	<0.005	0.037	0.004	0.010			
A	0.17	0.014	0.031	0.25	0.78	0.02	<0.01	<0.01	0.03	<0.005	0.047	0.003	0.008		7200/63,29,35,37	
DH34	0.19	0.021	0.015	0.01	1.33	0.07	0.05	0.02	0.14	0.033	0.006	0.011	0.002		7200/41	20000/1,10,3
133J	0.20	0.005	0.021	0.40	1.19	0.02	0.01	0.01	0.01	0.031	0.027	0.001	0.005			20000/7

Table III Chemical compositions of wires

Wire		Element, wt.%															Welds for which used		
Code	Type	C	S	P	Si	Mn	Ni	Cr	Mo	Cu	Ti	Al	B	Sn	O	N	Multipass	Two-pass	Triple arc, two-pass
s	S4	0.14	0.026	0.021	0.15	2.20	0.03	0.01	0.01	0.36	<0.01	<0.005	<0.001	<0.01	0.025	0.006	7200/109		
i	S3	0.10	0.016	0.014	0.26	1.52	0.06	0.05	<0.01	0.14	<0.01	<0.005	<0.001	0.01	-	-	7200/11,9	7200/41	
b	S1	0.07	0.035	0.009	<0.01	0.42	0.02	<0.01	<0.01	0.22	<0.01	0.006	<0.001	<0.01	-	-	7200/7		
a	S1	0.06	0.030	0.008	<0.01	0.42	0.02	0.01	<0.01	0.22	<0.01	0.006	<0.001	<0.01	-	-	7200/3		
e	S2	0.12	0.023	0.025	0.21	0.89	0.03	0.03	<0.01	0.11	<0.01	<0.005	<0.001	<0.01	-	-	7200/15		
f	S2	0.10	0.022	0.022	0.24	0.95	0.05	0.03	<0.01	0.26	<0.01	<0.005	<0.001	<0.01	-	-	7200/13		
n	S3	0.12	0.014	0.020	0.21	1.57	0.04	0.07	<0.01	0.11	<0.01	0.007	<0.001	<0.01	0.013	0.009	7200/55,53		
k	S3	0.11	0.014	0.012	0.32	1.79	0.06	0.06	0.04	0.14	<0.01	0.010	<0.001	<0.01	0.007	0.008	7200/21		
u	S3	0.15	0.008	0.006	0.20	1.46	0.03	0.01	<0.01	0.23	<0.01	0.010	<0.001	0.02	0.010	0.012	7200/113		
q	S3	0.14	0.013	0.010	0.21	1.53	0.11	0.07	0.02	0.31	<0.01	0.006	<0.001	0.02	0.010	0.012		7200/63	
r	S3	0.08	0.011	0.015	0.11	1.72	0.11	0.05	0.02	0.24	<0.01	0.027	<0.001	0.01	-	-		7200/29	
d	S1	0.07	0.038	0.028	<0.01	0.43	0.03	0.04	<0.01	0.40	<0.01	0.006	<0.001	<0.01	-	-		7200/27	
g	S2	0.10	0.027	0.021	0.24	0.92	0.06	0.02	0.02	0.36	<0.01	<0.005	<0.001	<0.01	-	-		7200/35	
h	S2	0.12	0.025	0.027	0.24	0.90	0.03	0.02	<0.01	0.17	<0.01	<0.005	<0.001	<0.01	-	-		7200/37	
x	S3	0.12	0.008	0.012	0.21	1.56	0.04	0.04	<0.01	0.12	<0.01	0.015	<0.001	0.02	0.012	0.012		7319/3	
-	Ti-Mo-B	0.10	0.006	0.005	0.11	1.31	0.08	0.05	0.25	0.13	0.05	0.040	0.003	<0.01	0.011	0.012			20000/1
-	Ti-Mo-B	0.07	0.008	0.011	0.07	1.37	0.09	0.10	0.29	0.16	0.04	0.026	0.005	<0.01	0.006	0.009			20000/10
-	S3Mo	0.12	0.011	0.007	0.21	1.53	0.06	0.05	0.47	0.19	<0.01	0.017	<0.001	0.03	0.008	0.009			20000/3,7

All wires contained <0.01%V and Pb; <0.02%Co; <0.005%Nb

0.025wt.% sulphur content, but weld 7200/27 was made on a plain carbon plate containing <0.01wt.%Si and 0.047wt.% sulphur (Table II).

Results

A summary of the major elements in the weld metals examined is given in Table IV, and the results of quantitative metallography (on unreheated weld metal) in Table V. The proportion of inclusions by volume, calculated from the composition (3) is also included in this Table for comparison with the value measured by PASEM. There is generally good agreement between these two sets of results, although the PASEM values do appear to be below the calculated values at inclusion contents above about 0.6%.

PASEM analyses

The results of selected PASEM analyses are presented as distribution curves in Fig.1a-d. These curves are intended only to facilitate visual discrimination between the distributions. Micrographs of as-deposited material from the same regions as were examined by PASEM are presented alongside each set of distribution curves in Fig.1. The total number of each inclusion type found in each weld is tabulated in Table VI, where the symbols for inclusion types used in Fig.1 are also displayed.

Size distribution

In all cases the inclusion size distribution appeared to peak between about 0.25 and 1.0 microns, in agreement with data from other sources for submerged arc welds (7-9,12-13). Overall average diameters, that is the average for all inclusions not separated according to type, are given in Table VI. Differences are small, but it appears that inclusion sizes rise with both increasing weld metal oxygen content and increasing heat input. Weld metal oxygen content is plotted against average inclusion diameter in Fig.2; the effect of oxygen content is more marked at the higher heat inputs, and at low oxygen contents the lines converge so that there is little difference between the average diameters over the range of heat inputs (3-6kJ/mm) studied. Oxygen content would be expected to affect both nucleation and growth rate of inclusions from the melt, and slower cooling rates would tend to reduce the nucleation rate and allow more time for growth. These effects could also be explained if it is assumed that the inclusions grow partly by impingement and coalescence, in which case both higher inclusion contents in the melt and slower cooling rates would lead to larger inclusions.

It is of some concern that the limit of detection of PASEM (0.25 microns) is so close to the peaks in the size distributions, but support for these curves is given by quantitative electron microscopy carried out by Barritte (7) in which extraction replicas were taken from submerged-arc welds made with basic and calcium silicate fluxes (F22 and F1) at a heat input of 4.2kJ/mm. The resulting size distributions of inclusions are shown in Fig.3. These data cannot be matched accurately to the size distributions produced by PASEM, as in the electron microscope study full diameters have been measured, while, particularly in the larger size ranges, it is

Table IV   Chemical compositions of welds

Weld Number	Element, wt.%										Inclusions, Vol.%		IIW carbon equivalent
	C	S	P	Si	Mn	Ti	Al	Mo	B	O	Calculated*	PASEM	
7200/109	0.04	0.053	0.024	0.94	0.56	<0.01	<0.005	<0.01	<0.001	0.126	1.09	0.98	0.16
7200/11	0.09	0.021	0.018	0.47	1.17	<0.01	<0.005	<0.01	<0.001	0.061	0.52	0.47	0.30
7200/9	0.09	0.009	0.016	0.32	0.82	<0.01	0.006	0.01	<0.001	0.022	0.28	0.16	0.25
7200/7	0.04	0.024	0.024	0.49	1.29	<0.01	<0.005	<0.01	<0.001	0.126	0.95	0.75	0.27
7200/3	0.07	0.027	0.025	0.24	1.01	<0.01	0.007	<0.01	<0.001	0.155	1.14	0.87	0.25
7200/15	0.06	0.030	0.036	0.26	1.01	<0.01	0.023	<0.01	<0.001	0.092	0.77	0.61	0.24
7200/13	0.08	0.027	0.026	0.22	1.03	<0.01	0.013	<0.01	<0.001	0.066	0.55	0.58	0.27
7200/53	0.07	0.013	0.031	0.39	1.80	<0.01	0.010	<0.01	0.002	0.044	0.31	0.21	0.39
7200/21	0.10	0.005	0.020	0.17	1.30	<0.01	0.015	0.02	<0.001	0.025	0.19	0.12	0.34
7200/113	0.09	<0.005	0.011	0.10	1.27	<0.01	<0.005	<0.01	<0.001	0.019	0.12	0.15	0.32
7200/55	0.07	0.006	0.025	0.20	1.27	<0.01	0.013	<0.01	<0.001	0.022	0.15	0.14	0.30
7200/63	0.16	0.019	0.028	0.30	1.09	<0.01	0.012	<0.01	<0.001	0.053	0.42	0.47	0.36
7200/29	0.16	0.019	0.032	0.26	0.80	<0.01	0.012	<0.01	<0.001	0.039	0.33	0.40	0.30
7200/27	0.10	0.038	0.032	0.19	0.98	<0.01	<0.005	<0.01	<0.001	0.064	0.61	0.58	0.28
7200/35	0.13	0.026	0.030	0.21	0.94	0.02	0.028	0.01	<0.001	0.055	0.43	0.27	0.31
7200/37	0.14	0.036	0.028	0.18	1.10	<0.01	0.010	<0.01	<0.001	0.043	0.45	0.53	0.30
7200/41	0.11	0.008	0.023	0.22	0.93	<0.01	0.026	<0.01	<0.001	0.022	0.15	0.20	0.27
7319/3	0.13	0.006	0.015	0.21	1.09	<0.01	0.019	<0.01	<0.001	0.024	0.15	0.16	0.32
20000/1	0.15	0.010	0.013	0.07	1.21	<0.01	0.019	0.12	<0.001	0.020	0.15	0.15	0.40
20000/10	0.14	0.006	0.012	0.05	1.18	<0.01	0.013	0.15	<0.001	0.027	0.18	0.17	0.40
20000/3	0.15	0.012	0.011	0.14	1.36	<0.01	0.015	0.22	<0.001	0.028	0.22	0.29	0.45
20000/7	0.17	0.006	0.016	0.33	1.35	<0.01	0.022	0.21	<0.001	0.021	0.13	0.14	0.45

* See Ref.3
+ MP = multipass; 2P = two-pass; 2P3 = two-pass, triple arc.   See Table 1 for details.
a See Table VII
b See Table III
c See Table II

Table V

Table V Summary of available quantitative microstructural data

Weld Number	Type	Flux[a] Type	Code	Wire[b] Type	Code	Plate[c] Type	Code	% constituents* Grain boundary ferrite	Ferrite with aligned second phase	Acicular ferrite	Ferrite carbide aggregate	Prior austenite grain size, average dia. (µm)
7200/109	MP	CSHS	F15	S4	s	C-Si-Al	J	67	21	12	0	–
7200/11	MP	CS	F1	S3	l	C	B	24	37	27	11	144
7200/9	MP	CSLS	F2	S3	l	C	B	41	18	41	0	112
7200/7	MP	MS	F5	S1	b	C	B	29	67	4	0	~188
7200/3	MP	MS	F3	S1	a	C	C	25	64	11	0	160
7200/15	MP	AR	F7	S2	e	C	C	38	56	5	0	153
7200/13	MP	AR	F6	S2	f	C	E	33	11	26	28	109
7200/53	MP	AB	F8	S3	n	C-Si-Al	G	16	2	82	0	100
7200/21	MP	B	F9	S3	k	C	D	22	4	74	0	90
7200/113	MP	B	F18	S3	u	C-Si-Al	J	22	8	70	0	93
7200/55	MP	B	F16	S3	n	C-Si-Al	G	25	3	72	0	132
7200/63	2P	CS	F1	S3	q	C-Si-Al	F	42	31	25	0	
7200/29	2P	CSLS	F2	S3	r	C-Si-Al	F	38	8	52	0	
7200/27	2P	MS	F5	S1	d	C	C	35	43	20	0	
7200/35	2P	AR	F7	S2	g	C-Si-Al	F	35	15	48	0	
7200/37	2P	AB	F8	S2	h	C-Si-Al	F	35	25	27	11	
7200/41	2P	B	F9	S3	l	C-Si-Al	A	40	16	44	0	
7319/3	2P	B	F16	S3	x	C-Si-Al	J	32	21	47	0	

a See Table VII
b See Table III
c See Table II

*Data taken from Ref.10.

399

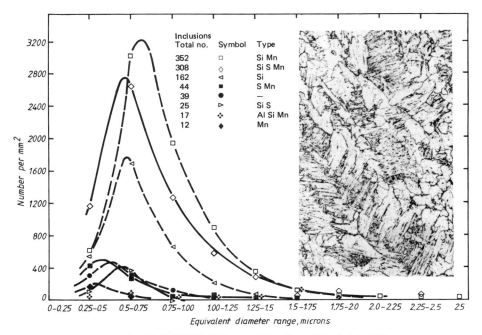

1 a) *Weld 7200/109. Multipass; calcium silicate (high silica) flux (F15).*

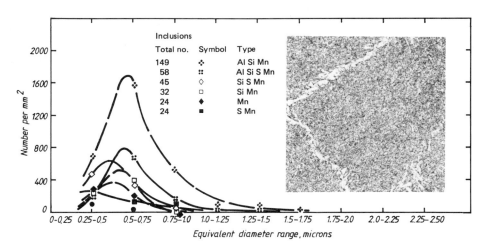

1 b) *Weld 7200/53. Multipass weld; alumina basic flux (F8).*

400

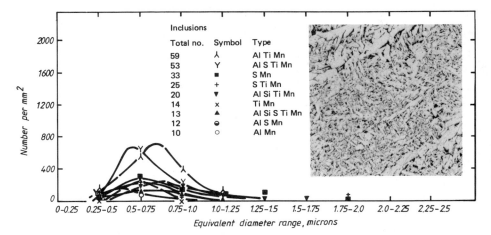

Inclusions

Total no.	Symbol	Type
59	λ	Al Ti Mn
53	Y	Al S Ti Mn
33	■	S Mn
25	+	S Ti Mn
20	▼	Al Si Ti Mn
14	x	Ti Mn
13	▲	Al Si S Ti Mn
12	⊖	Al S Mn
10	○	Al Mn

1 c) *Weld 7200/35. Two-pass; alumina flux (F7).*

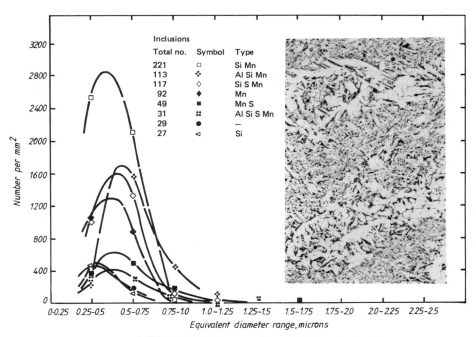

Inclusions

Total no.	Symbol	Type
221	□	Si Mn
113	❖	Al Si Mn
117	◇	Si S Mn
92	◆	Mn
49	■	Mn S
31	⠿	Al Si S Mn
29	●	—
27	◁	Si

1 d) *Weld 7200/29. Two-pass; calcium silicate, low silica flux (F2).*

Table VI   Summary of results of PASEM analyses

Weld Number	Type*	Flux Type[a]	Flux Code	Wire Code[b]	Plate Code[c]	Total	Overall average diameter (microns)
7200/109	MP	CSHS	F15	s	J	985	0.81
7200/11	MP	CS	F1	1	B	603	0.72
7200/9	MP	CSLS	F2	1	B	260	0.65
7200/7	MP	MS	F5	b	B	802	0.79
7200/3	MP	MS	F3	a	C	924	0.79
7200/15	MP	AR	F7	e	C	981	0.64
7200/13	MP	AR	F6	f	E	796	0.69
7200/53	MP	AB	F8	n	G	349	0.64
7200/21	MP	B	F9	k	D	177	0.65
7200/113	MP	B	F18	u	J	292	0.59
7200/55	MP	B	F16	n	G	225	0.64
7200/63	2P	CS	F1	q	F	379	0.90
7200/29	2P	CSLS	F2	r	F	722	0.58
7200/27	MS	MS	F5	d	C	498	0.88
7200/35	2P	AR	F7	g	F	272	0.81
7200/37	2P	AB	F8	h	F	608	0.76
7200/41	2P	B	F9	1	A	357	0.61
7319/3	2P	B	F16	x	J	213	0.70
20000/1	2P3	B	F16	-	DH04	212	0.68
20000/16	2P3	B	F16	-	DH04	221	0.71
20000/3	2P3	B	F17	-	DH04	404	0.68
20000/7	2P3	B	F17	-	1336J	198	0.68

* MP = multipass; 2P = two-pass; 2P3 = two-pass, triple arc. See Table 1 for details.

a See Table VII
b See Table III
c See Table II

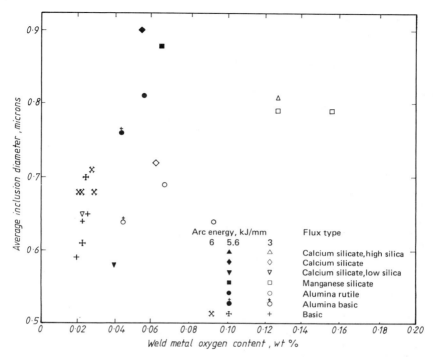

2 *Relationship between weld metal oxygen content and average inclusion diameter.*

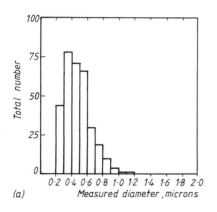

(a)

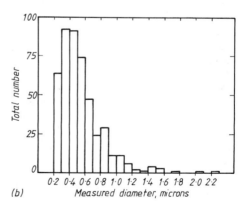

(b)

3 *Size distributions of inclusions from Ref.8. 50 fields of view, ×8000:*
a) *Basic flux;* b) *Calcium silicate flux.*

probable that section diameters are observed in PASEM. It is, however, reassuring that the majority of the inclusion diameters are shown to lie above the PASEM detection limit in these cases. In the electron microscope study particles traced from micrographs at 8000 times magnification were examined using a Quantimet grey level image analyser. The resolution (0.2 microns) was set by the image analysis and data above this threshold are considered to be reliable. The good correlation between the total volume fraction of inclusions measured by PASEM and that calculated from the composition is further indication that the majority of inclusions have been assessed, although even relatively large numbers of very small inclusions will not make a significant contribution to the total volume fraction.

Inclusion type

Inclusions of one type are defined as those which have been found to contain the same group of elements. A total of 32 are listed in Table VI, of which 27 occur in significant numbers, although this may be reduced by about half in reality (see discussion of sulphur and pairing of distributions later in this section). It is of interest that Ca was not found in any of the inclusions despite its presence in nearly all the fluxes. This is in agreement with inclusion analyses reported by other workers who were also examining welds made with Ca bearing fluxes (14-16).

From Table VI, the smallest number of inclusion types is given by the manganese silicate fluxes, an example of the size distributions being shown in Fig.1a. In all three cases, the peaks for inclusions containing Mn and Si, and Mn, Si and S are dominant. Silicon is present in the inclusions, despite the very low plate and wire contents in all cases. No Ti- or Al-bearing inclusions were observed, which, in view of the absence of these elements in plate and wire, only shows that they cannot be produced by the flux. It cannot be said whether they would be present in welds on Ti- or Al-bearing plate or made with Ti- or Al-bearing wire, and this should be investigated.

Similar inclusion distributions to the manganese silicate deposits were found for the calcium silicate and high silica calcium silicate fluxes with the exception that there was also a small number of Al-bearing inclusions. The low silica calcium silicate flux (F2) produced distributions similar to the other calcium and manganese silicate fluxes, although a number of Ti-bearing inclusions were observed, consistent with the higher TiO_2 content of the flux (3.8%cf. <0.7%, Table VII). There were virtually no Al-bearing inclusions in the multipass weld (7200/9), and those in the two-pass weld (7200/29) can be attributed to the plate and wire; this corresponded to a very low flux alumina content of 2.6% (Table VII).

The high alumina content of the alumina rutile and alumina basic fluxes is reflected by the high number of Al-bearing inclusions in these welds. The majority of inclusions in the multipass weld No.7200/15, made with both plate and wire containing <0.005% Al, are Al-bearing, as is the case in the two-pass welds made with fluxes F7 and F8 (welds 7200/53, 35 and 37), although Al was present in plate and/or wire in these cases. Weld 7200/13 appears to be slightly anomalous: the number of Al-bearing

404

inclusions was the least out of this group of welds, while the flux alumina content (42.5%) was very high, and the weld metal Al content (0.013wt.%) was mid range.

It can also be seen from Table VII that the two alumina fluxes, F6 and F7, contain 10% TiO_2, whereas none was detected in the alumina basic flux F8. This is reflected by the inclusion types. The majority of those in welds 7200/15 and 35 (F7) and a significant number in weld 7200/13 (F6) contained Ti, whereas very few did in welds 7200/53 and 37 (F8).

Table VII   Calculated oxide compositions of fluxes and basicity indices

Flux	Type/manufacture*	SiO_2	TiO_2	ZrO_2	Al_2O_3	FeO	MnO	CaO	MgO	K_2O	Na_2O	CaF_2	CO_2	Basicity index[0]
F15	CSHS/F	58.8	0.1		4.9	1.0	0.7	38.2	0.2	0.2		0.4		0.8
F1	CS/F	35.7	0.5		12.1	0.6	5.1	26.7	12.6	0.2	0.1	10.0		1.2
F2	CSLS/F+	34.9	3.8		2.6	0.4	0.5	37.5	1.0	0.1	5.1	13.9	1.0	1.5
F3	MS/F	39.4			10.0	1.8	42.2	6.7	1.8	0.5	0.5	3.3		0.8
F5	MS/F	44.1	0.7		3.8	1.2	42.5	(2.3x)		0.8		7.3		0.7
F6	AR/F	13.9	10.5	6.6	42.5	3.0	14.2	2.0	0.8	0.2	0.1	10.9		0.5
F7	AR/A	12.6	13.3	0.2	45.2	1.2	12.2	3.2		1.3	0.1	12.3		0.6
F8	AB/A	19.0		0.1	28.5	1.9	10.7	3.1	21.6	0.1	1.0	15.2		1.4
F9	B/A	12.4	0.5	0.2	13.5			11.6	36.0			26.0		3.8
F16	B/A	10.9	0.6		17.0	2.1	0.8	4.2	40.5	0.4		26.5	1.5	3.7
F17	B/A	15.6	4.0	0.1	21.4	1.3	3.2	3.2	24.4	0.1		26.9		1.5
F18	B/F	16.0	3.8	5.1	20.0	0.9	3.2	19.6	26.5	0.5		3.1		1.7

x   BaO, not CaO
0   Ref.11
*   Flux types: AR alumina rutile   CS   calcium silicate   Manufacture: A   agglomerated
                  AB alumina basic   CSHS calcium silicate,   F   fused
                  B   basic   high silica   F+ dressed, i.e. additions
                                           CSLS calcium silicate,   made after fusion
                                                 low silica
                                           MS   manganese silicate

The basic fluxes used in this work contained between 13 and 22% Al_2O_3 (Table VII), and significant numbers of Al-bearing inclusions were present in all the welds. Titanium-bearing inclusions were also observed in weld 7200/113, made using flux F18 which contained 3.8% TiO_2, but very few were evident in the other single arc basic flux welds, where flux TiO_2 contents were <0.6%. The triple arc welds 20000/3 and 7 were made with flux F17, containing 3.8% TiO_2, and both gave a significant number of Ti-bearing inclusions. Many inclusions in welds 20000/1 and 10 also contained Ti, but, as the flux contained only 0.6% TiO_2, which in the case of weld 7319/3 had not been sufficient to produce Ti-bearing inclusions, this was more probably due to the 0.04-0.05%Ti in the wire.

Although it has been shown (17) that flux type can affect weld metal sulphur content, it appears to have no bearing on the presence of sulphur in inclusions. Furthermore, when inclusions containing sulphur were identified, inclusions containing the same other elements but without sulphur were also present. There were only four exceptions to this

405

observation, the most significant of which arose with weld 7200/15, and even then concerned only 10 inclusions (Al,S,Ti,Mn type). In some cases it was noted that the distribution curves for inclusions with and without sulphur followed each other closely, clear examples of this pairing being shown in Figs.1a and b. It would seem probable that this is more than coincidental, but it is not possible to say whether it is due to some sectioning of duplex inclusions, or whether it is an instrumental effect caused, for example, by the position of the sulphur threshold, or in fact a combination of both. It may be, however, that in many cases inclusions with and without sulphur should be treated as one species, and this should be borne in mind when examining the distribution curves. For this reason, the symbols used for the distribution curves have been paired (see Table VI), and solid lines have been used for curves with sulphur and discontinuous lines for those without. Inclusion types encompassing both those with and without sulphur, are designated by putting S in brackets: e.g. Mn,Si(S).

Microstructure

Where there was a dominant peak for inclusions containing Si and Mn, with and without sulphur, the microstructure contained coarse grain boundary ferrite, usually associated with coarse ferrite with aligned second phase (18) (welds 7200/109, 11, 9, 7, 3 and 27 eg. Fig.1a). On the other hand, where the peaks for Al-, Si- and Mn-bearing inclusions, with and without sulphur, were clearly dominant, a significant proportion of fine acicular ferrite was found in the microstructure (welds 7200/53, 21, 113, 55 and 7319/3 eg. Fig.1b). Titanium-bearing inclusions were rarely dominant, but appeared to be associated with acicular ferrite of mixed grain size (welds 7200/35 and 20000/1 eg. Fig.1c). Where no one type of inclusion was clearly dominant, a mixture of microstructural constituents was observed, none of which accounted for the majority of the area (welds 7200/15, 63, 29, 37 and 20000/10 eg. Fig.1d).

These apparent associations of inclusion types and microstructural constituents are not, however, necessarily evidence of cause and effect. For example, all the welds with predominant Si,Mn(S) inclusions had carbon equivalents of <0.30 in the weld metal, and all those with predominant Al, Si,Mn(S) inclusions had carbon equivalents of >0.30, which will undoubtedly have affected the microstructure. The presence of Al bearing inclusions gives no indication of the amount of Al in solid solution, this being better derived by comparison of Al, oxygen and nitrogen levels, and it is most unlikely that it is Al in solid solution which is producing acicular ferrite where Al,Si,Mn(S) inclusions are found.

The triple arc welds were included in this study because the significant microstructural differences between welds 20000/1 and 10, and those between 20000/3 and 7 could not be attributed to the very slight compositional differences in terms of these elements normally cited in consumable specifications. Weld 1 had a very fine, acicular ferrite microstructure, while 10 consisted principally of ferrite with aligned second phase. The total numbers of inclusions (212 and 221 respectively) and the volume fractions (0.15 and 0.18) were similar, but there were 162 Al-bearing inclusions in weld 1, compared with only 88 in weld 10. Welds 3 and 7 had similar differences in microstructure, weld 3 being predominantly ferrite with aligned second phase, and weld 7 principally acicular ferrite. In this case the total numbers of inclusions were significantly different (404 and 198 respectively) but the numbers of Al-bearing inclusions were closely similar (140 and 157 respectively). However, whereas weld 3 contained 91 Si,Mn(S) inclusions, which have been associated with ferrite with aligned second phase, weld 7 only contained one.

Discussion

Application of the PASEM techniques

The major advantage of PASEM is clearly its ability to collect data from a large number of inclusions in an acceptable time period. However, the limitations of the techniques employed (particularly scanning electron microscopy with X-ray microanalysis) must be considered. The resolution of the microscope (0.25 microns at the magnification employed) has already been discussed (section 4.1.1.). A lower value would be preferable, but the transmission electron microscopy indicated that size data down to this level were reliable (7). Furthermore, X-ray data from smaller inclusions would be even less reliable than those obtained in this study. The diameter of the volume of material from which X-rays are produced by an incident electron beam of this energy (25kV) is approximately 10 times the beam diameter (14), equivalent to about 1.25 microns in this case. The concentrations in the matrix of the elements analysed are very low, however, compared with the concentrations of those elements in the inclusions, and an allowance has been made for Mn and Si in nearly all cases by using a higher background threshold. A more serious limitation of the analysis is X-ray absorption and/or fluorescence within both the inclusion and the surrounding matrix. Absorption would have a particularly marked effect on the elements of lower atomic weight (Al, Si, S). These effects (volume of analysis, X-ray absorption and fluorescence) mean that quantitative analysis is not possible, and for this reason inclusions are classified only according to the presence or absence of an element.

The level of the threshold remains a problem, and it is possible that in some inclusions certain elements will give peaks below the threshold because of strong absorption. This is one possible explanation for the pairing of some inclusion distributions with and without sulphur (section 5.2.2.); it may be that all such inclusions contain sulphur, and that the threshold level is producing an artificial distinction. It is also possible that these inclusions are complex in nature, as reported by several authors (14,20,21), and that they are sectioned by the polishing process. Certainly it would be reasonable to expect multiphase inclusions.

A number of refinements could be made to the techniques used in this study; a larger area could be covered by examining more fields of view; a higher magnification could be used (with more fields of view to compensate for the smaller area covered in each); problems of matrix X-ray absorption and fluorescence could be avoided by examining extraction replicas. Overall, the approach has been useful in providing data on flux/microstructure relationships, and in indicating which areas are worthy of further study: these now need to be explored in more detail, but by using techniques permitting closer examination of individual inclusions.

407

Relationship between flux and inclusions

Bearing in mind the above-mentioned limitations of the technique, it was nevertheless found that there were significant differences between the inclusion distributions arising from the use of different flux types. In the Mn silicate flux welds the Mn,Si(S) inclusions were most numerous, and there were very few of any other type. Similar distributions were found with Ca silicate fluxes, although Al bearing inclusions were present with these. Manganese and to a lesser extent Si, were identified in most of the inclusions from the more basic flux welds as well, but Al and Ti were also present to a varying degree, apparently principally dependent on the Al_2O_3 and TiO_2 contents of the fluxes. It is, in fact, clear that Al, Si and Ti may all be found in inclusions, as a result of their presence in the flux. In the case of Al, this was demonstrated with fluxes F1 and F7 (welds 7200/11 and 15) and, while it is possible that some metallic Al was present in the agglomerated flux F7, this could not be so in the fused flux, F1. Similarly, Ti was found in inclusions in welds made with the fused fluxes F2 and F18, and Si in welds made with F3 and F5, both of which are fused. Calcium, on the other hand, is present in significant proportions in all the fluxes, and has not been found in any of the inclusions.

It might be expected that, in something approaching an equilibrium situation, the amount of an element in the inclusions would be dependent on its concentration in the molten pool. The number of inclusions containing an unspecified amount of an element is not an accurate measure of the amount of that element in inclusions, but, if a relationship between certain inclusion types and microstructure is postulated, it is a useful value. This number for Al-bearing inclusions has been plotted in Fig.4 against total weld metal Al content, and against % Al_2O_3 in the flux. The correlation between the number of inclusions and weld metal Al content is not good, but, particularly in the multipass welds, where the effects of differences in plate composition are minimised, there is a more satisfactory relationship with flux alumina content. Thus it is possible that, if Al bearing inclusions are desirable in weld metal, they would be more satisfactorily introduced through the flux than by wire alloying. Indeed, it may be that the better relationship with flux Al_2O_3 content is due to a buffering effect of the flux, which would help to accommodate variations in plate Al content in high dilution welds, as suggested by Abson (18).

Relationship between inclusions and microstructure

It is suggested above that a good acicular ferrite microstructure may be associated with Al-bearing inclusions. Although it has also been shown, in other work, (23) that excess Al in weld metal can have a very strong deleterious effect on microstructure and toughness, a small amount of Al was nevertheless found necessary for optimum microstructure and toughness. Particularly low toughness appears to be experienced when the oxygen content of the weld metal is balanced or exceeded by the Al content in the stoichiometry Al_2O_3(24), suggesting that inclusions with a high Al_2O_3 content do not promote acicular ferrite nucleation. This was indeed observed by Devillers et. al. (20) who nevertheless found that acicular ferrite did nucleate on AlMn silicates. Mills et.al.(22) noted a similar effect, and commented that Galaxite (59%Al_2O_3, 41%MnO by weight) has a

408

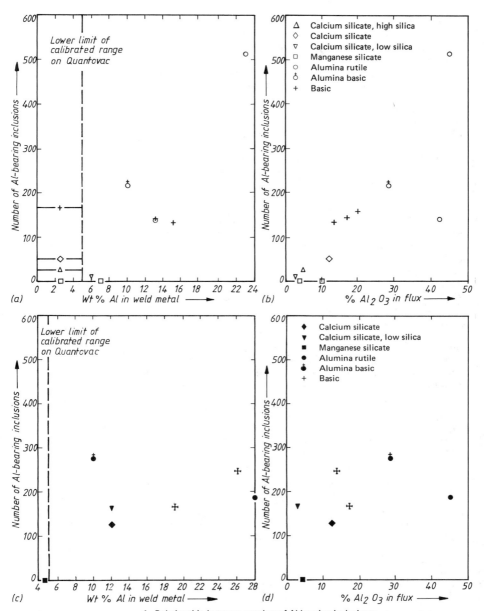

4 *Relationship between number of Al-bearing inclusions:*

a) *Weld metal Al content (multipass welds);*
b) *Flux Al₂O₃ content (multipass welds);*
c) *Weld metal Al content (two-pass welds);*
d) *Flux Al₂O₃ content (two-pass welds).*

409

particularly low lattice mismatch with ferrite. Al_2MnO_4 was also found in high acicular ferrite welds by Dowling et al (14). They, furthermore, concluded that there was no apparent relationship between TiO inclusions and acicular ferrite. Although it is known that Ti can have a powerful effect on weld metal microstructures, (23,24) no explanation of this phenomenon is possible from the present data.

From Tables V and VI, there seems to be an apparent association between Mn,Si(S) inclusions, and coarse grain boundary ferrite and ferrite with aligned second phase. This would correlate with the work of Cochrane and Kirkwood (1) where it was shown that manganese silicate inclusions in prior austenite grain boundaries could lead to the early nucleation of grain boundary proeutectoid ferrite, and the growth of Widmanst*tten side plates across the grains.

PASEM analysis has also been carried out by Terashima and Hart (23) and it is of interest to compare their results, reproduced in Table VIII, with those of the present work. The values presented are numbers/mm^2, and should be divided by 19.4 to correspond with the absolute numbers presented in Table VI. Microstructures were classified as A, B and C, corresponding to predominantly acicular ferrite, mixed, and predominantly ferrite with aligned second phase respectively. Three welds with large proportions of Al, Mn and Si bearing inclusions (W111, W34 and W9) have type A microstructures, and welds W16, W11 and W116, with significantly fewer such inclusions, have type C microstructures. In this case the poor microstructures are, however, probably a lower temperature transformation product due to a paucity of ferrite nucleants rather than a sideplate structure due to early grain boundary nucleation on Mn silicate inclusions. The reason for the poor microstructure of W100 is not clear, but in considering the good microstructures of W4 and W107, it must be recognised that these welds contain a large number of inclusions of other types which may well be good ferrite nucleants. Thus the data from Terashima and Hart, while unable to provide any information on possible competition between grain boundary and intragranularly nucleated products, do not conflict with the postulate that AlMn silicate inclusions are good nucleants for acicular ferrite.

Overall, a competitive situation may be envisaged whereby when both Mn silicate and Al,Mn silicate inclusions are present, the relative proportions and sizes control whether the grain boundary or intragranular transformation starts first and therefore predominates. For example, the effects of plate deoxidation practice have not been clearly defined by this work, but it is of interest to note that in the pair of triple arc welds where the major difference was plate deoxidation (20000/3 and 7), the numbers of Al-bearing inclusions were closely similar, and it was the Mn,Si (S) inclusions which appeared to be controlling the microstructure.

The reasons why one inclusion type should encourage intragranular nucleation, and another grain boundary nucleation are not immediately apparent, as both are nucleating ferrite from austenite. The explanation may be associated with the temperature of formation, and hence the spatial distribution of the inclusions within the microstructure, and possibly the orientation relationship with δ ferrite, if the same inclusions have

Table VIII Results of PASEM analyses from ref.23

The inclusion-type columns are headed by stacked element groups (combinations of Al, Si, S, Ti, Mn). The summary columns on the right are S, +, S, Others, Total and Microstructural type. The measured values for each weld are given below in column order (left to right of the original table).

Weld code	Measured values (composition columns, in order)	Total	Microstructural type [**]
W3	1408, 299, 70, 18, 18, 968, 810, 35, 88, 35, 18, 35, 18, 211, 493, 1180, 88, 70	5863	B
W111	978, 102, 122, 41, 183, 102, 102, 1936, 591, 999, 204, 20, 61, 41, 795, 530, 795, 163, 102	7967	A
W34	20, 20, 163, 41, 1120, 550, 20, 61, 20, 693, 143, 2303, 713, 20, 326, 122	6338	A
W16	600, 39, 445, 19, 77, 155, 77, 58, 39, 310, 19, 717, 194, 19, 155	2963	C
W4	61, 40, 40, 61, 242, 121, 121, 1170, 545, 121, 20, 40, 2865, 1372, 2764, 40, 161	9847	A
W9	484, 97, 19, 19, 116, 213, 988, 445, 77, 19, 19, 407, 19, 368, 736, 794, 58, 116	4687	A
W11	1220, 39, 39, 19, 77, 19, 213, 407, 19, 58, 19, 77, 542, 349, 58, 136	3390	C
W116	39, 813, 19, 19, 116, 19, 252, 523, 58, 1453, 600, 116, 39, 77, 1298, 659, 736, 39, 291	7438	C
W100	581, 155, 155, 19, 252, 39, 967, 194, 426, 97, 77, 232, 39, 19, 77, 19, 775, 581, 1569, 97, 291	6547	C
W107	1278, 174, 19, 19, 136, 426, 426, 39, 19, 58, 39, 755, 717, 97, 77	3731	A

[**] A = predominantly acicular ferrite
B = mixed
C = predominantly ferrite with aligned second phase

nucleated this constituent from the melt, as suggested by Cochrane and Kirkwood (1). However, it must be recognised that, if in one weld Mn silicate inclusions encourage grain boundary nucleation, under different conditions of, for example, alloying, or cooling rate, or if grain boundary nucleation is hindered, by for example boron, intragranular Mn silicate inclusions may also encourage the formation of acicular ferrite.

Summary and Conclusions

The PASEM has been used to examine non-metallic inclusions in submerged-arc weld metal. It has been shown that inclusions containing manganese and silicon, with and without sulphur, predominate in welds made with manganese silicate and calcium silicate fluxes. With other more basic flux types, aluminium and titanium were also found in some of the inclusions, apparently principally depending on the Al_2O_3 and TiO_2 contents of the fluxes. The majority of inclusions contained manganese in all cases.

Inclusion types and sizes have been considered with respect to both the final weld metal composition and plate, wire and flux compositions. The relationship between microstructure and inclusion type has also been examined.

It has been shown that aluminium, silicon and titanium can all be transferred from the flux by some means to the inclusions, but that calcium is not transferred.

An apparent association of inclusions containing only manganese and silicon, with and without sulphur, with coarse grain boundary ferrite and coarse ferrite with aligned second phase and aluminium-bearing inclusions with acicular ferrite was observed.

REFERENCES

1. COCHRANE, R.C. and KIRKWOOD, P.R. 'The effect of oxygen on weld metal microstructure'. Welding Institute conference on Trends in steel and consumables for welding, London 14-16 November 1978.

2. ABSON, D.J., DOLBY, R.E. and HART, P.H.M. 'The role of non-metallic inclusions in ferrite nucleation in carbon steel weld metals'. Ibid.

3. BAILEY, N. and PARGETER, R.J. 'The influence of flux type on the strength and fracture toughness of mild steel weld metal - Part 1 - two pass welds'. Welding Institute Research Report 70/1978/M, July 1978.

4. BAILEY, N. and PARGETER, R.J. 'The influence of flux type on the strength and fracture toughness of mild steel weld metal - Part 2 - multipass welds'. Welding Institute Research Report 100/1979, August 1979.

5. BAILEY, N. and PARGETER, R.J. 'The influence of flux type on the strength and fracture toughness of mild steel weld metal - Part 3 - further tests on two pass welds'. Welding Institute Research Report 101/1979, August 1979

6. EKELUND, S. and WERLEFORS, T. 'A system for the quantitative characterisation of microstructures by combined image analysis and X-ray discrimination in the scanning electron microscope'. Proceedings of the symposium on scanning electron microscopy. IIT Research Institute, Vol. 1, 1970.p. 417.

7. BARRITTE, G.S. "The microstructure of weld metals in low alloy steels". PhD thesis University of Cambridge 1982.

8. KIRKWOOD, P.R. The role and application of boron in submerged arc welding consumables for the offshore oil industry PhD thesis Teeside Polytechnic. September 1978.

9. FARRAR, R.A. and TULIANI, S.S. "The influence of high oxygen contents on the fracture toughness of submerged arc mild steel weld metals" Welding Research International 5(4) April 1975.

10. BAILEY, N. and PARGETER, R.J. 'The influence of flux type on the strength and toughness of mild steel weld metal - Part 4 Additional test data'. Welding Institute Research Report No. 124/1980.

11. TULIANI, S.S., BONISZEWSKI, T. and EATON, N.F. 'Notch toughness of commercial submerged-arc weld metal'. Weld and Metal Fabrication 37 (8), August 1969, pp. 327-339.

12. FERRANTE, M. and FARRAR, R.A. "The role of oxygen rich inclusions in determining the microstructure of weld metal deposits". J. Mat.Sci. 17 1981pp.3293-3298.

13. LIU, S. and OLSON, D.L. "The role of inclusions in controlling HSLA steel weld microstructures" Welding Journal, Research Supplement, 65 (6), June 1986. pp 139s - 149s.

14. DOWLING, J.M., CORBETT, J.M. and KERR, H.W. "Effects of inclusion compositions and size distributions on the microstructure and properties of submerged arc welds" Proc. conf. on Inclusions and Residuals in Steels: Effects on Fabrication and Service Behaviour. Ottawa March 1985 CANMET/CSIRA.

15. SAGGESE, M.E., BHATTI, A.R., HAWKINS, D.N., and WHITEMAN, J.A. "Factors influencing inclusion chemistry and microstructure in submerged arc welds". Proc.Conf. on The Effects of Residual, Impurity and Microalloying Elements on Weldability and Weld Properties. London, 15-17 November 1983. The Welding Institute.

16. BHATTI, A.R., SAGGESE, M.E., HAWKINS, D.N., WHITEMAN, J.A., and GOLDING, M.S. "Analysis of inclusions in submerged arc welds in microalloyed steels." Welding Journal Research Supplement 63 (7) July 1984 pp224s - 230s.

17. DAVIS, M.L.E. and BAILEY, N. 'How submerged-arc flux composition influences element transfer'. Weld pool chemistry and metallurgy.

International Conference, London, April 1980. The Welding Institute.

18. "Guidelines for classification of ferritic steel weld metal microstructural constituents using the light microscope" IIW Doc. No. IX-1377-85.

19. ANDERSON, C.A. (ed). 'Microprobe analysis'. Wiley, London, 1973.

20. DEVILLERS, L., KAPLAN, D., MARANDET, B., RIBES, A. and RIBOUD, P.V. "The effect of low level concentrations of some elements on the toughness of submerged arc welded C-Mn steel welds". The effects of residual, impurity and microalloying elements on weldability and weld properties. International Conference, London November 1983. The Welding Institute.

21. ABSON, D.J., Unpublished work, 1986.

22. BARRITTE, G.S., RICKS, R.A., and HOWELL, P.R. "Application of STEM/EDS to study of microstructural development in HSLA steel weld metals" proc.conf. on Quantitative Microanalysis with High Spatial Resolution Manchester 25-27 March 1981. UMIST/ The Metals Society.

23. TERASHIMA, H. and HART, P.H.M. 'Effect of flux TiO_2 content on tolerance to high Al content of submerged arc welds made with basic fluxes'. The effects of residual, impurity and microalloying elements on weldability and weld properties. International Conference, London November 1983 The Welding Institute.

24. BAILEY, N. "Ferritic steel weld metal microstructures and toughness" Perspectives in metallurgical development. Metals Society, London 1984 pp276-281.

25. MILLS, A.1., THEWLIS, G. and WHITEMAN, J.A. "The nature of inclusions in steel weld metals and their influence on the formation of acicular ferrite". Submitted for publication, Materials Science and Technology.

THE ROLE OF INCLUSIONS IN THE FRACTURE

OF AUSTENITIC STAINLESS STEEL WELDS AT 4 K*

T. A. Siewert and C. N. McCowan

Fracture and Deformation Division, MC 430
National Bureau of Standards
325 Broadway
Boulder, Colorado 80303

Abstract

Inclusion densities were measured for three types of austenitic stainless steel welds and compared to the 4-K yield strengths, 76-K Charpy V-notch absorbed energies, and the ductile dimple densities on the respective fracture surfaces. The welds included shielded metal arc (SMA) welds and gas metal arc (GMA) welds. The inclusion density was consistently a factor of 8 to 10 less than the fracture surface dimple density. Inclusion and dimple densities ranged from 3.9×10^4 inclusions$\cdot$mm^{-2} and 3.2×10^5 dimples$\cdot$mm^{-2} for one SMA specimen to 1.1×10^4 inclusions$\cdot$mm^{-2} and 1.2×10^5 dimples$\cdot$mm^{-2} for the fully austenitic GMA specimen. Both ductile dimple density and dimple morphology varied with specimen type. The inclusion data for the welds agreed with a linear relationship between fracture toughness and inclusion spacing that had been developed for base metals.

*Contribution of NBS; not subject to copyright.

Introduction

Type 316LN stainless steel is emerging as the preferred structural material for superconducting magnet casings designed to operate at 4 K because it has a higher strength than the older 316L grade and comparable toughness at cryogenic temperatures.[1-3] Unfortunately, welding electrodes matching this composition and strength have substantially poorer toughness at 4 K than the stainless steel plate. This is illustrated in figure 1 for type 304LN stainless steel (similar in strength and toughness to type 316LN) and various welds produced with type 308LN and type 316LN electrodes.[4] Matching the electrode strength to the base metal strength results in much lower toughness. This forces the designer to move the welds into regions of the structure where the lower toughness can be tolerated. Where high toughness welds are required, the corresponding weld strength is substantially reduced, and the structural design becomes less efficient. This study sought the reason for the lower weld toughness and also looked at parameters that can be used to predict the 4-K fracture toughness.

Experimental Procedures

Two shielded metal arc (SMA) welds and a gas metal arc (GMA) weld were selected for detailed microstructural evaluation. These welds were specifically selected on the basis of previous mechanical property evaluations, so that their range of toughness and strength would reveal any microstructural basis for variations in toughness.[5,6] Table 1 lists the mechanical property data and composition of these welds. Welds SMA-1 and SMA-21 differ primarily in nitrogen and manganese content. Data from previous studies indicate a reduction in Charpy V-notch (CVN) absorbed energy as nitrogen additions increase the strength.[6] Because GMA has twice the nickel content of SMA-21, the importance of the nickel content in increasing the CVN absorbed energy and its effect on the fracture surface appearance could be examined. The delta-ferrite content is also known to affect toughness, but the compositions in this study vary over a small range (0-4 FN), and Szumachowski and Reid have shown that the effect of delta-ferrite content is less than that of the nitrogen content.[7]

Specimens for microstructural evaluation were selected from the 25-mm-thick butt welds used for mechanical property tests. Fracture surfaces that were analyzed were those of specimens fractured at 4 K, CVN specimens fractured at 76 K, and compact tension (CT) specimens fractured at 4 K. The surfaces chosen to be polished were parallel to the direction of the welding and normal to the plate surface, at least 1 cm in each dimension, and they included several weld passes. They were polished until scratches were no longer visible at 3000X, but the polishing time was minimized to reduce relief effects and selective removal of inclusions. This finish enabled measurement of the inclusion size and density with a scanning electron microscope (SEM) having a quantitative image analyzer.

Examination of both the fractured and polished surfaces at 3000X enabled easy identification of the micrometer-scale inclusions characteristic of these welds. This magnification was quite useful for measuring the surface dimple dimensions and densities. At least 500 inclusions and dimples were counted to provide statistically significant data.

The all-weld-metal specimens for tensile testing were 6 mm in diameter and oriented parallel to the direction of welding; their fracture surfaces were, therefore, normal to the direction of welding. Tensile tests were performed in accordance with ASTM standard A370 at a strain rate of 2×10^{-4} s^{-1}. Details of the fixture and the procedure for tensile testing at 4 K have been described elsewhere.[8]

Specimens for CVN tests were standard 10-mm square, centered in the weld and notched so that the crack propagated in the weld direction; they were prepared according to ASTM E23. Because physical property constraints (adiabatic heating and low heat capacity) preclude obtaining meaningful data at 4 K, the CVN testing was performed at 76 K.

The 25-mm-thick CT specimens were notched so that the fatigue crack propagated down the center of the weld in the direction of the welding; they were prepared according to ASTM E813. The $K_{Ic}(J)$ value was calculated from the fracture energy, J, using the single-specimen unloading-compliance technique. Details of the fixture and testing procedure have been described elsewhere.[9] The CT data (average of two specimens) were available for only the GMA weld.

Results and Discussion

Role of Inclusions

Neither inclusion density nor the average inclusion size was affected by the differences in nitrogen content of the SMA-21 and SMA-1 welds (see table 2). The SMA-21 weld had greater yield strength owing to its higher nitrogen content, but the inverse relationship between strength and toughness observed in a previous study[6] is not evident here. The similar inclusion densities and sizes of SMA-21 and SMA-1 may be the reason for the similarity in their CVN absorbed energies. This suggests that the factors controlling strength and toughness might be considered independently, at least for these two welds.

The GMA weld metal has more than twice the CVN absorbed energy of the SMA welds, yet its strength is similar to SMA-21. Therefore, the difference in toughness is due to another factor, possibly the nickel content, inclusion content, inclusion size, or a combination of these factors. Unfortunately, these factors are confounded in these data.

A study by Reed and Simon separated the two factors of nickel content and inclusion size.[3] Although their study was of base metals, the stainless steel compositions were similar to these weld-metal compositions. Through a statistical regression analysis of NBS test data, they determined that the fracture toughness at 4 K could be predicted by a linear combination of strength, nickel content, and inclusion spacing (calculated from inverting and taking the square root of the inclusion density):

$$K_{Ic}(MPa \cdot m^{\frac{1}{2}}) = 130 - 0.34\sigma_y + 20[Ni] + 2.25\lambda, \tag{1}$$

where σ_y is the yield strength (MPa), [Ni] is the nickel content (wt.%), and λ is the inclusion spacing (μm). The standard deviation for the fit of this equation to 19 data points was 31 MPa·m$^{\frac{1}{2}}$. This equation was developed from data with a range of 550 to 1100 MPa, a [Ni] range of 8 to 12 wt.%, and a λ range of 40 to 110 μm.

The 4-K fracture toughness of the GMA weld (table 1) is predicted well by equation 1, even though it was developed from data on alloy 316-type base metals. Since fracture toughness data at 4 K are not available for the SMA welds in table 1, they cannot be used to evaluate the fit of the equation. However, 4-K fracture toughness has been evaluated in previous studies of austenitic stainless steel welds.[10-12] Therefore, inclusion densities were

417

measured on three of these welds. These data, listed in table 3, also fit the equation fairly well, although they are outside the range of composition (Ni, Mn) of alloy 316 for which the equation was developed. This is surprising because welds and base metal have different cooling rates and solidification structures and an order-of-magnitude difference in inclusion density. This emphasizes the importance of inclusion density in controlling fracture toughness. When the weld metal data are added to the base metal data, the predictive equation for fracture toughness can be revised to cover this expanded inclusion density range.

When revising this predictive equation, it is also appropriate to reassess the form. Because the fundamental relationships among toughness, strength, and composition are only beginning to be developed, it is difficult to choose the optimum form for a predictive equation. For this reason, a stepwise multiple linear regression program was used to choose the statistically most significant form from a selection of over thirty combinations of strength, nickel content, and inclusion spacing. The combinations included linear, product, and quotient forms with various exponents. The revised predictive equation selected by the regression program is

$$K_{Ic} = 30 + 6000 \frac{[Ni]}{\sigma_y}(1 + 0.02\lambda) \, , \tag{2}$$

where the units are the same as in equation 1. This equation, predicting the toughness for both weld and base metal compositions, was developed from the data used for equation 1 and four weld data with a σ_y range of 550 to 1100 MPa, a [Ni] range of 8 to 20 wt.%, and a λ range of 5 to 110 μm. It has a standard deviation of 27 MPa·m$^{1/2}$, an R^2 of 0.87, and an F ratio of 159. This standard deviation is comparable to that found for equation 1, which was developed for the restricted data set of alloy 316-type base metals.

Role of Dimple Density

All fracture surfaces were composed of ductile dimples. The inclusions that nucleated the dimples were still at the bottom of some of the dimples. Microprobe analysis indicated they were all of the manganese silicate type. The fracture-surface dimple measurements are given in table 4. For each weld, we see less than 25 percent variation when the dimple densities were measured on tensile, CVN, and CT (for one weld) fracture surfaces. This variation is small considering that testing was done at two temperatures, at strain rates ranging from 2×10^{-4} s^{-1} for the tensile and CT specimens to approximately 1×10^3 s^{-1} for the CVN specimens, and that the specimen types had different loading configurations. However, the CVN data are consistently higher than those for the other fracture surfaces, which may indicate a small effect. More detailed studies are needed to confirm this trend. The good agreement found here indicates that the dimple densities of the various welds can be characterized with any of these specimen types.

The higher CVN absorbed energy of the GMA weld could be explained by the lower density of inclusions measured on the polished surface section. Assuming all ductile dimples observed on fracture surfaces were nucleated at inclusions, one can correlate the higher CVN absorbed energy to the lower dimple density.

In a study of C-Mn welds, Passoja and Hill found very good agreement between the inclusion density measured on an electropolished planar surface and the number of dimples on a fracture surface.[13] In fact, they found that the 1.18 factor in an equation by Kochs,

$$D_S = 1.18 \ N_S^{-1/2}, \tag{3}$$

was sufficient to correct the results when the inclusion density was calculated in terms of a three-dimensional nearest-neighbor spacing.[14] Here D_S is the nearest-neighbor spacing, and N_S is the average planar surface inclusion density.

Comparison of tables 2 and 4 reveals about an order-of-magnitude difference between the inclusion density on planar surfaces and the fracture surface dimple density for this study. This difference in density is clearly visible in figures 2 and 3. The largest difference predicted by the fracture models for these two surfaces (corrected for the difference between plan views of a 2-and 3-dimensional surface) is about a factor of 2.[15] The unexplained difference (about a factor of 5) could be due to undercounting the inclusions on the polished surface, overcounting the dimples on the fracture surface, or a surface roughness greater than that predicted by general fracture models.

The first possible cause, undercounting inclusions on the polished surface, is unlikely since the surfaces were polished with care and the final polishing time was minimized. Pits indicating that inclusions had been removed from the surface were not evident. The electropolishing technique employed by Passoja and Hill, may, in fact, have led to an overestimate of inclusion content in their study owing to selective removal of the matrix material. The second possible cause, overcounting of the dimples, is also unlikely. Whether the small indentations in figure 3 are facets of one large inclusion or several smaller inclusions is somewhat uncertain, but the possibility of miscounting by a factor of 5 is very small. This leaves the roughness of the fracture surface as the most likely reason for the differences between inclusion counts and dimple density. Grain orientations that favor shear on other than a 45-degree angle, nonuniform inclusion distribution, and other microstructural heterogeneity could cause increased surface roughness.

Comparison of the inclusion sizes of the polished and fracture surfaces provides further insight into the fracture process. Table 2 includes average inclusion diameters measured on the polished surfaces. Figures 4 through 6 are plots of inclusion diameters measured on the fracture surfaces of these three welds versus the average dimple diameters. In some cases, the dimples had major and minor axes, and the data are represented with a bar. In other cases, the dimples were circular and are represented by an X within a circle. Notice that there were relatively few inclusion-ductile dimple pairs that could be measured. Many dimples had lost their inclusions during the fracture process. Also, notice that the average diameter of the fracture-surface inclusions is much greater than that of the polished-surface inclusions. This may indicate that the small inclusions were lost more frequently than the large ones or that large inclusions are more likely to initiate fracture and, therefore, they are more commonly observed on the fracture surface. We found that another possibility, the correction to the true inclusion diameter from random sections of a polished surface, increased the average inclusion diameter only slightly.

The measurements shown in figures 4 through 6 also indicate that larger inclusions form larger dimples. Since the higher toughness GMA weld had larger dimples, perhaps further improvements in toughness can be achieved by reducing the inclusion density through better refining or agglomeration of the manganese silicate into larger, widely spaced inclusions.

Conclusions

1. The fracture-surface dimple density and inclusion density were inversely related to the Charpy V-notch absorbed energy, indicating that inclusion spacing is a primary factor in determining the toughness.

2. The fracture-surface dimple density was found to be almost independent of the strain rate during fracture and specimen type. It varied less than 25 percent when measured on a tensile specimen fractured at 4 K, a Charpy V-notch specimen fractured at 76 K, and a compact tension specimen fractured at 4 K.

3. The fracture-surface dimple density was a factor of 5 greater than that predicted by fracture models based on inclusion density. This indicates either shear on an angle other than 45 degrees, a nonuniform inclusion distribution, or microstructural inhomogeneity.

4. The weld metal fracture toughness can be predicted from an equation developed for the base metal. The applicability of one equation for both weld and base metal indicates a role of strength, Ni content, and inclusion density in determining the toughness of the austenitic stainless steels.

Acknowledgments

This study was supported by the Office of Fusion Energy, U.S. Department of Energy.

References

(1) G. M. Goodwin: Weld. J., vol. 64, 1985, pp. 19-25.

(2) S. Yamamoto, N. Yamagami, and C. Ouichi: Advances in Cryogenic Engineering--Materials, vol. 32, Plenum Press, New York, 1985, pp. 57-64.

(3) N. J. Simon and R. P. Reed: J. Nucl. Mater., in press.

(4) R. L. Tobler, T. A. Siewert, and H. I. McHenry: Cryogenics, 1986, vol. 26, pp. 392-396.

(5) C. N. McCowan, T. A. Siewert, and R. L. Tobler: Trans ASME, 1986, vol. 108, pp. 340-343.

(6) C. N. McCowan, T. A. Siewert, R. P. Reed, and F. B. Lake: Weld. J., in press.

(7) E. R. Szumachowski and H. F. Reid: Welding J., 1979, vol. 58, pp. 34-44.

(8) D. T. Read and R. L. Tobler: Advances in Cryogenic Engineering-- Materials, vol. 28, Plenum Press, New York, 1982, pp. 17-28.

(9) R. L. Tobler, D. T. Read, and R. P. Reed: Fracture Mechanics: Thirteenth Conference, Richard Roberts, Ed., ASTM STP 743, American Society for Testing and Materials, Philadelphia, 1981, pp. 250-268.

(10) D. J. Kotecki: Materials Studies for Magnetic Fusion Energy Applications at Low Temperatures--III, R. P. Reed, Ed., NBSIR 80-1627, National Bureau of Standards, Washington, 1980, pp. 197-204.

(11) T. A. Whipple and D. J. Kotecki: Materials Studies for Magnetic Fusion Energy Applications at Low Temperatures--IV, R. P. Reed, Ed., NBSIR 81-1645, National Bureau of Standards, Washington, 1981, pp. 303-321.

(12) H. I. McHenry and T. A. Whipple: Materials Studies for Magnetic Fusion Energy Applications at Low Temperatures--III, R. P. Reed, Ed., NBSIR 80-1627, National Bureau of Standards, Washington, 1980, pp. 155-165.

(13) E. Passoja and D. C. Hill: Fractography--Microscopic Cracking Processes, ASTM STP 600, American Society for Testing and Materials, Philadelphia, 1976, pp. 30-46.

(14) U. F. Kochs: Philos. Mag., 1966, vol. 13, pp. 541-544.

(15) E. E. Underwood: J. Met., 1986, vol. 38, pp. 30-32.

Table I. Selected Data for the Welds[*]

	SMA-1	SMA-21	GMA
Yield Strength (MPa at 4 K)	460	880	1020
CVN Absorbed Energy (J at 76 K)	45	42	102
$K_{Ic}(J)$ (MPa·m$^{\frac{1}{2}}$)	--	--	203
Ferrite Number (FN)	4	0.5	0
Composition (wt.%)			
C	0.03	0.03	0.03
Mn	1.6	3.5	5.4
Si	0.29	0.37	0.26
Cr	17.6	17.8	18.1
Ni	9.2	9.1	20.4
N	0.05	0.15	0.16

[*]This study.

Table II.   Inclusion Measurements on Polished Surfaces

	Weld		
Measurement	SMA-1	SMA-21	GMA
Inclusion Density, n (mm^{-2})	3.9×10^4	4.0×10^4	1.1×10^4
Average Diameter (μm)	0.36	0.37	0.45
Volume Fraction	4.8×10^{-3}	4.9×10^{-3}	2.3×10^{-3}
Mean Spacing, λ (μm)	5.1	5.0	9.5

Table III.   SMA Fracture Toughness Data at 4 K Developed in Previous Studies

Weld Identification	K_{Ic} (MPa$\cdot$m$^{\frac{1}{2}}$)	σ_y (MPa)	Ni (wt.%)	n_{-2} (mm)	λ^* (μm)	Reference
497-5 FCA	159	608	15.2	4×10^4	5	10,11
508-2 GMA	163	779	11.8	4.2×10^4	4.9	10,11
1616-2 FCA	235	607	16	3.2×10^4	5.6	12

$^*\lambda$ (the inclusion spacing) is defined as the inverse root of n (the inclusion density).

Table IV.   Dimple Measurement on the Fracture Surfaces

	Weld		
Specimen	SMA-1	SMA-21	GMA
4-K Tensile Specimen Dimple Density (mm^{-2})	3.2×10^5	3.5×10^5	1.2×10^5
76-K CVN Specimen Dimple Density (mm^{-2})	4.1×10^5	4.6×10^5	1.6×10^5
4-K CT Specimen Dimple Density (mm^{-2})	---	---	1.2×10^5

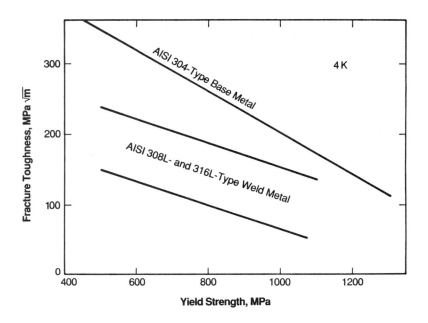

Figure 1.   Comparison of the strength-toughness relationships for type 304LN
austenitic stainless steel and type 308LN and 316LN welds.

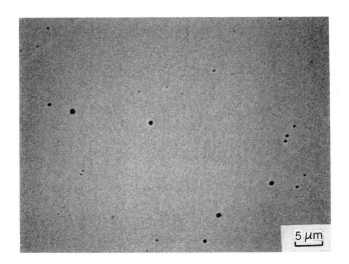

Figure 2.   Polished surface of GMA weld.

423

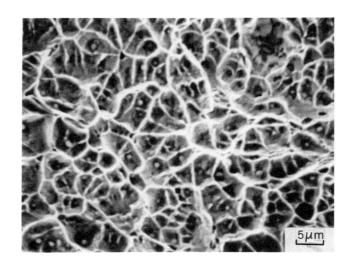

Figure 3.   Fracture surface of GMA tensile specimen.

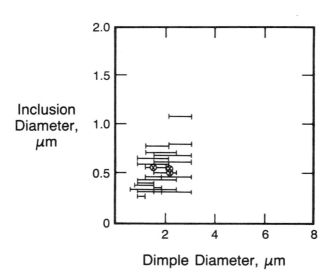

Figure 4.   Dimple diameter versus inclusion diameter for weld SMA-1.

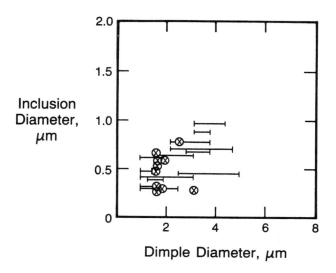

Figure 5.   Dimple diameter versus inclusion diameter for weld SMA-21.

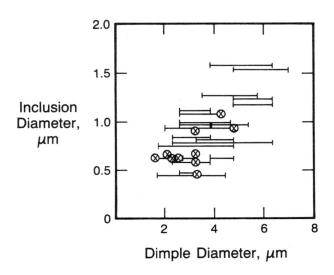

Figure 6.   Dimple diameter versus inclusion diameter for weld GMA.

INFLUENCE OF MOLYBDENUM ON THE STRENGTH AND

TOUGHNESS OF STAINLESS STEEL WELDS FOR CRYOGENIC SERVICE

C. N. McCowan and T. A. Siewert

Fracture and Deformation Division, MC 430
National Bureau of Standards
325 Broadway
Boulder, CO   80303

E. Kivineva

Department of Metallurgy
University of Oulu
90100 Oulu 10, Finland

Molybdenum additions to austenitic stainless welds were found to increase the 4-K yield strength by approximately 30 MPa per weight percent. Molybdenum additions had little effect on the 76-k Charpy V-Notch impact energy, with one exception: when the molybdenum content was raised from 1.7 to 3.8 wt.% in an otherwise equivalent stainless steel composition of approximately 17Cr-9Ni-6.6Mn-0.17N, the absorbed energy decreased from 33 to 16 J. At a nickel content of 14 wt.%, the higher molybdenum contents did not reduce the impact toughness. The loss in impact toughness for the 9 wt.% nickel, 3.8 wt.% molybdenum composition was linked to an increased number of both small and large inclusion sizes in the weld. The effects of nickel and manganese on the cryogenic strength and toughness are also reported.

427

Introduction

Recently there has been an interest in the development of higher strength austenitic stainless steels for lower cost and more compact magnetic fusion structural designs. Fusion magnet components are expected to operate at temperatures near 4 K. Austenitic stainless steels are one of the common structural materials chosen for service in this environment because unlike many high strength materials, these alloys have good toughness at temperatures approaching absolute zero. For example, high strength type 316LN stainless steel base materials have 4-K yield strengths of 1020 MPa, 76-K Charpy V-notch absorbed energies of 89 J, and 4-K fracture toughnesses [$K_{Ic}(J)$] of 224 MPa√m.[1] The properties of these nitrogen-strengthened base materials cannot be totally utilized, however, until welds of comparable toughness at these elevated strengths are also developed.

Nitrogen additions can increase the yield strength of 300-series stainless steel weld metals to levels comparable to those of the nitrogen strengthened base materials. The problem is that toughness values for the welds at cryogenic temperatures have been consistently lower than those for base materials of equal strength.[2] For this reason, other strengthening elements that have a less detrimental effect than nitrogen on toughness are being investigated.

The effect of molybdenum on the strength and toughness in shielded metal arc (SMA) welds is reported in this study, and interactive effects with nickel, nitrogen, and manganese are also investigated.

Background

Much of the data pertaining to the effects of alloying element additions on the strength and toughness of austenitic stainless steels at 4 K are from studies of base materials. In types 304 and 316 stainless steel base materials whether the strength is increased by nitrogen, carbon, or molybdenum additions the strength-toughness relationship remains essentially the same.[3,4] The toughness decreases as the strength increases. From this perspective, nitrogen, which increases the 4-K yield strength in both 300-series weld and base metal by approximately 3000 MPa per weight percent, would therefore be expected to have a more detrimental effect on toughness than molybdenum.[5-8] Molybdenum reportedly increases the yield strength at 4 K by only 40 to 80 MPa per weight percent in 300 series base materials.[3,6,9] The only alloying element that has been found to improve the strength-toughness relationship in 300 series stainless steels at cryogenic temperatures is nickel. Increased nickel contents increase the toughness while having very little effect on the strength.[10]

Although, substitutional alloying by molybdenum in 300-series stainless steel base materials has been found to predictively increase strength and decrease toughness, as molybdenum contents are increased, unexpected decreases in toughness have been observed.[3] One cause for the loss of toughness in molybdenum-alloyed stainless steels is the formation of molybdenum-rich chi and laves phases. Formation of these brittle intermetallic phases reduced the 76-K Charpy V-notch energy by more than 100 J in a 12Cr-12Ni-0.1N-Mo base metal alloy when the molybdenum content was increased from 5 to 7 weight percent.[9] However, not all reported toughness losses for molybdenum alloyed stainless steels can be clearly related to embrittlement by these intermetallic phases. A type 316 stainless steel base metal, for example, had unexpectedly low 4-K fracture toughness values at 4 weight percent molybdenum contents.[3]

428

Presumably, the molybdenum had not been uniformly distributed in this alloy. A second phase was present in the microstructure, but it was not rich in molybdenum.

In welds, nonuniform distributions of alloying elements are common owing to the dendrictic solidification process. Molybdenum, in particular, segregates extensively in austenitic stainless steel welds.[11] Therefore, decreases in weld metal toughness, similar to the decreases observed in base metals, may also result from an inhomogeneous distribution of molybdenum. In the case of a 25Cr-16Ni-0.34N-Mo weld, ductilities were found to decrease rapidly after the molybdenum content exceeded 3.5 weight percent.[12]

Materials and Procedure

The test matrix was designed to evaluate welds with molybdenum contents of approximately 0.02, 2, and 4 weight percent. At each level of molybdenum content, there were two possible nickel and nitrogen levels. In addition, several manganese additions were included in the test matrix. The alloying elements were added to the shielded metal arc (SMA) electrode coatings. The weld compositions were determined by optical emission spectroscopy, and the ferrite number (FN) of the weld deposits was measured magnetically; both are given in table 1.

Test welds were produced by using a series of 3.2-mm-diameter electrodes that had core wires from a single heat of type 308 stainless steel. The base material was 13-mm-thick by 305-mm-long mild steel plate. To overcome the effect of dilution by the base material, the exposed faces and backing strip were buttered with two layers of weld metal prior to welding as specified in AWS A5.4-78. The same welder, power supply, and welding parameters (110 A, 22 V, identical bead sequences, and a heat input of approximately 1.6 kJ/mm) were used to minimize variations in the weldments. The interpass temperature was maintained at 93°C.

Two all-weld-metal 6-mm-diameter tensile specimens, oriented along the axis of the weld, were machined from each weld. The second specimen, reserved as a spare, was tested in a number of cases to check the repeatability of the results; in these cases, the average value is reported. Tensile specimens were tested at 4 K (in liquid helium) with a strain rate of 9.5×10^{-5} s^{-1}. A 25-mm gauge length was used. Specific details on the 4-K tensile testing equipment and procedures have been reported previously.[13]

Five 10-mm-square Charpy V-notch specimens, with their notches oriented perpendicular to the plate surface and located along the weld centerline, were removed from each weld and tested at 76 K (in liquid nitrogen).

Scanning electron microscopy was employed in the acquisition of inclusion distribution data from polished weld metal samples. In addition, wavelength dispersive spectroscopy (WDS) was used to evaluate molybdenum segregation between the ferrite and austenite phases, and energy dispersive spectroscopy (EDS) was used to evaluate the composition of the inclusions on the fracture surfaces.

Welds 6 and 13 were evaluated with x-ray diffraction techniques. They were step-scanned at 0.02-degree intervals of 2θ for 25 seconds per step. The tube voltage was 44 kV at 40 mA. The radiation was $CuK\beta$ ($\lambda Cu\beta = 1.39225$ Å).

Results and Discussion

Strength

The yield strength of the weld metal at 4 K increased as a function of molybdenum, manganese, and nitrogen content. Relationships between alloy content and the ultimate strength could not be resolved. The strengths determined for the welds are listed in table 2.

The ultimate strength of the welds had a mean value of 1400 MPa, which is consistent with previous results obtained for SMA weld metals with compositions similar to the 9 wt.% nickel welds tested in this study.[5] Inhomogeneities in the weld structures and variations in ferrite contents evidently contributed to the scatter in the ultimate strength values of this study.

To evaluate the contributions of the alloying elements to the yield strengths of these welds, the data were analyzed with stepwise linear regression techniques. The 4-K yield strength (in MPa) was best predicted by the equation:

$$\text{Yield Strength} = 180 + 3200[\text{N}] + 33[\text{Mo}] + 32[\text{Mn}] + 13[\text{Ni}] \tag{1}$$

This equation (elements in wt.%) had an F value of 920, a coefficient of regression (R^2) of 0.98, and a standard error of estimate of 31 MPa.

As expected, the interstitial strengthening of nitrogen had a much greater effect on the yield strength at 4 K than the solid-solution strengthening contributions of molybdenum, manganese, or nickel. Equation 1 shows the strengthening of nitrogen (3200 MPa/wt.% N) to be two orders of magnitude greater than that of the solid solution elements. This nitrogen coefficient agrees well with those previously determined for both weld and base metals.[5-8] The effect of nitrogen on the yield strength at 4 K is shown in figure 1.

Much of the variation in yield strength, however, can be attributed to the solid-solution strengthening of molybdenum and manganese. To better show the strengthening contributions made by these elements, the yield strength data were normalized to a 0.05 nitrogen content using the 3200 MPa per weight percent strengthening coefficient from equation 1 (figure 2). The slopes of the dashed lines in figure 2 indicate the increases in yield strength due to molybdenum additions at the three manganese contents present in this study. The difference in the yield strength between respective lines, indicates the strengthening by manganese. Clearly, molybdenum and manganese also contribute to the 4-K yield strength of these welds.

Although both figure 2 and equation 1 indicate the relative magnitude of the strengthening contributions made by the alloying elements, singular strengthening coefficients to represent the effect of molybdenum or manganese could not be defined. The coefficients for these elements did not remain constant over the compositional range considered in this study. For example, when the 14 weight percent nickel data were evaluated separately, the molybdenum coefficient decreased to 25 and the manganese coefficient decreased to 15. Further manipulation of the data indicated that the substitutional strengthening effects of manganese and molybdenum may decrease as the nickel content, nitrogen content, or both increase. The coefficients in equation 1, therefore, represent the average effects of these elements.

<u>Toughness</u>

The Charpy V-notch absorbed energy (AE) and lateral expansion (LE) values determined for the welds are reported in table 2. Impact toughness at 76 K decreased as the interstitial alloy content of the welds increased. The effects of substitutional alloy content on the impact toughness varied.

Nickel increased both AE and LE for alloys having low nitrogen contents (figures 3 and 4). At higher nitrogen levels (≥ 0.15 wt.% N), however, the beneficial effect of increased nickel content on the impact toughness was reduced.

The effect of molybdenum on the impact energy was related to the nickel content. At nickel contents of 14 weight percent, molybdenum did not adversely effect either the AE or LE. Increased molybdenum contents, in fact, may have been slightly beneficial to the toughness in the higher nickel content welds. In the 9 weight percent nickel welds, however, the AE was much lower for the 3.8 weight percent molybdenum alloy (weld 6). To verify this finding, a new weld was made with another electrode of similar composition. Again, low AE values (17 J) were found for the 4Mo-9Ni type stainless steel weld metal. These AE values are approximately 50 percent lower than that of the 2Mo-9Ni alloy (weld 13). Although the chromium content in welds 6 and 13 also differs, otherwise they have quite similar compositions and the decrease in AE is felt to be directly related to 3.8 weight percent molybdenum content. Lower molybdenum contents had no adverse effect on impact toughness of the 9 weight percent nickel welds. Possible explanations for the decrease in toughness associated with the 3.8 weight percent molybdenum welds are given later in this report.

Regression analysis was applied to the AE and LE data, but equations that were significant at the 99% confidence level could not be developed. Nevertheless, the analysis consistently suggested that: (1) nitrogen reduced the AE and LE, (2) nickel increased the AE and LE, and (3) manganese and ferrite decreased the LE. Because the chromium content was lowered as the molybdenum content was increased to maintain a constant ferrite potential, the effects of these elements on toughness is uncertain. With the exception of the 4Mo-9Ni weld, however, the effect of molybdenum on toughness was small in comparison with the nitrogen effect. The decrease in LE associated with manganese indicated here agrees well with a previous study performed on 9 weight percent nickel SMA welds.[5] In general agreement with past studies, nitrogen[3] and ferrite[14] reduce the toughness, and nickel[10] increases the toughness.

<u>Evaluation of the 3.8 Weight Percent Molybdenum Alloy</u>

In an effort to determine the reason for the low impact toughness of the 4Mo-9Ni alloy (weld 6), the microstructure of the weld and the fracture surfaces of the test specimens were evaluated and compared to those of weld 13. As previously mentioned, welds 6 and 13 had very similar compositions with the exception of their chromium and molybdenum contents. The reduction in absorbed energy from 33 J (weld 13) to 16 J is attributed to the increase in molybdenum content from 1.7 weight percent in weld 13 to 3.8 weight percent in weld 6.

The two welds were found to have similar ferrite contents and primary ferritic solidification morphologies during light microscope evaluations of mechanically polished and etched weld metal specimens. The extent of molybdenum segregation between the ferritic and austenitic phases of welds 6 and 13 was also similar. Line scans using wavelength dispersive spectroscopy showed

431

molybdenum contents to be approximately 1 weight percent higher in the ferritic phase than in the austenitic matrix.

Evaluations of the weld metals to determine the presence of intermetallic phases were inconclusive. No evidence of intermetallic phases was found during light microscope examination of the respective weld metals. X-ray analysis showed that both welds 6 and 13 had very similar types and quantities of phases present in their structures. These findings would indicate that the lower absorbed energy found for the 3.8 weight percent molybdenum alloy cannot be attributed to the formation of brittle intermetallic phases.

The only resolvable differences found between the two welds were related to inclusion sizes. Large inclusions were found on the Charpy V-notch fracture surfaces of the 3.8 weight percent molybdenum weld. These inclusions (figure 5) had (a) diameters ranging from 20 to 100 μm; (b) compositions rich in calcium, manganese, silicon, and titanium; and (c) glassy, conchoidal fractures associated with them. Although large inclusions were also found on lower molybdenum content fracture surfaces, they were observed less frequently and generally were not as large. In addition to finding large inclusions on the fracture surfaces of weld 6 specimens, inclusion counts on polished specimens showed a difference between the size distributions for the smaller inclusions in weld 6, figure 6. Weld 6 was found to have a greater number of small diameter inclusion than the other welds. Whether or not the presence of more small and large inclusions in alloy 6 was solely responsible for lowering the toughness in these samples is not known.

Inclusions have, however, long been recognized as void nucleation sites in materials having dimpled rupture fracture modes and do affect the resistance of engineering alloys to fracture.[15] Attempts to correlate inclusion content and fracture toughness [$K_{Ic}(J)$] at cryogenic temperatures are not directly applicable to the Charpy V-notch toughness results of this study, but they indicate that as inclusion spacing decreases, the toughness decreases.[16] Although more complete investigations of inclusion morphology and spacing were beyond the scope of this study, ductile dimple density counts taken on the fractured tensile specimen surfaces of welds 6 and 1 support the argument that decreased inclusion spacing lowers the toughness. The weld 6 specimens had a substantially higher dimple density (4.4 x 10^5 dimples/mm^2) than the weld 1 specimens (3.2 x 10^5 dimples/mm^2). This indicates that the inclusions nucleating the voids in weld 6 were more closely spaced. Further study is needed to determine the quantitative relation between the presence of the large number of small inclusions (figure 6) observed in weld 6 and inclusion spacing. The very large inclusions, found on weld 6 fracture specimen surfaces, could also be speculated to reduce toughness. As a general observation, when the inclusion size is increased, the strain required to nucleate a void is decreased.[17,18]

Conclusions

1) The solid-solution strengthening contribution of molybdenum to the welds at 4 K is approximately 30 MPa per weight percent.

2) The effect of molybdenum on the 76-K Charpy V-notch absorbed energy and lateral expansion is small at molybdenum contents of two weight percent or less.

3) In 18Cr-9Ni stainless steel SMA welds, molybdenum contents approaching 4 weight percent cause substantial decreases in Charpy V-notch absorbed energy and lateral expansion at 76 K.

4)   In 18Cr-14Ni stainless steel SMA welds, increasing the molybdenum content
     from approximately 0 to 4 weight percent did not adversely affect the
     Charpy V-notch toughness at 76 K.

Acknowledgment

This work was supported by the U.S. Department of Energy, Office of Fusion
Energy.

REFERENCES

1)   Mazandarany, F. N., Parker, D. M., Koenig, R. F., Read, D. T., "A
     Nitrogen-Strengthened Austenitic Stainless Steel for Cryogenic Magnet
     Structures," Advances in Cryogenic Engineering - Materials 26, Plenum
     Press, New York, 58-170 (1980).

2)   McHenry, H. I., Read, D. T., Steinmeyer, P. A., "Evaluation of Stainless
     Steel Weld Metals at Cryogenic Temperatures," Material Studies for
     Magnetic Fusion Energy Applications at Low Temperatures - II, NBSIR
     79-1609, National Bureau of Standards, Boulder, CO, 299 (1979).

3)   Purtscher, P. T., Reed, R. P., "Effect of Molybdenum on the Strength and
     Toughness of Stainless Steels at Cryogenic Temperatures," Advances in
     Cryogenic Engineering - Materials 33, Plenum Press, New York, (1987) to
     be published 1987.

4)   Tobler, R. L., Read, D. T., Reed, R. P., "Strength/Toughness Relationship
     for Interstitially Strengthened AISI 304 Stainless Steels at 4 K,"
     Fracture Mechanics: Thirteenth Conference, Richard Roberts, Ed., ASTM
     STP 743, American Society for Testing and Materials, Philadelphia, PA,
     250-268 (1981).

5)   McCowan, C. N., Siewert, T. A., Reed, R. P., Lake, F. B., "Manganese and
     Nitrogen in Stainless Steel SMA Welds for Cryogenic Service, Welding
     Journal, in press (1987).

6)   Yamamoto, S., Yamagami, N., Ouchi, C., "Effect of Metallurgical Variables
     on Strength and Toughness of Mn-Cr and Ni-Cr Stainless Steels at 4.2 K,"
     Advances in Cryogenic Engineering - Materials 32, Plenum Press, New York,
     57-64 (1986).

7)   Reed, R. P., Simon, N. J., "Low Temprature Strengthening of Austenitic
     Stainless Steels with Nitrogen and Carbon," Advances in Cryogenic
     Engineering - Materials 30, Plenum Press, New York, 127-136 (1984).

8)   Simon, N. J., Reed, R. P., "Strength and Toughness of AISI 304 and 316 at
     4 K," Material Studies for Magnetic Fusion Energy Applications at Low
     Temperatures - IX, NBSIR 86-3050, National Bureau of Standards, Boulder,
     CO, 27-42 (1986).

9)   Miura, R., "Development of High Strength Structural Alloys for
     Superconducting Magnets in Fusion Reactor," unpublished.

10) Reed, R. P., Purtscher, P. T., and Yushchenko, K. A., "Nickel and
     Nitrogen Alloying Effects on the Strength and Toughness of Austenitic
     Stainless Steels at 4 K," in Advances in Cryogenic Engineering -
     Materials 32, Plenum Press, New York, 43-50 (1986).

11) Ogawa, T., Aoki, S., Sakamoto, T., and Zaizen, T., "The Weldability of Nitrogen-Containing Austenitic Stainless Steel: Part I - Chloride Pitting Corrosion Resistance," Welding Journal 62(5), 139s-148s (1982).

12) Sakamoto, T., Abo, H., Okazaki, T., Ogawa, T., Ogawa, H., Zaizen, T., "Corrosion Resistant Nitrogen-Containing Stainless Steels for Use by the Chemical Insustry," Alloys for the Eighties, Climax Molybdenum Company, 269-279 (1980).

13) Read, D. T., Tobler, R. L., "Mechanical Property Measurements at Low Temperatures," Advances in Cryogenic Engineering - Materials 28, Plenum Press, New York, 17-28 (1982).

14) Szumachowski, E. R., Reid, H. F., "Cryogenic Toughness of SMA Austenitic Stainless Steel Weld Metals: Part 1 - Role of Ferrite," Welding Journal 57(11), 325s-333s (1978).

15) Van Stone, R. H., Cox, J. R., Low, J. R., Jr., Psioda, J. A., "Microstructural Aspects of Fracture by Dimpled Rupture," International Metals Reviews 30(4), 157-179 (1985).

16) Reed, R. P., Simon, N. J., "Strength and Toughness of AISI 304 and 316 at 4k," Journal of Nuclear Materials 44, 141-143 (1986).

17) Tauoka, J. P., Pampillo, C. A., Low, J. R., Jr., "Review of Developments in Plane Strain Fracture Toughness Testing," Ed W. F. Brown Jr., STP 463, American Society for Testing and Materials, Philadelphia, PA, 191 (1970).

18) Van Stone, R. H., Merchant, R. H., and Low, J. R., Jr., "Fatigue and Fracture Toughness - Cryogenic Behavior," Eds. C. F. Hickey and R. G. Broadwell, STP 556, American Society for Testing and Materials, Philadelphia, PA, 93 (1973).

Table I. Weld Compositions and Ferrite Contents.

Weld	C	Mn	Si	P	S	Cr	Ni	Mo	N	FN Plate
1	0.033	1.57	0.29	0.013	0.006	17.58	9.19	0.02	0.047	4.9
2	0.034	1.49	0.34	0.013	0.006	15.37	9.06	2.03	0.034	4.4
5	0.039	6.31	0.34	0.013	0.007	15.39	9.14	1.99	0.047	3.6
6	0.036	6.27	0.34	0.013	0.007	15.17	9.14	3.84	0.162	2.9
12	0.037	6.62	0.38	0.013	0.008	19.51	9.06	0.02	0.151	3.6
13	0.035	6.61	0.32	0.013	0.007	17.77	9.10	1.66	0.166	3.5
28	0.028	3.23	0.33	0.014	0.007	19.56	9.31	0.02	0.186	1.4
29	0.029	3.21	0.34	0.019	0.007	18.18	9.24	2.17	0.179	1.8
60	0.036	2.38	0.29	0.024	0.005	19.89	9.40	0.02	0.187	0.9
23	0.037	3.10	0.29	0.012	0.011	19.51	14.61	0.03	0.044	0.4
24	0.038	3.01	0.30	0.016	0.010	14.29	14.29	2.01	0.039	0.5
25	0.040	2.97	0.28	0.020	0.010	15.68	14.33	3.90	0.036	0.4
26	0.040	6.51	0.29	0.017	0.013	17.80	14.49	1.92	0.038	0.0
27	0.044	6.71	0.35	0.021	0.013	15.94	14.57	3.84	0.042	0.4
68	0.028	3.20	0.27	0.012	0.010	19.36	14.67	0.02	0.191	0.3
69	0.031	3.21	0.27	0.013	0.009	18.76	14.69	1.06	0.182	0.3
70	0.028	3.10	0.27	0.015	0.009	18.03	14.70	2.02	0.183	0.3
71	0.027	3.05	0.25	0.018	0.009	16.57	14.74	3.98	0.178	0.4

Table II. Mechanical Properties of the Welds.

Alloy	Yield Strength MPa (4 K)	Ultimate Strength MPa (4 K)	Absorbed Energy J (76 K)	Lateral Expansion mm (76 K)
1	460	1598	46	0.69
2	523	1570	39	0.71
5	665	1489	37	0.53
6	1169	1432	16	0.13
12	1027	1376	37	0.31
13	1129	1409	33	0.23
28	999	1335	41	0.38
29	1076	1520	41	0.41
60	998	1227	37	0.41
23	638	1152	51	0.84
24	675	1269	60	0.89
25	752	1359	43	0.84
26	758	1299	59	0.91
27	836	1385	56	0.81
68	1093	1413	34	0.36
69	1098	1493	33	0.38
70	1086	1471	37	0.38
71	1124	1453	42	0.43

435

Figure 1. The effect of nitrogen on the 4-K strength.

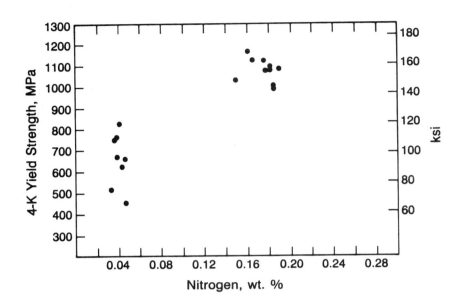

Figure 2. The 4-K yield strength data, normalized (3200 MPa/wt.% nitrogen) to a 0.5 weight percent nitrogen content to show the strengthening contributions of molybdenum and manganese.

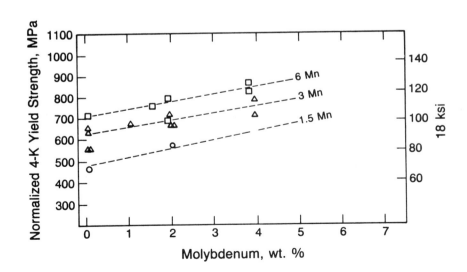

Figure 3. The effect of molybdenum and nickel content on the 76-K Charpy V-notch absorbed energy.

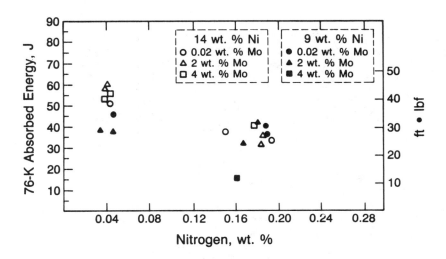

Figure 4. The effect of molybdenum and nickel content on the 76-K Charpy V-notch lateral expansion.

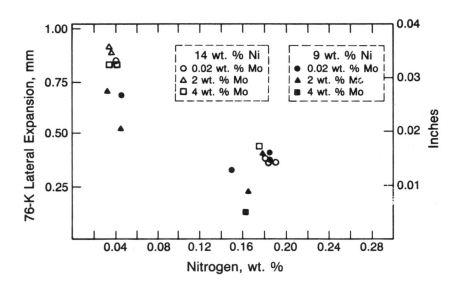

Figure 5.   SEM fractograph of a Charpy V-notch specimen from weld 6 showing one of the large inclusions on the fracture surface.

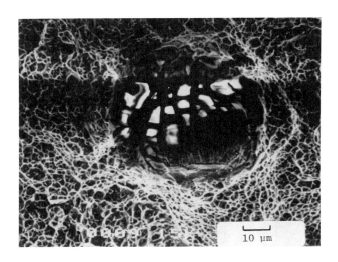

Figure 6.   Inclusion size distribution results of welds 1, 6, 12, and 27 acquired at 3000X.

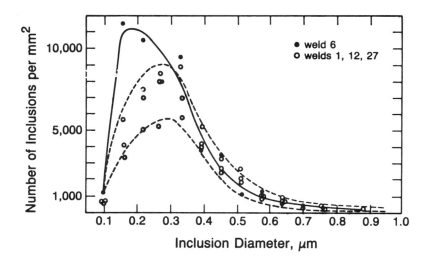

GRAIN REFINEMENT IN THE HAZ OF HIGH SPEED ELECTROSLAG WELDMENTS

STEPHEN LIU & CHWEN TZENG SU

DEPARTMENT OF INDUSTRIAL ENGINEERING
THE PENNSYLVANIA STATE UNIVERSITY
207 HAMMOND BUILDING
UNIVERSITY PARK, PA 16802

Abstract

The long thermal cycle of slow heating and cooling rate in electroslag welding results in very coarse primary solidification structure. The heat absorbed into the base metal creates an extremely large heat affected zone (HAZ), which reduces the applicability of electroslag welded structures in more critical applications. In this work, results of electroslag welding ASTM A588 grade structural steel with metal powder cored strip electrodes showed that it is possible to refine the HAZ size and obtain grain refinement in the coarse grained HAZ (CGHAZ). This is accomplished by increasing the welding current which decreases the specific heat input of the process to below one kJ/mm^2. At current levels below 1200-1300 Amperes, the decrease in grain size was observed to be gradual. Above the particular current range, however, the average size of the grains in the CGHAZ was found to reduce from approximately six hundred to less than three hundred microns which represented a grain size reduction by a factor of approximately two. Weld temperature profiles indicated the influence of electrode geometry on the heat flow conditions. In spite of similar peak temperatures, higher current welds showed significantly shorter thermal cycles with higher heating and cooling rates, restraining grain growth in the CGHAZ. The correlation of welding parameters (welding current and travel speed), thermal cycle (heat flow conditions) and HAZ microstructure was determined for high speed electroslag welding with metal powder cored strip electrodes.

439

Introduction

Due to the excessive coarse grained structure and low toughness found in the HAZ of electroslag weldments, the application of electroslag welding is currently very limited. Recognizing the detrimental effects of HAZ degradation, many high speed electroslag welding processes were developed and reported in the literature (1-10). There are essentially two approaches to achieve the goal of obtaining electroslag weldments with refined structure and improved mechanical properties. One is to narrow the root gap decreasing the need of large volume of metal deposit to form the weld joint. The other is to increase the deposition rate and welding speed such that the heat content of the process would be distributed throughout a longer section of the weld lowering the effective heat input.

Avramenko, Lebedev, and Bozhko (1) suggested that an expedient and simple way of increasing the productivity of the electroslag welding process is the reduction of the root gap size with a simultaneous increase of the electrode wire feed rate. This procedure decreases mainly the amount of weld metal deposited in the weld joint and the time needed to complete the weld. However, the extent to which the cross sectional area of the gap can be reduced is limited by the stability of the process and the size of the electrode and guide tube inserted into the gap. Too close root gap may cause arcing between the electrode and the base metal.

Using solid flux blocks to avoid arcing between a solid strip electrode and the base metal, Watanabe, Sejima, Kokura, Taki, and Miyake (2) could weld with root gap as small as 15 mm. For welds made with 800 Amperes, 34 Volts and a root gap distance of 15 mm, the welding speed was approximately 0.25 mm per second. The specific heat input was $10^5 J/cm^2$ ($10^3 J/mm^2$). Slight improvement in weld mechanical properties was observed. Atteridge, Venkataraman, and Wood (3) investigated also the effects of root gap and reported that for a similar specific heat input level, the narrow gap welds resulted in higher base metal dilution than standard gap ones.

The concept of adding powdered metal filler, which can be easily melted by the excess process heat of the arc, was introduced in the early 1960s to increase the deposition rate in the Bulkweld process (4). Reynolds and Kachelmeier (4) observed increased welding speed, decreased heat input, reduced HAZ size, and lower consumption of fluxes in hardfacing and cladding with a powder to wire weight ratio of 1:1.

Smirnov and Efimenko (5) reported that simultaneous powder metal filler addition with an electrode wire refined the electroslag weld metal microstructure. Low temperature impact strength of the weldments in the as welded condition increased by a factor of two. Deposition rate was approximately thirty to fifty pounds per hour. Slight reduction in specific heat input was also observed.

Eichhorn, Remmel, and Wubbels (6-8) added powder metal filler without electric current and increased the deposition rate while reducing the overheating of the weld pool. The metal powder was supplied to the weld by a current of argon gas. Once the metal powder reach the root gap, it would adhere to the electrode and move towards the weld pool. The powder to electrode weight ratio was 1.2:1. They used single and dual electrode configurations with different powder delivery locations. The welding speed was increased to approximately nine inches per minute with simultaneous reduction of energy input. Good Charpy impact toughness was observed in HAZ samples with notches made at a distance of one to two millimeters from the fusion line.

In spite of the amount of success reported in these high speed elect-roslag techniques, the general position of the welding industry is to wait for the development of improved steels which will retain the original fine grain structure even at the presence of the high heat content of the process. Additionally, Threadgill (9) pointed out that the good impact toughness reported in the high speed electroslag weldments (6-8) could not be verified with fracture mechanics techniques such as CTOD testing (crack tip opening displacement). Eagar (10) also indicated recently that although HAZ size reduction could be achieved, it would only improve slightly the mechanical properties of an electroslag weld. Furthermore, he suggested that HAZ grain refinement was unlikely to be accomplished.

Responding to the need of energy efficient welding processes that can operate economically while generating large volumes of good quality deposited metal, the concept of high speed electroslag welding is reviewed. It is the objective of this research to develop a high speed welding technique that will refine the HAZ and weld metal grain structure. It involves the design of a special cored strip electrode which add metal powder to the weld pool in a more efficient way, eliminating the need of a special inert gas metal powder delivery system. It also incorporates the ideas of reduced root gap distance, electrode geometry, chemical composition modification, and high electrode melt rate.

It is the intent of this paper to report and discuss the effects of the metal powder cored strip electrode on the solidification characteristics and the final HAZ and weld metal microstructure of an electroslag weldment.

Experimental Procedure

ASTM A588 grade structural steel was chosen for the consumable guide electroslag welding experiments. Plates of 305 mm x 305 mm x 38 mm (12 in. x 12 in. x 1-1/2 in.) were used. With cross-sectional dimensions of 35.4 mm x 4.5 mm, the metal powder cored strip electrode was made to carry high current and achieve high melt rate. The hollow low carbon manganese steel strip electrode is filled with metal powder with a compacting factor of approximat-ely fifty percent. The chemical composition of the base metal and electrode are given in Table I. Figure 1 shows the electrode to joint geometry. The

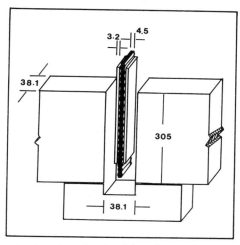

Figure 1. Schematic drawing showing the Electrode to Joint Geometry.

441

width of the electrode covers almost the entire thickness of the base plate promoting more uniform heating of the weld pool. In addition, a pair of low carbon steel consumable guide plates were also used.

The welds were produced at a constant potential of forty volts with the welding current varying from 1000 to 1500 Amperes. The root gap distance was held constant at 1-1/2 inch (38.1 mm) for all welds. Details of the process parameters selection have been presented previously (11-12) and will not be discussed in this paper. Commercial starting and running fluxes were used to carry out the welding. The chemical composition and basicity index (BI) of the running flux is shown in Table II.

Table I. Chemical composition of the electrode and base metal used in the welding experiments (in weight percent).

	C	Mn	Si	P	S	Cr	Ni	Mo	Al	Ti
Electrode	0.022	1.05	0.34	0.028	0.018	0.08	0.04	0.01	0.007	0.002
Base Metal	0.26	1.05	0.21	0.022	0.008	0.04	0.03	0.01	0.032	0.001

	Co	V	Cu	Nb	B
	0.013	0.002	0.065	0.005	0.0005
	0.004	0.0002	0.012	0.002	0.0003

Table II. Typical chemical composition of a running flux for electroslag welding (in weight percent).

CaO	MgO	MnO	CaF_2	SiO_2	Al_2O_3	TiO_2	K_2O	Na_2O	FeO	BI
12.2	2.3	22.5	8.6	33.0	8.3	8.0	0.9	0.6	1.8	0.9

Extensive metallographic analyses were performed to study the macro- and microstructure of the weldments. Chemical analyses of the welds were also carried out using an emission spectrometer and the composition given in Table III. For the temperature distribution and heat extraction conditions in the HAZ, thermocouples were inserted into drilled holes in the plates and monitored throughout the heating and cooling cycle. Temperature profiles of the welds were obtained and compared to data available in the literature.

Table III. Chemical Analyses of Some Representative Weldments.

WELD	C	Mn	Si	P	S	Cr	Ni	Mo	Al	Ti	Co
1000A	0.15	0.89	0.19	0.016	0.012	0.05	0.04	0.01	0.011	0.001	0.004
1200A	0.11	0.87	0.17	0.014	0.015	0.07	0.04	0.01	0.022	0.002	0.006
1300A	0.10	0.79	0.16	0.010	0.014	0.07	0.04	0.01	0.015	0.001	0.005
1400A	0.11	0.90	0.18	0.012	0.014	0.07	0.03	0.01	0.012	0.001	0.005
1500A	0.10	0.88	0.18	0.011	0.013	0.06	0.04	0.01	0.014	0.001	0.005

	V	Cu	Nb	B
	0.001	0.041	0.002	0.0003
	0.001	0.051	0.003	0.0003
	0.0005	0.048	0.003	0.0003
	0.001	0.054	0.002	0.0003
	0.001	0.048	0.002	0.0003

Results and Discussion

Solidification Characteristics

The structure of a weld metal is a deciding factor with regard to its mechanical properties, resistance to intergranular and transgranular crack propagation, and corrosion resistance. Therefore, metallographic techniques were used to study the solidification and solid state transformation structures of the welds produced. As expected, coarse columnar grains were observed in the region adjacent to the weld fusion line. At a certain distance from the fusion line, the columnar grains gradually gave way to the fine columnar grains. Only small amounts of equiaxed grains were observed at the center of the welds. As the welding current was increased, the volume fraction of coarse columnar grains decreased. This is shown in Figure 2. On the other hand, the volume fraction of fine columnar grains increased with current. However, the increase became more gradual with current levels above 1300 Amperes, with the fine columnar grains replaced by the thickening of the

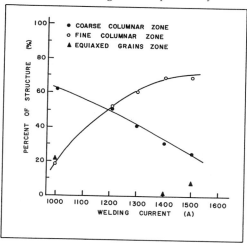

Figure 2. Variation of Solidification Structure with Welding Current.

equiaxed grained zone. A minimum of volume fraction of equiaxed grains was observed in the 1200 and 1300 Amperes welds.

The morphology of electroslag welded steel deposits was first categorized by Paton (13). The classification included four types of weld structures. The first group showed coarse and refined columnar grain zones. The coarse columnar grains protrude from the fusion line towards the center of the welds. At some distance from the fusion line, fine columnar grains would be observed. The second type was composed of three zones, the coarse and refined columnar grain, and the equiaxed grain zone. The third group showed only coarse columnar grains meeting in a very sharp acute angle at the center line of the weld and the fourth type, only refined columnar grains meeting again in acute angle at the center line. In a more recent publication, Yu, Ann, Devletian, and Wood (14) further detailed the characterization of the solidification structure of low carbon structural steel electroslag weldments and related them with the welding current and weld pool form factor. They observed both cellular and columnar dendrites (named previously as coarse and fine columnar grains by Paton) in the solidification structures. The location of cellular to columnar dendrite transition and the angle of inclination that the grains meet at the weld center were some of the criteria used in the classification. Figure 3 shows the variation of solidification mode in carbon

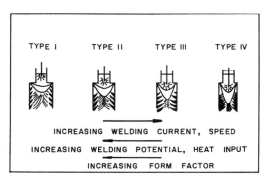

Figure 3. Variation of Solidification Mode in Carbon Steel Electroslag Weldments.

steel electroslag weld metal. Low current welds show shallow molten metal pools with a high form factor. High current and fast travel speed welds will generally exhibit deep pools and low form factor. Yu et al. (14) also observed that by increasing the welding current to the 1300/1500 Amperes range and speeding up the welding process, cracking developed along the center line of the welds or along the interfaces of the fine dendrites. In contrast to the weld defect reported, no cracking was observed, even though the welds in this experiment were also made in a similar current and voltage range. In addition, the welds did not exhibit type III nor IV solidification structure. Even at the higher current levels and high travel speeds, the solidification pattern resembled solidification mode I and II. This indicates the importance of electrode to joint geometry in the electroslag welding operation, affecting particularly the molten pool formation, heating and cooling conditions of the weldment. Together with heat input, the heat generation pattern and temperature distribution are also essential to produce a sound weldment.

HAZ and Weld Metal Microstructures

HAZ Size Reduction. Macro- and micrographic techniques were applied to determine the effects of the metal powder cored strip electrode and welding current on the welds. The CGHAZ and the fine grained HAZ (FGHAZ) sizes were measured and compared with data obtained in the literature. Under similar operating conditions, the present technique produced welds with reduced HAZ. In addition, the thickness of the heat affected zone varied with the current level. At 1500 Amperes, the CGHAZ and FGHAZ were approximately 5 mm and 2 mm, respectively. This represented a reduction of almost fifty percent from the 1000 Amperes weld. Figure 4 shows the HAZ reduction as a function of welding current.

CGHAZ Grain Size Reduction. Contrary to that indicated by Eagar (10), grain size reduction in the HAZ was possible and indeed observed. Figures 5 and 6 are representative HAZ micrographs. At 1000 Amperes, the average CGHAZ grain size was approximately 550 microns. Grain refinement occurred with increasing welding current which reduced the average grain size in the 1500 Amperes weld to approximately 250 microns. Figure 7 illustrates graphically the relationship between CGHAZ grain size and the welding current. When compared to conventional electroslag welds which generally exhibit grains of six hundred to eight hundred microns, the present technique produced welds with much refined grain size in the CGHAZ. It can also be noticed that grain

444

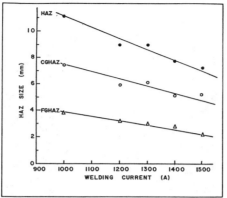

Figure 4. HAZ Size Reduction with Welding Current

size variation was more significant at current levels around 1200 and 1300
Amperes.

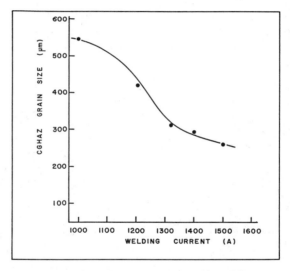

Figure 7. Variation of CGHAZ Grain Size with Welding Current.

Weld Metal Microstructures. Figures 8 and 9 are representative weld
metal micrographs of low and high current welds taken at equivalent positions. As expected, the high current weld exhibited finer prior austenite
grain size. On the other hand, the lower current weld showed microstructure
comparable to conventional welds with extremely large prior austenite grains.
Figure 10 shows the variation of prior austenite grain size with the welding
current. Again, a sharp decrease in austenite grain size was observed in the
1200 to 1300 Amperes range welds. Both Paton (13) and Yu (14) indicated that
the coarse columnar grain zone usually contain more acicular ferrite while
fine columnar grains zone exhibit higher volume fraction of grain boundary
ferrite. Yu et al. (14) also outlined the sequence of morphological change
in low carbon steel welds. During the cooling cycle, the cellular dendrites
will transform into coarse columnar austenite grains and result in an acicu-

445

200 UM

Figure 5. HAZ Light Micrograph of the 1000 Amperes, 40 Volts Weld.

Figure 6. HAZ Light Micrograph of the 1500 Amperes, 40 Volts Weld.

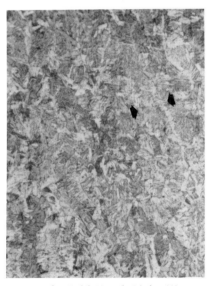

200 UM

Figure 8. Weld Metal Light Micrograph of the 1000 Amperes, 40 Volts Weld, at 1/2 Weld Depth and 1/4 Weld Width.

Figure 9. Weld Metal Light Micrograph of the 1500 Amperes, 40 Volts Weld, at 1/2 Weld Depth and 1/4 Weld Width.

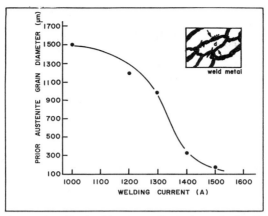

Figure 10. Prior Austenite Grain Size as a function of the Welding Current.

lar ferrite dominant room temperature microstructure. Columnar dendrites
will generally transform into fine columnar austenite grains. Grain boundary
and sideplate ferrite will then be the predominant structure. This is also
verified in the present work. Increasing current led to an increase in the
volume fraction of fine columnar grains. Subsequent transformations of the
finer austenite grains resulted in high volume fractions of grain boundary
ferrite.

Heating and Cooling Rate Effect

Welding conditions play an important role in the heat distribution in
electroslag welding. A major portion of the heat is carried away into the
workpart. The high heat transfer into the base metal explains the wide HAZ
observed in electroslag weldments and can also be attributed to the lower
concentration of heat of the electroslag welding process. DebRoy, Szekely,
and Eagar (15-16) developed mathematical models to represent the three dimensional temperature fields in the slag, metal pool, and base metal in the
electroslag weldments. They indicated that the heat generating patterns are
highly sensitive to the geometric location of the electrode in the slag.
Similarly, the shape of the electrode and the electrode to joint geometry
will also affect significantly the temperature distribution in the weldment.
As in the case of solid plate electrodes (13), the cored strip electrodes
used in the present work also promoted more uniform heat distribution in the
weld pool.

For the sake of clarity, only one set of two temperature profiles is
shown in Figure 11. The lowest and highest current welds were chosen to
illustrate the effects of heating and cooling rate on the HAZ grain size
control. The readings represented measurements taken from the mid section of
the weldments at 1.5 mm from the fusion line (CGHAZ). The 1500 Amperes weld
temperature profile showed a steep temperature increase with time during the
heating cycle and reached the peak temperature of approximately 1400°C. The
lower current weld (1000 Amperes) experienced a more sluggish temperature
increase and the peak temperature was slightly below 1400°C. During the
cooling cycle, similar observation was obtained that the higher current welds
showed faster cooling rate than the lower current samples.

Excessive grain growth in the CGHAZ occurs mainly as a result of the
coarsening and dissolution of non-metallic particles such as carbides and
nitrides (17). The presence of these particles is essential to the stability

447

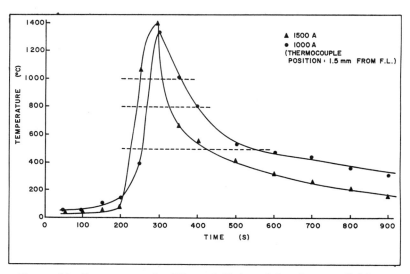

Figure 11. Temperature Profiles of High and Low Current Welds.

of the structure, since they can interact and pin the grain boundaries. In the case of an aluminum killed fine grain steel, the grain coarsening temperature is related to the aluminum nitride dissolution temperature. In the base metal used in the present experiments, the aluminum content is 0.032 weight percent, and the grain coarsening temperature was estimated to be approximately 1000°C (17). However, with heating and cooling rates greater than those corresponding to the equilibrium conditions, the dissolution of these particles in the HAZ will also depend on the peak temperature, the concentration of aluminum and nitrogen in the alloy and the time duration that the base metal is subjected to the severe thermal cycles during welding. In terms of peak temperature, the low and high current welds had experienced quite similar temperature readings, 1340 and 1400°C, respectively. However, the two welds showed distinct heating and cooling characteristics. The lower current weld had a heating rate of 9.9°C/s and its cooling rate was 6°C/s. On the other hand, the high current weld showed 14.5°C/s in the heating cycle and 30°C/s during the cooling cycle. This observation is particularly important in the interpretation of the grain refinement event in the higher current welds. The higher heating rate does not allow for the equilibrium heating and dissolution of the aluminum nitride particles, even up to a very high temperature. During the entire thermal cycle, particle pinning of the austenite grain boundaries was never ceased. In addition, the time that the CGHAZ experiences temperatures above 1000°C for the lower current weld was twenty five percent longer than the higher current one. Therefore, one can conclude that the higher current weld experienced a shorter thermal cycle above 1000°C with faster heating and cooling rate which limited particle coarsening or dissolution. This also explains the resulting finer CGHAZ grain size observed in the higher current welds. However, $\Delta t_{8/5}$ in the low current weld was not observed to be much longer (approximately 150 seconds, as compared with the higher current weld reading of 120 seconds). This is confirmed by the observation that the CGHAZ structure of both the welds were composed of proeutectoid ferrite networks along the prior austenite grain boundaries, intragranular ferrite and pearlite matrix with carbide and bainite.

Figure 12 shows the drawing of an ideal thermal cycle for the prevention

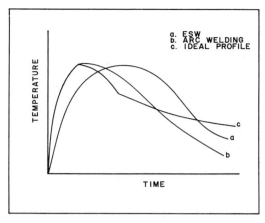

Figure 12. Ideal Thermal Cycle for the prevention of Cold Cracking.

of cold cracking (19). For comparison, thermal profiles of typical electroslag and arc weldments are also included. It can be seen that the higher
current weld thermal profile determined in the present work resembles more
the arc welding curve and the ideal curve. Since cold cracking is also
related to the overheating of the base metal, faster heating rate as achieved
in the present work would also be beneficial in weld quality improvement.

Specific Heat Input Effect

Grain refinement in electroslag weldments cannot be achieved by changing
one single parameter. As discussed in the previous paragraphs, the faster
heating and cooling rate of the electroslag weldments certainly has a strong
influence on the HAZ grain growth restriction. However, the temperature
profile itself is due to a combination of other factors such as the welding
current, welding speed, specific heat input, the geometry of the strip electrode, and the thermal properties of the materials used. The effect of specific heat input and welding speed on CGHAZ grain size is illustrated in Figures 13 to 15. The sharp decrease in grain size at specific heat input of
approximately 1.3 kJ/mm^2 in Figure 13 follows the trend described previously,
for it corresponds to the 1200 to 1300 Amperes range. The non-linear relationship between welding speed and current is shown in Figure 14. The same
relationship can also be seen in Figure 15 when wire feed rate (deposition
rate) is examined. Even though it is more appropriate to relate welding
current as a function of wire feed rate, the inverse function was plotted
purposely in Figure 15 to demonstrate the fast increase in deposition rate
with current at the 1200 to 1300 Amperes range. A significantly larger
amount of electrode is deposited into the weld joint which accelerates the
welding process. Consequently, heat input into the molten metal pool is
spread out to a longer section of the weld decreasing the specific heat
input. This also leads to a faster heating and cooling cycle resulting in
effective control of HAZ grain growth.

Hardness Measurements and Chemical Composition Variation

Hardness measurement of the weld specimens are shown in Figure 16.
When considering the highest and lowest current welds prepared, only slight
difference in the HAZ hardness readings was observed. This is consistent
with the previous observation that even though the HAZ size and the grains

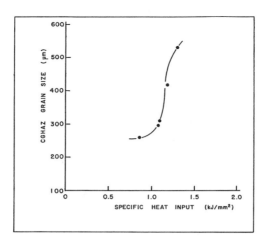

Figure 13. Variation of CGHAZ Grain Size with Specific Heat Input.

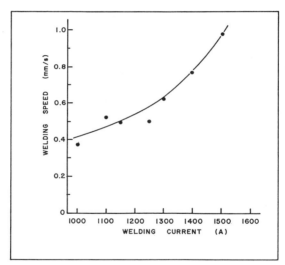

Figure 14. Variation of Welding Speed with Welding Current.

were refined in the higher current welds, the constituents found in the microstructures were generally the same. Therefore, the increase in HAZ hardness with increasing current was minimum. However, the average hardness of the weld metal with the refined solidification structure was approximately ten percent lower than that of the coarse structure weld. Venkataraman et al. (20) reported similar findings that the hardness of the quartz shroud refined weld metal was lower than that of the standard electroslag weld. The decrease in hardness is related to the microstructure discussed previously. Since higher current welds showed smaller austenite grains with large amount of grain boundary ferrite, the average hardness was observed to be lower. A weld with microstructure of coarser austenite grains and higher volume fraction of acicular ferrite will result in higher hardness.

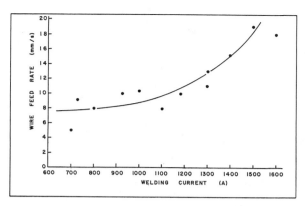

Figure 15. Variation of Wire Feed Rate with Welding Current.

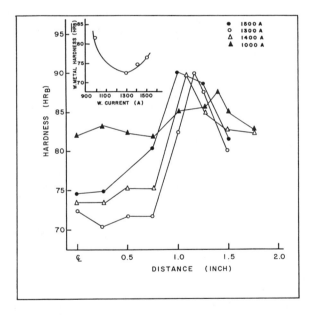

Figure 16. Hardness Profiles of High Speed Electroslag Weldments with the
Metal Powder Cored Strip Electrode.

An interesting feature observed, however, is the variation of weld metal hardness with the welding current. A minimum in hardness was observed in the specimens at welding current around 1200 to 1300 Amperes. This shows that the decrease in weld metal hardness is not due to the grain size variation alone. Process characteristics of the metal powder cored strip electrode technique can also have an important role in the observed behavior. It was suspected that the slag/metal interaction could also be distinct within the 1200 to 1300 Amperes current range. Therefore, relative chemical element

451

loss was determined for each of the major alloying elements. The loss can be
expressed as the difference between the calculated and measured composition
of a particular element in the weld. It was observed that at 1200 to 1300
Amperes range, larger loss of manganese, silicon, and phosphorus occurred,
Figure 17. A lower carbon pickup was also observed within the same weld.

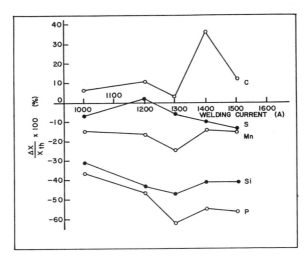

Figure 17. Variation of Relative Chemical Element Loss with Welding Current.

Lower chemical hardenability contributed to the lower weld metal hardness.
The loss observed at this particular current range is very interesting since
it follows the same trend as observed in CGHAZ grain size and weld metal
prior austenite grain size decrease with welding current. This indicates
that the observed behavior is originated from the thermal history of the
process, which include the heating and cooling cycle, temperature distribution and molten metal pool chemistry.

Conclusions

1. The metal powder cored strip electrode increased the travel speed and
   decreased the specific heat input of electroslag welding.
2. Even at high current levels and welding speed, type I and II solidification structures containing essentially cellular and columnar dendrites
   were observed, with no signs of hot or cold cracking.
3. With the metal powder cored strip electrode, both CGHAZ and FGHAZ size
   were reduced with increasing welding current.
4. Substantial grain refinement in the CGHAZ was achieved with the metal
   powder cored strip electrode. Average grain size was reduced from above
   six hundred to approximately two hundred and fifty microns.
5. With the metal powder cored strip electrode, it was possible to produce
   welds with faster heating and cooling rates than the conventional
   electroslag welds. This provided adequate conditions for grain growth
   control in the CGHAZ.
6. The metal powder cored strip electrode affected the heat distribution in
   the molten weld pool as well as the chemistry resulting in distinct
   carbon, manganese, silicon, and phosphorus transfer behavior within the
   range of 1200 to 1300 Amperes range.

ACKNOWLEDGMENT

The authors acknowledge the research support of the National Science Foundation under Grant DMC 8505204, the equipment and material support of The Hobart Brothers Company and Alloy Rods, Inc. The assistance of PSU IE Lab technicians during the different phases of the welding experiments is much appreciated.

REFERENCES

1. V.I. Avramenko, B.F. Lebedev & V.I. Bozhko. "Some Ways of Increasing the Productivity of Electroslag Welding," Svar. Proiz., 10(1973), 16-17.

2. K. Watanabe et al. "Problems and Improvement of Large Heat Input Electroslag Welding," Journal of Japan Welding Society, (1975), 519-524.

3. D. Atteridge, S. Venkataraman & W.E. Wood. "Improving the Reliability and Integrity of Consumable Guide Electroslag Weldments in Bridge Structures" (Research Report - Oregon Graduate Center, 1982).

4. G.H. Reynolds & E.J. Kachelmeier. "Adding Powdered Metal Filler speeds Deposition Rate," Metal Construction, (1978), 426-430.

5. A. Smirnov & L.A. Efimenko. "Special Structural Features and Mechanical Properties of Electroslag Welded Joints made using Powdered Filler Metal," Avt. Svarka, 9(1973), 46-50.

6. F. Eichhorn, P. Hirsch & B. Wubbels. "Use of Metal Powder Additions to improve the Strength and Toughness of High Speed Electroslag and Electrogas Welds in Microalloyed and Low Alloyed Steels" (Report IIW-DOC XII-76-80, 1980).

7. F. Eichhorn, J. Remmel & B. Wubbels. "High Speed Electroslag Welding," Welding Journal 62(1)(1984), 37-41.

8. F. Eichhorn & J. Remmel. "Efficient Fillet Welding in the Vertical Welding Position with Electrogas and Electroslag Welding Methods" (Report IIW DOC XII-908-85, 1985).

9. P. Threadgill. "Applications and Trends for Electroslag and Related Welding Processes in Western Europe," in Electroslag Welding for Marine Applications, ed. W.A. Palko, D.F. Hasson and C.A. Zanis (USNA, 1985), 4/11-4/17.

10. T. Eagar. "The Physics and Chemistry of Welding Processes," in Advances in Welding Science and Technology, ed. S.A. David (ASM, 1986), 291-298.

11. S. Liu & C.T. Su. "Performance Evaluation of a Metal Powder Cored Strip Electrode in High Speed Electroslag Welding," in Advances in Welding Science and Technology, ed. S.A. David (ASM, 1986), 401-412.

12. S. Liu & C.T. Su. "Innovations in Electroslag Welding," in Manufacturing Processes, Systems and Machines, ed. by J. Tlusty (NSF/SME, 1986), 29-35.

13. B.E. Paton. Electroslag Welding, (American Welding Society, 1962).

14. D. Yu et al. "Solidification Study of Narrow Gap Electroslag Welding," in Welding Research: The State of the Art, ed. E. Nippes and D. Ball,

(ASM, 1985), 21-32.

15. T. DebRoy, J. Szekely & T. Eagar. "Heat Generation Patterns and Temperature Profiles in Electroslag Welding," Met. Trans. B, 11B(1980), 593-605.

16. T. DebRoy, J. Szekely & T. Eagar. "Temperature Profiles, the Size of the Heat Affected Zone and Dilution in Electroslag Welding," Materials Science and Engineering, 56(1982), 181-193.

17. K. Easterling. Introduction to the Physical Metallurgy of Welding (Butterworths, 1983).

18. T. Gladman. "The Effect of Aluminum Nitride on the Grain Coarsening Behavior of Austenite" (The Iron and Steel Institute Special Report 81- Metallurgical Developments in Carbon Steels, 1963, 68-70).

19. B.E. Paton. Electroslag Welding and Surfacing - V. I (Mir Publishers, 1983).

20. S. Venkataraman et al. "Grain Refinement Dependence on Solidification and Solid State Reactions in Electroslag Welds," in Grain Refinement in Castings and Welds, ed. G.J. Abbaschian and S.A. David (AIME, 1982), 275-288.

RESIDUAL STRAINS IN GIRTH-WELDED LINEPIPE

S.R. MacEwen, T.M. Holden, B.M. Powell

Atomic Energy of Canada Limited, Chalk River Nuclear Laboratories
Chalk River, Ontario, K0J 1J0   Canada

and

R.B. Lazor
The Welding Institute of Canada, 391 Burnhamthorpe Road East,
Oakville, Ontario, L6J 6C9   Canada

ABSTRACT

High resolution neutron diffraction has been used to measure the axial residual strains in and adjacent to a multipass girth weld in a complete section of 914 mm (36") diameter, 16 mm (5/8") wall, linepipe. The experiments were carried out at the NRU reactor, Chalk River using the L3 triple-axis spectrometer. The through-wall distribution of axial residual strain was measured at 0, 4, 8, 20 and 50 mm from the weld centerline; the axial variation was determined 1, 5, 8, and 13 mm from the inside surface of the pipe wall.

The results have been compared with strain gauge measurements on the weld surface and with through-wall residual stress distributions determined using the block-layering and removal technique.

Residual stresses in engineering components are of considerable technological significance, and their measurement has been achieved using a variety of techniques. Surface stresses can be determined by X-ray diffraction, and near-surface stresses by blind-hole drilling; both can be considered as non-destructive. The determination of residual stress gradients through thick components generally involves measuring the change in surface strains (using strain gauge rosettes) that result when the component is slit, drilled, or trepanned. The procedure is clearly tedious and labour intensive, and the results are subject to considerable uncertainty, especially in anisotropic materials. Here we will describe a relatively new non-destructive technique for determining residual strains at known positions in a large component, and will demonstrate how it can be applied to measuring residual strains near the girth weld in linepipes.

The field-welding of lengths of standard pipe has always been the most important aspect of pipeline construction. The girth welding operation is subject to welding defects such as cracking, lack of fusion, undercut, slag intrusions, and porosity. Depending on the nature of the defect and its extent, it can be either left in place, repaired, or cut out. While flaw acceptance limits for pipeline girth welds based on workmanship standards (CSA Z184, API 1104) have been used successfully for years, recent test results have shown that alternative flaw acceptance criteria can be used based on "engineering critical assessments", ECA. Fracture mechanics can be used to reduce the repair rate and associated costs by conducting a fitness-for-purpose, or engineering critical assessment of the defect.

The acceptance of ECA methods requires extensive fracture toughness testing, periodic random inspection of production welds, and an analysis of the maximum stress levels imposed on the pipeline. The first two items are obtained by mechanical testing of weld samples, which is relatively straightforward, while the stress levels are more difficult to determine due to uncertainty in the levels of residual stresses.

The girth-welding operation induces long-range reaction stresses, as well as localized residual stresses in the weldment. The peak stresses are generally considered to be of yield magnitude, but marked differences in the growth rate of fatigue cracks suggests that the residual stresses vary around the weld and are less than tensile yield. It has been shown under certain conditions that axial residual stresses can be less than yield, or compressive.

The influence of welding procedure on the contraction behaviour of a series of girth welds prepared using standard field welding conditions has recently been completed in a program sponsored by the American Gas Association (1). The overall project included eight welds using 610 and 914 mm diameter linepipes, with various combinations of preheat, weld metal strength, wall thickness, and welding technique. One objective of the study was to measure the residual stress distribution in pipeline girth welds to determine whether welding procedure could reduce the level of residual stresses produced by welding. If these stresses are in fact less than yield, then ECA may justify a significant increase in the allowable defect size. Through-wall distributions of axial residual stresses were obtained using the "block removal and layering" technique (2). Here strain gauges mounted on the inner and outer surfaces of the wall of the pipe are used to monitor the changes in strain that are produced when thin, relaxation blocks are cut out. Subsequently, one of the gauges is used to monitor further changes in strain as layers are removed from the opposite face of the block. Separate blocks are used to measure strains parallel

456

and normal to the weld, and it is assumed that the original stress distributions at the positions of the two blocks are identical. The strain relaxation is analyzed assuming simple beam theory to determine residual stresses.

Neutron diffraction provides an alternative and non-destructive method for mapping out the variation of the strain within and adjacent to weldments. The basis of the technique is the same as for X-ray measurements; the distance between planes of atoms is used as a miniature, directional, internal strain gauge. However, the penetration of neutrons into metals is from 1000 to 10,000 times greater than that of X-rays, and thus neutrons provide a true bulk probe. Figure 1 is a schematic of the experimental setup used at CRNL. The use of slits in both the incident and diffracted beams defines a column with a diamond-shaped cross-section (dependent on the diffraction angle). By moving the sample on a computer controlled X-Y drive, the column can be positioned at will in the component, and all diffraction information comes from the volume of the column only. With the configuration shown in Figure 1, the through-wall variation of the radial component of residual strain could be determined.

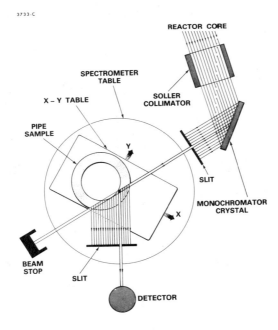

Figure 1 - Schematic of the apparatus used for residual strain measurements.

A comparison of the interplanar spacing in the component with that in a reference sample allows the residual strain in the direction of the scattering vector (bisection of the incident and diffracted beams) to be calculated. In general, six strain measurements must be made before the residual stress tensor can be determined, however the number of measurements required is reduced to three if it is assumed that no shear terms are present, and to two if one assumes either plane strain or plane stress conditions.

457

Experimental studies of weldments by neutron diffraction have been reported for welded plates and for a tube welded to a plate [3], but the present measurements are the first to be reported on axial strains on a linepipe girth weld.

Experimental

Linepipe

Two short lengths of Grade 448 N/mm² linepipe with a diameter of 914 mm (36") and a wall thickness of 16 mm (5/8") were prepared for welding with a V-preparation using a 75° included angle and a 1.6 mm root face. An internal clamp was used to remove misalignment and to maintain root spacing during deposition of the root pass. Welding was performed vertically down, with the pipe axis in the horizontal position, by four welders using conventional stovepipe welding techniques. Six passes were required to complete the girth weld.

Neutron Diffraction

The experiments were carried out using the N5 spectrometer at the NRU reactor, CRNL, employing the (331) planes of a squeezed silicon crystal monochromator to provide a neutron beam of wavelength 2.0411Å. The resolution of the spectrometer allowed interplanar spacings to be determined with precision of $\pm$ 2 x 10^{-4}. Soller slit collimators were used to achieve horizontal angular collimation before and after the linepipe of 0.4° and 0.3° respectively. Slits in absorbing cadmium sheet, each 1 mm wide and 25 mm high, defined the volume in the sample where the measurements were made. The diffracting volume was positioned with the 25 mm column in the hoop direction of the pipe, and the diamond-shaped cross-section in the axial-radial plane with the long diagonal (2.0 mm) in the radial direction and the short diagonal (1.2 mm) in the axial direction. Each diffraction measurement gave the average interplanar spacing of (110) crystal planes whose normals were parallel to the axial direction of the pipe, at the centroid of the diffracting volume.

The linepipe, weighing 100 kg, was suspended above the spectrometer but attached to a stage on the X-Y drive so that one wall of the pipe at its equatorial diameter was at the centre of the sample table, as shown in Figure 2. The linepipe was oriented so that its axis bisected the incident and scattered neutron beams, and hence the strains measured were axial. Positioning in the horizontal plane was achieved with an X-Y translator with a precision of $\pm$ 0.05 mm. Measurements of the through-wall distribution of axial residual strain were made at 0, 4, 8, 20 and 50 mm from the weld centerline. The axial variation of residual strain was determined 1, 5, 8, and 13 mm from the inside surface of the pipe wall.

The instrumental line-shape for neutron diffraction is Gaussian, so the precise position, $2\theta_{hkl}$, of a diffraction peak was obtained by fitting a Gaussian profile on a sloping background to the experimental data. The parameters of importance are the angular position of the peak, 2_{hkl}, the full-width at half-maximum and the integrated intensity. From the angular position, the interplanar spacing, d_{hkl}, for the planes with Miller indices (h,k,l) may be deduced from Bragg's Law,

$$\lambda = 2d_{hkl}\sin\theta_{hkl} \qquad (1)$$

Figure 2 - Section of girth-welded linepipe in position
for the measurement of axial residual strains.

The lattice spacing is for that subset of grains which has its hkl
plane normal aligned along the axis of the pipe. The measurement gives no
information about grains oriented in any other direction of the pipe, even
though they may be within the volume defined by the slits.

The reference lattice spacing was obtained in a subsequent experiment
on a cylinder of material 10 mm in diameter, cut at a position well away
from the weld and annealed at 675°C for eight hours.

Results and Discussion

The through-wall distribution of axial residual strain, as a function
of the distance from the weld centerline, are shown in Figures 3a-3e; the
axial variation, at fixed through-wall positions, are shown in Figures 4a-
4d. Since the strain values have been calculated from an annealed
reference, they represent a superposition of the girth-welding strains and
those introduced by the forming and seam-welding used to make the pipe from
the rolled plate, and by the hydrostatic testing of the welded section.

459

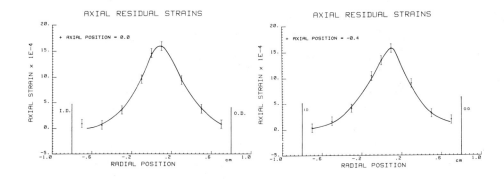

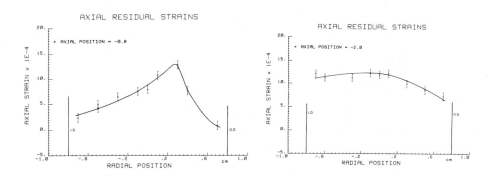

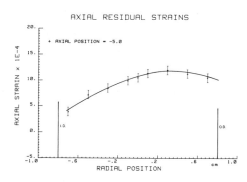

Figure 3 - Through-wall variation of axial residual strains.

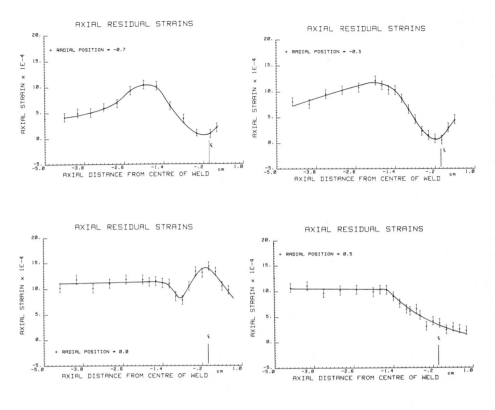

Figure 4 - Axial variation of axial residual strains.

The results show that the axial lattice strain on the weld centreline is approximately symmetric about the mid point of the wall, near zero at the inside and outside surfaces and having a peak tensile strain at a position 9 mm from the inside surface. A similar pattern appears on the through-wall scan 4 mm from the centerline, where the locus of points is still mostly within the weld metal. An asymmetric peak is observed at an offset of 8 mm where the loci of scans pass through the parent pipe as well as the weld. At offsets of 20 and 50 mm, the welding strains are expected to be small, and the data reflect primarily the residual strains in the original pipe.

The axial variation depends sensitively on the through-wall position. At the wall mid-thickness, the strain decreases from the high tensile value evident on the weld centreline in Figure 3a, passes through a local minimum just outside the weld in the heat-effected zone, HAZ, and then increases to a broad maximum centred about 16-20 mm from the weld centreline. Near the surfaces of the wall the behaviour is significantly different, with the residual strains increasing from a minimum in the HAZ to a peak value about 20 mm from the centerline.

The through-wall variation of axial residual stress deduced from the layering method (full line), and surface stresses measured by rosette gauges on the weld centreline in a similar linepipe are shown in Figure 5.

461

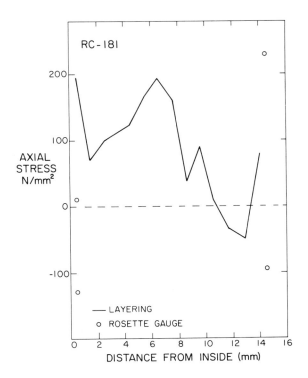

Figure 5 - Residual stresses obtained by block-layering
and removal technique.

At both surfaces, the stresses have a range from -100 to +200 N/mm², depending on the circumferential position. Figure 6, which shows the circumferential variation of the surface stresses in a similar pipe in more detail, indicates a periodic variation of the axial component with position around the pipe. The layering results, Figure 5, show a peak tensile stress at 6.5 mm from the ID, in reasonable agreement with results from the neutron diffraction experiments where a maximum in residual strain was observed 9 mm from the inner wall. A direct comparison of residual stresses determined by diffraction and strain gauge techniques is not, in this case, possible. The neutron diffraction measurements were of one component of residual strain only, and thus a calculation of residual stresses is clearly not possible without further measurements of the hoop and radial components of strain.

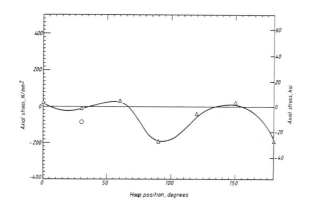

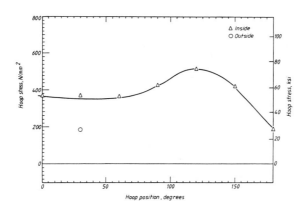

Figure 6 - Circumferential variation of surface
residual stresses. (Ref.1).

Conclusion

The results of these experiments demonstrate that neutron diffraction
is a viable non-destructive technique for measuring the distributions of
residual strains in welded components.

Acknowledgements

We acknowledge the expert technical assistance of H.F. Nieman,
A.H. Hewitt, M.F. Potter and D. Tennant.

References

1. R.B. Lazor, R.H. Leggatt and A.G. Glover, AGA PR-140-169/PR-164-170 (1985).

2. R. Leggatt, Residual Stress Measurements at Repair Welds in Pressure Vessel Steels in the As-Welded Condition, (Welding Institute report October 1986),315/1986.

[3] A.J. Allen, M.T. Hutchings, C.G. Windsor and C. Andreani, Advances in Physics (1985),34:445.

464

A LONG–LIFE REGIME FATIGUE DESIGN METHOD FOR WELDMENTS

F. V. Lawrence, J.-Y. Yung*, S.-K. Park

Department of Materials Science
University of Illinois at Urbana–Champaign
Urbana, IL 61801
*Mechanics Section of Battelle
505 King Avenue
Columbus, OH 43201

Abstract

A fatigue design methodology based on fatigue crack initiation and
early growth is presented which estimates the fatigue strength of weldments
at long lives when fatigue crack initiation becomes the dominant fraction of
the total fatigue life. The design method is based upon an analytical model
for the fatigue crack initiation life [1] which considers the four important
attributes of weldments which, together with the magnitude of the
fluctuating stresses applied, determine their resistance to fatigue: the
ratio of the applied or self-induced axial and bending stresses; the
severity of the discontinuity or notch which is an inherent property of the
geometry of the joint; the notch-root residual stresses which result from
fabrication and subsequent use of the weldment; and the mechanical
properties of the notch-root material in which fatigue crack initiation and
growth takes place. The described deterministic model gives estimates of
the mean fatigue strength under constant amplitude loading but can also be
used as a basis for probability based design of weldments. A method of
generating the expected probability distribution function of the weldment
fatigue strength is suggested in which the fatigue variables are treated as
random variables and the probability distribution function of the fatigue
strength is generated using a Monte Carlo computer simulation.

Estimates of Weldment Fatigue Resistance Based on Crack Initiation

The total fatigue life of a weldment (N_T) is comprised of a period devoted to fatigue crack initiation and early growth (N_I) and a period devoted to the growth of a dominant crack (N_p), i.e., $N_T = N_I + N_p$. It is controversial whether N_I is a significant portion of weldment fatigue life; however, there is increasing evidence [2,3] that the initiation and early growth of fatigue cracks becomes the controlling portion of the fatigue life at long lives. Hence, it is proposed that the fatigue design of weldments for long-life applications can be carried out using the relatively simple models for the estimation of N_I.

The key assumption of the design method described is that, despite the observation that cracks appear early in the life of weldments exhibiting short to intermediate fatigue lives, the initiation and early growth of (short) cracks is the largest portion of arc weldment fatigue life at long lives. This assumption is strengthened by observed crack growth behavior of the resistance spot welds discussed below and by the fatigue damage maps of Socie (2).

The spot weld should be a severe test of importance of fatigue crack initiation since the spot weld provides a very high stress concentration notch ($K_t \approx 30$). McMahon [3] recently studied fatigue crack initiation and propagation in SAE 960x tensile-shear spot welds using two direct techniques: post-test sectioning of companion specimens and replicating the exposed polished surfaces of pretest-sectioned spot weld specimens: see Fig 1. The results of this study are summarized in Fig 2 in which the fraction of total life devoted to establishing a crack of a given length is plotted against the total life here defined as the cycles to propagate a crack through the sheet thickness (0.055 in. or 1.4 mm). For both R = -1 stress ratio, creating a 0.01-in. (0.0254 mm) fatigue crack required 50% or more of the total fatigue life in the life range studied (2×10^4 to 4×10^6 cycles) and apparently even greater percentages at longer lives.

Thus, it is assumed that the fatigue strength or fatigue life of a weldment at long lives can be estimated by considering only crack initiation and early growth through the Basquin equation with the Morrow mean stress correction [4]:

$$\sigma_a = (\sigma_f' - \sigma_m)(2N_I)^b \qquad [1]$$

where σ_a is the stress amplitude, σ_f' is the fatigue strength coefficient, σ_m is the mean stress which includes the residual and local mean stress after the first cycle of load (set-up cycle), $2N_I$, is the reversals devoted to

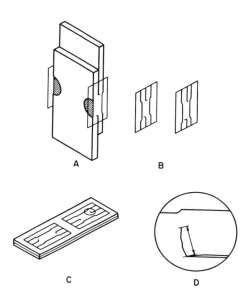

Figure 1 - Pre-test sectioned specimens of SAE tensile shear
spots. Two spot welds of each specimen were sectioned at
their mid-sections to permit the continuous observation of
fatigue crack development through surface replicas [3].

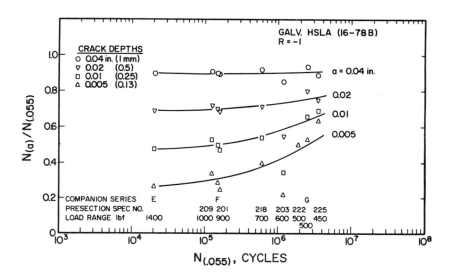

Figure 2 - Fraction of total life devoted to establishing
a crack of a given length plotted as a function the total
life of SAE 960X tensile shear-spot welds (R=-1) [3].

crack initiation and early growth (one cycle equals two reversals), and b is
the fatigue strength exponent. The notch-root stress amplitude, the stress
at the weld discontinuity can be related to the remote stress using the
fatigue notch factor so that Eq. 1 becomes:

$$(\Delta S/2)K_f = (\sigma_f' - \sigma_m)(2N_I)^b \qquad [2]$$

where ΔS is the remote stress range and K_f is the fatigue notch factor.

The Worst-Case Notch

The fatigue notch factor K_f quantifies the severity of a discontinuity.
To cope with the variable nature of the weld toe, we have developed the
concept of the "worst-case notch" in which a radius giving the highest
possible value of fatigue notch factor is presumed to occur somewhere at a
weld toe or other similar notch. Our experience with the notch-size effect
for steels has led us to conclude that Peterson's equation correctly
interrelates the fatigue notch and elastic stress concentration factors:

$$K_f = 1 + (K_t - 1)/(1 + a/r) \qquad [3]$$

where K_t is the elastic stress concentration factor for a given notch
geometry, notch-root radius, and remote stress state, a is Peterson's
material parameter which may be approximated by the expression 1.087×10^5
S_u^{-2} (mm) for steel, r is the notch root radius (mm), and S_u is the ultimate
strength of the notch-root material. A general form which can be used to
describe the K_t of most welds is [1]:

$$K_t = \beta[1 + \alpha(t/r)^\lambda] \qquad [4]$$

where α, β, and λ are constants whose values are determined by the weld
geometry and the nature of the remote stresses, t and r are the plate
thickness and notch-root radius. The constants β and λ are usually 1 and
1/2, respectively, so the K_t is often simply $1 + \alpha(t/r)^{1/2}$. The worst-case
notch value of the fatigue notch factor, $K_{f\ max}$, can be found by
substituting Eq. 4 into Eq. 3 and differentiating with respect to r to find
the value of notch root radius for which the fatigue notch factor is
maximum. Because λ is usually 1/2, $K_{f\ max}$ occurs at notch root radii
numerically equal to Peterson's parameter a ($r_c = a$). The value of $K_{f\ max}$
depends upon: the nature of the remote stresses (axial or bending) and the
geometry of the joint through the constant (α), the ultimate strength of the
material at the notch root (S_u), and the absolute size of the weldment
through the dimension (t). The use of the worst-case notch concept leads to
models which predict that the fatigue strength of a weldment depends upon
its size as well as its shape, material properties and manner of loading.

468

$$K^A_{f\ max} = 1 + 0.0015\alpha_A\ S_u t^{1/2} \qquad [5]$$
$$K^B_{f\ max} = 1 + 0.0015\alpha_B\ S_u t^{1/2}$$

where A and B represent the axial and bending loading conditions, respectively. Reference 1 lists K_t and $K_{f\ max}$ values for several weldments.

An Empirical Expression For Weldment Fatigue Strength

An empirical expression for the fatigue strength of weldments subjected to either axial or bending loads has been suggested by Lawrence and Yung [1]. Weldments are generally subjected to both axial and bending loads; the latter often result from the straightening of fabrication induced distortions under load. Thus, the combined effects of axial and bending loads must be considered in estimating weldment fatigue strength.

For fatigue lives greater than 10^5 cycles, the local stress-strain response to the applied remote stress amplitude (S_a) and the nominal mean stress (S_m) can be assumed to be elastic. Thus, the local mean stress (σ_m) and stress amplitude (σ_m) at the notch root can be expressed as:

$$\sigma_m = \sigma_r + (K^A_{f\ max}\ S^A_m + K^B_{f\ max}\ S^B_m) \qquad [6]$$

$$\sigma_m = \sigma_r + \frac{1+R}{1-R}(K^A_{f\ max}\ S^A_a + K^B_{f\ max}\ S^B_a),\ \text{since}\ S_m = (\frac{1+R}{1-R})\ S_a$$

where σ_r is the notch-root residual stress and R is the stress ratio. Also,

$$\sigma_a = (K^A_{f\ max}\ S^A_a + K^B_{f\ max}\ S^B_a) \qquad [7]$$

From Basquin's relationship (Eq. 1) and Eqs. 6 and 7.

$$\sigma_a = (K^A_{f\ max}\ S^A_a + K^B_{f\ max}\ S^A_a) \qquad [8]$$

$$\bar{\ }_a = [\sigma'_f - \sigma_r - \frac{1+R}{1-R}(K^A_{f\ max}\ S^A_a + K^B_{f\ max}\ S^B_a)](2N_I)^b$$

Rearranging Eq. 8, the fatigue strength of a weldment at long lives ($N_T > 10^5$ cycles) is:

$$S^T_a = \frac{(\sigma'_f - \sigma_r)(2N_I)^b}{K^{eff}_{f\ max}[1 + \frac{1+R}{1-R}(2N_I)^b]} \qquad [9]$$

$$K^{eff}_{f\ max} = (1 - x)K^A_{f\ max} + xK^B_{f\ max} \qquad [10]$$

$$x = S^B_a/S^T_a$$
$$S^T_a = S^A_a + S^B_a$$

469

A comparison of fatigue strength predictions made using Eq. 9 and experimental data for both as-welded and post-weld treated steel weldments [5] is given in Fig. 3.

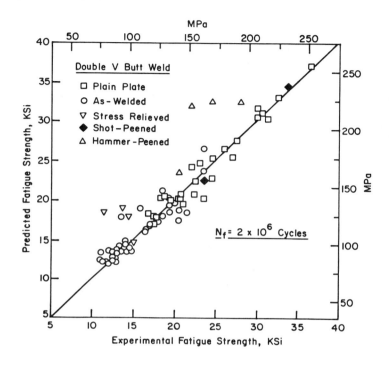

Figure 3 - Comparison of weldment fatigue strength
predicted using Eq. 11 with experimental data
principally from the UIUC weldment fatigue data bank [5].

Graphical Aids for the Fatigue Design of Weldments

Since the fatigue strength coefficient (σ_f'), the fatigue strength exponent (b), the residual stress (σ_r), and $K_{f\ max}$ all depend upon or can be correlated with hardness or ultimate strength of the base metal [1,6], Eq. 9 can be expressed as a function of ultimate strength and constants which depend upon ultimate strength and the type of post-weld treatment:

$$S_a^T = \frac{AS_u + B}{C(K_{f\ max}^{eff} - 1) + 1} \cdot \frac{(2N_I)^b}{1 + \frac{1+R}{1-R}(2N_I)^b} \qquad [11]$$

where: S_u = tensile strength of base metal
           $b = -1/6 \log [2(1+D/S_u)]$

A, B, C, D = coefficients given in Table I and below

$$AS_u + B = CS_u + 344 - \sigma_r$$

$$\sigma_r = \pm S_y(BM) = 5/9\, S_u \qquad\qquad \text{Hot Rolled}$$

$$= 7/9\, S_u - 138 \qquad \text{Normalized}$$

$$= 1.2\, S_u - 345 \qquad \text{Quenched and Tempered}$$

$$\sigma_r = 0 \qquad\qquad\qquad\qquad \text{Stress Relieved}$$

$$\sigma_r = .6\, S_u \qquad\qquad\qquad \text{Shot Peened } (S_u < 862)$$

$$\sigma_r = -\,(.21\, S_u + 551) \qquad \text{Shot Peened } (S_u > 862)$$

$$C = 1 \qquad\qquad\qquad\qquad \text{Plain Plate}$$

$$= 1.5 \qquad\qquad\qquad \text{HAZ (stress relief might reduce this value)}$$

$$= 1.5 \times 1.2 = 1.8 \qquad \text{Peened HAZ}$$

$$D = 344/C$$

Table I. Coefficients of Eq. 11 for Various
Post-weld Treatments and Base Metals

Post-Weld Treatment	Base Metal Heat-Treatment	A	B	C	D
Plain Plate	-	1	345	1.0	345
As-Welded	Hot-Rolled	0.94	345	1.5	230
	Normalized	0.72	483	1.5	230
	Q & T	0.30	690	1.5	230
Stress-Relief	-	1.50	345	1.5	230
Over-Stressed	Hot-Rolled	2.06	345	1.5	230
	Normalized	2.28	207	1.5	230
	Q & T	2.70	0	1.5	230
Shot-Peening	S_u(HAZ) < 862 MPa	2.12	896	1.8	191
	S_u(HAZ) > 862 MPa	2.12	896	1.8	191

Units: t(mm), S_u(MPa)

Figure 4 gives an example of the graphical determination of the fatigue strength of weldments based upon Eq. 11 for as-welded ASTM A36 steel. Comparison of the conditions described by lines A → A''' and B → B''' show that welds with more favorable geometries (A → A''') may have lower fatigue strengths than weldments having worse geometries but smaller thicknesses, having smaller flank angles, and having a smaller R ratio. Comparison of line B → B''' with line C → C''' shows that weldments subjected to bending (C → C''') give higher fatigue lives than smaller weldments subjected to more nearly axial loading conditions (B → B'''). Nomographs for other steels and post-weld treatments can be constructed in a similar way.

The accuracy of predictions based on Eq. 11 requires further study, but comparison of predictions made using Eq. 11 and available test data is given in Fig. 3. If one discounts the data for stress-relieved and hammer-peened

471

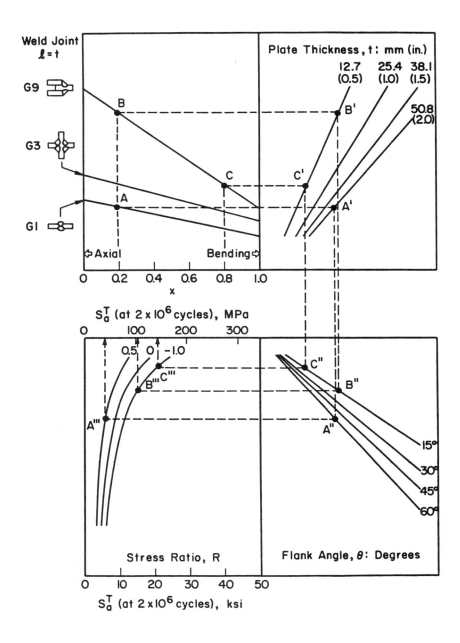

Figure 4. Nomograph for the fatigue design of as—welded ASTM A36 steel.

472

weldments (treatments which may not be as effective as hoped), then Eq. 11
and Fig. 4 would seem to predict the fatigue strength of steel weldments
with an accuracy of about 25%.

A Proposed Extension of the Initiation Model to Probability Based Design

Munse, et al [7,8] have suggested a simple but effective method of
fatigue design which deals with the complex geometries, the variable load
histories, and the variability in these and other factors encountered in the
fatigue design of weldments. The Munse method fits weldment fatigue data
with the basic S-N relationship shown in Fig. 5.

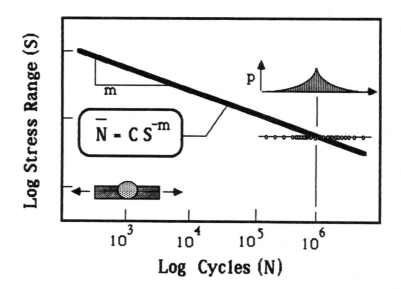

Figure 5 - Basic S-N relationship for fatigue and the
distribution of fatigue life at a given stress level [8].

When stress histories other than constant-amplitude are used, different
S-N diagrams result if the data are plotted against the maximum stress
amplitude or range: see Fig. 6. The Munse method accounts for this effect
by introducing a term ξ calculated using Miner's Rule which when multiplied
by the constant amplitude fatigue strength at a given life will predict the
fatigue strength for the variable load history at the same life: see Fig.
6.

473

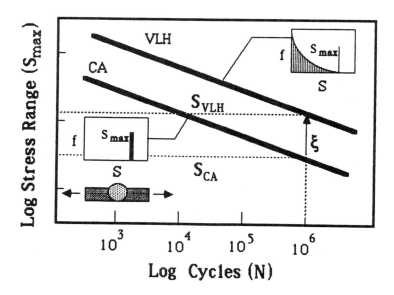

Figure 6 - Relationship between maximum stress range of variable
load history and the equivalent constant-amplitude stress
range [8,9]. Miner's rule is used to calculate the factor (ϵ)
which predicts the VLH from the CA S-N diagram.

Similarly, the natural scatter in fatigue data shown in Fig. 5 together
with the uncertainties in fabrication and stress analysis are dealt with by
the Munse method through the concept of total uncertainty.

$$\Omega_n^2 \approx \Omega_f^2 + m^2 \, \Omega_S^2 + \Omega_c^2 \qquad [12]$$

where, Ω_n = the <u>total</u> uncertainty in fatigue life.

Ω_f = the uncertainty in the fatigue data life.

m = reciprocal of the slope to the component S-N diagram.

Ω_S = measure of total uncertainty in mean stress range, including the
effects of impact and error of stress analysis and stress
determination.

Ω_c = the uncertainty in the mean intercept of S-N regression lines
and includes in particular the effects of workmanship and
fabrication.

Having estimated the total uncertainty in fatigue life Ω_n, the
reliability factor R_f is calculated after assuming an appropriate
distribution to characterize the load history and after specifying a desired
level of reliability [7].

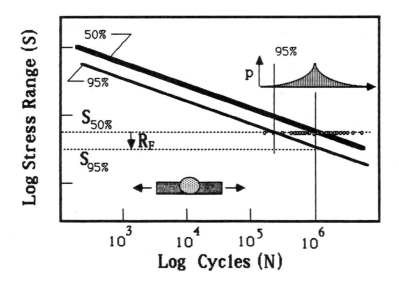

Figure 7 - Application of the reliability factor R_f to the mean resistance from either constant amplitude or variable load histories to determine the design stress i.e. $S_{90\%}$ at a given life or the useful life for the service stresses $S_{50\%}$ at a given value of reliability (95%) [8].

The Munse method estimates the maximum allowable design stress range ΔS_D from the weldment S-N diagram by determining the average fatigue strength at the desired design life ΔS_N and multiplying this value by the random load correction factor ξ and the reliability factor R_F in a manner similar to that shown in Figs. 7 and 8:

$$\Delta S_D = {}^{\wedge}S_N \ (\xi) \ (R_f) \tag{13}$$

One difficulty in applying the simple design method outlined above is that it relies on weldment fatigue data banks in which constant-amplitude fatigue data for welded components have been collected [5,9]. There are several problems with such data banks: they seldom have sufficient data for detail and material of interest; they generally do not permit one to deal with the individual factors which actually control the fatigue resistance of a weldment; they do not allow the sources of scatter in the S-N diagrams to be separated and identified; and they therefore probably force an unnecessary conservatism.

As shown in the preceding section, the mean fatigue strength of weldments can be successfully estimated using a model based on fatigue crack

475

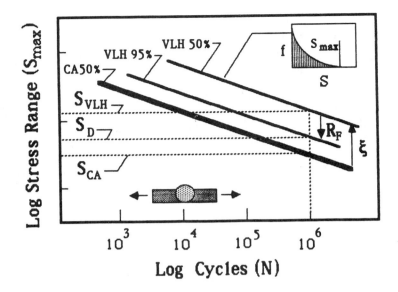

Figure 8 - The design stress S_D is calculated from the
constant amplitude S-N diagram using the factors (ξ) and (R_F).

initiation (Eq. 11). As finite element stress analysis methods become
easier and cheaper to use, it will be more economical and much faster to
generate welded component S-N diagrams using Eq. 11. Moreover, Eq. 11 can
be used to create a stochastic model for the fatigue strength of weldments
and to provide estimates of scatter due to random events occurring during
the fabrication of weldments (Ω_c in Eq. 12).

The variables in Eq. 11 can be divided into constants or known
quantities (S_a^A, S_a^B, $2N_I$) and random variables ($\alpha, t, S_u, b, R, \sigma_r, x$). The random
variables include the effects of weld geometry (α, t), material properties
and welding process (S_u, b), mean and residual stress (R, σ_r), induced
stresses due to joint distortions (x). Data of the means and variances of
each of these random variables must be collected and fitted with an
appropriate probability distribution function. In the instance in which
nothing is known concerning the variance of a given random variable, one can
assume very pessimistic values and later refine them through experience and
subsequent measurement.

Having modelled each of the random variables in Eq. 11, the mean and
variance of the fatigue strength of the weldment can be estimated using a
Monte Carlo computer simulation of Eq. 11. It is proposed that the results
of this Monte Carlo simulation, that is the estimated mean fatigue strength
and estimated value of uncertainty due to weld fabrication variables (Ω_c in

476

Eq. 12), be used as input to Eq. 13 (ΔS_D and R_f) and the probability based design stress range (ΔS_D) could then be determined in the manner suggested by Munse et al. [7,8].

Experience to date with the Monte-Carlo simulation using Eq. 11 is that the simulation predicts the experimental average value of log C but an order of magnitude less variation (this is scatter in the S-N diagram) than suggested by fatigue data bank compilations for similar details. This difference in scatter between the experimental S-N diagrams and the simulation which considers the effects of geometry and induced bending stresses may reflect the unavoidable variation in results inherent in the fatigue testing of weldments, but it may also reflect the penalty in uncertainty paid for our current inability to quantify and thus ultimately control the effects of residual stress, mean stress, induced secondary member stresses, and specimen size on the fatigue strength of weldments. Even if the conservatism forced upon the designer by the current scatter in S-N diagrams cannot be avoided, the use of the initiation model and the Monte Carlo simulation would permit the contributions to the overall uncertainty to be factorized, and the designer can then decide the economic merits of reducing the variation in a particular fatigue variable such as fabrication distortions or weld geometry.

References

1. J.-Y. Yung, F. V. Lawrence. "Analytical and Graphical Aids for the Fatigue Design of Weldments." Fatigue Fract Engng Mater Struct. 8 (3) (1985) 223-241.

2. D. F. Socie, "Fatigue Damage Maps" Fatigue 87. The Third International Conference on Fatigue and Fatigue Thresholds, Charlottesville, VA, June 1987.

3. J. McMahon, F. V. Lawrence, "Fatigue Crack Initiation and Early Growth in Tensile-Shear Spot Weldments." (Report 131 UILU-ENG 86-3608 Univ. of Ill. Fracture Control Program 1986).

4. R. W. Landgraf, "Effect of Mean Stress on the Fatigue Behaviour of a Hard Steel." (M.S. Thesis, University of Illinois at Urbana-Champaign).

5. J. B. Radziminski et al., "Fatigue Data Bank and Data Analysis Investigation." (Structural Research Series No. 405, Civil Engineering Studies, University of Illinois at Urbana-Champaign, June, 1973).

6. J. C. McMahon and F. V. Lawrence, "Predicting Fatigue Properties Through Hardness Measurements." FCP Report No. 105, University of Illinois at Urbana-Champaign.

7. A. H.-S. Ang, W. H. Munse, "Practical Reliability Basis for Structural Fatigue," (preprint no. 2494 ASCE Nat Struct Eng Conf 1975).

8. W. H. Munse et al., "Fatigue Characterization of Fabricated Ship Details for Design." (Report SSC-318 Ship Structure Committee SSC-318 1983).

9. The Welding Institute: Proceedings of the Conference on Fatigue of Welded Structures, July 6-9, 1970, The Welding Institute, Cambridge, England, 1971.

Microstructural Evolution in Weldments

MICROSTRUCTURES OF LASER BEAM WELDING OF STEELS

E.A. Metzbower and D.W. Moon
Naval Research Laboratory
Washington, DC   20375-5000   USA

P.E. Denney
Westinghouse Research &
Development Center
Pittsburgh, PA   15235

R.H. Phillips
Materials Research Laboratories
Melbourne, Victoria, Australia

Abstract

Laser beam welding is a high energy density, low heat input process which results in rapid solidification of the small welding pool and consequently a narrow heat-affected zone. Usually laser beam welds are fabricated in a single pass without the addition of filler metal, although preplaced shims of a complimentary alloy have resulted in improved properties. Several different steels have been laser beam welded using a 15 kilowatt, continuous wave, carbon dioxide laser. ASTM A36 steel has a microstructure of ferrite and pearlite. The fusion zone of the laser beam weld is bainite which when given a stress relief heat treatment has significantly improved fracture toughness. Several high strength low alloy steels (ASTM A633, A710, and A737) have also been laser beam welded. The properties of the weldments will be discussed. Quench and tempered steels of high strength (550 to 900) have also been laser beam welded with various successes.

Introduction

Laser beam welding is a high energy density, low heat input process that can be utilized to join a wide variety of metals and alloys. The significant advantages of laser beam welding include: fast travel speed; simple geometry of the weld joint; lack of a need for preheat, post heat or interpass temperature usually lack of a need for filler metals; low distortion and consequently very little postweld straightening; and the need to automate the process.

In the interaction between the laser beam and the surface of the piece to be welded, part of the laser energy is absorbed and part of it is reflected. If the energy density is sufficient, melting occurs. In the case of high power laser beams, deep penetration welding with a high depth-to-width ratio occurs a result of keyholing in which the laser beam drills a thin cylindrical volume through the thickness of the metal. A column of vapor is produced in this volume which is surrounded by a liquid pool. As the column is moved along is surrounded by a liquid pool. As the column is movedalong between the two plates to be joined, the material on the advancing side of the vapor column is melted throughout its depth. The molten metal flows around the vapor column and solidifies along the rear. The vapor column is stabilized by a balance between the energy density of the laser beam and the welding speed. Thus it is important that the energy density beam at the workpiece and the welding speed be chosen so as to compliment each other. An energy density that is too high will result in an unstable molten pool which can drop throughwhereas in energy density that is too low will not permit vaporization and the formation of a keyhole. A welding speed that is too fast will result in incomplete fusion, whereas a welding speed that is too slow will results in a very wide fusion zone and possibly undercut and/or drop through.

When the high power laser beam interacts with the workpiece, vaporization occurs and a plasma is formed. This plasma, consisting of vaporized metal ions and electrons, is opaque to the laser beam, moves over the surface, and effectively decouples the laser beam from the workpiece. In order to weld with a laser beam it is necessary to minimize the absorption of the laser beam by the plasma. This is usually accomplished by directing a high velocity jet of inert gas into the area of the interaction and displacing the plasma continuously from the top of the workpiece.

Laser beam welding results in a high depth-to-width ratio (typically greater than 4) and a fast solidification rate. Very little direct experimental data is available on the solidification and cooling rates of laser welds, but knowledge of the solidification mode and the fact that the heat affected zone narrow leads to the conclusion that the solidification and cooling rate is very fast.

The interaction of the laser beam with the workpiece is often non-uniform, in that fusion zone purification occurs. A general characteristic of the 10.6 μm radiation of the CO_2 laser is that it is highly reflected by metals and absorbed absorbed by insulators. Preferential absorption occurs if, during the movement of the movement of the keyhole, the beam encounters an impurity. The impurity can be vaporized out of the fusion zone. A reduction in impurities in the fusion zone has been measured (1).

The intent of this paper is to describe the changes in microstructural features that result from laser beam welding. A basic understanding of these processes rests on a knowledge of the laser beam welding process, fusion zone purification, and the relationship between solidification modes and cooling rates. In order to systematically elucidate the process, laser beam welding of several alloys will be described.

The alloys are: ASTM A36, HY80/100, ASTM A710, A633, and A737.

Laser Beam Welding

Using a 15 KW, continuous wave, CO_2 laser, 12mm (0.5 in) thick plates were welded.   The laser power was varied from 12 to 14 KW, and the travel speed varied from 9.7mm/s (23 in/min) to 12.7mm/s (30 in/min). Consequently the heat input ranged from 1.02 KJ/mm to 1.23 KJ/mm.   Although several different plasma "suppression" techniques were tried, the most successful was the use of a gas lens through which a tube with an inside diameter of about 1mm was inserted. Helium gas flowed through both the tube and around it.   The gas lens was positioned in the plane in which the weld was to be made and at an approximate 45° angle to the weldment.   The gas lens was aimed at the intersection of the laser beam and the plates.

ASTM A36

The most widely used carbon steel commercially available for welded struc- tures is ASTM A36, which is a steel with optimum composition for providing strength and weldability (2).   The minimum yield strength is 36 ksi (249 MPa). The microstructure of the steel is a mixture of ferrite and pearlite.   The heat-affected zone consists of a refined ferrite/pearlite microstructures, whereas the fusion zone consists of fine ferrite plus plus bainite (Fig. 1). A significant increase in the energy absorbed by Charpy V-notch (CVN) specimens as a function of temperature occurred after a stress relief heat treatment of 600°C (1112°F) for one hour.   No noticeable change in the microstructure was observed optically between the as-welded and stress reliefed fusion zones of the weldments.   Transmission electron microscopy of the fusion zones showed that in that in the as-welded fusion zones the carbides were continuous along the ferrite laths.   Small discrete carbide particles were also found in the laths.   In the case of the heat treated fusion zones, the carbides particles were found to have spherodized into discrete spherical particles between the ferrite laths.   In addition the small carbide particles observed in the ferrite coarsened (3) and began to spherodize (Fig. 2).

HY Steels

The HY steels are quenched and temper products with good strengths and toughness.   The compositions of these steels are similar in that the carbon content varies from .20 to .14 w/o depending on the strength level and the nickel content varies from 2 w/o tp 4 w/o over the same strength range.   The microstructure is martensite, with some bainite allowed in thick plates. Autogenous laser beam weldments of these steels result in good mechanical properties.   The welds are usually tested in the transverse direction.   The hardness travels across the base plate, heat-affected and fusion zones show a considerable increase in the hardness level from a Rc value of 20 in the base plate to nearly 40 in the heat-affected and fusion zones.   The microstructure of the fusion zone is untempered martensite. The CVN values of these weldments at low temperatures were quite low and unsatisfactory.   Fractographic studies (4) of these specimens indicated a considerable amount of solidification or hot cracking.   Autogenous laser beam welding procedures were modified over the extent applicable for a full penetration weldment with no success.   A shim of Inconel 600 was placed in the but joint and an alloyed weld was made (5).   This technique resulted in welds with excellent CVN values.   The hardness level of the fusion and heat-affected zones of the alloyed welds was essentially the same as in the case of the autogenous welds.   The microstructure was a refined untempered martensite (Fig. 3).   Transmission electron microscopy of the fusion zone of both the autogenous and the alloyed weldments showed a highly dislocated and twinned martensite structure.   No discernable differences were found in the thin foil examinations of the two conditions.

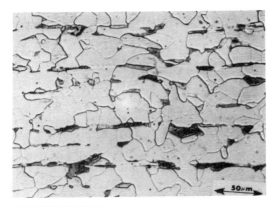

a) BASE PLATE

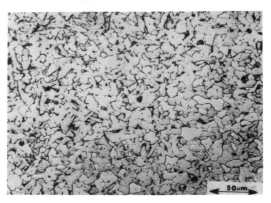

b) HEAT AFFECTED ZONE

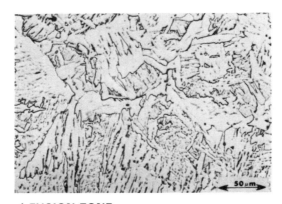

c) FUSION ZONE

Figure 1 -Micrographs of ASTM A36. (a) base plate, (b) HAZ of laser beam weld, and (c) fusion zone of laser beam weld. Note the segregation in the base plate materials (a) and how it is not present in the HAZ or fusion zone.   (Etch 1% Nital).

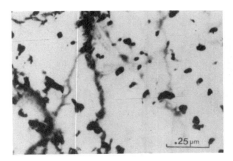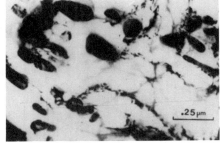

Figure 2 - TEM electron micrographs of ASTM A-36 which indicate the spheroidization of the carbides in the stress relieved materials.

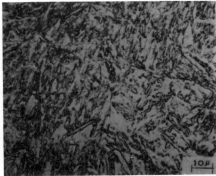

Figure 3 - Photomicrographs of the fusion zone material of HY-80 welds without Inconel 600 insert (Etchant 1% Nital)

The solidification mode of an autogenous laser beam weldments varies from the top of the weldment to the bottom (6). At the top of the weldment (Fig. 4) the solidification mode is cellular; midway through the thickness of the 12mm (0.5 in) weld the solidification mode is cellular/cellular dendritic; whereas at the bottom of the weldment the solidification mode is cellular dendritic.

Considering the keyholing process of the laser beam with a plasma on top of the workpiece, the solidification modes are in the correct order for heat transfer. The top of the weld solidifies slowest, (keyhole conduction plus plasma), whereas the bottom of the weld solidifies fastest (narrow keyhole conduction plus some convection). The differences in solidification mode also affect the crack propagation mechanism of a stress corrosion crack (SCC) through the weld-ment. In the top of the weldment the SCC crack propagates through the cells, in the center along the walls and cell walls and through the dendrites. From a fractographic point of view the top failed by microvoid coalescence; the center by a combination of microvoid coalescence and cleavage; the bottom by intergranular and some cleavage.

485

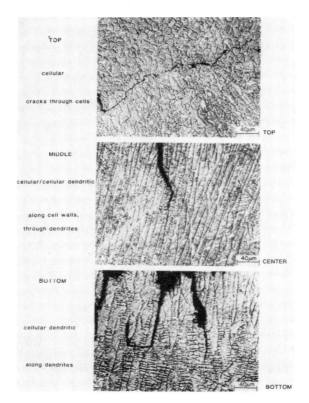

Figure 4 - Photomicrographs of the solidification structure in the different areas of the Fatigue weld and the path that fatigue cracks take passing through those regions. (Etchant 10% Ammonium Persulfate).

HSLA Steels

Several HSLA steels were laser beam welded. The microstructures and mechanical properties were determined for each steel. The microstructures were classified according to a scheme proposed by Dolby (7).

A633, Grade E - The base plate material of A633, Grade E, was a banded blocky ferrite structure which readily indicated the rolling direction of the plate. The banding was assumed to have been a carbide structure, possibly a very fine pearlite, which would be expected in an alloy with the given carbon and alloy content. The region of the HAZ furthest from the weld was a refined base plate structure. Traversing the HAZ toward the weld the apparent carbide conglomerated into large "islands." These "islands" appear to form while the amount of banding in the structure decreases. Approaching the fusion zone, the microstructure ucture of the HAZ became a mixture of ferrite with second phase, and polygonal ferrite. The microstructure of an autogenous A633 fusion zone was identical to that of an HAZ but having a more refined structure size (Fig. 5). The microstructure of the Inconel insert welds was ferrite with aligned second phase and intergranular ferrite.

486

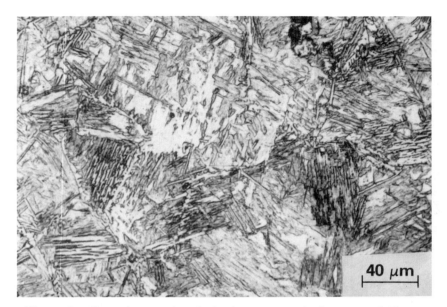

Figure 5 - Photomicrograph of the Autogeneous fusion zone of the A633 laser weld. Microstructure consists of grain boundary ferrite with aligned MAC (Nital etch).

A737, Grade B - The microstructure of the A737 is very similar to that found in A633. The base plate material is again a blocky ferrite but in the case of the A737 there is no banding and the grain size is somewhat smaller. In the HAZ near the parent material the ferrite grain size is reduced while the carbide aggregate spherodized. Adjacent to the fusion zone the HAZ comprised of ferrite with second phase. The autogenous weld microstructure included ferrite with second phase and grain boundary ferrite (Fig. 6). In the Inconel insert welds of the A737, as was the case in the A633, the microstructure was ferrite with second phase and polygonal ferrite. In this material the grain size was larger in the Inconel insert than in the autogenous weld.

A710/736, Grade A - In the base plate of the A710, Grade A, Class 1, the microstructure is a large grain blocky ferrite with carbide regions which were banded to some extent (Fig. 7). In the Class 1 material that was laser welded the grain size appeared to vary extensively. The HAZ of the Class 1 weld ranged from a refined polygonal ferrite near the parent material to a large grain martensitic structure near the fusion zone interface. The fusion zone microstructure of the laser welded Class 1 was a combination of ferrite with second phase, acicular ferrite, and intergranular polygonal ferrite.

In the A710, Grade A, Class 3, the base plate was a fine grain polygonal ferrite with some small carbide aggregate regions located along the grain boundaries (Fig. 8). The HAZ was a mixture of martensite and ferrite with aligned MAC. The microstructure of the fusion zone of the autogenous weld was a combination of ferrite with second phase, acicular ferrite and intergranular polygonal ferrite (Fig. 9). In the Inconel 600 and pure nickel insert welds, the microstructure was an extremely large grain martensitic structure.

Figure 6 - Microstructure of the autogeneous laser weld of A737 which consists of grain boundary ferrite with aligned martensite-austenite-carbide (MAC) structure (Nital etch).

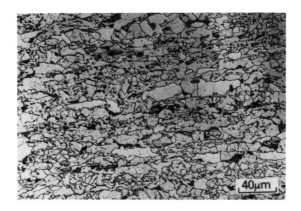

Figure 7 - Base plate A710, Grade A, Class 1 material used in laser welding. The microstructure is a blocky ferrite with some carbide regions and shows some banding due to production methods (Nital etch).

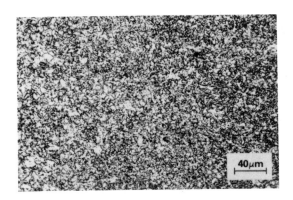

Figure 8 - Fine grain polygonal ferrite with some small carbide aggregates are found in the base plate of A710. Grade A, Class 3. Compare grain size to that of the Class 1 material (Nital etch).

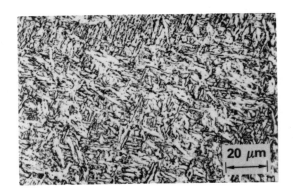

Figure 9 - Fusion zone of the autogeneous zone of laser welded A710, Grade A, Class 3. The photomicrograph shows the structure to be a mixture of acicular ferrite, ferrite with aligned MAC, and grain boundary ferrite (Nital etch).

Base plate materials were received with two different compositions. There were significant variations in the amounts of sulfur and oxygen. A chemical analysis of the base plate and fusion zone of these materials was carried out for sulfur, oxygen and nitrogen. The alloys that are designated "high S/O" had a sulfur content of 60 ppm, oxygen content of 44 ppm and a nitrogen content of 114 ppm. The sulfur level was unchanged in the fusion zone whereas the oxygen content was 98 ppm and the nitrogen content was 127 ppm. The alloy that was designated "low S/O" had a sulfur content of 8 ppm and an oxygen content of 21 ppm and a nitrogen content of 80 ppm. In the fusion zone the sulfur content was unchanged whereas the oxygen content increases to 101 ppm and the nitrogen content was 99 ppm. The increase in gas content occurred, but below the amount anticipated, indicating that the shielding techniques used worked well.

Charpy V-notch (CVN) specimens were tested as a function of temperature for foreach of the base plates and the alloyed welds. The results of the autogenous welds are shown in Fig. 10 and the results of the nickel alloyed welds are shown in Fig. 11.

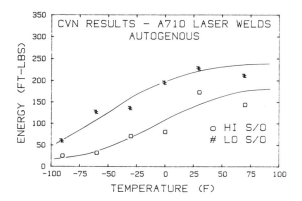

Figure 10 - Toughness versus temperature for A710 autogeneous laser welds with two levels of sulfur and oxygen.

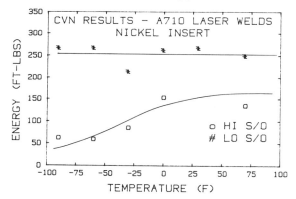

Figure 11 - Charpy V-notch results for ASTM A710 laser welds with pure nickel inserts. Welds were made with two different heats of steel containing two levels of sulfur and oxygen.

No significant difference in the microstructures of fusion zone of the high and low sulfur/oxygen were observed. The presence of acicular ferrite in the fusion zone was surprising since its occurrence is found to coincide with slowercooling rates and higher oxygen contents. Oxides are considered to be the dominant nucleation points for this fine structure in the austenite grain (7).

Discussion

The high power density of the laser welding process and the resulting rapidcooling rate have a strong influence on the microstructure and in turn the properties of the laser beam weldments. This was especially true in the HY steels in which the resulting weldments met the mechanical properties but because of rapid cooling, the toughness values for the welds were below specifi cations. This was related to the fact that the microstructure of the fusion metal was in untempered martensite. However, in comparison is the results of the ASTM A710 material which has similar mechanical properties to the HY steels and was successfully welded autogenously and met both mechanical requirements. The differences in the materials are related to the mechanisms used to achieve the properties desired. In the HY steel base plate the high strength is a result of a martensitic structure which has been tempered to increase the toughness. In the ASTM A710 the strength is attained by small ferrite grain size accomplished through alloy additions and rolling procedures and through precipitation of copper particles throughout the matrix (9). The toughness is related to these factors and to the "cleanliness" of the steel especially in the form of lower amounts of sulfide particles which have been shape controlled (10, 11). The A710 also forms a martensitic structure upon rapid cooling but because of its composition (low carbon, nickel, and chromium content) the structure has an acceptable toughness. In both of these materials the fusion zone microstructure can be altered to produce structures which have lower brittle transition temperatures (12, 13). This is done with the addition of nickel which preliminary transmission electron microscopy results indicate decreases the lath spacing in the martensite of the A710.

In the A710 the carbon, copper, nickel, and sulfur/oxygen levels are the major factors that result in the high toughness values in the base plate material. As reported by Pickering (14), and Porter, Manganello, Dabkowski, and Gross (15) the effect of lowering the carbon level is to increase the upper shelf energy and to some extent lower the transition point on the impact-transition temperature curve for a material. The lowering of the carbon decreases the amount of carbides in the material such as in pearlite, which are normally detrimental to the impact energy.

Copper is used in the A710 as a precipitated material to increase the strength of the alloy and to retain a high impact toughness shelf energy. To achieve the higher impact energy values the material must be quenched and aged. In the aging of the steel a trade off, between the strength of the material and the toughness of the material must be made as reported by Cutler and Krishnadev (16) and Hicho, Fields, and Singhal (12). The as rolled condition of the material has the higher strength (Class 1) but also has the lower toughness . This lack of aging could explain the lower CVN values in the autogenous welds.

The addition of nickel to the A710 is to prevent the copper from causing hot-shortness in the steel (8) and to lower the CVN transition temperature (8-17). The lowering of the transition temperature is the reason that additions of nickel were made to the laser welds. The exact method by which the nickel accomplishes this is unclear but it is believed to be related to the fact that nickel additions increase cross slip of screw dislocations or by spreading of shear misfit of edge dislocations and thereby making them less effective as crack nuclei (18).

In the literature there are many references given to the effects of sulfur and oxygen on the toughness of HSLA steels (14, 16). The influence of sulfur and oxygen is that their presence in the form of inclusions act as nucleation sites for the information of acicular ferrite. In many of these papers it is also stated that acicular ferrite is the preferred microstructure for high CVN results (9, 10, 11).

Both of these statements appear to conflict with the results that were achieved with the lower sulfur/oxygen material and the laser welding. However, in support of the laser beam results it has also been found that lower sulfur levels especially when shape controlled increase the toughness of the material. This relationship is related to the number of sites for microvoid nucleation and the distance between such voids.

In many of these papers it is also stated that acicular ferrite is the preferred microstructure for high CVN results (9, 10, 11). Both of these statements appear to conflict with the results that were achieved with the lower sulfur/oxygen material and the laser welding. However, in support of the laser beam results it has also been found that lower sulfur levels especially when shape controlled increase the toughness of the material. This relationship is related to the number of sites for microvoid nucleation and the distance between such voids.

From the CVN results of the low sulfur/oxygen A710 it is apparent that either andoptimum level of sulfur and oxygen has been reached or that the act of laser welding has in some way altered the recommended inclusion levels and still produced a microstructure that has a high impact shelf energy. When the microstructure of the A710 GMA weld is compared to that of the laser weld it is apparent that the conventional weld has a much higher percentage of acicular ferrite. However, the CVN curves indicate that the low sulfur/oxygen laser weld has substantially higher values. A possible explanation for this higher toughness value in the laser welded material could be related to the microstructure which is predominantly parallel lath ferrite with aligned MAC. This structure is randomly oriented throughout the fusion zone and appears to be an interlocking structure similar to normal acicular ferrite. This possible relationship between the laser weld microstructure and acicular ferrite is presently being studied by TEM.

The nickel, as stated before, reduces the transition temperature in the CVN curves and does so for both chemistries of A710. The most effective addition apparently was in the form of pure nickel but increases in the CVN values with the Inconel 600 inserts were also noted. Microprobe work and simple weight calculations indicated that the level of nickel in the fusion zone increased from approximately 1% to 8% total weight of the weld material depending on the size of the melt pool, the exact thickness of the insert, and the type of insert used.

In materials such as ASTM A36, the production procedures are such that a center line segregation of manganese-sulfide is formed. Segregation of this type lowers the toughness properties of this material. In laser welding, the segregation is broken up and redistributed throughout the fusion zone which results in an increase in the toughness of the weld (2).

In addition to the break up and redistribution of inclusions in the A36, laser welding may not maximize the properties of a material. In the as-welded state of the ASTM A36 the carbides are small and discreet. After heat treating, these carbides coarsen and spherodize which explains the increase in the toughness values (3).

In the above cases no general statement can summarize the relationship between the microstructure of a material and the laser welding process. Most of the property characteristics are related to the composition which determine the effect that rapid cooling has on the microstructure and in turn the properties of the weld. In some cases the rapid solidification and cooling is beneficial, such as in the A710, while in the HY steels the resulting microstructure is a brittle structure. Microstructural variations through a fusion zone may or may not effect the properties of a weld.

In the HY steels the mode and path of the SCC cracks were determined by the solidification structure while in the Ti-6-4 there was no apparent effect on the tensile properties due to the prior beta grain orientation. The lased weldment process can alter the microstructure through its mixing ability as in the alloying of the HY steels and A710 or through the redistribution of damaging structures as in the sulfides of the A36. In addition to redistribution the weld can vaporize alloys which results in concentration variations and microstructure and property changes (1).

Conclusion

The type and physical properties of the microstructure resulting from laser beam welding is highly dependent on the characteristics of the material following rapid solidification and cooling. The degree to which this affects the system is compositionally related but may not be the only factor or the major factor in determining the properties of a fusion zone. In many cases the composition, in both the base plate and the resulting fusion zone, is a parameter of the microstructure and in turn the properties of the weld joint; however, each case must be examined to determine whether other factors such as material redistribution or vaporization will alter the expected microstructure and properties of the weld.

Acknowledgements

1. D.W. Moon and E.A. Metzbower, Welding Journal: Welding Research Supplemental, 62 (2) (1983) pp. 53S-58S.

2. E.A. Metzbower and D.W. Moon, in Lasers in Metallurgy, p. 255-262 AIME, New York, NY, 1981.

3. R. Strychor, D.W. Moon, and E.A. Metzbower, Journal of Metals, 36 (5) (1984) pp. 59-61.

4. E.A. Metzbower, Naval Engineers Journal, 93 (4)(1981) pp. 49-58.

5. D.W. Moon and E.A. Metzbower, in Lasers in Materials Processing, p. 248-253, ASM, Metals Park, OH, 1983.

6. F.W. Fraser and E.A. Metzbower, in Lasers in Materials Processing, p. 196-207, ASM, Metals Park, OH, 1983.

7. R. E. Dolby, "Advances Metals Tech. 10, (9), pp. 349-362, (1983).

8. R. E. Kelley, Master of Science Thesis, University of Texas at El Paso, 1974.

9. R. A. Ricks, P. R. Howell, and G. S. Barritte, J. of Materials Science 17 (1982), pp. 732-740.

10. L. E. Samuels, Optical Microscopy of Carbon Steel (ASM, Metals Park, OH, 1980), pp. 422-427, 454-463.

11. P. L. Harrison and R. A. Farrar, J. of Materials Science 16, pp. 2218-2226 1981.

12. G. E. Hicho, S. Singhal, and R. J. Fields, Int. Conf. Tech. Appl. HSLA Steels, pp ASM, Metals Park, OH. 1983.

13. A. D. Wilson, Metals Progress, 121 (5) (1982) pp. 41-46.

14. F. B. Pickering, Proceedings Int. Symp. HSLA Steels, Microalloying '75, pp. 9-31, ASM, Metals Park, OH 1975.

15. L. F. Porter, S. J. Manganello, D. S. Dabkowski, and J. H. Gross, Metal Engineering Quarterly 6 (3), pp. 17-32 (1966).

16. L. R. Cutler and M. R. Krishnadev, Microstructural Science 10, pp. 79-90 (1982).

17. W. C. Leslie, Met. Trans. 3 (1), pp. 5-26 (1972).

18. W. C. Leslie, The Physical Metallurgy of Steel (Hemisphere Publishing, Washington, 1981) pp. 122-124, 204-209, 300-303.

SEGREGATION, OXYGEN CONTENT AND THE TRANSFORMATION

START TEMPERATURE IN STEEL WELD DEPOSITS

M. Strangwood and H. K. D. H. Bhadeshia

University of Cambridge
Department of Materials Science and Metallurgy
Pembroke Street, Cambridge CB2 3QZ, U.K.

Abstract

The high cooling rates associated with most arc-welding processes causes non-equilibrium solidification so that the final weld is compositionally heterogeneous. The effect of such segregation on the temperature at which the transformation from austenite to ferrite first occurs on cooling from the austenite phase field (i.e., T_h), is investigated both experimentally and theoretically for reheated Fe-C-Mn-Si weld deposits and pure alloys. The role of oxygen concentration in influencing the transformation temperatures is also investigated. It is found that the effect of alloy segregation is to elevate T_h, to an extent depending on the average alloy concentration. For homogeneous alloys, the agreement between experiment and theory for T_h is reasonable, the former being a little higher than expected, but it is demonstrated that the discrepancy has only a small effect on calculations of weld microstructure. It is found that the oxygen content has no noticeable effect on the transformation start temperature of allotriomorphic ferrite in welds at the levels studied ($\approx$ 200 ppm).

495

Introduction

The microstructure of the fusion zone of a weld, which is not influenced by any subsequent heat-treatment (for example, by the deposition of further passes) is called the primary microstructure. A major step in the evolution of the primary microstructure occurs during the continuous cooling of the weld from the austenite phase field, and is a complex process, which may be summarised as follows.

During cooling of a low-alloy steel weld deposit, the austenite (γ) first begins to decompose by a diffusional transformation mechanism to allotriomorphic ferrite (α) at a temperature T_h which depends on the cooling rate, alloy composition and the austenite grain size and shape. In order to avoid cracking and to optimise toughness, the alloying of welds with respect to both substitutional and interstitial elements is low. The rate of α growth is therefore very high, and the original γ grain boundaries rapidly become decorated with fairly uniform layer of allotriomorphic ferrite. This ferrite continues to thicken in a direction normal to the γ grain boundaries, until at some temperature T_l, diffusional transformations become relatively sluggish and the austenite then transforms by a displacive mechanism to Widmanstätten ferrite, acicular ferrite or martensite/degenerate pearlite.

Considerable work has recently been done on the prediction of the primary microstructure using detailed phase transformation theory (1-4), and this has met with reasonable success; the method is now being used commercially in the design of welding consumables and procedures. A small but significant part of the calculation requires the temperature T_h which is calculated by first computing (5) an isothermal transformation diagram (as a function of the alloy chemistry). This gives, for a particular transformation, a plot of the incubation time τ_i required to obtain a detectable degree of transformation after quenching from the γ phase field to an isothermal transformation temperature T_i. Of course, transformations during welding occur over a range of temperatures, and this is taken into account by treating the continuous cooling curve as a series of isothermal transformations, so that if the specimen spends a time Δt_i at a temperature T_i, then transformation begins when:

$$\sum_i \Delta t_i / \tau_i = 1 \qquad \qquad .....(1)$$

When the sum reaches a value of unity, $T_i = T_h$, when τ_i are for allotriomorphic ferrite (Ref.6 includes a detailed discussion of eq.1). Equation 1 is essentially an *additive reaction rule* and works approximately in converting an isothermal transformation curve into a continuous cooling curve, at least for the beginning of transformation.

During solidification, the cooling rate is sufficiently high to cause considerable chemical segregation. The austenite, when it begins to transform, is therefore far from homogeneous, and this must in turn influence its transformation kinetics. Gretoft *et al.* (7) took this into account by calculating the liquid/delta-ferrite partitioning coefficients of substitutional alloying elements, and hence obtaining the composition of the *solute-depleted* regions of a weld. A separate isothermal transformation diagram was then calculated for the solute-depleted regions, and since austenite transforms more rapidly in these regions, it was used to calculate T_h, which turned out to be higher than that expected for a homogeneous alloy. This also means that the volume fraction of allotriomorphic ferrite obtained is higher for a heterogeneous weld.

Although the procedure discussed above works well, it was decided to investigate further the factors influencing T_h for three reasons:

(a) It is now necessary to get very accurate predictions of the allotriomorphic ferrite content of a weld; too large a volume fraction of this phase is known to be detrimental to the toughness of a weld, acicular ferrite being better in this respect. However, it is possible to obtain virtually 100% acicular ferrite (given that the cost, strength etc. can be tolerated), but the microstructure then does not give optimum toughness (8). In general, a microstructure containing a small amount of allotriomorphic ferrite with the rest being acicular ferrite is known to give the best toughness. The reason for this is not clear, but could be related to the fact that as the acicular ferrite content increases, so does the yield strength and the level of hard microphases (martensite, degenerate pearlite); also, undecorated prior austenite grain boundaries may be more susceptible to trace impurity element embrittlement.

(b) It has been suggested (9) that the presence of oxide rich inclusions, for oxygen levels between 320 and 600 ppm, in a weld raises T_h, even though α nucleates at austenite grain boundaries, because the γ grain size is influenced by inclusion content. However, the issue is confused because recent work, at oxygen levels between 220 and 420 ppm, suggests that the columnar γ grains of weld deposits are not pinned by inclusions (10), because they evolve from δ-ferrite, and the driving force for the $\delta \rightarrow \gamma$ transformation increases indefinitely with undercooling below the equilibrium transformation temperature. Inclusions may however have a pinning effect on grain growth in *reheated* weld deposits; the driving force for grain growth is approximately independent of temperature and in the present context is rather small compared with that for phase transformations. Furthermore, the original work which claimed a significant effect of inclusions on T_h was based on a doubtful interpretation of length versus temperature dilatometric curves, which were significantly non-linear *before* transformation; these experiments also did not take any account of surface effects and other details to be discussed later.

(c) It should be interesting to compare T_h for as-deposited welds with welds which have been homogenised to remove compositional variations.

The purpose of the present work was therefore to study the effect of oxygen concentration and of alloying element segregation on the temperature at which austenite first transforms to allotriomorphic ferrite. It was hoped to compare transformation start temperatures obtained from three submerged arc weld deposits with three corresponding high-purity wrought alloys and calculated data; the pure alloys serve to check theory and any special effects that occur in welds as a direct consequence of impurities such as oxide inclusions. The present work is a part of a programme of research aimed at the accurate calculation of weld microstructure using phase transformation theory.

Experimental Procedures

Dilatometric measurements were carried out using a high-speed Theta Industries dilatometer (which has a RF furnace of virtually zero thermal mass, allowing rapid changes in specimen temperature), the temperature being recorded with a Pt/Pt-10wt.%Rh thermocouple attached directly to the specimen. The dilatometer length transducer was calibrated using a pure platinum specimen of known thermal expansivity; the dilatometer was programmed to give a constant cooling rate during any given experiment; the cooling rates were varied to span a wide variety of welding conditions. Some of the very rapid cooling rates were obtained using a helium gas quenching facility available. Austenitisation was carried out at $1000^\circ C$ for 10 min with the specimen chamber flooded with helium.

Dilatometer specimens were in the form of 3mm diameter rods of 20mm length, some of which had a 1mm diameter hole along the cylinder axis, to cope with high cooling rates necessary for some of the experiments. In the case of the welds, the specimens were machined from various positions, as illustrated in Fig. 1, and each specimen was identified accordingly. For the pure alloys, specimens were prepared by swaging, as described below. It was felt that the presence of a free surface might enhance the nucleation of ferrite, and this was confirmed by cutting one specimen (from each weld) into two 10mm lengths, and plating one of them with a thin layer of nickel; subsequent experiments showed that the specimen not plated with nickel had a transformation start temperature T_h which was 40-100°C higher than the plated specimen, the actual difference increasing slightly with cooling rate but considerably with austenite stabilising alloy content. This is reasonable because nucleation becomes more difficult with alloy content and with fast cooling; hence any enhanced nucleation due to surface effects would be most noticeable in these circumstances. All subsequent experiments were carried out using nickel plated specimens.

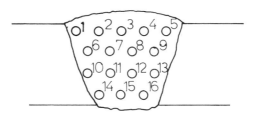

Figure 1 - Identification of dilatometer specimens
extracted from submerged arc welds.

The compositions of the alloys used are presented in Table 1. The welds were all prepared by a multipass submerged arc process with about 30 runs in an ISO2560 design joint.

Alloys P1-3 were prepared from high-purity elements as 65g melts in an argon arc furnace on a water-cooled copper mould. The ingots (and some of the specimens machined from welds) were sealed in a quartz tube under a partial pressure of pure argon, and homogenised at 1200°C for 3 days before cold swaging down to 4mm diameter rods. Further cold swaging down to 3mm diameter rod was done in two stages, each preceded by annealing treatments at 600°C for 30 min. The specimens were protected with inert argon during the annealing treatment. The pure alloys were prepared to correspond in C, Si, and Mn concentration to the weld deposits, but their impurity content was meant to be much lower than that of the welds; P1 and P2 have virtually the same C, Si and Mn concentration as W1 and W2 respectively, but unfortunately, P3 has a somewhat lower C concentration than its counterpart W3. This latter discrepancy has been allowed for in the theoretical calculations.

Table 1: Compositions, in wt.%, of welds (W1-3) and pure alloys (P1-3). The concentrations of oxygen and nitrogen are given in parts per million by weight.

No.	C	Si	Mn	Al	S	P	O	N
W1	0.10	0.19	1.10	0.015	0.012	0.015	188	115
W2	0.11	0.25	1.57	0.012	0.012	0.018	196	57
W3	0.14	0.22	1.81	0.012	0.013	0.030	138	107
P1	0.10	0.18	1.09	<0.003	<0.003	<0.003	<10	<10
P2	0.11	0.22	1.56	<0.003	<0.003	<0.003	<10	<10
P3	0.09	0.25	1.82	<0.003	<0.003	<0.003	<10	<10

Linear intercept measurements needed to characterise the austenite grain size were carried out using a Quantimet 720 Image Analysing Computer; the very large number of measurements ($\approx$ 200 grains per sample) that this made possible makes the statistical error negligible.

Results and Discussion

Austenite Grain Size

Measurements of the mean lineal intercept are presented in Table 2 below: they refer to the austenite grain structure obtained following *reaustenitisation* at 1000°C for 10 min. Reaustenitisation involves the nucleation and growth of austenite from a mixture of ferrite and cementite, and after the alloy becomes fully austenitic, the grains may coarsen in order to reduce the total amount of interface present. We note again that this process is not to be compared with the formation of columnar austenite grains from δ-ferrite during the evolution of the primary microstructure of welds.

All of the pure alloys and welds W1 and W3 have approximately the same austenite grain size, but weld W2 for some reason which is not clear shows a significantly higher grain size.

Specimen	Lineal intercept, μm
W1	27
W2	42
W3	17
P1	22
P2	20
P3	21

Transformation Start Temperatures

A typical dilatometric curve is presented in Fig. 2. During cooling from the austenite phase field, the slope of the curve simply represents the expansion coefficient of austenite, which is, found to be constant over the temperature range of interest. By fitting a least squares line to the high temperature part of this curve, the slightest deviation from linearity, caused by the formation of ferrite, can easily be detected; we estimate that this allows the transformation start temperature to be recorded reproducibly to an accuracy of better than 5°C.

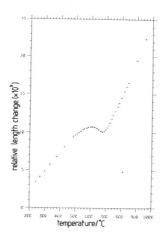

Figure 2 - A typical dilatometric plot of relative length change versus specimen temperature.

During an early stage in the experiments it was found that weld specimens extracted from near the fusion boundary of the parent plate (specimens 1, 5, 6, 9, 10,13, 16 of Fig. 1) had considerably higher T_h temperatures than all of the other specimens. This is because of dilution of the weld by the parent material. Hence, only data from regions of the weld suffering low dilution are considered further.

The data on transformation start temperatures as a function of cooling rate are presented in Figs. 3-5. Calculated curves, derived from a combination of the additive reaction rule and computed isothermal transformation curves (assuming homogeneous alloys) are also presented in Figs. 3-5.

The results are encouraging; experimental curves for the homogenised welds in all cases correspond well to the calculated curves, although they consistently underestimate T_h by about 10-20°C, the difference decreasing at high cooling rates. The reason for this could be the approximations made in the calculation of the isothermal transformation diagrams or indeed in the use of the additive reaction rule for converting such diagrams to continuous cooling transformation curves. The error caused by an underestimation of T_h by 20°C can be calculated using the microstructure model mentioned earlier (1-4). For a manual metal arc weld of compositions identical to W1, welded at 170A, 21V, 3.6mm/s, interpass temperature 250°C, arc transfer efficiency 0.775, austenite grain size $L_{tn}=50\mu m$, the volume fraction of allortiomorphic ferrite changes from 0.38 to 0.37 as T_h changes from 780-760°C. This small change is not surprising since the rate of formation of ferrite is small at T_h.

Comparison of the data from homogenised welds W1 and W2 with alloys P1 and P2 respectively shows that there is excellent agreement between these samples, even though the welds contain considerably higher oxygen levels. This proves that oxygen concentration, upto 200 ppm, does not influence the transformation start temperature of austenite. This is different from published work which we believe is based on an incorrect interpretation of dilatometric curves, from experiments in which the specimens are not protected from effects such as surface nucleation.

The data from homogenised specimen W3 and alloy P3 cannot be compared directly, but when allowance is made for the difference in carbon concentration (Fig. 5), it is clear that both show equally good agreement with calculated data, consistent with the above conclusion.

The effect of alloying element segregation is evident from Figs. 3-5. The transformation start temperatures for the unhomogenised welds are typically 30-60°C higher than those for the homogenised welds, the difference increasing with average alloy content. The amplitude of composition differences in a given alloy must increase with average alloy content, since it is given by the product of the average composition and the partition coefficient during solidification. Thus, the influence of segregation should be most pronounced for more heavily alloyed materials. We note that one important effect of segregation is to increase the amount of allotriomorphic ferrite content of welds, by raising the temperature at which this ferrite first forms. It can be seen from Figs. 3-5 that the procedure for calculating the effect of chemical segregation (7) on T_h works well; the experimental data from the unhomogenised weld agree with the CCT curves calculated for the solute-depleted regions.

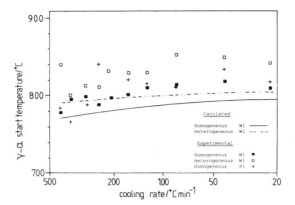

Figure 3 - Calculated and experimental continuous cooling
transformation diagrams for specimens W1 (in both the
heterogeneous and homogeneous conditions) and P1.

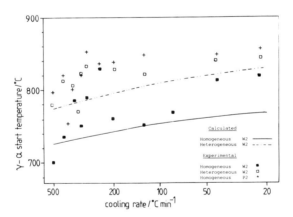

Figure 4 - Calculated and experimental continuous cooling
transformation diagrams for specimens W2 (in both the
heterogeneous and homogeneous conditions) and P2.

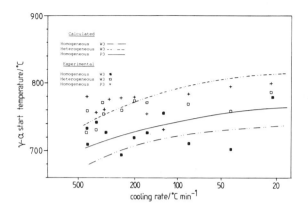

Figure 5 - Calculated and experimental continuous cooling transformation diagrams for specimens W3 (in both the heterogeneous and homogeneous conditions) and P3.

Conclusions

The temperature at which austenite begins to transform diffusionally to allotriomorphic ferrite (T_h) can be measured accurately using high-speed dilatometry, as long as precautions are taken to avoid surface nucleation effects, and effects due to the dilution of the weld by parent material.

A comparison of pure alloys with reheated welds indicates that the presence of oxide inclusions in the latter has no significant effect on T_h.

Continuous cooling curves calculated using thermodynamically predicted isothermal transformation curves combined with an additive reaction rule have been found to be in reasonable agreement with those experimentally obtained from homogenised welds and pure alloys, although the calculations systematically underestimate the transformation start temperatures by some 10-20°C, the discrepancy decreasing with increasing cooling rate.

The presence of alloying element segregation in welds has been shown to raise T_h relative to homogeneous alloys, the difference increasing with average alloy content. The effect can be predicted using a model developed by Gretoft et al. (7).

Acknowledgments

The authors are grateful to the Science and Engineering Research Council and to ESAB (UK) for financial support, and to Professor D. Hull for the provision of laboratory facilities at the University of Cambridge. During the course of this work, the authors had many helpful discussions with G. Barritte, P. Judson and L. -E. Svensson.

References

1. H. K. D. H. Bhadeshia, L.-E. Svensson and B. Gretoft, "A Model for the Development of Microstructure in Low-Alloy Steel Weld Deposits," Acta Metall., 33 (1985) 1271-1283.

2. L.-E. Svensson, B. Gretoft and H. K. D. H. Bhadeshia, "An Analysis of Cooling Curves from the Fusion Zone of Steel Weld Deposits," Scand. J. Metall., 15 (1986) 97-103.

3. L.-E. Svensson, B. Gretoft and H. K. D. H. Bhadeshia,"Computer-Aided Design of Electrodes for Manual Metal Arc Welding," Computer Technology in Welding, (Welding Institute, Abington, Cambridge), in press.

4. H. K. D. H. Bhadeshia, L.-E. Svensson and B. Gretoft, "Prediction of the Microstructure of the Fusion Zone of Multicomponent Steel Weld Deposits, Advances in Welding Science and Technology, ed. S. A. David, (Metals Park, OH: American Society of Metals, 1987), 225-229.

5. H. K. D. H. Bhadeshia,"A Thermodynamic Analysis of Isothermal Transformation Diagrams," Metal Science, 16 (1982) 159-165.

6. J. W. Christian, Theory of Transformations in Metals and Alloys, Part 1, 2nd ed.,(Pergamon Press, Oxford, U. K., 1975) 489.

7. B. Gretoft, H. K. D. H. Bhadeshia and L.-E. Svensson, "Development of Microstructure in the Fusion Zone of Steel Weld Deposits," Acta Stereologica, 5 (1986) 365-371.

8. G. Snieder, and H. W. Kerr, "Effects of Chromium Additions and Flux types on the Structure and Properties of HSLA Steel Submerged Arc Weld Metal,"Canad. Metall. Q., 23 (1984) 315-325.

9. R. C. Cochrane and P. R. Kirkwood,"The Effect of Oxygen on Weld Metal Microstructure,"Trends in Steels and Consumables for Welding(Welding Institute, Cambridge, U. K., 1978), 103-121.

10. H. K. D. H. Bhadeshia, L.-E. Svensson and B. Gretoft, "The Austenite Grain Structure of Low-Alloy Steel Weld Deposits," J. Mat. Sci., 21 (1986) 3947-3951.

THE DEVELOPMENT OF MICROSTRUCTURE
IN C-Mn STEEL WELD DEPOSITS

Stephen A. Court* and Geoffrey Pollard**

* Department of Materials Science and Engineering, University of Illinois,
1304 W. Green St., Urbana, IL 61801, U.S.A.

** Department of Metallurgy, University of Leeds,
Leeds, LS2 9JT, West Yorkshire, ENGLAND.

Abstract

A series of shielded metal arc (SMA), bead-on-plate welds has been carried out on 0.15C-1.0Mn-0.3Si (wt.%) steel plates using AWS E7016 type electrodes, deposited at two arc energies. After welding, the deposits were either air cooled or quenched. Thermocouples lanced into the molten weld pool provided a temperature profile along the welds immediately prior to quenching, and enabled determination of the quench rate. The quenching resulted in varying degrees of $\gamma \rightarrow \alpha$ transformation along the weld, which could be related to the temperature prior to quenching.

A detailed optical and transmission electron microscopy (TEM) study has been made of the microstructures found within the air cooled welds, and of sections along the quenched welds. In addition, analysis of the weld cooling curves and hardness measurements enabled determination of the temperature range of the acicular ferrite transformation. Non-metallic inclusions containing surface sulphide globules were frequently found to be the sites for the nucleation of acicular ferrite, and there was evidence of sympathetic nucleation on pre-existing ferrite laths, which led to the interlocking acicular ferrite structure.

Introduction

There is general agreement on the sequence of transformation in as-deposited weld metal, and in order of decreasing temperature of transformation, the micro-constituents are primary ferrite (prior austenite grain boundary or intragranularly nucleated), Widmanstatten ferrite sideplates, acicular ferrite, bainite and the microphases (ferrite/carbide aggregates, martensite and retained austenite), and this has been the subject of a recent review [1].

There have been a considerable number of investigations into the relationship between weld metal microstructure and toughness, and as a result, it is generally accepted that to obtain strong and tough steel weld metal it is necessary to develop a microstructure which contains a high proportion of acicular ferrite [e.g. 2,3]. Acicular ferrite forms intragranularly during the austenite to ferrite transformation and which, by virtue of its fine grain size and high angle boundaries, hinders the propagation of cleavage cracks. However, an acicular ferrite microstructure is not always easy to produce and this has necessitated a better understanding of the factors which influence microstructural development in ferritic weld metal. It is widely recognised [2-8] that the factors which influence such development are: weld metal composition, cooling rate, the prior austenite grain size and the nature of the non-metallic inclusion population. All are inter-related.

Of particular recent interest has been the way in which inclusions may affect the nucleation of acicular ferrite. The primary effect of the oxygen-containing inclusions is considered to be the provision of intragranular nucleation sites for acicular ferrite [3-5], although austenite grain size control [6-9], and matrix solute depletion of elements such as Mn [10] may also be important. It has been proposed that acicular ferrite is a form of proeutectoid ferrite which nucleates intragranularly on inclusions [4,5], although some workers have suggested that it is bainitic in nature [e.g.11,12]. The relationship between inclusions and acicular ferrite has been the subject of a recent review [13]. To date, however, the precise mechanisms involved in the nucleation of acicular ferrite have not been elucidated. In the present study, the early stages of acicular ferrite formation, and the role of non-metallic inclusions in microstructural development have been investigated.

Experimental

The welds were made using the SMA welding technique using a commercial, basic coated, low hydrogen electrode deposited onto 15cm x 8cm coupons of 12mm thick steel plate (BS 4360 43E, see Table 1) using 4mm diameter "*Ferex 7016*" electrodes (manufactured by Murex Welding Products, Ltd, U.K.). The welding current and voltage used were 180A and 25V (DC +ve), respectively, and the travel speed was adjusted to give arc energies of ≈4.7kJ/mm and ≈1.8kJ/mm.

Weld cooling curves were obtained by lancing a Pt/Pt-Rh thermocouple into the centre of the molten weld pool. The welded plate, plus thermocouple, was then either air cooled or quenched into a solution of brine and liquid nitrogen maintained at -15°C. The low arc energy (low heat input) welds were quenched when the thermocouple indicated a temperature of approximately 500°C, whereas the high arc energy (high heat input) welds were quenched when the electrode "ran-out", which was generally at some temperature greater than 500°C. The quenching resulted in varying degrees of partial ($\gamma \rightarrow \alpha$) transformation along the weld, which from the weld cooling curve, could then be related to a temperature prior to quenching. The quench rate of each weld was also determined from the weld cooling curves.

Weld metal chemical analyses from a representative low arc energy (W101) and high arc energy (W105) deposit are presented in Table 1. These analyses were assumed to be typical of all the welds produced at the respective arc energies, whether air cooled or quenched. Values of hardness (HV10) were determined for the air cooled welds designated W106 (arc energy ≈4.6kJ/mm) and W103 (arc energy ≈1.8kJ/mm), and hardness profiles were determined from sections of the quenched welds designated W105 (arc energy ≈4.7kJ/mm) and W102 (arc energy ≈1.7kJ/mm). Thin foil and carbon extraction replica specimens taken from sections of the same welds were prepared using standard techniques and were examined in a JEOL 200CX scanning transmission electron microscope (STEM), operating at 200kV, interfaced with a Link EDX system.

Table 1. Base plate and weld metal chemical analyses (wt.%).

								p.p.m.	
Identity	C	S	Mn	Si	Ti	Al	Cu	O	N
Base Plate	0.15	0.025	1.02	0.32	<0.01	0.048	0.02	N.D.*	N.D.
Weld W101	0.10	0.120	1.25	0.25	0.02	0.017	0.03	341	83
Weld W105	0.08	0.090	1.40	0.26	0.02	0.014	0.03	N.D.	N.D.

* N.D. = Not Determined.

Results and Discussion

The results have been divided into two sections, corresponding to the analyses of: (A) the air cooled welds (W106 and W103), and (B) the quenched welds (W105 and W102). These results will be described in turn.

(A) Air cooled welds

A typical photomicrograph from weld W106 is shown in Fig.1(a). The microstructure contains a large proportion of coarse, high temperature grain boundary transformation products and a coarse form of the lower temperature acicular ferrite. A typical transmission electron micrograph of the acicular ferrite is shown in Fig.1(b); the acicular ferrite does not appear to be highly acicular, the individual laths being poorly defined, their width typically being ≈1.5-2.5 μm. The scale of the microstructure is a consequence of the slow cooling rate ($\Delta t_{800°C-500°C}$ ≈50 seconds) resulting from the high arc energy employed. Electron optical examination generally upholds the light optical observations, revealing large areas of grain boundary ferrite, often containing parallel arrays of interphase cementite precipitation (Fig.2), and colonies of ferrite sideplates (Fig.3). The majority of the microphase islands, estimated to amount to ≈7% by volume of the total microstructure, appear to be ferrite/carbide (cementite) aggregates, Fig.4 (see also Fig.3).

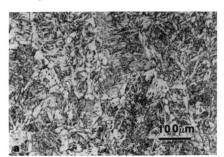 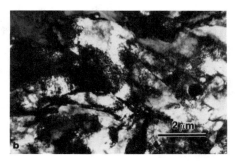

Fig. 1. (a) Optical micrograph, and (b) bright field electron micrograph taken from sections from weld W106.

The more rapid rate of cooling ($\Delta t_{800°C-500°C}$ ≈12 seconds) associated with the lower arc energy, and to some extent, the slightly lower hardenability of weld W103 (see Table 1) has resulted in a much finer microstructure than that of weld W106 (see Fig.5(a)), although TEM observation of samples from weld W103 generally revealed similar microstructural features to those previously observed in weld W106. Figure 5(b) shows an area of acicular ferrite, in which, by comparison with Fig.1(b), it can be seen that the acicular ferrite structure in weld W103 is finer than that of weld W106, although it possesses the same overall, irregular appearance. The microphase regions in weld W103 were also found to be predominantly ferrite/carbide aggregates and were estimated to account for ≈10% by volume of the total microstructure.

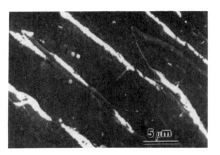

Fig. 2.   Bright field electron micrograph taken from a section from weld W106, showing an area of grain boundary ferrite and interphase cementite precipitation.

Fig. 3.   Scanning electron micrograph taken from a section from weld W106, showing an area of ferrite sideplates.

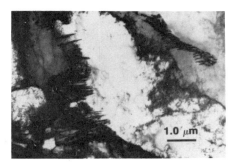

Fig. 4.   Bright field electron micrograph taken from a section from weld W106, showing ferrite/carbide microphase islands.

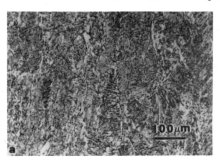

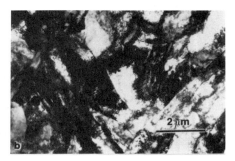

Fig. 5.   (a) Optical micrograph, and (b) bright field electron micrograph taken from sections from weld W103.

The non-metallic inclusions in both welds were typically 0.1-1.0μm in diameter and were distributed quite randomly throughout the microstructure. There did not appear to be any significant differences in inclusion composition between the two welds, and microanalysis showed the inclusions from both welds to be rich in Ti, Mn, Si and Al (see Fig.6), of varying proportions, often with smaller amounts of Cu and S. Light element analysis was not possible, but it was assumed that the major elements were combined with oxygen (say 30-40 wt.%) to give inclusion compositions within the TiO-MnO-SiO$_2$-Al$_2$O$_3$ system. The compositions of the inclusions in the quenched welds were found to be essentially the same as those in the air cooled welds.

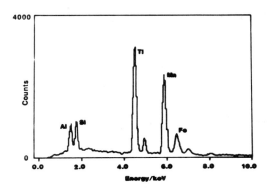

<u>Fig. 6.</u>   Typical EDX output from an inclusion found in weld W103.

(B) Quenched welds

The temperature when quenched versus distance along the weld for weld W105 is shown graphically in Fig.7, together with the results of the hardness testing. The curve represents the pre-quench weld cooling curve, with distance having been substituted for time. From this cooling curve, it is evident that some phase transformation was initiated at approximately 780-770°C, probably corresponding to the onset of the proeutectoid ferrite reaction. The hardness data also shows the onset of a quite distinct softening stage occurring at ≈740°C, which probably corresponds to the reduction in hardness resulting from the increase in the proportion of the proeutectoid ferrite morphologies and the concomitant decrease in the proportion of lath martensite. Figure 8(a) shows an optical micrograph taken from a section corresponding to a temperature of ≈815°C prior to quenching. The intragranular microstructure is predominantly one of lath martensite (untransformed austenite when quenched); however, from Fig.8(a), it can also be seen that the quench had not been sufficiently rapid to suppress the formation of grain boundary nucleated ferrite, particularly Widmanstatten sideplates, and intragranular plates are also visible. The quench rate was fairly slack (approximately 115°C/s), and this probably accounts for the discrepancies between the "expected" (i.e. fully martensitic) microstructure and the pre-quench temperature. In a section taken from a temperature of ≈700°C prior to quenching (Fig.8(b)), it can be seen that further intragranular ferrite nucleation and growth has occured and the grain boundary ferrite structure is well-defined. Subsequent coarsening of the microstructure produces a microstructure similar to that of the equivalent fully transformed air cooled weld.

TEM examination of sections corresponding to a temperature of ≈800°C prior to quenching showed the microstructure to be almost completely martensitic, Fig.9, although the grain boundary and Widmanstatten ferrite morphologies were often present at the prior austenite grain boundaries, and some intragranular plates were also observed. In sections taken from a temperature of ≈750°C prior to quenching, the microstructure again consists primarily of lath martensite. Frequently, however, within the intragranular regions, areas of narrow, interlocking ferrite laths (≈0.5-1.0μm in width) were observed, Fig.10. Furthermore, small steps at the ferrite/martensite interfaces were often evident (Fig.11), which is consistent with the findings of Ricks et al. [5], of acicular ferrite growth by a ledge advance mechanism. Many examples of a close association between inclusions and individual ferrite laths were also observed, Fig.12.

509

Figure 12(a) is a STEM image taken from a thick area of a foil and demonstrates two examples of ferrite lath nucleation associated with pairs of inclusions. Examples of multiple lath nucleation by single inclusions were also observed (see Fig.12(b)). It should also be noted at this point that there were many areas in which no such inclusion/ferrite association was evident. However, this may be accounted for if the degree of undercooling necessary for the operation of the nucleation mechanism is considered for individual inclusions; such that some (inoperative) inclusions may be consumed by the ferrite nucleated at adjacent (operative) inclusions at low undercoolings. Alternatively, it may simply be the result of sectioning effects.

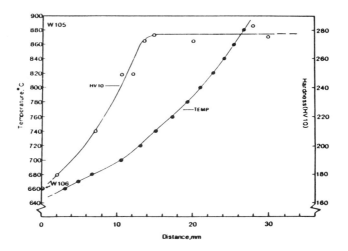

Fig. 7.   Plot of Temperature prior to quenching versus distance along the weld for weld W105, together with the results of the hardness testing. See text.

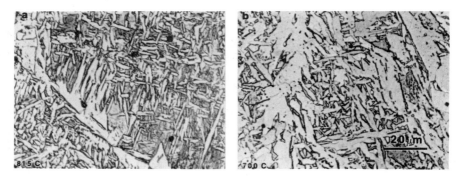

Fig. 8.   Optical micrographs of sections along weld W105, recorded from sections at a temperature of (a) ≈815°C, and (b) ≈700°C, prior to quenching. See text.

510

Fig. 9. Bright field electron micrograph taken from a section at ≈800°C prior to quenching, from weld W105, showing lath martensite.

Fig. 10. Bright field electron micrograph taken from a section at ≈750°C prior to quenching, from weld W105, showing interlocking ferrite laths.

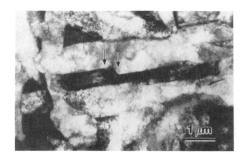

Fig. 11. Bright field electron micrograph taken from a section at ≈750°C prior to quenching, from weld W105, showing the stepped nature of the acicular ferrite/martensite interface.

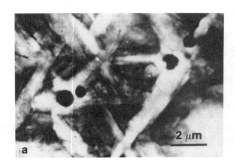

Fig. 12. (a) STEM micrograph, and (b) TEM micrograph taken from sections at ≈750°C prior to quenching, from weld W105, showing an inclusion/acicular ferrite association.

511

Sections taken from a temperature of ≈700°C prior to quenching reveal a well developed intragranular microstructure, and the lath-like nature of the acicular ferrite is more apparent (Fig.13), although the average lath width is of the order of 0.8-1.2 μm, i.e. smaller than that typically observed in the fully transformed structure. Again, examples of an association between the acicular ferrite and some inclusions was observed. Frequently, ferrite "arms" could be seen growing from pre-existing laths, which is consistent with the development of an acicular ferrite microstructure by a multiple sympathetic nucleation mechanism [5]. The sympathetic nucleation mechanism, supported by experimental observations [5], has been invoked to account for the apparent lack of a 1:1 inclusion/acicular ferrite lath correspondence noted by Cochrane [14]; although Ahblom [15], in a criticism of the work of Ricks et al., suggested that they had failed to take into account three dimensional effects, such that inclusions outside the plane of the foil may have nucleated the "sympathetically nucleated" laths. However, it seems unlikely that the symmetry exhibited by the two pairs of secondary laths illustrated in Fig.13 could be the result of lath nucleation on inclusions outside the plane of the foil. Finally at this temperature, Fig. 14 would appear to provide some evidence to support the proposal [5] that the nucleation and growth of acicular ferrite occurs at temperatures similar to those associated with Widmanstatten ferrite formation, and is therefore a proeutectoid morphology, through the observation of inclusion nucleated ferrite, which can be seen to have disturbed the growth of an area of aligned (Widmanstatten) ferrite laths.

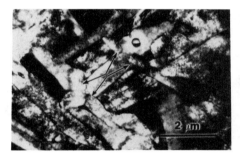

Fig. 13. Bright field electron micrograph taken from a section at ≈700°C prior to quenching, from weld W105, showing sympathetic nucleation of acicular ferrite laths.

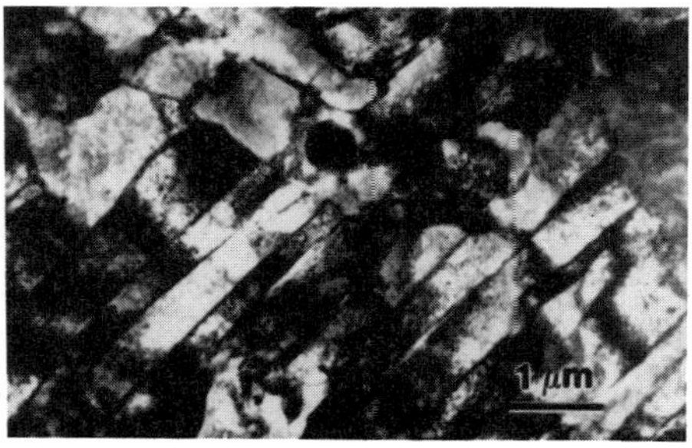

Fig. 14. Bright field electron micrograph taken from a section at ≈700°C prior to quenching, from weld W105, showing inclusion nucleated acicular ferrite.

The temperature when quenched versus distance along the weld for weld W102 is shown graphically in Fig.15, together with the results of the hardness testing. The hardness results indicate the onset of two distinct softening stages occurring at ≈740°C and at ≈680-670°C, which probably correspond to the growth of proeutectoid and acicular ferrite morphologies, respectively, although no inference of the respective transformation start temperatures can be made. The onset of the lower temperature softening stage, in this case, coincides with the mid-point region of separate inflexions in the weld cooling curve. Points of inflexion in the cooling curves of all the welds produced at the low arc energy were apparent, and enabled a thermal analysis of the acicular ferrite transformation. The high temperature inflexion was considered to be representative of the acicular ferrite start temperature (AF_s), and the lower temperature inflexion of the acicular ferrite finish temperature (AF_f). The results of this analysis are presented in Table 2. The results suggest that the acicular ferrite transformation occurs over a temperature range of approximately 120°C between ≈670°C and ≈550°C, although the results will be specific to the hardenability, cooling rate and prior austenite grain size of these deposits, and the characteristics of the non-metallic inclusion population.

Optical micrographs representing two areas along the section of weld W102 are shown in Figs.16(a) and (b). Once again, the slack quench (≈77°C/s) has resulted in discrepancies between the temperature when quenched and the "expected" microstructure. A section taken from a temperature of ≈740°C prior to quenching (Fig.16(a)) reveals a well-defined grain boundary structure. The intragranular regions are predominantly martensitic, although examples of

512

intragranular ferrite plates can also be seen. A section taken from a temperature of ≈570°C prior to quenching (Fig.16(b)), reveals a microstructure similar to that previously observed in the equivalent fully transformed weld.

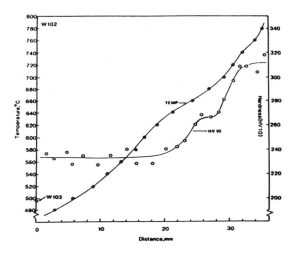

Fig. 15. Plot of Temperature prior to quenching versus distance along the weld for weld W102, together with the results of the hardness testing. See text.

Table 2. Thermal analysis of the acicular ferrite transformation.

Weld Identity	Arc Energy (kJ/mm)	$\Delta t_{800°C-500°C}$ (seconds)	AF_s (°C)	AF_f (°C)
W101	1.93	12.6	670	560
W102	1.73	11.7	680	540
W103	1.84	12.1	660	540
W107	1.71	11.1	660	560
W108	2.14	19.4	670	560

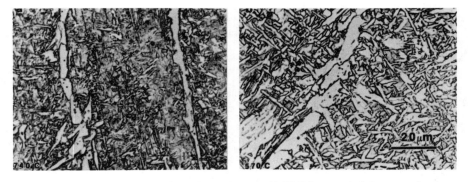

Fig. 16. Optical micrographs of sections along weld W102, recorded from sections at a temperature of (a) ≈740°C, and (b) ≈570°C, prior to quenching. See text.

513

As observed previously in weld W105, many examples of an inclusion/acicular ferrite association were evident in weld W102, and an example recorded from a section corresponding to a temperature of ≈690°C prior to quenching is shown in Fig.17(a). This inclusion was found to contain a sulphur-rich globule (Area A in Fig.17(b)), which was also rich in copper with a lower level of Mn. Figures 17 (c) and (d) show the corresponding EDX "dot-map" images for S and Cu, respectively. Area B indicated the presence of high levels of Mn, Ti and Si, and is presumably a mixed oxide or silicate. The presence of similar copper and sulphur-rich inclusion regions has been reported by other authors [16-19]. In this investigation, these globular sulphide regions were found to be a surface feature of many of the inclusions examined, and were frequently found to be associated with acicular ferrite nucleation events. This result is consistent with the observations of these authors [16] for inclusions within a range of SMA weldments, although other workers have suggested that inclusions covered with a layer of sulphide are inefficient nucleants [18,19].

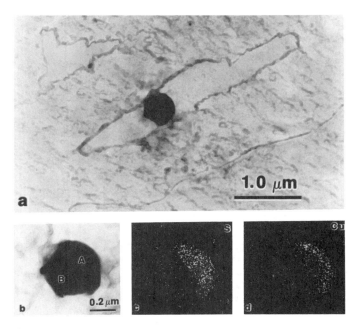

Fig. 17.   A nucleant inclusion from weld W102, showing: (a) a bright field electron micrograph of the inclusion and the surrounding ferrite; (b) the duplex nature of the inclusion, and (c) and (d) EDX "dot-map" images for S and Cu, respectively.

Considering a mechanism for acicular ferrite nucleation, it has previously been suggested that the differential thermal contraction between the elastically soft austenite matrix and the hard inclusion will make a significant difference to the stress distribution in the vicinity of the inclusion, which could lead to an increase in the austenite dislocation density and so accelerate its transformation to ferrite [6,20]. Another proposal is that of nucleation via an inclusion/ferrite lattice matching mechanism [21]; although, the results of a recent investigation [17] would suggest that an epitaxial relationship between the ferrite and the inclusions is unlikely. However, in this respect, bulk inclusion chemistry may not be the controlling factor, but instead a low lattice mismatch between certain inclusion regions, such as the sulphide globules observed in this investigation, or surface films not easily detected by the microanalysis techniques used may be important. Alternatively, the inclusions may simply act as inert substrates [17,22].

It is suggested here that nucleation may be favoured by the introduction of a matrix strain associated with the elastically soft sulphide regions [16], i.e. a shape-enhanced strain; although it is recognised that these globular sulphide areas may simply provide high energy inclusion surface features, which favour the nucleation of acicular ferrite by lowering the energy barrier for nucleation [17]. However, Dowling et al. [17] have suggested that low melting point phases such as copper sulphide would have low surface energies as compared with the more refractory oxides or spinels (e.g. Al_2O_3) and would, therefore, be less efficient nucleants. This suggestion does not appear to be consistent with the findings of this and a previous investigation by these authors [16], in which copper and sulphur rich globules were often found to be the sites for acicular ferrite nucleation.

Summary Remarks

In the air cooled welds, the differences in microstructure between the two welds could generally be accounted for in terms of the arc energy (heat input) employed and, to a lesser extent, the slight compositional variations between the two welds (see Table 1), a reduction in the arc energy leading to a refinement of the microstructure. In both welds the acicular ferrite appeared very irregular, and the overall appearance could not be described as being highly acicular.

In the quenched welds difficulties arose in correlating the microstructure and the temperature prior to quenching, because of the slack nature of the quenches, which had failed to suppress the $\gamma \rightarrow \alpha$ transformation, particularly at the prior austenite grain boundaries. However, in both welds the sequence of transformation was in agreement with that observed in previous investigations [1].

Thermal analysis of the weld cooling curves of deposits produced at the low arc energy (≈ 1.8kJ/mm) suggested that the acicular ferrite transformation occurred over a temperature range of $\approx 120°C$, between $\approx 670°C$ and $\approx 550°C$, although these results will be specific to the welding procedure and consumables employed in this investigation.

Non-metallic inclusions were frequently found to be sites for the primary nucleation of acicular ferrite. Furthermore, evidence was provided to support the proposal [5] that subsequent growth occurs by a ledge advance mechanism, as in the growth of the proeutectoid ferrite morphologies. However, many of the inclusions appeared to have played no part in microstructural development. This may be the result of sectioning effects although, as mentioned previously, some inclusions may be consumed by the ferrite nucleated at adjacent inclusions before they become "operative" nucleating agents. Evidence of sympathetic nucleation on pre-existing ferrite laths has also been provided, which adequately accounts for the lack of a 1:1 inclusion/ferrite correspondence. The globular sulphide regions, present as surface features of many of the inclusions examined, were frequently found to be associated with the acicular ferrite nucleation events. A mechanism for acicular ferrite nucleation is proposed which involves a local, shape-enhanced matrix strain associated with the presence of the surface sulphide globules. However, it is recognised also that inclusions may simply act as inert substrates, thereby lowering the energy barrier for acicular ferrite nucleation.

Acknowledgements

At the time the work was carried out, both authors were in the Department of Metallurgy at the University of Leeds. One of us (S.A.C.) wishes to thank The Welding Institute, U.K. and the Science and Engineering Research Council (SERC) for financial support, and Professor J. Nutting for provision of the research facilities.

References

1. D.J. Abson and R.J. Pargeter, Int. Met. Rev., 1986, vol. 31 (4), pp. 141-194.
2. R.E. Dolby, Factors Controlling Weld Toughness - The Present Position. Part II - Weld Metals, The Welding Institute Research Report, 1976, # 14/1976/M.
3. Y. Ito and M. Nakanishi, The Sumitomo Search, 1976, vol. 15, pp. 42-62.
4. D.J. Abson, R.E. Dolby and P.H.M. Hart, in Proc. Conf. "Trends in Steels and Consumables for Welding", The Welding Institute, London, 1978, pp. 75-101.
5. R.A. Ricks, P.R. Howell and G.S. Barritte, J. Mater. Sci., 1982, vol. 17, pp. 732-740.

6. P.L. Harrison and R.A. Farrar, J. Mater. Sci., 1981, vol. 16, pp. 2218-2226.
7. R.C. Cochrane and P.R. Kirkwood, in Proc. Conf. "Trends in Steels and Consumables for Welding", The Welding Institute, London, 1978, pp. 103-121.
8. G.S. Barritte and D.V. Edmonds, in Proc. Conf. "Advances in the Physical Metallurgy and Applications of Steels", The Metals Society, London, 1982, pp. 126-135.
9. S. Liu and D. L. Olson, Weld. J. Res. Supp., 1986, vol. 65 (6), pp. 139-s-149-s.
10. R.A. Farrar and M.N. Watson, Metal Construction, 1979, vol. 11 (6), pp. 285-286.
11. Y. Ito, M. Nakanishi and Y. Komizo, Metal Construction, 1982, vol. 14 (9), pp. 472-478.
12. M. Strangwood and H.K.D.H. Bhadeshia, in Proc. Conf. "Trends in Welding Research", ASM, Metals Park, Ohio, 1986.
13. R.J. Pargeter, The Welding Institute Research Bulletin, 1983, vol. 24 (7), pp. 215-220.
14. R.C. Cochrane, in Written Discussion to Proc. Conf. "Trends in Steels and Consumables for Welding", The Welding Institute, London, 1978.
15. B. Ahblom, International Institute of Welding Document, 1984, IIW Doc. IXJ-81-84.
16. S.A. Court and G. Pollard, J. Mater. Sci. Letts., 1985, vol. 4, pp. 427-430.
17. J.M. Dowling, J.M. Corbett and H.W. Kerr, Metall. Trans. A, vol. 17A, pp. 1611-1623.
18. G.M. Evans, International Institute of Welding Document, 1985, IIW Doc. II-A-640-85.
19. L. Devillers, D. Kaplan, B. Marandet, A. Ribes and P.V. Riboud, in Proc. Conf. "The Effects of Residual, Impurity and Microalloying Elements on Weldability and Weld Properties", The Welding Institute, London, 1983, p. 1.
20. M. Ferrante and K. Akune, in Proc. Conf. "Phase Transformations in Solids", (T. Tsakalakos, ed.) MRS Symposia Proceedings, vol. 21, Elsevier Science Publishing Co, Inc, New York, 1984, pp. 817-822.
21. N. Mori, H. Homma, S. Okita and M. Wakabayashi, International Institute of Welding Document, 1981, IIW Doc. IX-1196-81.
22. R.A. Ricks, G.S. Barritte and P.R. Howell, in Proc. Conf. "Solid→Solid Phase Transformations", (H.I. Aaronson, D.E. Laughlin, R.F. Sekerka and C.M. Wayman, eds.), TMS-AIME, Warrendale, Pennsylvania, 1982, pp. 463-468.

THEORY FOR ALLOTRIOMORPHIC FERRITE FORMATION

IN STEEL WELD DEPOSITS

H. K. D. H. Bhadeshia[*], L. -E. Svensson[**] and B. Gretoft[**]

[*]University of Cambridge
Department of Materials Science and Metallurgy
Pembroke Street, Cambridge CB2 3QZ, U.K.
[**]ESAB AB, Gothenburg, Sweden

Abstract

Recent theoretical work has enabled the rationalisation and prediction of the primary microstructure of steel welds as a function of chemical composition, welding conditions and other variables. There are, however, systematic discrepancies in the estimation of the volume fraction of allotriomorphic ferrite. In this work, we present a detailed theoretical analysis of allotriomorphic ferrite formation, which avoids some of the approximations of the earlier method. The new theory accounts also for factors influencing the nucleation of ferrite, and hence can in principle be used for high-alloy welds and for welds containing boron as a minor addition. The results are compared (in a limited way) with some published experimental data.

Throughout this work, braces are used exclusively to indicate functional relations. For example, $D\{x\}$ implies that x is the argument of the function D.

a	Side-length of the cross-section of a hexagonal prism
c	Length of a hexagonal prism
D	Diffusivity of carbon in γ
$\bar{D}$	Weighted average diffusivity of carbon in γ
I_B	Grain boundary nucleation rate per unit area
O	Total area of a plane surface
O_b	Area of a particular grain boundary
O_B	Area of all grain boundaries
O_α^e	Extended area due to the intersection of an α allotriomorph with a test plane
O_α	Actual area due to intersection of an α allotriomorph with a test plane
q	Half thickness of an allotriomorph of ferrite
S_v	Grain boundary surface area per unit volume
T_h	Temperature at which γ first transforms to α during continuous cooling
T_l	Temperature at which $\gamma \rightarrow \alpha$ transformation stops during continuous cooling
t_s	Time at which soft impingment becomes significant
V_v^α	Volume fraction of α
V_α	Actual volume of α
V_α^e	Extended volume of α
$\bar{x}$	Average carbon content of alloy
$x^{\alpha\gamma}$	Equilibrium or paraequilibrium C content of α
$x^{\gamma\alpha}$	Equilibrium or paraequilibrium C content of γ
α	Allotriomorphic ferrite
α_a	Acicular ferrite
α_w	Widmanstätten ferrite
α_1	One-dimensional parabolic thickening rate constant
α_2	Parabolic thickening rate constant for oblate ellipsoid
δ	Delta-ferrite
γ	Austenite
η	Ratio of length to thickness of allotriomorph
τ	Incubation time for the nucleation of one particle
ϕ	Equilibrium volume fraction of α

Introduction

The primary microstructure of a weld is that part which is not influenced by any subsequent heat input. It consists of allotriomorphic ferrite (α), Widmanstätten ferrite (α_w), acicular ferrite (α_a) and "microphases" (i.e., small amounts of martensite, degenerate pearlite, retained austenite). Recent work (1-7) has enabled the estimation of the primary microstructure of steel weld deposits to a reasonable degree of accuracy.

All of the constituents of the primary microstructure are important in determining the properties of a weld deposit. One of the more important properties is toughness, and it is generally recognised that thick layers of allotriomorphic ferrite at the austenite grain boundaries are detrimental to toughness because they offer little resistance to cleavage crack propagation (8). It is therefore necessary to control carefully the volume fraction V_v^α of allotriomorphic ferrite. The theory (1-7) for the prediction of the allotriomorphic ferrite has a number of unsatisfactory approximations, whose significance can best be understood after discussing briefly the way in which microstructure is calculated (1-7):

(a) The input data for the calculations consist of chemical composition, which may include C, Si, Mn, Ni, Cr, Mo and V in any reasonable combination. A weld cooling curve, either measured or estimated (7) using heat-flow theory and welding parameters is also necessary. The austenite grains of welds can be represented (9) as a honeycomb of space-filling hexagonal prisms of length "c" and hexagon side-length "a"; since c>>a, and since the model ignores α formation on the ends of the hexagons, only the parameter "a" is needed as an input. This can either be measured or estimated empirically (7) as a function of chemical composition and heat-input.

(b) The chemical composition is used to calculate a paraequilibrium (10-12) phase diagram which includes the Ae3' $\alpha/\alpha+\gamma$ phase boundary, the T_0 curve, and the same curves after appropriate modifications for strain energy effects due to the displacive character of some of the phase transformations that arise in welds.

The Ae3' curve on a plot of temperature versus carbon concentration represents the *paraequilibrium* $\alpha/\alpha+\gamma$ phase boundary, where paraequilibrium is a form of constrained equilibrium in which substitutional atoms do not redistribute during transformation, but carbon partitions such that it has equal chemical potential in both phases at the interface. The T_0 curve on the same plot represents the locus of all points where α and γ of the same composition have equal free energy. Diffusionless transformation is thermodynamically possible only if the γ composition falls to the left of the T_0 curve.

The phase diagram is needed in order to calculate the compositions at the α/γ interface during α growth, which is assumed to occur at a rate controlled by the diffusion of carbon in the γ ahead of the interface. Since the diffusivity D of carbon in γ is a function of its carbon concentration, a reliable method (13,14) of representing this function is necessary so that a weighted average diffusivity (15) $\bar{D}$ can be substituted for D in the rate equations. This is strictly justified only when growth occurs at a constant rate, but numerical calculations (3) have demonstrated the approximation is extremely good for typical weld deposits. Finally, the Ae1 curve, which represents the composition of α in equilibrium with γ is calculated, as a good approximation (since the carbon concentration of ferrite is very small) to the paraequilibrium Ae1' curve.

(c) Having calculated the phase diagram, an isothermal transformation (TTT) curve is also computed using methods described elsewhere (16). However, welds are known to be chemically heterogeneous and this can cause an underestimation in the calculated value of V_v^α. To allow for this, the weld is divided into solute-rich and solute-depleted regions (with respect to substitutional elements only, since C diffuses very rapidly so that its activity should become

519

uniform during cooling) (5). It is assumed that ferrite formation first begins (at a temperature T_h) in the solute-depleted regions, whose composition is calculated by computing the partition coefficients for substitutional elements, for partitioning between liquid and δ (it is solidification which gives rise to segregation in the first place). Thus, a new TTT curve is calculated for the solute-depleted region, and in combination with the weld cooling curve, can be used to compute T_h using an additive reaction rule (reviewed in ref.17), which treats continuous cooling as a series of short isothermal transformations, each occurring at a successively lower temperature until the transformation eventually stops at the temperature T_l. This is the temperature at which diffusional transformations become sluggish and give way to displacive reactions (e.g., α_w), when the diffusional C-curve of the TTT diagram crosses the displacive C-curve.

(d) The half-thickness q of the allotriomorph is calculated using the equation (1):

$$q = \int_{t=0}^{t_1} 0.5\alpha_1 t^{-0.5} \, dt \tag{1}$$

where t=0 when the alloy reaches the temperature T_h, and $t=t_1$ at T_l. Note that α_1 is the one-dimensional parabolic thickening rate constant, and for continuous cooling transformation is a complicated function of temperature and hence of time; the equation is therefore numerically integrated. The analysis assumes that at T_h, the γ grain boundaries become decorated instantaneously with an infinitesimally thin layer of α, so that α formation is (below T_h) a growth problem involving the one-dimensional thickening of allotriomorphs.

(e) Having calculated q, V_v^α is given, from geometrical considerations by (1):

$$V_v^\alpha = 4qC_3(a - qC_3)/a^2 \tag{2}$$

where $C_3 = \tan\{30^0\}$.

This procedure works well if the calculated volume fraction is multiplied by a factor of about two (1); the theory *consistently* underestimates the amount of allotriomorphic ferrite. This is clearly not satisfactory, but is encouraging since the calculated volume fraction correlates extremely well (correlation coefficient typically 0.95) with the experimental data, and bearing in mind the factor of two, the model can, and is being used extensively in the design of weld deposits (18). In this work, we present new theory which should help eliminate many of the approximations made in the earlier model. The new theory is more general in the sense that it can also be applied to more heavily alloyed welds and to welds where α formation is nucleation limited.

Theory

It is believed that the major assumption (illustrated in Fig. 1) in our earlier model, which causes a consistent underestimation of V_v^α, is that at T_h the γ boundaries become decorated uniformly with a thin layer of α, the α subsequently thickening normal to the γ grain boundary. In some heavily alloyed welds this is unjustified since discontinuous layers of α are observed (3). Even when uniform and continuous layers of α are observed, the early stages must involve a faster rate of transformation since growth is initially not confined to one dimension. It is only when site-saturation occurs that one-dimensional allotriomorph thickening should be a good approximation.

520

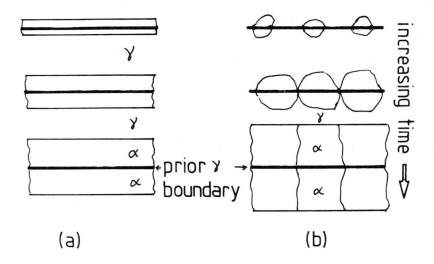

Figure 1 - Diagram illustrating allotriomorphic ferrite as
theoretially modelled (a), and as is the case in reality (b).

Allotriomorphs as discs

The allotriomorphs prior to site-saturation are modelled as discs parallel to the austenite grain boundary planes, of half-thickness q, radius ηq. An analysis will also be presented for allotriomorphs with are oblate ellipsoids, but for reasons discussed later, the disc model is probably a better approximation. The aspect ratio η of the allotriomorphs is considered constant because in reality, lengthening and thickening processes are coupled. The method takes account of hard impingement; the problem of soft-impingement is discussed later.

In the derivation that follows, we use the method of Cahn and Avrami (19,20), for the calculation of isothermal reaction kinetics for nuclei forming preferentially at grain boundary faces. Nucleation on grain edges and corners is ignored because the γ grain size for welds is large, and because ferrite formation in welds starts at relatively high supersaturations (1). Cahn's method is here modified for the parabolic growth kinetics of α. The general theory of overall transformation kinetics has been thoroughly reviewed by Christian (17), and for the sake of brevity, an understanding of the concept of extended area or volume is assumed.

Consider a plane surface of total area O parallel to a particular boundary; the extended area O_α^e is defined as the sum of the areas of intersection of the discs with this plane. It follows that the change dO_α^e in O_α^e due to disc nucleated between $t=\tau$ and $t=\tau+d\tau$ is:

$$dO^e_\alpha = \pi O_b I_B[(\eta\alpha_1)^2(t - \tau)]d\tau \qquad (3)$$

$$\text{for } \alpha_1(t - \tau)^{0.5} > y$$

and

$$dO^e_\alpha = 0 \qquad \text{for } \alpha_1(t - \tau)^{0.5} < y$$

where y is the distance between the boundary and an arbitrary plane parallel to the boundary.

Bearing in mind that only particles nucleated for $\tau > [(y/\alpha_1)^2]$ can contribute to the extended area intersected by the plane at y, the whole extended area is given by:

$$O^e_\alpha = \int_0^{t-(y/\alpha_1)^2} (\eta\alpha_1)^2 \pi O_b I_B(t - \tau)d\tau$$

$$= 0.5\pi O_b I_B(\eta\alpha_1)^2 t^2[1 + \theta^4 - 2\theta^2] \qquad (4)$$

where $\theta = y/(\alpha_1 t^{0.5})$.

Given that the relationship between extended area and actual area O_α is given by (17,19):

$$O^e_\alpha / O = -\ln\{1 - [O_\alpha / O]\} \qquad (5)$$

and assuming that there is no interference with allotriomorphs from other boundaries, the total volume V_b of material originating from this grain boundary is given by integrating for all y between negative and positive infinity; in terms of θ, the integral amounts to:

$$V_b = \int_0^1 2O_b (\alpha_1 t^{0.5})(1-\exp\{-O^e_\alpha/O_b\}) \, d\theta$$

$$= \int_0^1 2O_b (\alpha_1 t^{0.5})(1-\exp\{-0.5\pi I_B(\eta\alpha_1)^2 t^2[1 - \theta^4]\}) \, d\theta$$

$$= 2O_b(\alpha_1 t^{0.5})f\{\theta, \eta\alpha_1, I_B, t\}$$

where

$$f\{\theta, \eta\alpha_1, I_B, t\} = \int_0^1 (1-\exp\{-0.5\pi I_B(\eta\alpha_1)^2 t^2[1 - \theta^4]\}) \, d\theta \qquad (6)$$

If the total grain boundary area is $O_B = \Sigma O_b$, then by substituting O_B for O_b in the above equation the total extended volume V^e_α of material emanating from all boundaries is found; this is an *extended* volume because allowance was not made for impingement of discs originating from different boundaries. Thus,

$$V_\alpha^e = 2O_B(\alpha_1 t^{0.5})f\{\theta, \eta\alpha_1, I_B, t\}$$

and if V is the total volume, and S_v the γ grain surface area per unit volume, then:

$$V_\alpha^e / V = 2S_v(\alpha_1 t^{0.5})f\{\theta, \eta\alpha_1, I_B, t\}.$$

This can be converted into the actual volume V_α using the equation:

$$V_\alpha / (V\phi) = 1 - \exp\{-V_\alpha^e / (V\phi)\}$$

where $\phi = (x^{\gamma\alpha} - \bar{x})/(x^{\gamma\alpha} - x^{\alpha\gamma})$

It follows that:

$$-\ln\{1 - \zeta\} = (2S_v/\phi)(\alpha_1 t^{0.5})f\{\theta, \eta\alpha_1, I_B, t\} \qquad (7)$$

where $\zeta = V_\alpha / (V\phi)$, i.e., it is the volume of α divided by its equilibrium volume.

Finally, it should be noted that the value of the integral f tends to unity as I_B increases or time increases, since site saturation occurs. In the limit that the integral is unity, eq.7 simplifies to one dimensional thickening as in the earlier model:

$$-\ln\{1 - \zeta\} = (2S_v / \phi)(\alpha_1 t^{0.5}) \qquad (8)$$

Allotriomorphs as Oblate Ellipsoids

Obara *et al.* (21) have stated an equation for an oblate ellipsoid representation of allotriomorphs, although the derivation was not given and the factor ϕ was not included in their analysis. From geometrical considerations and by comparison with eq.3, it follows that for oblate ellipsoids, the change in the extended area at the plane located at distance y from the boundary, due to particles nucleated between $t=\tau$ and $t=\tau+d\tau$ is given by:

$$dO_\alpha^e = \pi O_b I_B[(\eta\alpha_2)^2(t - \tau)(1 - (y^2/(\alpha_2^2(t - \tau))))]d\tau \qquad (9)$$

The function f is now replaced by the new function f_2 as follows:

$$f_2\{\theta, \eta\alpha_2, I_B, t\} = \int_0^1 (1-\exp\{-0.5\pi I_B(\eta\alpha_2)^2 t^2[1 - 2\theta^2 + \theta^4]\}) \, d\theta \qquad (10)$$

where α_2 is the parabolic thickening rate constant for oblate ellipsoids; α_2 is in general larger than α_1. It follows that:

$$-\ln\{1 - \zeta\} = (2S_v \ / \ \phi)(\alpha_2 t^{0.5}) \ f_2\{\theta, \eta\alpha_2, I_B, t\} \tag{11}$$

There is a disadvantage in using the oblate ellipsoid model for allotriomorphic ferrite formation in situations where site saturation occurs at an early stage of transformation, as is the case for welds. The oblate ellipsoid parabolic thickening rate constant α_2 only applies while the shape of the ferrite remains in the form of oblate ellipsoids. When the γ boundaries become decorated with α, the allotriomorphs grow in just one dimension, and at that stage, the one-dimensional parabolic thickening rate constant α_1 applies because the boundary conditions for the growth problem are then different.

Non-isothermal growth

The equations derived above may be used for the non-isothermal growth of ferrite in welds, if it is assumed that the reaction is isokinetic; this is probably a good assumption for welds since site-saturation occurs at an early stage of transformation.

The continuous cooling curve can then be divided into a series of isothermal steps. The procedure is not discussed here but has been reviewed by Christian (17).

Comparison with Experimental Data

The model derived above cannot be compared fully with experimental data; it is not possible to predict from first principles, the nucleation rate of ferrite.

It is however possible to test the *form* of the equations. For example, under both isothermal and non-isothermal conditions, if everything else is kept fixed, then $\ln\{1-\zeta\}$ should be directly proportional to S_v. Thewlis (22) has recently published a large amount of data (some 34 tandem submerged arc welds), which includes details on the austenite grain size, volume fractions of all the phases and chemical composition. The data can, for the present purposes be divided into two sets, one containing welds with negligible boron content, and another with welds containing about 30ppm by wt. of boron. These have to be treated separately since I_B should be much lower in the latter case; boron is known to reduce the austenite grain boundary energy by segregation (23), thereby causing an increase in the activation energy for heterogeneous nucleation.

The boron-free welds have a typical composition (wt.%) Fe-0.06C-0.25Si-1.37Mn-0.04V-0.11Mo, the other welds containing a slightly higher Mo concentration ($\approx$ 0.21 wt.%) and about 32 ppm by wt. of boron. Thewlis also tabulated mean lineal intercept ($\overline{L}$) measurements for γ grains; S_v may be derived from the stereological rule (24):

$$S_v = 2 \ / \ \overline{L} \tag{12}$$

Because the chemical compositions of the welds in a given set are virtually identical, and since the same welding conditions are used for all welds, a plot of $\ln\{1-\zeta\}$ versus S_v should be

524

linear (even for non-isothermal conditions).

To obtain ζ, it is necessary to define ϕ, the equilibrium volume fraction of α. A phase diagram was calculated for each weld, taking into account the detailed composition of each weld, and ϕ was evaluated at the temperature T_1, where α formation ceases and gives way to more rapid displacive transformation. T_1 was determined from the calculated TTT curve for each weld, and the cooling curve was estimated using the welding conditions listed by Thewlis, and the heat flow parameters evaluated by Bhadeshia *et al.* (3)

A typical phase diagram and TTT curve is presented in Fig. 2. Fig. 3 shows that both sets of welds satisfy, to a reasonable degree, the linear relationship between $\ln\{1-\zeta\}$ and S_v, the scatter probably arising because: (a) the compositions of welds in each set are not strictly constant (Fe 0.05-0.07C 0.20-0.27Si 1.28-1.37Mn 0.09-0.15Mo 0.039-0.044V, wt. % for the boron free welds, and Fe 0.05-0.07C 0.20-0.27Si 1.36-1.42Mn 0.21-0.23Mo 0.038-0.044V wt.% for the other welds), (b) statistical error in the measurement of volume fraction.

Further testing of the theory requires a method for the reliable estimation of nucleation rates, and this is where our efforts will be concentrated in the future.

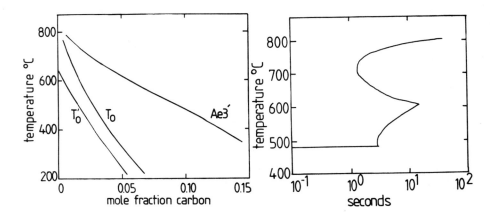

Figure 2 - Fe-0.06C-0.20Si-1.27Mn-0.041V-0.09Cr-0.11Mo-0.05Ni wt% alloy
(a) Calculated phase boundaries, (b) calculated TTT diagram.

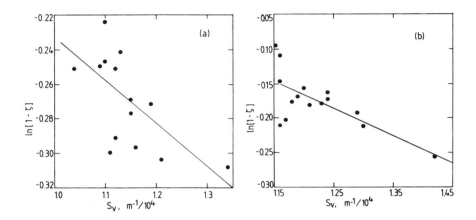

Figure 3 - Plots of ln{1-ζ} versus S_v, as discussed in the text
(data due to Thewlis). (a) Boron free welds, (b) boron containing welds.

Soft Impingement

Gilmour *et al.* (25) have treated the problem of soft impingement, i.e., the overlap of the diffusion fields of allotriomorphs growing from different boundaries. The diffusion profile ahead of the allotriomorph was approximated as being finite in extent, and soft impingement was considered to occur when the concentration at the end of the profile begins to rise above the average alloy carbon concentration. The time t_s before soft impingement is then given by:

$$t_s \approx (0.5a / \tan\{30^o\})^2 \, \bar{x} \, x^{\gamma\alpha} / [D\{x^{\gamma\alpha}\} \, (x^{\gamma\alpha} - \bar{x})] \tag{13}$$

where we have taken the extent of the γ grain ahead of the grain boundary to be $(0.5a/\tan\{30^o\})$, consistent with the hexagonal prism γ grain morphology.

Typical calculations for t_s are given in Table 1; it is evident that for arc welds, where allotriomorphic ferrite formation is complete within a few seconds, soft impingement between allotriomorphs growing from opposite grain boundaries is not a problem.

$x^{\gamma\alpha}$, mol. frac.	$D\{x^{\gamma\alpha}\}$, $cm^2/(10^{-8}s)$	t_s, s	Temp., ^oC
0.0004	1.470	5750	800
0.0113	0.926	2410	760
0.0196	0.591	1910	720
0.0303	0.389	1780	680
0.0435	0.255	1830	640
0.0521	0.140	2760	600

Conclusions

A new theory has been developed which allows the calculation of the volume fraction of allotriomorphic ferrite in welds, even when site saturation does not occur at an early stage of transformation. It has not been possible to fully test the model, since this awaits the development of at least a working hypothesis for the prediction of the nucleation rate of ferrite. However, it already provides an interpretation of effects such as the influence of austenite grain size on the volume fraction of ferrite. The model is incidentally, independent of the austenite grain shape, and as such should be applicable also to the heat-affected zone of welds or to the reheated microstructure of multipass welds.

Acknowledgments

The authors are grateful to ESAB AB for financial support and for the provision of laboratory facilities, and to Professor D. Hull for the provision of laboratory facilities at the University of Cambridge. Helpful discussions with A. A. B. Sugden are acknowledged with pleasure.

References

1. H. K. D. H. Bhadeshia, L.-E. Svensson and B. Gretoft,"A Model for the Development of Microstructure in Low-Alloy Steel Weld Deposits," Acta Metall., 33 (1985) 1271-1283.

2. H. K. D. H. Bhadeshia, L.-E. Svensson and B. Gretoft, "Prediction of the Microstructure of the Fusion Zone of Multicomponent Steel Weld Deposits," Trends in Welding Research, ed. S. A. David, (Metals Park, OH: American Society for Metals, 1987), 225-229.

3. H. K. D. H. Bhadeshia, L.-E. Svensson and B. Gretoft, "Prediction of the Microstructure of Submerged-Arc Linepipe Welds," Welding and Performance of Pipe Welds, (Welding Institute, Abington, Cambridge) in press.

4. L.-E. Svensson, B. Gretoft and H. K. D. H. Bhadeshia, "Computer-Aided Design of Electrodes for Manual Metal Arc Welding," Computer Technology in Welding, (Welding Institute, Abington, Cambridge), in press.

5. B. Gretoft, H. K. D. H. Bhadeshia and L.-E. Svensson, "Development of
    Microstructure in the Fusion Zone of Steel Weld Deposits,"
    Acta Stereologica, 5 (1986) 365-371.

6. H. K. D. H. Bhadeshia, L.-E. Svensson and B. Gretoft, "The Influence of
    Alloying Elements on the Formation of Allotriomorphic Ferrite in Low-Alloy
    Steel Weld Deposits," J. Mat. Sci., 4 (1985) 305-308.

7. L.-E. Svensson, B. Gretoft and H. K. D. H. Bhadeshia, "An Analysis
    of Cooling Curves from the Fusion Zone of Steel Weld Deposits,"
    Scand. J. of Metall., 15 (1986) 97-103.

8. J. H. Tweed and J. F. Knott, "Effect of Reheating on Microstructure and
    Toughness of C-Mn Weld Metal," Metal Science, 17 (1983) 45-53.

9. H. K. D. H. Bhadeshia, L.-E. Svensson and B. Gretoft, "The
    Austenite Grain Structure of Low-Alloy Steel Weld Deposits,"
    J. Mat. Sci., 21 (1986) 3947-3951.

10. A. Hultgren, Jernkontorets Ann., 135 (1951) 403.

11. M. Hillert, Jernkontorets Ann., 136 (1951) 25.

12. H. K. D. H. Bhadeshia, "Diffusional Formation of Ferrite in Iron
    and its Alloys," Progress in Mat. Sci., 29 (1985) 321-386.

13. R. H. Siller and R. B. McLellan, "Application of First Order
    Mixing Statistics to the Variation of the Diffusivity of Carbon in
    Austenite," Metall. Trans., 1 (1970) 985-995.

14. H. K. D. H. Bhadeshia, "The Diffusion of Carbon in Austenite,"
    Metal Science, 15 (1981) 477-479.

15. R. Trivedi and G. M. Pound, "Effect of Concentration-Dependent
    Diffusion Coefficient on the Migration of Interphase Boundaries,",
    J. Appl. Phys., 38 (1967) 3569-3576.

16. H. K. D. H. Bhadeshia, "A Thermodynamic Analysis of Isothermal
    Transformation Diagrams," Metal Science, 16 (1982) 159-165.

17. J. W. Christian, Theory of Transformations in Metals and Alloys,
    Part 1, 2nd Ed., (Oxford, Pergamon Press, 1975).

18. L.-E. Svensson, B. Gretoft and H. K. D. H. Bhadeshia, "Design
    of the Primary Microstructure of Arc Welds", Parts I-III,
    Internal Reports of ESAB AB and the University of Cambridge, 1986.

19. J. W. Cahn, "Kinetics of Grain Boundary Nucleated Reactions,"
    Acta Metall., 4 (1956) 449.

20. M. Avrami, J. Chem. Phys., 7 (1939) 1103.

21. T. Obara et al., Solid→Solid Phase Transformations, (Metals
    Park, Ohio: American Society for Metals, 1981) 1105.

22. G. Thewlis, <u>Welding and Performance of Pipe Welds</u>, (Cambridge: Welding Institute 1986), in press.

23. F. B. Pickering, <u>Phys. Metall. and the Design of Steels</u>, (London: Applied Science Publishers, 1978), 104-5.

24. R. T. DeHoff and F. N. Rhines, <u>Quantitative Microscopy</u>, (New York: McGraw Hill, 1968).

25. J. B. Gilmour, G. R. Purdy and J. S. Kirkaldy, "Partition of Manganese During the Proeutectoid Ferrite Transformation in Steels," <u>Metall. Trans.</u> 3 (1972) 3213-3222.

MICROSTRUCTURAL CHARACTERISATION IN TYPE 316 WELDS

Y. P. Lin, G. T. Finlan and J. W. Steeds

H. H. Wills Physics Laboratory
Tyndall Avenue
Bristol BS8 1TL
England, U.K.

Abstract

The microstructure of Type 316 stainless steel weld metal has been characterised. The columnar grain structure has been studied by orientation contrast in the SEM. In the TEM, columnar grain boundaries were located using convergent beam electron diffraction. Emphasis has been placed on the orientation relationship between austenite and ferrite. The results suggested that co-solidification of austenite and ferrite was the dominant process. Analysis of the ferrite composition indicated that substantial elemental partitioning had taken place in the solid state. During creep testing at 600°C or stress relieve treatment at 800°C, the formation of sigma phase was found to take place preferentially on or close to columnar grain boundaries. R phase particles were identified in samples after exposure at 600°C but not at 800°C. The R phase precipitates formed on dislocations in ferrite and the orientation relationship with ferrite has been determined. Closely spaced lathy ferrite was found to be more resistent to intermetallic formation, even though no compositional difference was detectable between lathy and vermicular ferrite.

Introduction

Type 316 Stainless Steel is widely used in the nuclear power generating industry. Of interest to the present project is the use of the steel for the above core structure of fast breeder reactors, where the structure experiences high temperature, low stress and a predominantly thermal neutron environment. In such structures, welded joints are unavoidable. The present work is part of a collaborative program to investigate the creep properties of Type 316 welds before and after thermal neutron irradiation.

The composition of the Type 316 weld metal, Table 1, when reduced to the ternary equivalent (1), intersects the liquidus surface close to but on the primary ferrite side of the 3 phase region of the Fe-Cr-Ni system, as indicated in Fig. 1 which shows the projections of the liquidus and solidus with the associated isotherms (2). The as-welded microstructure of the weld metal is essentially austenitic with 5-10% δ-ferrite, and exhibits a columnar grain structure with a cellular-dendritic infrastructure resulting from the solidification process. The ferrite decorates high as well as low angle austenite boundaries. In Type 316 and similar austenitic stainless steel weld metals, the retained ferrite often has a skeletal appearance. The skeletal ferrite is generally accepted to be due to primary ferrite solidification followed by partial transformation to austenite in the solid state (3-6). Depending on the composition, austenite can also be formed from the melt as the secondary phase (7,8). For primary ferrite solidification a controversy exists in the literature over the nature of the solid state ferrite to austenite transformation. Lippold and Savage (3) suggested a diffusionless massive mechanism for the transformation; while other workers (4-6) concluded that the transformation is diffusion controlled. In a recent study (9), a Kurdjumov-Sachs (K-S) relationship between the ferrite and austenite was found after slow continuous cooling; while a non-K-S relationship was found after gas quenching, and the latter relationship was attributed to massive transformation.

Table 1. Composition of weld metal (wt%)

Cr	Ni	Mo	C	Mn	Si	S	P	Fe
18.0	9.7	1.7	0.06	1.86	0.15	0.013	0.019	bal.

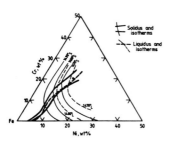

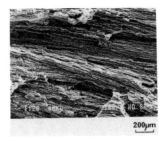

Fig. 1 Liquidus and solidus projections for the Fe-Cr-Ni ternary system(2). Equivalent composition is indicated.

Fig. 2 Fracture surface of low ductility creep failure, showing linear features.

For the present Type 316 welds, the creep properties at 600°C show a trend of decreasing ductility with decreasing stress (10) particularly after thermal neutron irradiation (11). Fracture surfaces of the low ductility failures, Fig. 2, show long linear features related to the cellular-dentritic solidification structure. Creep failures frequently occurred along ferrite/austenite interfaces or high angle columnar boundaries, but a full correspondence between the failure path and the weld microstructure remains unclear. The purpose of the present work was to fully characterise the weld microstructure and identify points of potential weakness, such as regions where brittle intermetallics might form.

In this paper, the results of characterisation of the as-welded structure as well as the transformation products after a stress relieving treatment at 800°C and after creep testing at 600°C are presented. The work involves the characterisation of the columnar structure, location of columnar grain boundaries, study of relative orientations between austenite and ferrite as well as phase identification. The results are discussed in terms of the solidification process. The techniques used range from conventional optical microscopy, through orientation contrast in a scanning electron microscope (SEM) to advanced analytical electron microscopy in a transmission electron microscope (TEM), combining energy dispersive spectroscopy (EDS) with convergent beam electron diffraction (CBED).

Materials and experimental details

The nominal composition of the Armex GT 17/8/2 weld metal is given in Table 1. Welds were deposited by the Mannual Metal Arc process and consisted of 13 passes, including the backing bead. The welds were examined in the as-deposited state or after a stress relieving treatment at 800°C for 10 hours. In addition, samples were examined after creep testing at 600°C for times up to 2500 hours (10,11). For general metallography, an electrolytic etch was used with 50% HNO_3 in water at 4V; while for selective etching of the sigma phase, a solution of 10% KOH in water at 4V was used. Thin foils for TEM were produced using Struers Tenupol twin jet polisher with a solution of 10% perchloric acid in methanol at -10°C and 35V. TEMs used were the Philips EM430 and EM400. The latter is equipped with a Link 860 EDS facility, and the k-factors have been reported elsewhere (12). SEMs used were the JEOL 840 and Cambridge Instrument Stereoscan.

Results

The columnar grain boundaries (CGB) are of special interest, since they may represent locations which were the last to solidify and hence potential locations for change of composition and sites for tramp element segregation. One of the prime tasks was thus to locate the CGBs on a microscopic scale for detailed analysis. The columnar structure in the multi-bead weld can be clearly seen with a suitable macro etch, Fig. 3. It should be noted that since austenite is the major phase, the term columnar grains refers to grains of austenite. In Fig. 3, the contrast arises from the etching of austenite/ferrite interfaces and the columnar nature does not necessarily represent austenite grains but rather the existence of columnar regions with the same ferrite morphology. At higher magnifications, the duplex microstructure becomes apparent and the ferrite morphologies can be classified as either being skeletal or vermicular, Figs. 4 and 5 respectively, with occasional patches with a lathy morphology, Fig. 6. The columnar structure, however, then becomes difficult to identify. Since solidification started from grains in the heat affected

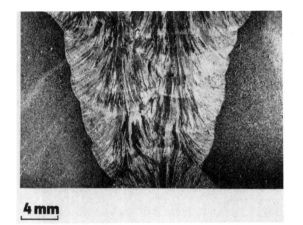

Fig. 3 Macro image of a multi-pass weldment showing columnar structure.

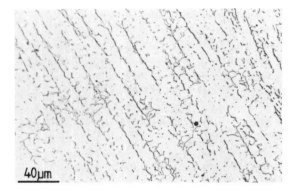

Fig. 4 Ferrite with skeletal morphology.

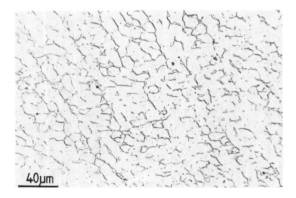

Fig. 5 Ferrite with vermicular morphology.

Fig. 6 Ferrite with lathy morphology.

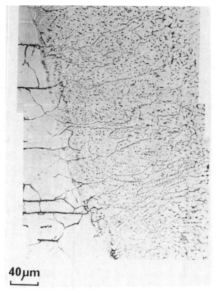

Fig. 7 Example of epitaxial solidification from HAZ grains.

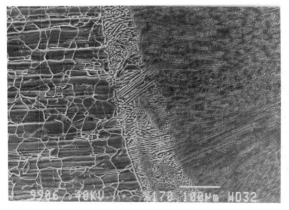

Fig. 8 Typical HAZ-weld interface showing the presence of a dilution zone. Taken from a stress relieved sample containing grain boundary $M_{23}C_6$.

zone (HAZ), in principle, it should be possible to follow columnar grains from the HAZ into the weld. An example of epitaxial growth on HAZ grains is shown in Fig. 7. However, this sort of epitaxial solidification was not found to be very common. Furthermore, the epitaxially grown grains soon petered out in the weld, to be replaced by grains which failed to give any clear evidence of epitaxial relationship. A more typical example of the microstructure near the HAZ is shown in Fig. 8. A dilution zone is present between the HAZ grains and the cellular-dendritic microstructure of the weld itself. No epitaxial relationship with the HAZ grain is apparent. However, by using oblique optical illumination, continuity of ferrite regions across bead interfaces was observed, Fig. 9, indicating an epitaxial relationship for the ferrite (at least) across the interface.

In order to follow the austenite grains, electron channelling or orientation contrast in the SEM was used. In the skeletal cellular-dendritic regions, long austenite grains were easily seen, Fig. 10. In the reheated regions, the columnar grains were sometimes visible at low magnifications, but the heat input from the subsequent bead had often resulted in relatively large mis-orientations between austenite subgrains, making the precise CGBs hard to locate. There were nontheless also regions where austenite grains extented across the bead interface, Fig. 11. Thus austenite grains can also show epitaxial relationships across bead interfaces. It was noted that the columnar grains were in general much smaller than expected from from the low magnification micrograph, Fig. 3. Furthermore, there was not always a one-to-one correspondence between a region with a particular ferrite morphology and specific austenite orientation; multiple austenite orientations were often observed in the same region. While it is likely that ferrite with the same lathy morphology has the same orientation, the possibility of multiple ferrite orientations cannot be ruled out, particularly in the vermicular region.

It was somewhat surprising to note that the CGBs were often quite ragged, and in some cases isolated regions of austenite grains or subgrains were observed with orientations quite different from the surrounding austenite, Fig. 12. When the second transverse section (plan view, horizontally perpendicular to the section of Fig. 3) was examined, the CGBs were also ragged, Fig. 13, but the grain shapes did not show preferential elongation in any particular direction. Texture measurement in the present work showed a <100> texture parallel to surface normal of this section. Such a texture has been reported for similar weld metals (13,14).

Using the CBED technique in the TEM, the orientations of austenite and ferrite can be determined precisely. Examination of as-welded samples showed that the austenite and ferrite were related, to within a few degrees, by the K-S relationship: $(\bar{1}11)_\gamma$ // $(1\bar{1}0)_\delta$ and $[110]_\gamma$ // $[111]_\delta$. It should be noted that the observed orientation relationship could be equally well described as being within a few degrees of the Nishiyama-Wasserman (N-W) relationship: $(111)_\gamma$ // $(110)_\delta$ and $[211]_\gamma$ // $[1\bar{1}0]_\delta$. With the K-S or N-W type of relationships, it might be expected that for a primary ferrite dominated solidification process, multiple austenite orientations would be present for a given ferrite orientation and vice versa for primary austenite solidification. In order to locate CGBs and study the ferrite/austenite orientation relationships, montages of thin area on foils have been made. Using CBED, high angle CGBs can then be mapped out on the montage. At the same time, ferrite pockets with the same orientation were identified. An example is shown in Fig. 14 which shows the presence of a CGB. In one of the austenite grains, multiple ferrite orientations were observed. By orienting that austenite grain onto [100], it was found that all of the ferrite variants had one particular <100> direction within the angular field of view, i.e. within about 10 of the

AR 200μm

Fig. 9 Optical micrographs using: a) reflected light, b) combined reflected light and oblique illumination and c) oblique illumination, showing continuity of ferrite morphology across a bead interface.

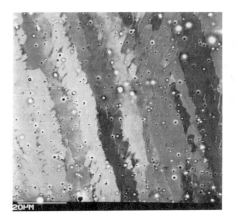

Fig. 10 Orientation contrast SEM micrograph showing long austenite grains in cellular-dendritic region.

Fig. 11 Orientation contrast SEM micrograph showing continuity of austenite across a bead interface.

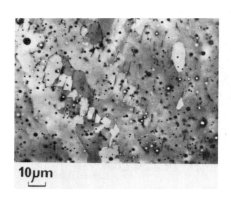

Fig. 12 Orientation contrast SEM micrograph showing isolated regions of differently oriented austenite.

Fig. 13 Orientation contrast SEM micrograph of a plan section perpendicular to the columnar grains.

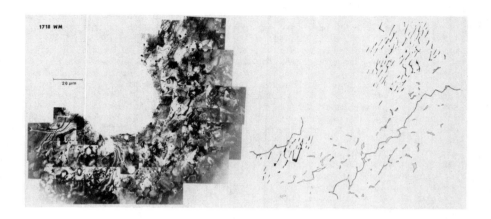

Fig. 14   Montage of thin area on a TEM foil. The line drawing shows   the
locations of a high angle CGB and ferrite pockets. Ferrite pockets   with
differnt orientations are shaded differently.

austenite [100]. Note that ferrite pockets with a lathy morphology had the
same orientation, and that one variant of the ferrite straddles the high
angle CGB. The latter observation is a case of ferrite with multiple
austenite orientations. Indeed, cases have been observed where several
austenite orientations but only one ferrite orientation was found in the
thin foil. Evidence for both primary ferrite and austenite have thus been
observed, often in close proximity, sugesting that neither is the dominant.

The composition of the various ferrite pockets have been monitored
using EDS. For a given foil, more than 15 ferrite pockets were analysed for
each category and the average taken. Typical results are shown in Table 2.
It should be noted that only those elements with atomic numbers >11 can be
detected using EDS. The results showed no significant differences between
ferrite on and off CGBs; nor was there any difference between regions with
different morphologies. The slight differences between the ferrite on CGBs
and those with different morphologies were probably due to the fact that
different foils were used.

Table 2. Typical ferrite compositions.

	Fe	Cr	Ni	Mo
lathy ferrite	: 66.8 +0.4	27.7 +0.7	3.5 +0.3	2.0 +0.15
vermicular ferrite	: 67.0 +0.7	27.5 +0.6	3.5 +0.3	2.0 +0.2
ferrite on CGB	: 67.4 +0.5	26.7 +0.6	3.6 +0.3	2.33 +0.1
ferrite off CGB	: 67.3 +0.6	26.6 +0.6	3.8 +0.4	2.33 +0.25

During creep testing at 600°C, the retained metastable ferrite underwent transformation mainly into austenite and $M_{23}C_6$ (15-19). The carbide generally decorated the ferrite/austenite boundary, and the ferrite pockets became progressively smaller as the transformation proceeded. In some cases the ferrite/austenite interface became completely divorced from its original position, as indicated by $M_{23}C_6$ precipitates, Fig. 15. The ferrite can also transform into intermetallic phases such as the sigma and R phases (17-19). In the present weld metal, the sigma phase was found to be formed preferentially on high angle CGBs, Fig. 16. Occasionally, sigma phase was observed in ferrite pockets off the CGBs themselves but in the near vicinity. This is interpreted as a ferrite pocket impinging on the CGB out of the sectioned surface. Sigma phase was generally observed in ferrite pockets which had formed little or no $M_{23}C_6$, as illustrated, for example, by comparison of the two ferrite pockets in Fig. 16. Such observations are consistent with the suggestion that the formation of $M_{23}C_6$ renders the ferrite more stable against transformation to sigma phase (16,20). In addition to the sigma phase, R phase (17-19,21) was also formed in some of the ferrite pockets, Fig. 16. The R phase tended to form in ferrite with a vermicular morphology and not in lathy ferrite. No clear pattern has been found for the location in relation to CGBs of ferrite pockets which were susceptible to R phase formation. The R phase formed preferentially on dislocations in the ferrite, Fig. 17a and often had a mottled appearance as shown by the dark field micrograph in Fig. 17b, indicating the presence of structural defects. The EDS spectrum was characterised by the presence of Si and traces of P. An EDS spectrum for the R phase is shown in Fig. 18 together with those for the sigma phase and $M_{23}C_6$. The presence of Si and particularly P in the R phase is unusual for an intermetallic compound, and may have been responsible for the observed faulting. The phase was identified by electron diffraction and Fig. 19 shows the $[11\bar{2}0]$ pattern with the $\bar{1}101$ and $\bar{2}202$ spots charateristic of the rhombohedral lattice. The crystallographic data for the R phase together with the other phases mentioned earlier is summarised in Table 3. The R phase shows a well defined orientation relationship with the ferrite:

$$(0003)_R \quad // \quad (22\bar{2})_\delta \quad \text{and}$$
$$[11\bar{2}0]_R \quad // \quad [123]_\delta \quad .$$

This orientation relationship differs by a rotation of 30 degrees, about the c-axis of the R phase, from that reported for R phase in a 12Cr-10Co-6Mo steel (21). In addition to these intermetallic phases, slag inclusions were present throughout the weld metal. The identity of the inclusion had been indentified and reported elsewhere (22).

Table 3. Phases in weld metal (excluding ferrite and austenite).

Phase	space group	lattice parameters (nm)
M C	Fm3m	a = 1.067
Sigma	P4 /mnm	a = 0.884, c = 0.466
R	R$\bar{3}$	a = 1.092, c = 1.965
		(a_R = 0.909, α = 73.82°)

During a 10 hours stress relieving treatment at 800°C, the main transformation products were $M_{23}C_6$ and sigma. The R phase was not observed. The sigma phase once again formed on, or close to, high angle CGBs. An interesting example is shown in Fig. 20. Three austenite grains are present within the montage. On this section, two of the grains are intertwined,

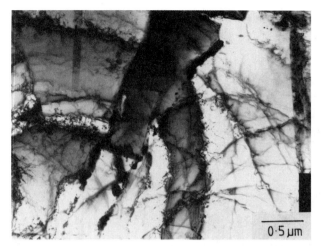

Fig. 15 TEM micrograph showing $M_{23}C_6$ precipitation at the austenite/
ferrite interface. Central region also shows the migration of the interface.
The original position is indicated by carbide decoration.

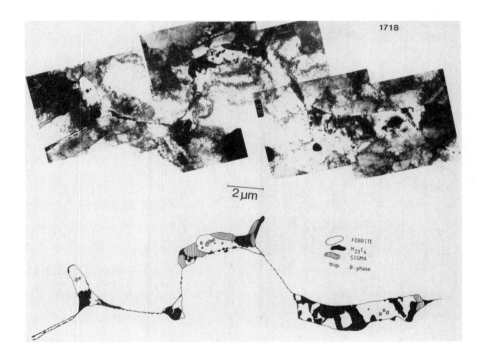

Fig. 16 Montage of a region near a CGB. The line drawing shows the location
of sigma, R and $M_{23}C_6$.

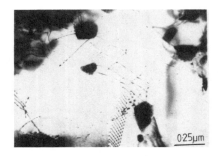

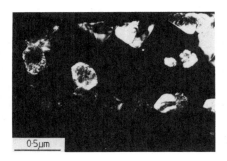

Fig. 17 a) Micrograph showing R phase particles on dislocations in ferrite. b) Dark field micrograph showing mottled appearance of the R phase. Some $M_{23}C_6$ particles are also in contrast.

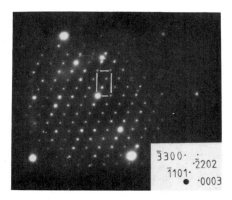

Fig. 19 Electron diffraction pattern for the $[11\bar{2}0]$ zone axis of the R phase, showing the $\bar{1}101$ and $\bar{2}202$ spots characteristic of the rhombohedral lattice. Ferrite zone axis is [123].

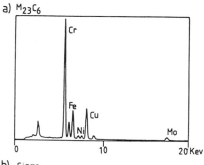

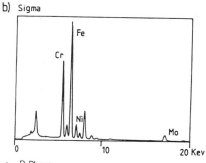

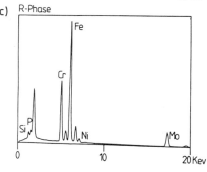

Fig. 18 EDS spectra for a) $M_{23}C_6$, b) sigma and c) R phase.

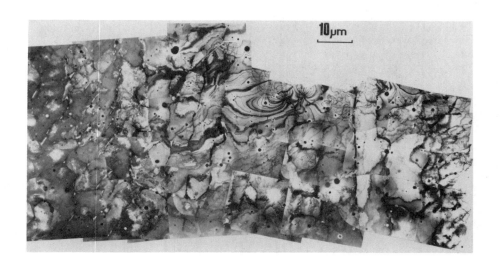

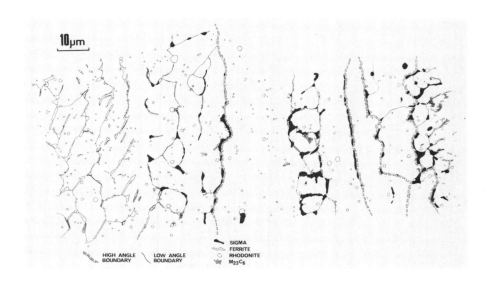

HIGH ANGLE   LOW ANGLE   SIGMA
BOUNDARY   BOUNDARY   FERRITE
                                           RHODONITE
                                           $M_{23}C_6$

Fig. 20 Montage and line drawing of a stres relieved sample. Sigma phase
is formed on or close to CGBs

resulting in several high angle CGBs. Sigma formation was restricted to these two grains plus some in the bordering region of the third grain. No sigma was formed inside the third grain away from the CGB region. The ferrite pockets in the latter grain have a lathy appearance with a narrow width and are of the same orientation, which probably distinguishes them from those which had transformed to sigma. The resistence of the lathy ferrite to sigma formation was also illustrated by using a selective etching technique as shown in Fig. 21.

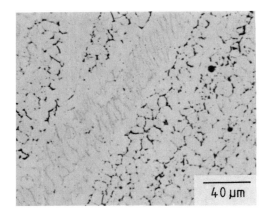

Fig. 21 Optical micrograph of a stress relieved sample selec-tively etched for sigma. Sigma phase is not formed in lathy ferrite.

Discussion

The study of relative orientations between regions of ferrite and austenite has shown that neither primary ferrite nor primary austenite was the dominant solidification mode. In Type 316 weld metal (23) as well as in a Nb stabilised stainless steel (24), regional variations in the solidification mode have been reported. In the present weld metal, variations were observed on a much finer scale. Such variations can be expected from a solidification process in which both the austenite and the ferrite solidify from the melt (7,8). The ternary equivalent composition of the weld metal lay close to the three phase (liquid + ferrite + austenite) field of the Fe-Cr-Ni system, Fig. 1 (2). Thus, after some primary ferrite solidification, the dominant mode of solidification could have changed to the decomposition of the liquid into ferrite and austenite. The ferrite and the austenite could exhibit a K-S or N-W orientation relationship, since the K-S relationship corresponded to matching of the close packed planes and directions, while the N-W relationship differs from the K-S by $5.3°$. It is worth noting that with either of K-S and N-W relationships, for a given austenite [100] direction, 8 out of the 24 distinct ferrite variants would have one <100> type direction at about 10° to the austenite [100] direction; and vice versa. It is thus perfectly feasible to have a number of ferrite and austenite grains with the direction of easy growth close to <100> while still maintaining a K-S or N-W relationship between all the ferrite/austenite interfaces. The observation shown in Fig. 14 is consistent with this conclusion. In the solid state, part of the ferrite transformed into austenite. The skeletal and vermicular ferrite morphologies resulted from the migration of the ferrite/austenite boundaries without generating new austenite orientations. The net result would then be a microstructure consisting of columnar grains of austenite interwoven with the remains of columnar ferrite, with an overall <100> texture along the direction of maximum heat flow.

In view of the sigma phase forming preferentially along CGBs, it is somewhat surprising that no significant compositional differences were noted between ferrite pockets located either on or off CGBs. However, the compositions of the various ferrite regions, determined using EDS, contained substantially less Ni than that expected from the ternary phase diagram, Fig. 1. This predicts that when the liquid composition first reached the three phase region, at around 1470°C, the ferrite would have a composition given by the ferrite apex of the three phase triangle for that temperature. This deviation suggests that, in addition to elemental partitioning during solidification, substantial partitioning had taken place in the solid state and masked any differences in composition due to different stages of the solidification process.

It is significant, however, that the measured ferrite compositions, Table 2, are very close to the sigma field boundary at 800 °C (2). The susceptibility to sigma formation would thus depend sensitively on the precise composition. The presence and effects of other minor elements may well be crucial; in particular, factors influencing the stability of the ferrite with respect to the formation of $M_{23}C_6$ would be important. These factors may include, for example, the diffusion of Mo (15,16). Both the sigma and the R phase were enriched in Mo as compared with ferrite, Fig. 18, and tended not to form in lathy ferrite which were often quite narrow and closely spaced as compared with vermicular ferrite, Figs. 5 and 6. It is likely that a smaller austenite volume is associated with each ferrite lath. There is hence less Mo available to diffuse to the lathy ferrite, insufficient to cause transformation to sigma or R phase. The tendency of ferrite on CGBs to form sigma may also be due to the higher boundary duffusion rates. In addition it may be that CGBs represent regions of a high degree of strain, retained from the last stages of the solidification process, rendering ferrite on boundaries more susceptible to sigma formation. During creep testing at 600°C, the strain may be enhanced by the applied stress. The susceptibility of ferrite on CGBs to sigma formation suggests that CGBs may represent locations where creep crack propagation is easiest.

Conclusions

The columnar structure in Type 316 welds has been analysed. The orientation relationships between ferrite and austenite suggest that co-solidification of ferrite and austenite was the dominant solidification process. The composition of the ferrite was found to be depleted in Ni, suggesting that a diffusional process was responsible for the partial transformation of ferrite to austenite. During the stress relieving treatment at 800°C or creep testing at 600°C, closely spaced lathy ferrite was found to be less susceptible to intermetallic formation than vermicular ferrite. Sigma phase formed preferentially on CGBs, even though no significant differences were detected in the as-welded samples for ferrite on and off the CGBs. During creep testing at 600°C, the intermetallic R phase formed on dislocations mainly in vermicular ferrite with an orientation relationship given by:

$$(0003)_R \quad // \quad (22\overline{2})_\delta \quad \text{and}$$
$$[11\overline{2}0]_R \quad // \quad [123]_\delta \quad .$$

Acknowledgements

The authors would like to thank the Science and Engineering Research Council and the Central Electricity Generating Board for providing financial support. We would also like to thank Drs. D.S. Wood and B.v.d. Schaaf for provision of samples, Drs. P. Marshall and R.G. Thomas for usefull discussions, S. Underwood for assistance with operation of the Stereoscan and Dr. C. Baker for help with specimen preparation.

References

1. A. Schaeffler, "Constitution Diagram for Stainless Steel Weld Metal," Met. Progress, 56(5) (1949) 680.

2. V.G. Rivlin and Raynor, "Critical Evaluation of Constitution of Cr-Fe-Ni system," Int. Met. Rev., 1 (1980) 21-38.

3. J.C. Lippold and W.F. Savage, "Solidification of Austenitic Stainless Steel Weldments: Part I-A Proposed Mechanism," Welding J., 58 (1979) 362s-374s.

4. H. Fredriksson, "The Solidification Sequence in an 18-8 Stainless Steel, Investigated by Directional Solidification," Met. Trans., 3 (1972) 2898-2997.

5. Y. Arata, F. Matsuda and S. Katayama, "Solidification Crack Susceptibility in Weld Metals of Fully Austenitic Stainless Steels (Report I)," Trans. Japanese Weld. Res. Inst., 5(2) (1976) 35-51.

6. J.A. Brooks, J.C. Williams and A.W. Thompson, "Microstructural Origin of the Skeletal Ferrite Morphology of Austenitic Stainless Welds," Met. Trans., 14A (1983) 1271-1281.

7. N. Suutala, T. Takalo and T. Moisio, "Ferritic-Austenitic Solidification in Austenitic Stainless Steel Welds," Met. Trans., 11A (1980) 717-725.

8. G.L. Leone and H.W. Kerr, " Ferrite to Austenite Transformation in Stainless Steels," Welding J., 61 (1982) 13s-21s.

9. J. Singh, G.R. Purdy and G.C. Weatherly, "Microstructure and Microchemical Aspects of the Solid State Decomposition of Delta Ferrite in Austenitic Stainless Steels," Met. Trans., 16A (1985) 1363-1369.

10. D.S. Wood, unpublished research, UKAEA Risley, U.K. 1986.

11. B.v.d. Schaaf, unpublished research, ECN Petten, The Netherlands, 1986.

12. R. Graham and J.W. Steeds, "Determination of Cliff-Lorimer k Factors by Analysis of Microdroplets," J. Micrsc., 133 (1984) 275.

13. B.L. Baikie and Yapp, "Oriented Structure and Properties in Type 316 Stainless Steel Weld Metal," Solidification and casting of metals, (London: The Metals Society, 1979) 438-443.

14. T. Tokalo, N. Suutala and T. Moisio, "Austenitic Solidification Mode in Austenitic Stainless Steel Welds," Met. Trans., 10A (1979) 1173-1181.

15. R.A. Farrar, C. Huelin and R.G. Thomas, "Phase Transformation and Impact Properties of Type 18-8-2 Austenitic Weld Metals," _J. Mat. Sci._, 20 (1985) 2828-2838.

16. R.A. Farrar, "Microstructure and Phase Transformations in Duplex 316 Submerged Arc Weld Metal, an Ageing Study at 700 C," _J. Mat. Sci._, 20 (1985) 4215-4231.

17. S.R. Keown and R.G. Thomas, "Role of Delta Ferrite in Thermal Ageing. of Type 316 Weld Metals," _Met. Sci._, 15 (1981) 386-392.

18. J.K. Lai and J.R. Haigh, "Delta Ferrite Transformation in a Type 316 Weld Metal," _Welding J._, 58 (1979) 1s-6s.

19. A.A. Tevassoli, A. Bisson and P. Soulat, "Ferrite Decomposition in Austenitic Stainless Steel Weld Metals," _Met. Sci._, 18 (1984) 345-350.

20. G.F. Slattery, S.R. Keown and M.E. Lambert, "Influence of Delta Ferrite Content on Transformation to Intermetallic Phases in Heat Treated Type 316 Austenitic Stainless Steel Weld Metal," _Met. Technol._, 10 (1983) 373-385.

21. D.J. Dyson and S.R. Keown, "A Study of Precipitation in a 12%Cr-Co-Mo Steel," _Acta Met._, 17 (1969) 1095-1107.

22. G.T. Finlan, Y.P. Lin and J.W. Steeds, "Microstructural Examination Around Weldments in AISI 316 Stainless Steel," _Inst. Phys. Conf. Ser. No. 78_, (1985) 257-260.

23. V.P. Kujanpaa, S.A. David and C.L. White, "Formation of Hot Cracks in Austenitic Stainless Steel Welds-Solidification Cracking," _Welding J._, August (1986) 203s-212s.

24. R.M. Boothby, "Solidification and Transformation Behaviour of Nb-Stabilized Stainless Steel Weld Metal," _Met. Sci. and Technol._ 2 (1986) 78-87.

REAUSTENITISATION IN STEEL WELD DEPOSITS

J. R. Yang and H. K. D. H. Bhadeshia

University of Cambridge
Department of Materials Science and Metallurgy
Pembroke Street, Cambridge CB2 3QZ, U.K.

Abstract

The process of reaustenitisation in weld deposits, beginning with a microstructure of acicular ferrite and austenite, has been studied in order to enable the prediction of the reheated microstructure of welds. The transformation mechanism by which the original acicular ferrite formed is found to strongly influence the reaustenitisation process. The reverse $\alpha \rightarrow \gamma$ transformation does not occur immediately when the temperature is raised, even though the alloy may be in the $\alpha + \gamma$ phase field. Reaustenitisation only begins when the carbon concentration of the residual austenite exceeds its equilibrium carbon concentration. This is a direct consequence of the fact that the acicular ferrite transformation ceases before the lever rule is satisfied. A theory has been developed, which explains the experimental data, including the fact that the degree of reaustenitisation varies with temperature above the Ae3 curve.

Braces are used exclusively to denote functional relations; $x\{T\}$ therefore implies that T is the argument of the function x.

k_i	Partitioning coefficient for alloying element i
Δl_m	Maximum relative length contraction observed during isothermal reaustenitisation
T	Temperature
T_a	Temperature at which acicular ferrite forms
T_γ	Temperature of isothermal reaustenitisation
$T_{\gamma 1}$	Minimum T at which reaustenitisation commences
$T_{\gamma 2}$	Minimum T at which alloy becomes fully austenitic
V_γ	Equilibrium volume fraction of austenite
$\bar{x}$	Average carbon concentration of alloy
x_γ	Carbon concentration of austenite
x'_γ	Carbon concentration of residual austenite when the formation of acicular ferrite ceases
$x_{T'_o}$	Carbon concentration given by the T'_o curve
x_{Ae3}	Carbon concentration given by the Ae3 curve
γ	Austenite
α	Ferrite

Introduction

The microstructure of the fusion zone of a multipass weld can be classified into two main components: (a) the primary microstructure obtained during cooling of the weld from the liquidus temperature which consists of allotriomorphic ferrite, Widmanstätten ferrite, acicular ferrite and very small amounts of "microphases" (i.e., martenisite, degenerate pearlite, retained austenite), and (b) the secondary microstructure, commonly called the reheated microstructure, obtained during the deposition of further weld metal. The thermal cycle experienced may just anneal the microstructure and cause recrystallisation and grain growth, or if the Ac_1 temperature is exceeded, it may reaustenitise to a degree depending on the thermodynamics and kinetics of reaustenitisation. The part which becomes austenitic then undergoes a series of transformations during the cooling part of the thermal cycle, the nature of the transformations depending on alloy chemistry, austenite grain size and morphology and cooling conditions amongst other factors.

Much work has recently been done on the prediction of the primary microstructure of welds using detailed phase transformation theory, and this has met with reasonable success (1-4).

550

Indeed, it is now possible to estimate quantitatively, the primary microstructure of arc welds as an aid to the design of welding consumables and procedures. However, there is no corresponding theory available for the reheated microstructure, which can form a substantial proportion of a multipass weld deposit.

This work is part of a programme on the theoretical design of welding consumables and deals specifically with the problem of reaustenitisation, and the factors influencing the temperature at which austenite first forms as a function of alloy chemistry, heating rate and starting microstructure. In order to establish a quantitative model for reaustenitisation, the work is initially limited to *isothermal* reaustenitisation experiments.

Experimental Procedures

Dilatometry

All dilatometry was performed on a Theta Industries high-speed dilatometer, which has a water cooled radio-frequency furnace of essentially zero thermal mass, since it is only the specimen which undergoes the programmed thermal cycle. The length transducer on the dilatometer was calibrated using pure platinum and pure nickel specimens of known thermal expansion characteristics. This enables rapid heating or cooling experiments. The dilatometer has been specially interfaced with a BBC/Acorn microcomputer so that length, time and temperature information can be recorded at microsecond intervals, and the data stored on a floppy disc. The information is then transferred to a mainframe IBM3081 computer for further analysis. Such a facility is essential for the present work since the transformations can occur at very rapid rates, and since a large number of experiments were conducted.

Specimens for dilatometry were in the form of 3mm diameter rods 15mm long, machined from the weld deposits with the cylinder axes parallel to the welding direction. The specimens were machined from regions far from the fusion line of the parent plate and are not affected by dilution from the parent material. To avoid surface nucleation and surface degradation, all specimens were plated with nickel (plating thickness $\approx$ 0.08mm) and all heat treatments were carried out in a helium gas environment. All experiments were conducted on specimens which had been homogenised at 1300°C for 3 days, while sealed in quartz tubes containing a partial pressure of pure argon. The microstructure of the specimens before reaustenitisation was always acicular ferrite in a matrix of austenite.

Weld metal preparation

The welds were deposited using manual metal arc welding. The electrodes (4mm diameter) used were of a AWS-E10016-G type, the joint geometry being compatible with British Standard BS639 (similar to ISO2560), a geometry which gives large regions of weld metal free from dilution by the parent plate. All samples for the present work are removed from regions not influenced by dilution. The welding was carried out in the flat position using the stringer bead technique, the parent plate thickness being 20mm. The welding current and voltage used were 180A 23V (DC+) respectively, with an electrical energy input of about 2kJ/mm, the weld consisting of some 27 runs with 3 runs per layer. The interpass temperature was typically 250°C.

The present study is connected with other work on high-strength weld deposits, and the electrodes for this purpose gave a deposit composition as follows:
Fe-0.06C-0.27Si-1.84Mn-2.48Ni-0.20Mo-0.0032O-0.01Al-0.02Ti, wt.%.

Transmission Electron Microscopy and Microanalysis

Thin foil specimens were prepared for transmission electron microscopy from 0.25mm thick discs slit from specimens used in the dilatometric experiments. The discs were thinned to 0.05mm by abrasion on silicon carbide paper and then electropolished in a twin jet electropolisher using a 5% perchloric acid, 25% glycerol and 70% ethanol mixture at ambient temperature, 60V.

The microscopy and microanalysis were conducted on a Philips EM400T transmission electron microscope operated at 120kV. An energy dispersive X-ray analysis facility was used for the microanalytical measurements, with the specimens held in a beryllium holder tilted from the normal by 35°, which is the take-off angle. The X-ray count rate was optimised to about 200 counts/s over a livetime of about 100s. The data were analysed using the *LINK RTS 2 FLS* program for thin foil microanalysis; this corrects that data for atomic number and absorption and accounts for overlapping peaks by fitting standard profiles. Even though the probe diameter used was about 3nm, beam spreading due to scattering of electrons within the thin foil gave an estimated broadened beam diameter of ≈ 20nm.

Results and Discussion

The Nature of Acicular Ferrite

Before discussing the results in detail, it is necessary to state briefly some recent work (5,6) on the mechanism of acicular ferrite formation, particularly with respect to the thermodynamics of transformation. This provides a rational explanation for the design of the experiments and a basis for the interpretation of the results.

The composition of the alloy used is such that after welding, it gives a microstructure consisting mainly of acicular ferrite (volume fraction ≈ 0.9), with very little allotriomorphic ferrite and Widmanstätten ferrite (Fig. 1). The latter constituents only arise because of the presence of chemical segregation in the weld, which cools under non-equilibrium conditions. The high hardenability of this alloy (Fig. 1) is a major advantage for the present study because it allows the starting microstructure to be controlled with ease.

A thermodynamic investigation (5,6) has shown that the mechanism of transformation from austenite to acicular ferrite is the same as that for bainite: the plates grow by a diffusionless and displacive transformation mechanism (nucleation, although displacive, involves partitioning of carbon - Ref.8), and immediately after plate growth, carbon is partitioned into the residual austenite. The transformation does not therefore obey the lever rule and exhibits a classical *incomplete reaction phenomenon* in which reaction ceases well before the residual austenite achieves its equilibrium carbon concentration. In fact, the transformation stops when the carbon concentration of the residual austenite (i.e., x_γ) reaches the T_0' curve on the phase diagram.

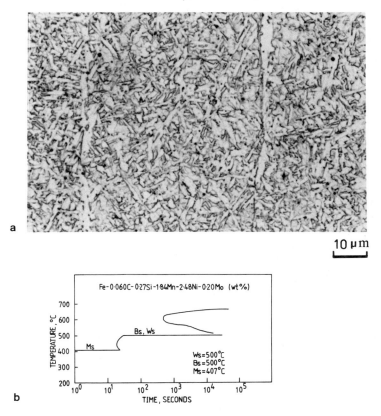

a

10 μm

Fe-0·060C-0·27Si-1·84Mn-2·48Ni-0·20Mo (wt%)

b

Figure 1 - (a) The primary microstructure of the weld deposit, consisting essentially of acicular ferrite, with very little allotriomorphic and Widmanstätten ferrite. (b) Calculated Time-Temperature-Transformation curve for a homogeneous alloy of the weld composition.

The T_0 curve, in a plot of temperature versus carbon concentration, represents all compositions where austenite and ferrite of the same composition have equal free energy (7-10). Diffusionless transformation is thermodynamically only possible if at the transformation temperature concerned, x_γ is less than the carbon concentration given by the T_0 curve. In fact, it is now known that the formation of acicular ferrite causes an invariant-plane strain (IPS) shape change in the transformed region. Consequently, allowance has to be made for the strain energy due to the IPS shape change. This strain energy amounts to about 400J/mol (5,6,8), and when this is accounted for, a T_0' curve is defined such that diffusionless transformation is only possible if x_γ is less than $x_{T_0'}$, the carbon concentration specified by the T_0' curve for the isothermal transformation temperature concerned.

This mechanism of acicular ferrite formation explains many of its other features, such as the plate morphology, and the fact that the plates never cross austenite grain boundaries. Without going into further details, which are already published (5), the work indicates that acicular ferrite is really bainite, whose morphology differs from classical sheaf like bainite simply because acicular ferrite nucleates intragranularly at point sites, and is limited also by hard impingement with other neighbouring plates. Classical bainite sheaves form in pure alloys not containing large numbers of inclusions, where the bainite nucleates from austenite grain boundaries. Acicular ferrite requires the presence of inclusions to enable intragranular nucleation, and will only form when the austenite grain size is relatively large. Intragranular nucleation on inclusions, which has a higher activation energy compared with grain boundary nucleation so that the number of grain boundary sites must be minimised to obtain acicular ferrite. In fact, it has been shown (5) using light microscopy that on reaustenitisation and subsequent isothermal transformation, classical bainite forms in the present alloy if the austenite grain size is small, but acicular ferrite forms when the grain size is large. Fig. 2 provides striking transmission electron micrographs from specimens with varying austenite grain size which confirm the earlier work.

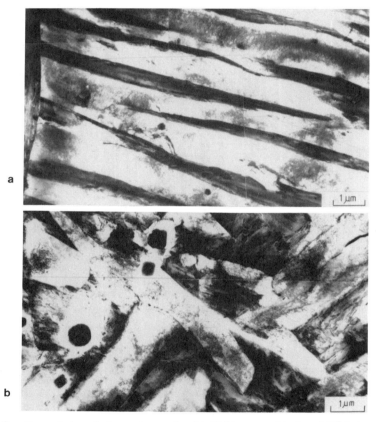

Figure 2 - Transmission electron micrographs. (a) Weld reaustenitised at 950°C for 10 min, isothermally transformed at 460°C for 30 min and water quenched. Shows classical sheaf of upper bainite. (b) Weld reaustenitised at 1200°C for 30 min and then transformed as in (a). Shows classical acicular ferrite.

Our aim was to study isothermal reaustenitisation, beginning with a microstructure consisting of acicular ferrite and austenite. For this reason, the experiments were carried out on weld specimens homogenised, austenitised at 1200°C for 30 min, isothermally transformed to acicular ferrite at a temperature $T_a = 460°C$ for 30 min and then rapidly up-quenched to a temperature T_γ for isothermal reaustenitisation. The specimens were not cooled below 460°C, in order to avoid the martensitic decomposition of some or all of the residual austenite at that temperature. It should be noted that 30 mins at 460°C is more than adequate to allow the acicular ferrite transformation to terminate.

In addition, to avoid the *diffusional* decomposition of austenite during the up-quench from T_a, a sufficiently high up-quench rate has to be used during heating to T_γ. In these circumstances, reaustenitisation does not require the nucleation of austenite, just the growth of austenite. In the present study, the computer link to the dilatometer allowed high resolution monitoring of events during up-quenching, and for all the results reported here, there was no transformation during heating. A study involving the nucleation austenite will be reported separately. We note that the present results are relevant in any case, because the primary microstructure of welds often contains substantial amounts of retained austenite.

Isothermal Reaustenitisation

The results obtained from dilatometric experiments are presented in Fig. 3; the first detectable growth of austenite was found to occur at $T_\gamma = 680°C$. In all cases the transformation rate was initially rapid, but decreased with time so that the specimen length eventually stopped changing, as transformation ceased. If the maximum relative length change obtained at any T_γ is designated Δl_m, then Fig 4 shows that Δl_m increases as T_γ increases from 680°C to 760°C and then stays essentially constant with further increase in T_γ. It should be noted that a small amount of transformation occurred during the up-quench, when attempts were made to isothermally reaustenitise specimens at temperatures above 735°C. The transformation during the up-quench was recorded by the computer; it contributes to specimen length change, and a correction for this has been included in the data of Fig. 4. The correction is easy to make. During continuous heating, the specimen length varies linearly with temperature. If the low-temperature part of the curve (in a length versus temperature plot) is extrapolated to T_γ, the vertical difference between the extrapolated curve and the actual curve gives the length change due to transformation during the up-quench, which should be added to any length change due to isothermal transformation at T_γ

Since the temperature range covered is not very large, the results indicate that below 760°C, reaustenitisation is incomplete, the alloy becoming fully austenitic only above this temperature. Furthermore, the maximum degree of transformation to austenite increases from nearly zero at 680°C to complete reverse transformation at 760°C and above.

It is also seen from Fig. 3 that the rate of the reverse $\alpha \rightarrow \gamma$ transformation increases with T_γ

555

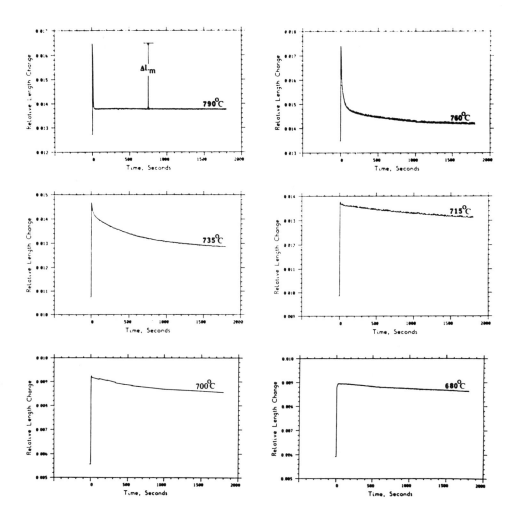

Figure 3 - Isothermal reaustenitisation experiments. The specimens were initially austenitised at 1200°C for 30 min, isothermally partially transformed to acicular ferrite at 460°C for 30 min and then rapidly up-quenched to an isothermal reaustenitisation temperature. Some transformation could not be avoided during the up-quench to temperatures above 735°C.

556

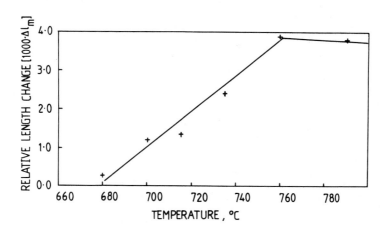

Figure 4 - A plot of the maximum relative length change Δl_m obtained at reaction termination, as a function of the isothermal reaustenitisation temperature T_γ

Table I. Microanalytical data (wt.%) for the γ and α during reaustenitisation.

Reaustenitisation		Mn	Mn	Ni	Ni
Temperature (°C)	Time (min)	γ	α	γ	α
680	10	2.83	1.79	2.86	2.31
735	10	2.46	1.64	2.69	2.12
760	2	1.93	1.97	2.58	2.50
715	120	2.73	1.39	2.87	1.89
735	120	2.82	1.42	3.05	1.85
760	120	1.97	1.96	2.44	2.44

Microanalytical data

The results (Fig. 5, Table I) cover a range of times (t) at each T_γ, and a range of isothermal reaustenitisation temperatures. They show that for low T_γ, substitutional alloying additions redistribute during the $\alpha \rightarrow \gamma$ transformation, even at low t. For low transformation times, the degree of partitioning of alloying elements (as indicated by the deviation of the partition coefficient k_i from unity, where $k_i = x_i^\gamma / x_i^\alpha$) increases with decreasing T_γ. This is consistent with the fact that at low-T_γ, the redistribution of substitutional alloying elements is a thermodynamic necessity (11,12). As the driving force for reaustenitisation increases, the transformation tends towards paraequilibrium or negligible-partitioning-local equilibrium; this is illustrated clearly by the data for 760°C, even for times as short as 2 minutes.

The concepts of paraequilibrium and local equilibrium have recently been reviewed (10). Paraequilibrium is a state of constrained equilibrium in which the substitutional lattice is configurationally frozen with respect to the transformation interface. Hence, even though the transformation is diffusional in nature, the ratio (atom fraction of substitutional element i/atom fraction of iron) is the same in α and γ. Thus, the chemical potentials of the substitutional elements are not equal in the two phases. Carbon which diffuses faster reaches equilibrium subject to this constraint. In negligible-partitioning-local-equilibrium (NPLE), equilibrium is maintained for all species at the transformation interface, but the concentration of substitutional elements is essentially the same in all phases.

The results also show that as t increases for a given T_γ, the partition coefficient k_i changes even though the volume fraction of austenite undergoes negligible change. This is strong evidence that the concentrations of alloying elements at the interface during the growth of austenite are not equilibrium concentrations.

These results are qualitatively consistent with the dilatometric data; one of the factors responsible for the increased rate of transformation at high T_γ is the fact that the degree of redistribution of alloying elements during transformation decreases with increasing T_γ. The results emphasize that although the states of local equilibrium or paraequilibrium are for convenience often assumed to define the compositions at the interface during diffusion-controlled growth, an infinite number of other possibilities exist for multicomponent alloys (10). All compositions of austenite which allow a reduction in free energy during growth can in principle form from ferrite, giving rise to situtations in which one or more of the elements is trapped by the moving interface; trapping implies an increase in he chemical potential of that element during transfer across the interface. The *prediction* of compositions at the interface is then a formidable theoretical problem which cannot be addressed until detailed information on interface mobility becomes available. On the other hand, any system must eventually tend towards equilibrium, so that the limiting composition of austenite can easily be calculated; this is the approach adopted in the present work.

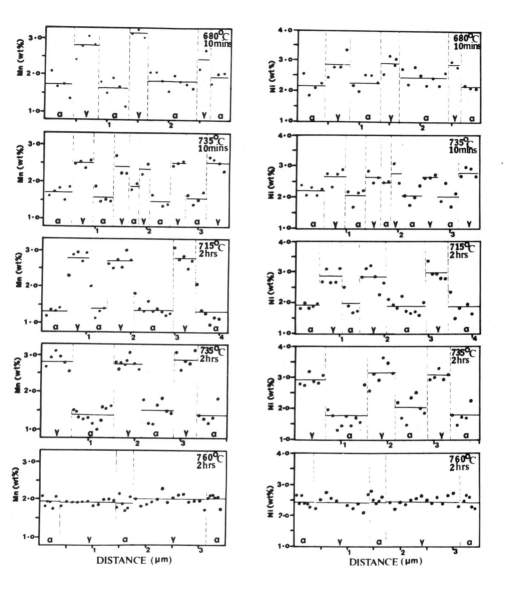

Figure 5 - Microanalytical data obtained using energy dispersive X-ray analysis on a transmission electron microscope.

<u>Theoretical analysis</u>

Here we present a model for the interpretation of the observations on reaustenitisation from a mixture of acicular ferrite and austenite. Since after the diffusionless growth of acicular ferrite, carbon is rapidly and spontaneously redistributed into the residual austenite with an accompanying reduction in free energy, the $\alpha_a \rightarrow \gamma$ transformation in its original form is irreversible. The problem of reaustenitisation is therefore considerably different from the case of reverse transformation from martensite to austenite in for example, shape memory alloys.

We have noted that the formation of acicular ferrite ceases prematurely during isothermal transformation when the carbon content of the residual austenite reaches the T_0' curve (the phase diagram for the present alloy is presented in Fig. 6). It follows that the carbon concentration x_γ' of the austenite when the formation of acicular ferrite ceases at T_a, is given by (point marked **a** in Fig. 6):

$$x_\gamma' = x_{T_0}\{T_a\} \qquad \qquad .....(1)$$

Furthermore, we note that:

$$x_\gamma' << x_{Ae3}\{T_a\} \qquad \qquad .....(2)$$

where $x_{Ae3}\{T_a\}$ is marked **b** in Fig. 6.

Thus, although the formation of *acicular ferrite* ceases at T_a, because $x_\gamma' << x_{Ae3}\{T_a\}$, the driving force for austenite to transform *diffusionally* to ferrite is still negative. Another way of expressing this is to say that the volume fraction of acicular ferrite present when its formation ceases at T_a is much less than required by the lever rule. In fact, this remains the case until the temperature T is high enough (i.e., $T=T_{\gamma 1}$) to satisfy the equation:

$$x_\gamma' = x_{Ae3}\{T_{\gamma 1}\} \qquad \qquad .....(3)$$

Hence, reaustenitisation will first occur at a temperature $T_{\gamma 1}$, as indicated on Fig. 6 (marked **c**), and as observed experimentally. Note that this is a direct consequence of the mechanism of the acicular ferrite transformation, which does not allow the transformation to reach completion. If this were not the case, then the lever rule demands that the temperature need only be raised infinitesimally above T_a in order for the reverse $\alpha \rightarrow \gamma$ transformation to be thermodynamically possible!

The theory goes further than explaining just the temperature at which the reverse transformation should begin. It also predicts that at any temperature T_γ greater than $T_{\gamma 1}$, the reverse $\alpha \rightarrow \gamma$ transformation should cease as soon as the residual austenite carbon concentration (initially x_γ') reaches the Ae3 curve, i.e., when

$$x_\gamma = x_{Ae3}\{T_\gamma\} \qquad\qquad .....(4)$$

with the equilibrium volume fraction of austenite (at the temperature T_γ), $V_\gamma\{T_\gamma\}$, being given by:

$$V_\gamma\{T_\gamma\} = \bar{x}/x_{Ae3}\{T_\gamma\} \qquad\qquad .....(5)$$

assuming that the carbon concentration of ferrite is negligible and that $x_{Ae3}\{T_\gamma\} > \bar{x}$. When $x_{Ae3}\{T_\gamma\} = \bar{x}$, the alloy eventually becomes fully austenitic (point **d**, Fig. 6), and if this condition is satisfied at $T_\gamma = T_{\gamma2}$, then for all $T_\gamma > T_{\gamma2}$, the alloy transforms completely to austenite.

These concepts immediately explain the dilatometric data in which the degree of $\alpha \rightarrow \gamma$ transformation increases (from $\approx$ zero at 680°C) with the temperature of isothermal reaustenitisation, until the temperature 760°C$=T_{\gamma2}$ where the alloy transforms completely to austenite.

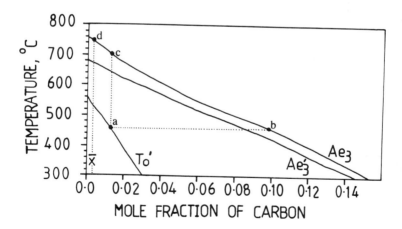

Figure 6 - Phase diagram showing the Ae3, Ae3', T_0 and T_0' curves for a Fe-0.27Si-1.84Mn-2.48Ni-0.20Mo wt.% alloy. The Ae3', T_0 and T_0' curves are calculated as in Refs.7,8, and the Ae3 curve is calculated as in Ref.13.

Conclusions

Reaustenitisation of a weld deposit, beginning with a microstructure of just acicular ferrite and bainite has been studied under isothermal conditions and in circumstances where the nucleation of austenite is not necessary. The reverse transformation from acicular ferrite to austenite does not happen immediately the temperature is raised (above that at which the acicular ferrite formed), even though the alloy is within the $\alpha + \gamma$ phase field.

This is because acicular ferrite plates grow by a diffusionless displacive transformation mechanism (similar to that of bainite) which ensures that the transformation ceases prematurely, before the residual austenite achieves its equilibrium composition. Hence, reaustenitisation only occurs when the alloy is heated to a temperature where the carbon concentration of the residual austenite exceeds its equilibrium concentration.

Complete transformation to austenite only occurs at a temperature where the alloy composition equals the austenite equilibrium composition. However, at all intermediate temperatures, the reverse $\alpha \rightarrow \gamma$ transformation terminates before the alloy becomes fully austenitic, with the volume fraction of austenite increasing with increasing T_γ.

The work has led to a theory for reaustenitisation which explains most of the experimental results, but which as yet does not address the detailed kinetics of reaustenitisation. The model predicts accurately the temperature at which reaustenitisation should begin and the degree of transformation to austenite as a function of the isothermal reaustenitisation temperature. All this is in turn a function of the temperature at which the original acicular ferrite itself formed.

Acknowledgments

The authors are grateful to the Chinese Ministry of Education for financial support, to ESAB for the provision of weld samples, to Professor D. Hull for the provision of laboratory facilities, and to G. Barritte, P. Judson and L. -E. Svensson for helpful discussions throughout the course of this work.

References

1. H. K. D. H. Bhadeshia, L.-E. Svensson and B. Gretoft, "A Model for the Development of Microstructure in Low-Alloy Steel Weld Deposits," Acta Metall., 33 (1985) 1271-1283.

2. L.-E. Svensson, B. Gretoft and H. K. D. H. Bhadeshia, "An Analysis of the Cooling Curves from the Fusion Zone of Steel Weld Deposits," Scand. J. Metall., 15 (1986) 97-103.

3. L.-E. Svensson, B. Gretoft and H. K. D. H. Bhadeshia, " Computer-Aided Design of Electrodes for Manual Metal Arc Welding," Computer Technology in Welding, (Welding Institute, Abington, Cambridge), in press.

4. H. K. D. H. Bhadeshia, L.-E. Svensson and B. Gretoft, "Prediction of the Microstructure of the Fusion Zone of Multicomponent Steel Weld Deposits," Advances in Welding Research, ed. S. A. David, (Metals Park, OH: American Society for Metals, 1987), 225-229.

5. J. R. Yang and H. K. D. H. Bhadeshia, "Thermodynamics of the Acicular Ferrite Transformation," Advances in Welding Research, ed. S. A. David,

(Metals Park, OH: American Society for Metals, 1987), 187-191

6. M. Strangwood and H. K. D. H. Bhadeshia, "Mechanism of Acicular Ferrite Formation in Steel Weld Deposits," Advances in Welding Research, ed. S. A. David, (Metals Park, OH: American Society for Metals, 1987), 209-213.

7. H. K. D. H. Bhadeshia and D. V. Edmonds, "The Mechanism of Bainite Formation in Steels," Acta Metall., 28 (1980) 1265-1273.

8. H. K. D. H. Bhadeshia, "A Rationalisation of Shear Transformations in Steels," Acta Metall., 29 (1981) 1117-1130.

9. H. K. D. H. Bhadeshia, "Bainite: Overall Transformation Kinetics," J. de Physique, C4-43 (1982) 443.

10. H. K. D. H. Bhadeshia, "The Diffusional Formation of Ferrite in Iron and its Alloys," Prog. in Mat. Sci., 29 (1985) 321-386.

11. G. R. Speich, V. A. Demarest and R. L. Miller, "Formation of Austenite During Intercritical Annealing of Dual-Phase Steels," Metall. Trans., 12A (1981) 1419-1428.

12. J. B. Gilmour, G. R. Purdy and J. S. Kirkaldy, "Thermodynamics Controlling the Proeutectiod Ferrite Transformation in Fe-C-Mn," Metall. Trans., 3 (1972) 1455-1469.

13. A. A. B. Sugden and H. K. D. H. Bhadeshia, Unpublished research based on the method of J. S. Kirkaldy, B. A. Thomson and E. Baganis, "Hardenability Concepts with Applications to Steel", Ed. D. V. Doane and J. S. Kirkaldy, pub. ASM, Warrendale, U. S. A., 1977, p. 82.

Steel Developments for Improved Weldability

THE INFLUENCE OF STEEL DEVELOPMENTS ON WELD PROPERTIES AND WELDABILITY

P. L. Threadgill
Edison Welding Institute
1100 Kinnear Road
Columbus, Ohio   43212

Introduction

Over the last 10 to 15 years, a major revolution in steelmaking technology has occurred, leading to many changes in steelmaking practice, and steel composition. The more significant advances that have been made include the implementation of continuous casting for plates of approximately three-inch thickness or below, the introduction of clean steels by improved desulphurization techniques, and a switch to the use of controlled rolling and other steel processing techniques to increase strength, rather than the use of expensive alloying elements. This has meant a reduction in the steel carbon levels and also in other indices such as carbon equivalent (CE) or the Japanese P_{cm}, which both reflect the influence of other alloying elements.

These changes have been brought about by two complementary factors. The first is a need by steelmakers to reduce their costs and maintain their profits in the face of a worldwide recession which has resulted in an over-capacity in steelmaking capability, and the second in ever more stringent demands being placed on the steels by the customers.

In this paper, some of the many developments will be discussed in more detail including:

(1) Clean steels
(2) Structural and linepipe steels of low carbon equivalent
(3) Steels for high heat input welding
(4) New high strength roller quenched steels.

These steels are considered individually below.

Low Sulphur Steels

In the late 1960's and early 1970's, the problems of lamellar tearing in steel structures, particularly for the oil industry, led steelmakers to develop steels containing low sulphur levels which would give high through-thickness ductility. The level of sulphur and the properties of sulphides were controlled by adding rare-earth metals (principally cerium rich mischmetal) and subsequently (and now more commonly) by calcium. These elements not only remove a great deal of sulphur in the refining ladle, but also combine with remaining sulphur to form hard sulphide or oxysulphide inclusions which are not deformed during subsequent rolling, and which led to excellent through-thickness ductility.

The increased cleanliness of the steel has also resulted in improved upper shelf toughness, and improved resistance to hydrogen pressure cracking in wet H_2S environments. While these improvements can only be considered beneficial, there is a substantial volume of evidence to suggest that low sulphur steels may have a substantially higher hardenability than conventional steels, leading to an increased risk of hydrogen cracking. The available literature has been reviewed by Hart (Ref. 1) and from this study it appears that the majority of researchers have found an adverse effect of low sulphur on the risk of HAZ hydrogen cracking. However, as Hart has noted all but one of the reviewed publications refer to laboratory trials rather than practical fabricator experience, although more than half of the investigations described have used commercial rather than experimental heats of steels. This distinction is of some importance since the morphology and size distribution of inclusions in steel are dependent on the size (and cooling rate) of the ingot in which they solidified.

Changing the sulphur content appears to have two effects. The first is that lowering the sulphur level increases the hardenability of the steel and, therefore, the hardness at a given cooling rate. The second is that sulphide inclusions can act as sinks for hydrogen. Reducing the sulphur content, therefore, means more hydrogen is free to embrittle the steel.

The metallurgical reasons for the effect of sulphur (and oxygen) on steel hardenability is reasonable well understood. It is well established that the austenite decomposition reaction can be initiated by inclusions, leading to increased transformation start temperatures, and the formation of softer microstructures. This effect is also well documented in weld metals. A typical example of preferential nucleation of a high temperature transformation product on an inclusion in an HAZ is shown in Figure 1. The effect of inclusions on transformation is not so significant in low alloy steels where the alloying has a greater influence on the transformation than the inclusions. However, in C-Mn steels where nonmartensitic products can be formed, the effects can be of major importance.

One of the more intriguing facts concerning the relationship between sulphur and hardenability is the wide variation in the magnitude of the effect. Although Hart (Ref. 1) has shown approximately 80 percent of published work shows an adverse influence of low sulphur, Yurioka (Ref. 2) has published data that show no effect of sulphur on hardenability between 0.002 percent and 0.020 percent. Other studies even indicate a reduction in hardenability due to sulphur (e.g., Suzuki) (Ref. 3). Typical data are shown in Figures 2-4. Figure 2 (from Glover, et al) (Ref. 2) shows the expected increase in preheat required as the sulphur level decreases. Data from Hirai, et al (Ref. 3) are reproduced in Figure 3, and this shows a general trend to increased preheat requirements below about 0.008 percent S, although certain heats of steel indicate the opposite trend. Data reviewed by Suzuki (Ref. 4) are reproduced in Figure 4 and shows both beneficial and adverse effects of low sulphur on preheat requirements.

One explanation for this wide variation in behaviour is that the hardenability is controlled not by the weight percentage of sulphide/oxide inclusions, i.e. the volume fraction, but more by the actual number of inclusions. Wide variations in the number of inclusions for a given volume fraction can be easily accounted for in different steelmaking processes, and

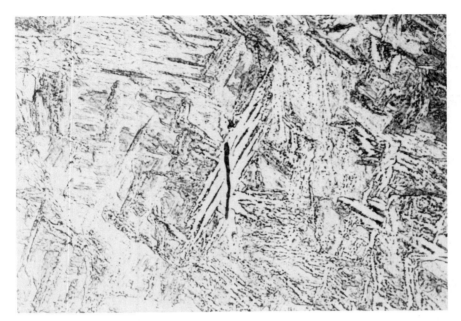

Figure 1. Manganese sulfide inclusion nucleating ferrite in a heat affected
zone   (OG 1451)   (from Hart, Ref. 1)(X1600)

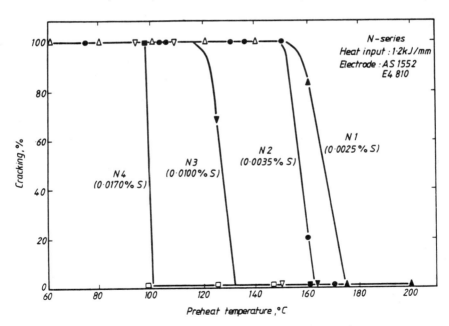

Figure 2.   Effect of sulfur on CTS tests on Nb microalloyed steels
(from Glover, Ref. 2)

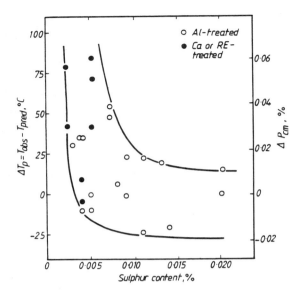

Figure 3. Effect of sulfur on charge in preheat temperature
(from Hirai, Ref. 3)

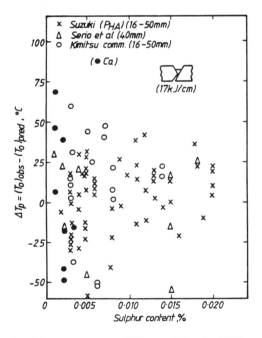

Figure 4. Effect of sulfur content on HAZ cracking in HT50 and HT60 steels
(from Suzuki, Ref. 4)

casting practices. Changes in the number of inclusions for a given volume fraction would result in changes in mean particle size and interparticle spacing, and also in the elastic mismatch between the inclusion and matrix. Since the elasticity energy associated with an inclusion (as well as its epitaxial properties) is likely to have a significant effect on subsequent transformation (which initiate preferentially in high energy areas), the range of effects of sulphur is perhaps not totally unexpected.

The main problem from the welding point of view is knowing how to estimate the magnitude of an effect which cannot be quantified. It is not possible to include a sulphur term in the carbon equivalent or other similar formula for the reasons outlined above, and therefore not possible to predict any change in hardenability from the analysis of the steel. Hart's examination of the data has shown that an approximate estimation of the increase in carbon equivalent in reducing the sulphur level from 0.025 percent to 0.005 percent would be approximately 0.03 percent, which could correspond to an increase in preheat to 100-200 F (50-100 C).

The most practical approach is to assume, in line with the bulk of experimental evidence, that low sulphur is detrimental. When devising procedures for a particular fabrication where several heats of steel of the same specification are to be used, the procedural trials should be run on the heat whose produce analysis shows the lowest sulphur level, using a correction to the carbon equivalent in line with the above relationship. Since the majority of welding procedures are necessarily conservative with respect to hydrogen cracking, fabricators will often find that the use of clean steels with existing procedures proves satisfactory. It should be noted that the margin of safety will be reduced.

A further point which can be easily overlooked is that although in certain circumstances the risk of hydrogen cracking may be low, the risk of increased hardness in clean steels may produce problems where strict control of hardness is exercised, e.g. for service in sour environments, or where hardness levels are specificed as part of procedure control.

However, although the introduction of clean steels can present some problems to the fabricator, it is felt that these are heavily outweighed by the advantages of such steel.

Low Carbon Equivalent Structual Steels

Over the last two decades, there has been a major revolution in steelmaking practices, leading to the introduction of vastly superior products in terms of weldability and mechanical properties. The use of ingot casting for thinner grades of plate (less than 2 inches) is rapidly diminishing, and the introduction of sophisticated thermo-mechanical rolling processes is becoming more common. These improvements in steelmaking practices have led to significant changes in steel composition, particularly in grades giving yield strengths of approximately 50 ksi.

These changes are:

(1) Lower carbon contents, giving improved weldability and HAZ toughness

(2) Cleaner steels, with sulphur less than 0.005 percent, and fre-quently even cleaner. Lower oxygen levels are also now common

(3) Reduced microalloying elements such as niobium

(4) Increased copper and nickel in normalized grades to compensate for reduced carbon and niobium

(5) Use of titanium additions to improve HAZ microstructures

(6) Lower levels of deoxidants such as aluminum

(7) Lower levels of nitrogen.

There is an increasing trend, which is likely to continue, to produce structural steels by controlled rolling, followed by accelerated cooling through much of the transformation range. This has been shown to result in steels of very high toughness, and produces a strength level which is considerably higher than would be found in a normalized steel of similar composition. Thus, accelerated cooled steels can have a lower carbon equivalent than conventional normalized steels of equivalent strength.

It is generally claimed by steelmakers that lower carbon equivalent values mean that the weldability of the steels is improved. Since the hardenability is lower, either the preheat and interpass temperature requirements can be relaxed during welding, or, in cases where little or not preheat is required, the requirements for drying of electrodes and fluxes become less severe.

These arguments must be treated with caution for two reasons. Firstly, there is a growing body of evidence which suggests that the critical hardness in the HAZ to cause hydrogen cracking is dependent on the carbon content and carbon equivalent of the steel. Lower carbon equivalent steels will crack at lower hardness levels. Some data is shown in Figure 5. Secondly, preheat and/or suitable preparation of consumables is also necessary to prevent weld metal hydrogen cracking. Unfortunately, weld metal cools at a comparatively slow rate compared to the high rate during transformation in accelerated cooled steels, and requires alloying elements to provide satisfactory strength, rather than accelerated cooling.

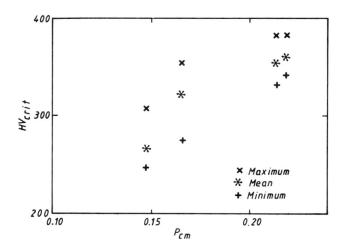

Figure 5. Dependence of critical hardness for cracking in a CTS test on composition   (from Matharu and Hart, Ref. 9)

A number of researchers have shown that the maximum hardness of a heat affected zone depends primarily on the carbon content of the steel, since it is this which determines the hardness of the martensitic constituent. This is reflected in empirical relationships between martensite hardness and composition, e.g.

$H_m = 812C + 293$ (Satoh) (Ref. 5)
$H_m = 802C + 305$ (Lorenz and Duren) (Ref. 6)

The critical hardness above which the risk of hydrogen cracking becomes unacceptable is not, as initially assumed 15-20 years ago when ways of predicting safe procedures were first being developed, independent of steel composition. There is now a growing volume of evidence to suggest that the onset of hydrogen cracking is controlled by the microstructure rather than by the hardness. In cases where the martensitic constituent is greater than about 75 percent of the volume fraction, the risk of hydrogen cracking becomes severe. The problem of predicting cold cracking susceptibility therefore reduces largely to one of understanding the hardenability of the steel, rather than the hardness. Hardenability can be defined as the ability of a steel to form a martensitic microstructure. For steel of a given composition, there exists a critical cooling rate between 800 C and 500 C above which martensite is formed. In earlier work, the critical cooling rate to just avoid a critical hardness level (dependent on the hydrogen level and joint restraint) has been used as the criterion for procedural development. These studies were on steels of comparatively high carbon equivalent (0.40-0.50 percent), where no great dependence of critical hardness on the risk of hydrogen crtacking was recognized, but it is apparent that this approach could be non conservative for lower carbon steels.

When using present procedures, high cooling rates can apparently be tolerated for low CE steels since rapid cooling will not lead to assumed critical hardness level for cracking. However, although this high cooling rate may well produce a hardness level which is below the assumed level, it could be above the actual critical hardness level, since this level will decrease with carbon equivalent.

This area of procedure prediction is at present under researched, although studies are in hand at a number of centers. Work published by Boothby (Refs. 7-8) shows that the susceptibility to cracking appears to increase as the carbon equivalent decreases in steels with CE_{IIw} 0.42. However, other studies (Refs. 9-10) have not found this effect and there is an obvious need for more work in this area to clarify this issue. At TWI preliminary results demonstrate that the present equations overestimate the reduction in risk of cracking in low C steels welded without preheat. The implication of this is that procedures leading to rapid cooling rates may appear to be safe, since the presumed critical hardness is not reached, whereas in fact the actual critical hardness may be exceeded.

A further complicating factor which must be recognized is that the inherent susceptibility to hydrogen embrittlement may be different for the same microstructures of different hardness levels, with a greater potential embrittlement being possible at higher hardness levels. It has been suggested (1) that the stress developed during welding may not be constant for a given geometry, but is probably dependent on the steel transformation behavior, which in turn depends on its composition. Increasing the transformation temperature (particularly M_s) may increase the contraction stress, since the temperature interval between transformation and ambient will increase. Increases in volume expansion during transformation (which would be expected with higher C levels) will offset the contraction stresses, leading to a reduction in overall stress. Thus, the situation may well arise where the softer, inherently less suceptible microstructures may experience a greater stress than the harder, more susceptible microstructures.

Assessments of data on new steels is still continuing, but from a practical point of view, it would appear that the preheat scheme for prediction of safe procedures is only conservative if the carbon equivalent value exceeds 0.40 percent. For lower steels, the present system may sometimes be non-conservative and, until work on the revised scheme for lean alloy steels is completed, it is recommended that C-Mn steels with a CE of below 0.40 percent are assumed to have a CE of 0.40 percent. This restriction may not apply to low C steels of 1 percent Mn.

An extreme case of low carbon steels exists in the so-called ULCB (ultra low carbon bainitic) steels which have been developed for linepipe applications. These are now available up to API X80 grades, and contain very low carbon levels (typically 0.02 percent, together with about 1.6 percent Mn, and small quantities of Nb, Ti and B to improve hardenability). Their strength is obtained primarily through accelerated cooling at the start of the austenite decomposition reaction. These rather unusual steels have been developed to satisfy hardness requirements in HAZ root regions in sour service in the as-welded condition of Rockwell 22C, equivalent to about 245 HV. In sweet conditions, higher values (275-350 HV, depending on process) may be permitted. Pipe steels are traditionally welded onsite using down hill manual welding with cellulosic electrodes, a process which gives very little heat input together with large weld metal hydrogen levels. Although the second pass (the hot pass) is supposed to be deposited before the root has cooled, this method of reducing hardness cannot be guaranteed and this has led to the very low carbon equivalent steels being developed. Increased interest is now being shown in mechanized GMA welding. Although this reduces the hydrogen to a minimal level, the heat input is still very low and again a requirement for low hardness must still be met, particularly for sour conditions. There is some evidence that low carbon weld metal produced as a consequence of welding these steels may be susceptible to soldification cracking.

Steels for High Heat Input Welding

It is generally recognized that the single factor which restricts the exploitation of more productive welding processes for fabricating C-Mn steels is the fall in toughness in the weld metal and HAZ as the arc energy increases. Although this can be overcome in the weld metal by the use of suitability micro-alloyed consumables, the problems in the heat affected zone have proved more challenging. It is well recognized that the main problem is the rapid rate of growth of the austenite grains in the HAZ, followed by the slow cooling rate through the transformation temperature range. These factors combine to produce a coarse HAZ microstructure of low toughness.

About 10 years ago (Ref. 12), it was noted that a suitable dispersion of TiN particles in the steel could restrict the austenite grain growth. In addition, there is considerable evidence that a suitable dispersion of such particles could also significantly influence the austenite decomposition reaction, by providing nuclei on which the transformation reaction could nucleate, thus increasing the number of sites from which transformation initiated, leading to a finer microstructure. The dispersion of precipitates required to control grain growth and also the austenite decomposition reaction are critical. It has been demonstrated that TiN particles above

574

about 0.05 m in diameter are ineffective. The particle size however is not a simple function of titanium and nitrogen content, but also depends on the thermal history of the material and in particular on the rate of cooling shortly after solidification. Since this cooling rate needs to be high, titanium treatment is really only effective on continuously cast steels of up to about 2 inches (50mm) in thickness. Nevertheless, a number of researches have indicated that the optimum levels are about 0.02-0.03 percent Ti, and approximately 0.008 percent N.

The need for continuous casting has already been noted, but this has a further indirect benefit in that the aluminum which is always added to such steels protects the titanium, and restricts TiO formation. Research has shown that this leads to the formation of a greater number of fine precipitates.

A problem which has been encountered in these steels is that of obtaining sufficient strength. Traditional methods for improving the strength (i.e. to give yield values of approximately $350N/mm^2$) by addition of niobium do not work, since this element has an adverse effect on the toughness of high heat input welds. The incompatability of miobium and titanium is probably related to their opposing effects on austenite transformation temperatures. Niobium tends to reduce the transformation temperature range, whereas titanium nitride precipitates raise these temperatures.

The approaches that are taken depend on the steel processing route. For normalized grades, small additions of copper and nickel (to about 0.3 to 0.4 percent total) can be made. Alternatively, the strength can be reaised by the use of accelerated cooling techniques at the higher end of the transformation temperature range.

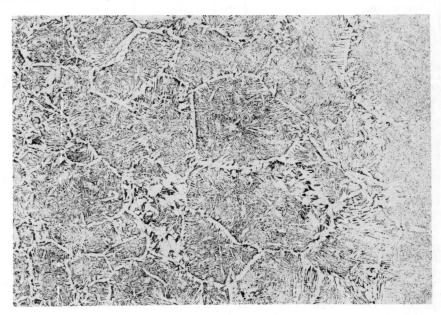

OM2842

Figure 6a.   HAZ regions in electroslag welds in titanium free C-Mn
             steels of similar carbon equivalent   X40

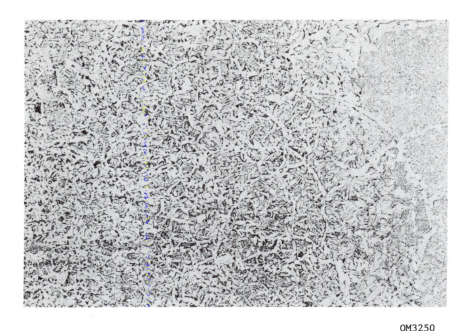

OM3250
Figure 6b.   HAZ regions in electroslag welds in titanium treated C-Mn
steels of similar carbon equivalent   X40

The properties of such steels when welded at high heat inputs are
generally impressive. Published literature (Ref. 13) generally shows
excellent HAZ properties and a reduction in the width of the HAZ over which
microstructural degradation, hardness increase and toughness reductions
occur. An example comparing the HAZ microstructures of titanium treated and
titanium free steels of similar carbon equivalent is shown in Figure 6. It
is evident that the titanium bearing steel has a much finer grain size,
although there is very little difference immediately adjacent to the fusion
boundary. However, examination of microstructures at and close to the
fusion boundary show that TiN precipitates are ineffective in these regions
due to their entering solid state solution at temperatures of approximately
1250 C. This, therefore, leaves a narrow band of coarse microstructure, and
it is now becoming apparent that the toughness of the fusion boundary region
in these steels can be poor. This has been demonstrated at The Welding
Institute in both CTOD and wide plate tests.

At present, the further development of such steels is still underway.
It seems probable that other second phase systems may lead to greater
success rates, particularly those using calcium oxysulphide particles which
are stable to much higher temperatures. Work by Nakanishi, et. al. (Ref.
14) has shown that synthetic HAZs treated with titanium and calcium provide
an improved level of HAZ toughness with temperatures up to 1450 C, although
control of the dispersion of the calcium-rich precipitates is believed to
present considerable problems to steelmakers.

Recent work by Homma, et. al. (Ref. 15) has shown that a suitable
dispersion of TiO particles can be effective almost to the melting point.

His work was influenced by observation of the behavior of reheated submerged arc weld metal containing TiO inclusions in which the beneficial effect of this precipitate species was noted. Experimental heats of TiO bearing steel were produced with favorable results. It is interesting to note that earlier researchers considered TiO to be an undesirable inclusion species.

An alternative approach that has been developed to production status is the use of very low nitrogen steels (approximately 0.002 percent) combined with high aluminum (approximately 0.06 percent) and a small amount of titanium (approximately 0.01 percent). Although in these steels grain growth is not prevented, the low nitrogen is claimed to provide good toughness at heat inputs of up to approximately 15kJ/mm, i.e. equivalent to a two-pass weld in one-inch-thick plate.

High heat input resistant steels were developed primarily for the Japanese shipbuilding industry. The use of such steels could perhaps be broadened to extend into areas where low hardness is required, e.g. for sour gas service. Substantially lower hardenability has been demonstrated in this type of steel (Ref. 13). Titanium treated steels have also been used extensively where good properties are required at low temperatures. In this case, the steels are welded at conventional heat input levels.

Welding of high heat input resistant steel generally causes few problems. Preheat is seldom required since the arc energies can be high and advantage can be taken of the more productive processes (e.g. multi-arc submerged arc welding, electroslage, etc.). A number of manufacturers are developing systems and consumables for use with such steels.

To date, widespread acceptance of these steels has been slow outside the Japanese shipbuilding industry, but there are a number of non-critical applications in which the steels and appropriate weld techniques could be used to advantage.

Roller Quenched Steels

In the higher yield strength steels (e.g. 70 ksi and above), the required properties have been traditionally achieved by suitable alloy additions followed by quenching and tempering. This has led to highly hardenable steels which had to be welded with some care to avoid hydrogen cracking.

In the last few years, a number of steelmakers throughout the world have introduced roller quenched steels. In these products, the steel is hot rolled normally to the required thickness, and is quenched from the austenite temperature range while traveling through straightening rolls. The quench is applied by very high pressure water jets resulting in an extremely efficient quench. Using this technique it is possible to generate much higher yield strengths than would normally be possible for steel of a given carbon equivalent. The steels are usually C-Mn steels containing small additions of Mo, Nb and B to improve hardenability. It is possible to obtain sufficient hardenability in thicknesses of about 2 to 2-1/2 inches (50-60 mm).

There are very little data available on welding these steels. That which is published by steelmakers indicates that lower preheats are required than would be necessary for competitive quenched and tempered grades, and that for thinner grades, preheat may not be necessary. The lean alloy contents mean that excessive heat inputs cannot be used without risk of softening outside the transformed HAZ, although this is unlikely to be a problem in the thicker grades where a better heat sink exists.

577

A point to note is that these steels must be welded with alloyed weld metal if matching strengths are required, and too great a reduction in preheat levels could induce hydrogen cracking in the weld metal.

Concluding Remarks

The rapid developments in steelmaking technology have generally progressed side by side with developments in welding. Indeed, the development of clean steels, ultra low carbon steels and steels for high heat input welding has been prompted by the needs of particular sections of the welding industry.

Although many of the newer steels may be unfamiliar to fabricators, in general the difficulties in welding them are not great, and major problems would not be expected.

The Welding Institute and Edison Welding Institute are currently active in assessing the weldability of many of the newer steels, as well as assessing the mechanical properties of weld and HAZ regions in such steels. Much of the data obtained, and its implications to fabricators and users will be published in due course.

References

1. Hart, P.H.M. The influence of steel cleanliness on HAZ hydrogen cracking: the present position. IIW Doc IX1308-84.

2. Glover, G., Chang, C.J. and Chipperfield, C.G. May, 1982. The influence of sulphur content on the weldability of structural steels. Proc. 35th Annual Conference of Australian Institute of Metals.

3. Hirai, Y., Minakawa, S. and Tsuboi, J. IIW Doc-IX-1160-80.

4. Suzuki, H. 1983. IIW Houdremont Lecture, Trans ISIJ, 23, pp 189-204.

5. Satoh, K. and Terasaki, T. 1979. General Assembly of JWS. No. 25, p 98.

6. Lorenz, K. and Owen, C. 1981. Evaluation of large diameter pipe steel weldability by means of carbon equivalent. Proc. "Steels for Line Pipe and Pipeline Fittings" to Metals Society, London, p 322.

7. Boothby, P.J. August, 1985. Weldability of offshore structural steels, part 1. Metal Construction, 7 (8).

8. Boothby, P.J. September, 1985. Weldability of offshore structural steels, part 2. Metal Construction, 7 (9).

9. Matharu, I.S. and Hart, P.H.M. Heat affected zone hydrogen cracking behavior of low carbon equivalent C-Mn structural steels. Welding Institute Members Report 290/1985.

10. Matharu, I.S. Heat affected zone hydrogen cracking behavior of low carbon equivalent C-Mn structural steels - further data. Welding Institute Members Report/1986.

11. Boniszewski, T and Watkinson, F. Effect of weld microstructures hydrogen induced cracking in transformable steels. Metals and Materials, February 1973, p 91 and March 1973, p 45.

12. Kanazawa, S. and Nakashima, A., Okamoto, K. and Kanaya, K. 1976. Improvement of weld fusion zone toughness by fine TIN. Trans. ISIJ, 16, p 486.

13. Threadgill, P.L. July, 1981. Titanium treated steels for high heat input welding. Welding Institute Research Bulletin, p 189.

14. Nakanishi, M., Komizo, Y. and Seta, I. 1983. Improvement of welded HAZ toughness by dispersion with nitride particles and oxide particles. Tetsu te Haganu, 52, p 117.

15. Homma, H., Ohkita, S., Matsuda, S., and Yamamoto, K. Improvement of HAZ toughness in HSLA steel by finely dispersed Ti oxide. Paper presented at 67th AWS Convention, Atlanta, 1986.

DEVELOPMENT OF STEEL PLATES

USED FOR OFFSHORE STRUCTURES AND THEIR MANUFACTURING TECHNOLOGY

S. Yano, K. Itoh, M. Katakami, H. Nakamura and T. Kusunoki

Yawata R & D Lab. Nippon Steel Corporation
1-1-1 Edamitsu Yawata-Higashi, Kitakyushu 805, Japan
Plate Technical Division Nippon Steel Corporation
2-6-3 Otemachi Chiyoda, Tokyo 100, Japan
Oita R & D Lab, Nippon Steel Corporation
1 Nishinosu, Oita 870, Japan
Plate Technical Division Nippon Steel Corporation
2-6-3 Otemach Chiyoda, Tokyo 100, Japan
Nagoya Works Nippon Steel Corporation
5-3 Tohkai, Tohkai 476, Japan

Abstract

In the present paper, the current development of steel plates used for offshore steel structures and their manufacturing technology are reported.
Firstly, the basic research on the leading principles for design of alloying elements and manufacturing conditions are reported.
(1) Mathematical model for prediction of effective strain governing grain size during and after hot rolling.
(2) Mechanism of strengthening and grain refining by accelerated cooling.
(3) Major metallurgical and mechanical factors affecting toughness of heat affected zones caused by welding.
(4) Metallurgical factors related to toughness of weld metals.
Secondly, the properties of developed steel plates (HT50-100) and welded joints are explained.
Thirdly, development of steel manufacturing technology is briefly summarized.
(1) Steel refining process and casting technology
(2) Plate rolling process including "The Accelerated Cooling"
(3) Computer aided control system for plate production process
Finaly, future scope for R & D activities on steel materials and related technologies taking account of trends of offshore technology will be mentioned.

Introduction

Though the development of offshore oil and gas has become stagnant throughout the world in recent years, it is expected that hydrocarbon fuels (petroleum, natural gas and coal) will remain the principal fuels in the next century. However, as a long time is required for the exploration and production of such fuels strenuous efforts must be continued for their development from a long-term viewpoint. The structures to be used for the development and production of offshore oil and gas are exposed to severe environmental conditions, such as strong winds, great waves, thick ice and low temperatures. Accordingly, the most advanced technologies are required for design, materials, fabrication, construction and operation of offshore structures (1,2). In particular steel products with high strength, high toughness and excellent weldability have made and will make a great contribution to the safety of offshore structures.

This report outlines the results of basic research conducted in recent years to provide criteria for improvement of the strength and toughness of heavy plates, heat-affected zones (HAZ) of such plates and weld metals. The report then describes the representative properties of various types of heavy plates and welding materials which have been developed on the basis of the results of basic research. Steel products of excellent quality cannot be produced on a commercial basis unless excellent manufacturing technologies are developed at the same time. In this sense, the present report introduces the recent advances made in refining and casting technologies, accelerated cooling after rolling and the control systems for plate manufacturing processes. Finally, the report describes the subjects for future research and development in connection with steel products and related engineering, paying attention to the recent trend in offshore development.

To maintain the technological consistency, this report is based on the results of R & D efforts made at Nippon Steel Corporation.

Basic research on the leading principles for design of alloying elements and manufacturing conditions

The required properties of steel plates for offshore structures are dependent upon the environment and conditions under which the structures are used. Steelmakers design the steel composition and manufacturing process so as to meet the given requirements. Throughout the design stage, safety of the structure is given prime consideration. It is particularly important to secure high strength and toughness at the welded joints. Accordingly, a guide to the production of steel plates must be established by elucidating the metallurgical and dynamic factors affecting the strength and toughness of the base metal and the welded joints. In recent years, the thermomechanical control process TMCP, in which the amounts of alloying elements are reduced to the lowest possible levels to prevent cracking at the welded joints and in which plates attain the designed strength and toughness through rolling alone, has become widespread. This process has been applied to the production of steel plates for offshore structures. Extensive study is underway for further development of this process in the search for better combinations of strength and toughness and aiming at the improvement of toughness at the welded joints.

This chapter describes the results of the basic research conducted in recent years to clarify what metallurgical and mechanical factors are related to the strength and toughness of the plates and at the welded joints.

Mathematical model for prediction of effective strain affecting the grain size during and after hot rolling

In the production of heavy plates, "controlled rolling" has found wide application to improve their strength and toughness. In recent years, the quantitative determination of the conditions for recrystallization of austenite (γ) in the rolling process has become an important problem in the design of properties, because the recent trend is toward rolling with heavy reduction in the temperature range in which the austenite is not recrystallized.

As shown by the chain lines in the schematic drawing of Fig. 1, the critical condition at which the static recrystallization of austenite occurs is considered to be a function of rolling temperature and rolling strain. In the so-called austenite recrystallization temperature range, the critical strain (ε_{cr}) at which static recrystallization takes place is smaller than the rolling strain in a normal single pass, and therefore the recrystallization of austenite is begun during the time interval before the next pass. If the rolling temperature falls, ε_{cr} increases and becomes larger than the rolling strain in a normal single pass. In such cases, therefore, ε_{cr} is exceeded and recrystallization is started only after the rolling strains in several passes have been accumulated. In this case, a decrease in accumulated strain due to the recovery during the interval before the next pass as described later should be taken into consideration. If the rolling temperature is further lowered, ε_{cr} increases significantly beyond the level which cannot be exceeded by the accumulated strain. Under this condition, rolling is ended without recrystallization of the austenite.

The recrystallization phenomenon during hot rolling, which is closely related to the properties and toughness of steel products, has often been treated in reports (3-7) as a function only of temperature and draft strain. This treatment, however, does not have universal applicability because the results are only those of statistical analysis and therefore are not applicable to a different rolling process or to the development of a new rolling schedule.

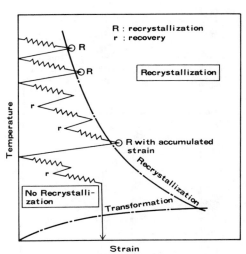

Figure 1 - Schematic illustration of recrystallization of γ in hot rolling process (8).

In recent years, the development of a mathematical model for relating the retained strain generated through the change in dislocation density to the recovery or recrystallization phenomenon has been in progress. The details are not yet published, but have recently been presented (8). Examples of the results obtained by this model which requires approximately 30 parameters are shown in Figures 2, 3, and 4. Fig. 2 shows the critical conditions of recrystallization of γ. The right upper sides of the lines are the recrystallization ranges and the left lower sides are the no recrystallization ranges. If the time interval during deformation is longer than incubation period (parameter in Fig. 2), recrystallization starts during the interval. Therefore, a short time interval extends the no recrystallization range to the higher temperature side. In Fig. 3, the change in the retained strain by each pass is shown for different rolling condition. In Fig. 4 the relationships are shown between the toughness of plates and the retained strain when rolled under the conditions indicated in Fig. 3.

These results of model calculation indicate that rolling should be ended in a short time, maintaining the rolling temperature at a low level, to increase the accumulated strain by controlling the recovery and recrystalliza-

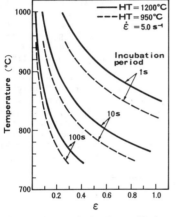

Figure 2 - Critical conditions of recrystallization of γ (8).

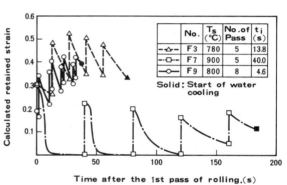

Figure 3 - Change of retained strain in hot rolling process (8).
$$\begin{pmatrix} T_S: \text{Start rolling temperature} \\ t_i: \text{Average interval time} \end{pmatrix}$$

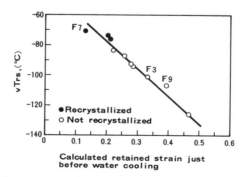

Figure 4 - Relation between retained strain and vTrs (8).

tion. This research is in progress today, and it is hoped that this model, when completed, will be applied in real time to the actual designing of rolling conditions.

Mechanism of strengthening and grain refinement by accelerated cooling

It is well known that the strength and toughness of steel are affected by various factors, such as grain size, dislocation density, precipitates, alloying elements in solid solution, second phase and microstructure. Of these, the refinement of grain size is the most important factor, because this refinement improves both the strength and the toughness. For this reason, controlled rolling has been developed mainly for the production of high -strength low-alloy steels. In recent years, extensive study has been made on a new technology for accelerated cooling after controlled rolling in the search for better combinations of strength and toughness (9-20). For the development of a new process, however, it is necessary to clarify the strengthening mechanism, which is closely related to other mechanical properties.

It is estimated that the strength of ferrite-pearlite steel is a function of the strengths and volume fractions of ferrite and pearlite. For example, Fig. 5 shows the results of measurement of macro-hardness, micro -hardness in the ferrite matrix, ferrite grain size and fraction of second phase (pearlite or pearlite + bainite) in HT50 which was cooled at various cooling rates after being heated to 950°C. The micro-hardness and $d^{-\frac{1}{2}}$ (d: ferrite grain size) of ferrite itself increase in proportion to the logarithm of cooling rate. The fraction of pearlite is constant when the cooling rate is small, but increases if the cooling rate increases beyond a certain level. In this HT50, bainite is observed when the cooling rate is higher than 25 °C/sec. This corresponds to increases in macro-hardness. Fig. 6 shows the macro-hardness and ferrite hardness of the same steel as shown in Fig. 5 which was heated to 950°C and was first cooled at a rate of 25°C/sec and then at a rate of 10°C/min from the temperatures shown on the X-axis. In all specimens, the microstructure consists of ferrite and pearlite and the ferrite grain size and the fraction of pearlite are nearly constant. However, when the temperatures at which the cooling rate was changed (accelerated cooling interrupted temperature) were lowered, the macro-hardness increased in proportion to increases in the micro-hardness of ferrite.

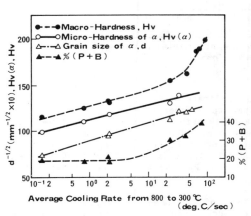

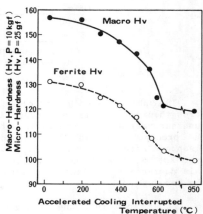

Figure 5 - Effect of cooling rate on strengthening factors of Steel 1 (0 15C- 0.17Si-0.66Mn-0.02Al-0.0056N) (21).

Figure 6 - Effect of accelerated cooling interrupted temperature on the macro-hardness of the steel and micro-hardness of its ferrite(Steel 1) (21).

Figs. 5 and 6 show that the grain refinement of ferrite and the volume fraction of pearlite are dependent upon the cooling rate but the hardness of ferrite depends on the cooling rate and accelerated cooling interrupted temperature. In other words, it may be concluded that there are three strengthening factors in the case of ferrite-pearlite microstructure obtained after accelerated cooling, i.e., the grain refinement of ferrite, the strengthening of ferrite itself and increases in volume fraction of pearlite. Here, let us cite an example in which the degrees of contribution of these strengthening factors to tensile strength were evaluated (21). According to Pickering (22), the tensile strength of low-carbon ferrite-pearlite steel is expressed by

$$TS \ (kgf/mm^2) = 30.0 + 2.83(\%Mn) + 8.48(\%Si) + 0.393(\% \ pearlite) \\ + 0.785d^{-\frac{1}{2}} \tag{1}$$

In the case of accelerated cooled steels, it is necessary to add the strengthening of ferrite to the first term of Eq. (1). This strengthening of ferrite will be about 4 kgf/mm² if TS is assumed as shown below, because the ferrite hardness of the accelerated cooled steel is greater by about 14 Hv than that of the steel made by controlled rolling:

$$TS = Hv/3.3 \tag{2}$$

The results of calculations are given in Table I. These results are in good agreement with the values obtained in the test, showing that the three factors mentioned above are the strengthening factors.

Table I. Analysis of Tensile Strength (21)

factors	(a) CR	(b) CR+AC	(b)−(a)
% Pearlite	14.7	21.0	6.3
N_α	8.0	10.7	2.7
Hv of ferrite	95	109	14
$30+2.83\%Mn+8.48\%Si$	33.31	33.31+4.0	4.0
$0.393\% Pearlite$	5.78	8.25	2.47
$0.785 d^{-1/2}$	5.28	8.43	3.15
TS (cal.)	44.37	53.99	9.62
TS (exp.)	44.7	54.0	9.3

CR : Controlled Rolling , AC : Accelerated Cooling
d : ferrite grain size (mm)

The above description clearly indicates the importance of ferrite strengthening. It is considered that this strengthening is caused by the supersaturation of carbon atoms, the dislocations induced by transformation, the precipitation of fine carbides, etc. However, these phenomena will disappear after tempering. As a matter of fact, the strength was decreased by 3 kgf/mm² from 54 to 51 kgf/mm² after treatment at 610°C for 10 minutes. This decrease approximates to a value of 4 kgf/mm² which was estimated from the hardness.

It is said that the strengthening of the steel in which the ferrite -bainite structure has been formed by accelerated cooling is dependent upon the formation of bainite. It should, however, be stressed that the strengthening of ferrite itself contributes to steel strengthening.

Major metallurgical and mechanical factors affecting the toughness at the heat-affected zone

As is seen in North Sea projects, recent construction of offshore structures has moved into deeper waters and colder regions. As these offshore structures are exposed to higher waves, requirements for the strength and toughness of steel plates to be used for such structures have become more severe. In particular, strength and toughness almost identical to those of the base metal are required at all positions of the welded joints. For testing the toughness of welded joints, therefore, the CTOD test has found wide application as a method for direct evaluation of brittle fracture initiation in addition to the Charpy impact test. As is empirically known, the CTOD values vary widely at the time when unstable fracture initiates. They also vary depending the type of welded joint, the type of notch, the position of notch in the welded joint, etc. As the specimens used for CTOD test have smaller radius of curvature at the notch tip than those used for Charpy impact test, the zone at the notch tip where high tension is applied is small. Accordingly, this zone is more likely to be affected by the localized brittle structure in the HAZ. Conversely speaking, the localized brittle structure does not readily coincide with the stress-concentration zone at the notch tip. Accordingly, the probability that the true minimum CTOD value is obtained is low. This is the reason why the CTOD values vary widely. As the brittle fracture resistance of the welded joint is evaluated in terms of the minimum CTOD value, it is not easy to satisfy the required specification in view of wide variations in CTOD value as described above.

Much work (23-25) has been conducted on the localized brittle structure at the HAZ which is the cause of variations in CTOD value. Research (26) on these variations and their cause by conducting multi-pass submerged arc welding using four types of HT50 as shown in Table I has obtained the following findings (see Figs. 7 and 8).

Table II. Chemical Compositions of Steels Tested (26)

Steel	C	Si	Mn	P	S	Cu	Ni	Nb	V	Al	Ti	N	Ca	Ceq
V	0.09	0.36	1.47	0.004	0.002	0.21	0.19	0.026	0.040	0.033	0.007	0.0026	0.0044	0.370
N	0.11	0.37	1.50	0.004	0.001	0.22	0.21	0.027	—	0.027	0.006	0.0024	0.0046	0.396
L	0.08	0.26	1.44	0.003	0.003	—	—	0.013	—	0.029	0.008	0.0027	0.0030	0.337
C	0.18	0.35	1.46	0.002	0.002	—	—	—	—	0.032	0.007	0.0027	0.0040	0.423

$$Ceq = C + \frac{Mn}{6} + \frac{Cu+Ni}{15} + \frac{Cr+Mo+V}{5}$$

a) The CTOD values in the as-welded state are low in steels V, N and L containing precipitation hardening elements such as V and Nb, but are high in steel C which does not contain precipitation hardening elements.

b) The CTOD values in all steels are improved by conducting PWHT (Post Weld Heat Treatment).

c) The examination of the fracture surface of the specimens with CTOD values smaller than 0.25 mm revealed that cracks initiate from the points in the grain-coarsened zone which was intercritically reheated by the succeeding weld beads.

They also elucidated the following by preparing Fig. 9, which schematically shows the microstructure and thermal cycle at each position of the HAZ, and by conducting the thermal cycle simulation test for the intercritically reheated zone and its vicinity (see Figs 10 and 11).

a) The specimens of all of four steels which were intercritically reheated in the double thermal cycle showed low CTOD values.

b) In the microstructures of these specimens, high-carbon martensite islands (M*) are observed mainly in the former austenite grain boundaries.

587

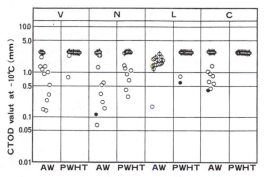

Figure 7 - CTOD values of as-welded and post-weld heat treated welded joints: submerged arc welding, heat input: 45 KJ/cm (26).

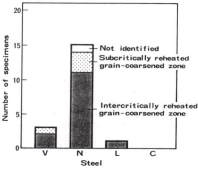

Figure 8 - Histogram of microstructures at cleavage crack initiation points: as welded $\delta_c \leq 0.25$mm (26).

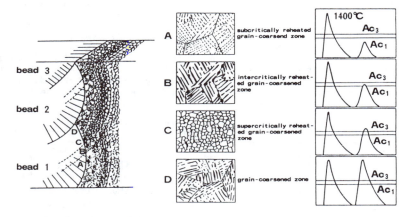

Figure 9 - Schematic illustration of the heat-affected zone of a multi-pass welding (26).

c) If the third thermal cycle is applied, the CTOD values of steels C and L which do not contain or contain few precipitation hardening elements are significantly increased.

The improvement of CTOD values by the third thermal cycle is attributable to the decomposition of M* in a manner similar to that causing the improvement by PWHT. A more detailed study (27) is being conducted on the effect of heat history on the formation of M*.

The CTOD value of the welded joint is affected not only by the toughness of the most embrittled zone but also by the strength distribution in the base metal, HAZ and weld metal. The effect of strength distribution on the critical CTOD value was studied (28) by conducting the CTOD test of specimens of model welded joints which were prepared using plates of different strengths (Table III and Fig. 12). For the specimen type A, the maximum tensile stress ahead of the crack tip (σ_m) was calculated as a function of crack-tip opening displacement (δ) by the two-dimensional, elastic-plastic

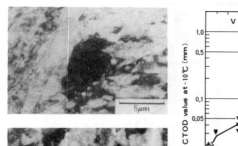

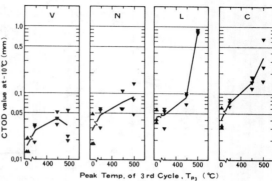

Figure 11 - Simulated HAZ CTOD value for triple thermal cycle: $T_{p1}=1400°C$, $T_{p2}=800°C$ (26).

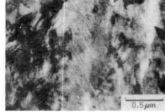

Figure 10 - Blocky second phase at prior austenite grain boundary observed by transmission electron microscopy:double thermal cycle, $T_{p1}=1400°C$, $T_{p2}=800°C$, $dt_{8/5}=20$ sec, steel V (26).

Table III. Chemical Compositions (wt%) and Yield Stress after Thermal Cycle (28)

No.	C	Si	Mn	P	S	Ni	Mo	Nb	Al	$\sigma_Y(N/mm^2)$
I	0.11	0.21	1.98	0.003	0.007	0.99	0.09	0.022	0.015	598 (-90 °C)
II	0.10	0.21	1.54	0.003	0.006	—	—	—	0.020	519 (-90 °C)

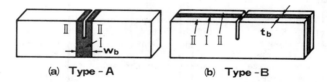

Figure 12 - COD specimen with strength distribution (28).

plane strain, finite element analysis method. For the type B, a reasonable distribution of stress through thickness direction was assumed. Changes in δ_c with changes (0.2 mm~4mm) in the thickness of high-strength part (w_b or t_b)were calculated, using $\sigma_m=\sigma_f$ (σ_f: critical tensile stress) as the condition of fracture. The result of the calculation for type A is shown in Fig. 13. It was confirmed by calculation that the plastic constraint ahead of the notch tip is increased with increases in w_b or t_b even if the toughness at the notch tip is locally the same, and σ_m reaches σ_f in the presence of small crack tip opening, resulting in decreases in δ_c with increasing w_b or t_b. It is important not only to decrease the brittle structure at the welded joint but also to minimize the unevenness of strength.

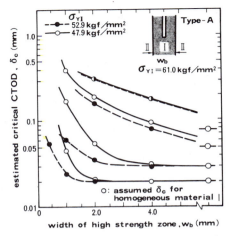

Figure 13 - Estimated change of δ_c (Type-A) (28).

Metallurgical factors affecting the toughness of weld metal

As described previously, the welded joints of steel plates for offshore structures to be used in severe environments are required to possess strength and toughness identical to or higher than those of the base metal. To satisfy these requirements, various new welding materials and welding processes have been developed. Of these new materials, the Ti-B type weld metal is known to have the highest toughness. The main reason why the Ti-B type weld metal shows the highest low-temperature toughness is that the solidified structure consists of acicular ferrite fine grains (29, 30). This structure is attributable to the suppression of the formation of polygonal ferrite at the grain boundaries by boron and intragranular transformation with many titanium oxides as the nuclei.

Fig. 14 shows the results of multi-pass submerged arc welding with a heat input of 33 KJ/cm. It will be noted that the Mn content has a large effect. Namely, when the Mn content is less than 1.2%, the amount of polygonal ferrite in the austenite (γ) grain boundaries increases, forming the lath-type, coarse ferrite structure within the austenite grains, and both the toughness in the as-welded (AW) state and the toughness after stress-relief (SR) decrease. If the Mn content exceeds 1.5%, a fine ferrite + upper bainite structure is obtained and the toughness after SR is decreased. In the microstructure with 1.2 ~ 1.5% Mn, a small amount of polygonal ferrite is observed at the austenite grain boundaries but the structure within the grains is fine acicular ferrite formed with titanium oxides as the nuclei. Accordingly, the toughness is high both in the AW state and after SR.

In the SEM image of the fracture surface formed by Charpy impact test of the SR embrittled specimen of steel with high Mn content, the propagation of cracks along the columnar γ-grain boundaries is observed. This indicates that the embrittlement of the grain boundaries is significant. In addition to Mn, Nb and V also participate in the SR embrittlement of the weld metal. The following formulae were proposed (31):

Structure in AW state: vTrs(°C) = 1430Nb(%) + 311V(%) - 67   (3)

590

Structure after SR: $vTrs(°C) = 990Nb(\%) + 387V(\%) - 71$ (4)

It will be noted that Nb has a greater effect than V and that it is necessary to minimize the Nb and V contents to enhance the toughness after SR.

On the other hand, the Ti-B type sealed metal arc welding rod has been developed (32) with a view to obtaining excellent CTOD characteristics. As a long arc is used in sealed metal arc welding, increases in [N] content in the weld metal are inevitable. Accordingly, the optimum composition range for obtaining the microstructure, consisting of small spheroidized polygonal ferrite formed in the austenite grain boundaries and the acicular ferrite formed from within the grain (this structure is the feature of Ti-B type weld metal as shown in Fig. 15), is enlarged as shown below: 1.45 ~ 1.8% Mn, [Ti]≧ 0.04%, and 30 ~ 70 ppm B. The CTOD value is correlated with the [N] content of the weld metal. To obtain a CTOD value ≧ 0.25 mm at -50°C, welding should be performed with a short arc length so that [N] is 100 ppm or less. For the rod connection, it is desirable to adopt the overlap method instead of the back step method. The dependence of CTOD value on the [N] content is observed not only in the Ti-B type weld metal but also in the 2.5%Ni type weld metal (E8016C1). [N] embrittlement is not a phenomenon peculiar to the Ti-B type weld metal only. The Ti-B type weld metal shows higher CTOD values than the 2.5%Ni type weld metal over the entire [N] range, including the normal [N] range (<120 ppm). [N] embrittlement is attributable mainly to [N] in solid solution when the tensile strength of the weld metal is less than 60 kg/mm². It is considered that the structure is changed by [N], resulting in increases in upper bainite or martensite islands, if the strength exceeds 60 kg/mm².

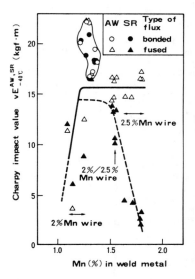

Figure 14 - Effect of Mn content on Charpy impact value of submerged arc welded metal with Ti-B type flux (31).

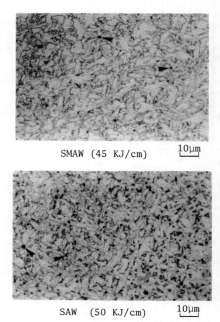

SMAW (45 KJ/cm)   10μm

SAW (50 KJ/cm)   10μm

Figure 15 - Microstructures of weld metals containing Ti-B.
(◄:Titanium oxides as the nuclei of intragranular ferrite plate)

591

Properties of developed steel plates (HT50, 70, and 100)

The movement of offshore oil exploration into deeper waters and colder regions will continue, though it may be considerably affected by the economic situation. It is one of the missions of the steel industry to supply the required quantity of high-quality steel products for offshore development projects. In this chapter, the steel plates HT50, 70 and 100 developed to meet increasingly severe requirements are introduced below.

HT50

HT50 with excellent CTOD value at the HAZ (33)   As a result of the study on the factors affecting the toughness at the HAZ as described in previous section, it was found that the CTOD values are low in the following three zones:
a) the grain-coarsened zone in the vicinity of the fusion line, which is not subjected to tempering,
b) the grain-coarsened zone which is reheated to a temperature between Ac_1 and Ac_3 transformation temperatures by the succeeding weld beads,
c) the boundary between the HAZ and the base metal which is reheated to a temperature between Ac_1 and Ac_3 transformation temperatures.

It was also made clear that the zone (a) is closely related to the hardness and [C] content and that the zone (b) is attributed to the existence of high -carbon martensite islands (M*) and that the formation of martensite islands is promoted by elements such as Nb and V. Furthermore, it was clarified that the zone (c) disappears if the toughness of the base metal is improved and M* is reduced. A 50 kgf/mm² tensile strength steel plate for offshore structures which ensures high toughness at the welded joint has been developed by reducing the contents of C, Nb and V based on the results of study described above and offsetting the reduction in strength due to the reduction of C, Nb and V contents by TMCP. The performance of this steel plate is described below.

The chemical composition of this steel plate is shown in Table Ⅳ. It will be noted that the Nb content is reduced to less than 0.01% and Cu and Ni are added to improve the toughness of the base metal. The steel plate was produced in thicknesses of 50 and 100 mm by water-cooled type TMCP. The tensile strengths and Charpy impact values of these steel plates are shown in Table Ⅴ. The yield and tensile strengths of the plates of both thicknesses satisfy the ABS EH36-060 even after PWHT. The Charpy transition temperature is below -80°C, and consequently the toughness is high. The weldability is excellent. For example, the maximum hardness of 50 mm thick plate is 289, and the crack arrest   temperature in the JIS y-groove cracking test is 25°C. The tensile strengths of the joints submerged arc welded with a heat input of 50 kJ/cm using the Ti-B type welding material well satisfies the specification both in the as-welded state and after PWHT. The results of CTOD test with fusion line notch are shown in Fig. 16. The Charpy impact values are high and the CTOD values are higher than 0.5 mm at -10°C. All specimens after PWHT could withstand the maximum load. Such excellent weldability and high toughness at the HAZ are unattainable by conventional steel plates.

Table Ⅳ. Chemical composition of newly developed steel (33)

t (mm)	C	Si	Mn	P	S	Cu	Ni	Nb	Ti	Al	Ceq.
50	0.09	0.17	1.43	0.005	0.001	0.25	0.27	0.007	0.007	0.025	0.36
100	0.08	0.21	1.47	0.003	0.001	0.24	0.40	0.009	0.006	0.031	0.37

Table V. Tensile and Charpy Impact Properties (33)

Thickness (mm)	PWHT	Thick Location	Direction	Tensile Properties			Impact Properties	
				YP (kgf/mm²)	TS (kgf/mm²)	Eℓ (%)	vE₋₄₀ (kgf-m)	vTrs (°C)
50	–	1/4 t	L	–	–	–	26.8	-85
			T	42.6	53.2	30	24.9	-90
		1/2 t	L	–	–	–	26.2	-85
			T	41.9	52.0	29	22.5	-80
	575 ℃ × 2 hr	1/4 t	L	–	–	–	28.4	-80
			T	42.6	50.4	31	27.3	-80
		1/2 t	L	–	–	–	27.6	-85
			T	39.9	49.4	31	25.1	-80
100	–	1/4 t	L	–	–	–	30.8	-80
			T	40.4	51.4	34	30.1	-100
		1/2 t	L	–	–	–	31.1	-95
			T	37.6	48.9	33	29.6	-85
	575 ℃ × 4 hr	1/4 t	L	–	–	–	31.2	-80
			T	39.7	50.9	33	30.6	-90
		1/2 t	L	–	–	–	32.1	-80
			T	36.3	49.3	35	32.5	-80

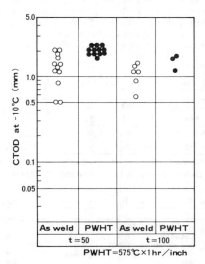

Figure 16 - CTOD test results with fusion line notch (33).

HT50 for low temperature use which can be welded with high heat input
(34) In the fabrication of large welded steel structures, high heat input welding is employed to enhance the welding efficiency. It is well known that the toughness at the welded joint is decreased by high heat input welding. The newly developed HT50 exhibits high toughness at the HAZ even if high heat input welding is employed, because this plate is produced by combining two new techniques, water-cooled type TMCP and IFP (intragranular ferrite plate) treatment utilizing trace elements. Not only the strength and toughness of the base metal but also the weldability and the toughness at the HAZ can be improved by accelerated cooling after controlled rolling. In other words, as the Ceq of 50 mm thick plate can be reduced to 0.30 ~ 0.35%, pre-

heating can be eliminated and the toughness can be improved. Moreover, very fine intragranular ferrite is formed in large quantities by IFP treatment during $\gamma-\alpha$ transformation in the welding heat cycle, as a result of which the toughness at the welded joint made by high heat input welding is improved. The nucleus in IFP is TiN-MnS-Fe$_{23}$(CB)$_6$. As the number of such nuclei is increased by adding small amounts of Ti and B at low N (= 0.0035%), the toughness at the HAZ of the newly developed HT50 satisfies the required Charpy value at -60°C even if high heat input welding is performed.

The chemical composition of this HT50 is shown in Table Ⅵ. It will be noted that this steel is produced by adding small amounts of Ti and B to Si-Mn steel and does not contain any expensive elements. The tensile and Charpy impact properties are shown in Table Ⅶ. The results of impact test of various types of welded joints made by high heat input welding and the dependence of these results on welding heat input are shown in Fig. 17. These results satisfy the target vE$_{-60}$ $\geq$ 3.5 kgf.m sufficiently. Namely, this HT50 plate in thicknesses up to 50 mm can be welded with high heat input and its performance as well as the performance of the welded joint satisfy ABS EH36-060. They have already found application in offshore structures, LPG tankers, etc., showing a great economic effect. Fig. 18 is an optical micrograph of a welded joint made by one-side one-pass welding with high heat input of 126 kJ/cm. The structure at the HAZ of low-N steel of the same composition with the exception of Ti and B is a mixture of polygonal ferrite and upper bainite, while that of the newly developed steel added with Ti and B is an acicular ferrite structure which is finely dispersed in the old austenite grain. Moreover, TiN-MnS-Fe$_{23}$(CB)$_6$ is observed as the nucleus of acicular ferrite.

Table VI. Chemical Compositions of Specimens (34)

Steel	Chemical composition (wt%)							Ceq*1	P$_{CM}$*2
	C	Si	Mn	P	S	N	Others		
A	0.08	0.26	1.38	0.007	0.001	0.0021	Ti-B-treated	0.32	0.158
B	0.09	0.26	1.45	0.007	0.002	0.0027	〃	0.34	0.171

*1 Ceq = C+Mn/6+Si/24+Ni/40+Cr/5+Mo/4+V/14

*2 P$_{CM}$ = C+Si/30+Mn/20+Cu/20+Ni/60+Cr/20+Mo/15+V/10+5B

Table VII. Mechanical Properties of Specimens (34)

Steel	Thickness (mm)	Specimen	Tensile tests					Charpy impact tests		
			Location and direction of specimen	0.2%P.S (kgf/mm²)	T.S (kgf/mm²)	Eℓ (%)	RA (%)	Location and direction of specimen	50% FATT (℃)	Absorbed energy at-60 ℃ (kgf-m)
A	25	Round-type specimen (JIS Z 2201 No. 4)	L	36.7	51.3	34	81	1/4 t	L - 90	Ave Min 24.1/19.8
			T	37.4	52.2	36	82		T - 85	21.4/20.3
	32	〃	1/4 t — L	38.2	52.2	35	77	1/4 t	L - 90	26.8/24.4
			1/4 t — T	39.6	53.5	32	76		T - 80	22.9/21.0
			1/2 t — L	39.2	52.7	36	75	1/2 t	L - 90	22.6/20.8
			1/2 t — T	39.9	53.1	35	75		T - 85	20.7/18.9
B	25	Flat-type specimen (JIS Z 2201 No. 1B)	Full thickness — T	42.2	55.2	23	—	1/4 t	T -100	24.3/24.0
	32	〃	〃	41.0	52.0	23	—	〃	〃 -100	25.9/25.5
	40	〃	〃	40.7	54.2	29	—	〃	〃 - 90	25.1/25.0

L : Longitudinal   T : Transverse

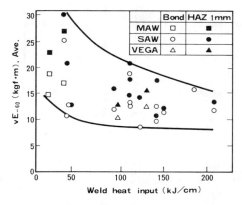

Figure 17 - Relationship between weld heat input and V-Charpy absorbed energy at -60°C for welded joints of CLC steel (34).

Fusion line

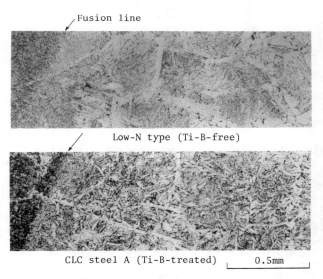

Low-N type (Ti-B-free)

CLC steel A (Ti-B-treated)   0.5mm

Figure 18 - Comparison of microstructures between CLC steel A and low-N steel in welded joints by one-side one-pass SAW with 126 KJ/cm (25mmt) (34).

HT70 (35)

HT70 has been developed for tendons of tension leg platforms (TLP) which are considered to be a highly promising means for oil exploration in deeper waters. To reduce the manufacturing cost, the tendon is fabricated by connecting the threaded forged connectors by girth welding to both ends of the pipe manufactured by bending plate into cylindrical shape and welding. Each length of tendon is 12 m in overall length, 500 mm in outside diameter at the straight portion, and 25.4 mm in wall thickness. Its proof stress is higher than 56 kgf/mm^2, tensile strength higher than 67 kgf/mm^2, and Charpy impact

value vE-20 higher than 6.9 kgf·m. To increase the reliability of the joint made by girth welding, the differential thickness pipe with increased thickness at the both ends is used. Furthermore, three kinds of heat treatment are applied to improve the toughness of the seam and girth welded joints (Fig. 19). For the development of such tendon, the following were indispensable:
a) a process for manufacture of differential thickness pipe,
b) a method for uniform quenching of steel pipe with differential wall thickness,
c) development of steel product and flux whose properties are kept stable even if the heating temperature and cooling rate are varied to some extent.

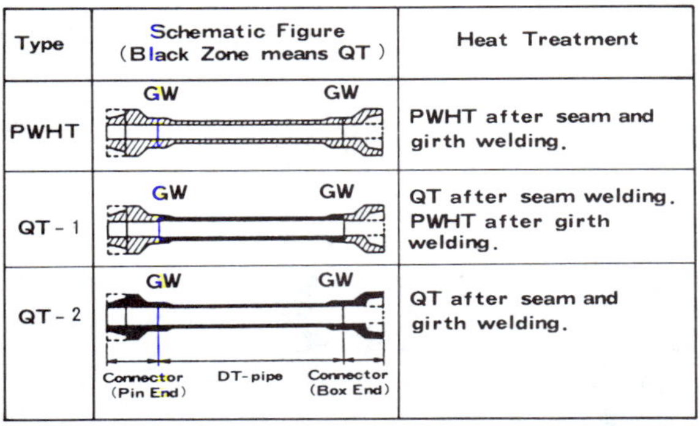

Type	Schematic Figure (Black Zone means QT)	Heat Treatment
PWHT	GW GW	PWHT after seam and girth welding.
QT-1	GW GW	QT after seam welding. PWHT after girth welding.
QT-2	GW GW Connector (Pin End) DT-pipe Connector (Box End)	QT after seam and girth welding.

Figure 19 - Tendon fabrication methods by three types of heat treatment (35).

In this section, the development of steel product and flux mentioned in (c) above is described in detail. As is well known, high-tension steel is produced by utilizing the high strength and high toughness of the tempered martensite and tempered lower bainite structures. It is, therefore, important to secure high hardenability. To improve the hardenability without decreasing the weldability, boron which enhances the hardenability remarkably is added in a very small amount. As a means for maximum utilization of the effect of boron, the following two techniques have been developed to secure the solid solution B.
a) To fix [N] which has a deleterious effect, Aℓ is added in a quantity twice that normally added.
b) The slab reheating temperature before rolling is limited to less than 1,100°C at which the AℓN precipitated during reheating is not dissolved.
If the slabs are slowly cooled in the cooling process, however, the coarse AℓN is precipitated in a reticulated manner in the austenite grain boundaries in the slab. If the temperature in succeeding reheating is low and the coarse AℓN is not dissolved as shown in Fig. 20, the toughness is drastically decreased. To obtain both excellent hardenability and high toughness, a slab water cooling - low temperature reheating process has been developed. This is called the CAP (Control of Aluminum Precipitate) treatment (36-38). This steel can be well quenched by the CAP treatment over a wide range of cooling rates. In particular, when the steel is rapidly reheated, the strength and toughness are stabilized at high levels (Fig. 21). Thus, the induction heating of differential thickness pipe has been made possible.

The aimed chemical composition and manufacturing sequence of this steel is given in Table VIII. This steel dose not have added with Nb and V, to avoid

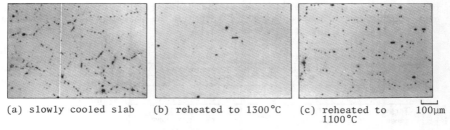

(a) slowly cooled slab   (b) reheated to 1300°C   (c) reheated to   100μm
                                                         1100°C

Figure 20 - Photomicrographs showing precipitates in slowly cooled
and reheated slabs (38).

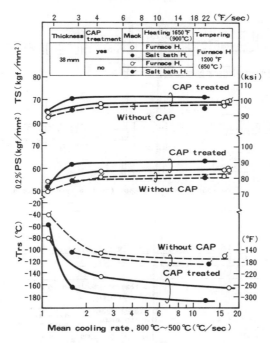

Figure 21 - Effect of CAP treatment and cooling
rate on mechanical properties (35).

SR embrittlement. The differential thickness plate is produced by the re-
versing rolling process with hydraulic AGC (Automatic Gauge Control). This
plate is then bent into cylindrical shape and seam-welded. The pipe thus
produced is quenched after induction heating and tempered in an ordinary
furnace. The mechanical properties of the pipe are uniform throughout its
entire length as shown in Table IX. The toughness at the welded joint is
excellent as shown in Fig. 22. Fig. 23 shows the fatigue properties of the
welded joint in synthetic seawater. As the properties of the welded joint
are nearly identical to those of the base metal, it may be said that welded
steel pipe type tendon with high reliability can be produced. With regard to
the welding materials, the wire composition was designed and the low-oxygen
SAW flux which does not cause weld defects was developed on the assumption

597

that quenching and tempering are carried out after welding. As a result, high toughness could be obtained by reducing the oxygen content of the weld metal to less than 200 ppm.

Table VIII. Chemical Compositions of a Ingot Cast Produced by LD Convertor (35)

	C	Si	Mn	P	S	Ni	Cr	Mo	Aℓ	B	Ceq. (%)	D$_I$ (in)
Target range	0.10 / 0.12	0.20 / 0.30	0.90 / 1.00	≦0.010	≦0.005	1.30 / 1.50	0.40 / 0.50	0.28 / 0.32	0.060 / 0.080	~14	Av. 0.48	Av. 3.26
Ladle analysis	0.116	0.28	0.97	0.006	0.002	1.41	0.46	0.30	0.076	0.0012	0.49	3.25

Ceq. D$_I$ ······ Refer to Table 2

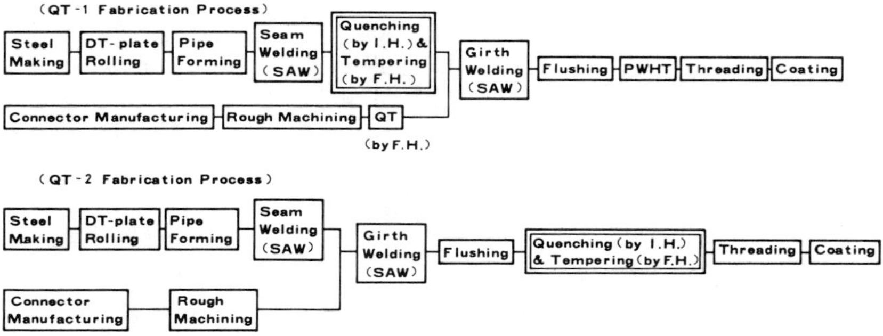

(QT-1 Fabrication Process)

(QT-2 Fabrication Process)

The fabrication processes for two types of tendon.

Table IX. Mechanical Properties of The Real Size DT Pipe after QT (35)

Tensile test specimen A 370	Direction	Distance from the top end			Distance from the bottom end
		100	190	300	100
		Thick zone	Tapered zone	Thin zone	Thick zone
P. S	L	57.8	—	59.9	60.8
(kgf/mm²)	C	58.5	60.2	60.1	59.7
T. S	L	67.7	—	69.3	70.1
(kgf/mm²)	C	66.9	68.1	68.3	69.4
Eℓ (%)	L	30.0	—	32.0	29.0
	C	28.0	24.0	29.0	32.0
vE$_{-20℃}$(kgf-m)	L	25.5	28.3	28.5	28.5
	C	27.6	—	28.0	28.3
vTrs (℃)	L	-115	-126	-122	-92.2
	C	-115	—	-108	-88.8

Position of specimen：¾ t,　Distance：mm

L ······ Longitudinal direction
C ······ Transverse direction

598

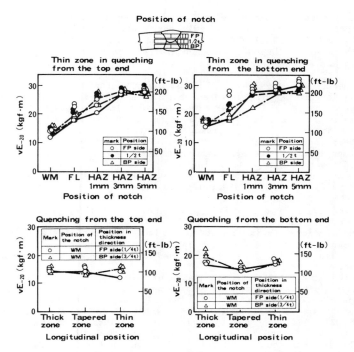

Figure 22 - Toughness in the weld metal and HAZ of a real size DT pipe for tendon after Q and T (35).

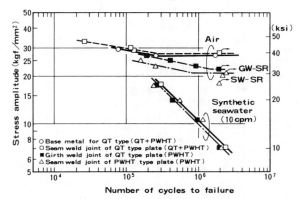

Figure 23 - Comparison of fatigue strength in the welded joint, mean stress = 50 ksi (35kgf/mm²) (35).

HT100 (39)

Of the many types of offshore structures for oil drilling, the jack-up rigs have found wide application. As the operation of these rigs has moved into deeper and colder waters, their sizes have been considerably enlarged. With such changes, the conventional HT80 steel for the rig chords and racks has been replaced by steel with higher strength, such as HT90 and HT100. This section introduces the newly developed HT100 of up to 4 inches in thickness, which has weldability identical to that of HT80, maintains the aimed strength and toughness throughout the thickness and exhibits excellent brittle crack arrest capability.

To secure the same level of weldability as that of HT80, the amounts of alloying elements are minimized so that Ceq and Pcm identical to those of HT 80 can be obtained, and the CAP treatment in which the effect of boron on the improvement of hardenability as described above is utilized to the utmost limit is employed. It is expected that if the steel plate is quenched in this manner, a structure consisting mainly of martensite will be formed at the surface layer of the plate, but the structure in the middle will consist mainly of lower bainite. Accordingly, the effects of γ grains and (Ni) content on the strength and toughness of these structures were investigated in the laboratory (Fig. 24). The following will be noted:
a) The vTrs of both the tempered martensite (TM) and the tempered lower bainite (TB$_L$) depend upon γ grain size, but the dependence of TM is greater than that of TB$_L$.
b) If the effective grain size d_e is defined in terms of the spacing between the tear lines on the cleavage brittle fracture surface, the relation between vTrs and $d_e^{-\frac{1}{2}}$ can be expressed by a straight line, regardless of whether the structure is TM or TB$_L$.
c) When the (Ni) content is in the range of 1 to 3%, little effect of (Ni) content on the toughness is observed in all structures.
d) The proof stress (0.2% PS) is dependent upon the γ grain size. Namely, the proof stress increases as the γ grain becomes finer, but the tensile strength (TS) is kept nearly constant, regardless of the grain size.
e) Both the 0.2%PS and TS increase with increasing (Ni) content. In the case of the TB$_L$ structure, however, the strength level decreases. At 1%Ni, both the 0.2%PS and TS become insufficient.

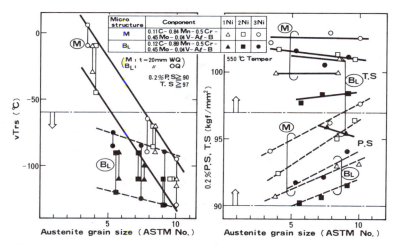

Figure 24 - The effect of austenite grain size and (Ni) contents on the mechanical properties of HT100 steel (39).

To produce HT100 without increasing the contents of alloying elements from those of HT80, it is essential to refine the γ grain to more than No.8 at the surface layer where the structure becomes martensite. At the middle in the thickness direction, however, the chemical composition should be designed so that the mixed structure of lower bainite and martensite can be obtained. Based on the results of study as described above, the CR–DQT (Controlled Rolling – Direct Quenching & Tempering) process has been developed, in which the uniform effective grain size is obtained from the surface layer up to the middle in the thickness direction by controlling the γ grain size and structure in the thickness direction. HT100 plates of 50 and 100 mm in thickness were produced by this process.

The chemical compositions and manufacturing process of HT100 are shown in Table X. It will be noted that the chemical composition of 50 mm thick plate is nearly the same as that of HT80. The mechanical properties are given in Tables XI and XII. The HT100 plates of both thicknesses sufficiently satisfy the aimed properties. Fig. 25 shows the γ grain size and brittle fracture surface of 50 mm thick plate. It will be noted that the effective grains in the plate produced by the CR–DQT process are considerably refined. The brittle fracture arrest capability of this plate is excellent. With regard to weldability, the maximum hardness and the results of JIS y groove cracking test are nearly the same as those of HT80 as shown in Figs. 26 and 27. The toughness of welded joint is also excellent (Fig. 28).

Table X.   Chemical Composition of a New HT100 Steel (38)

Thickness (mm)	C	Si	Mn	P	S	Cu	Ni	Cr	Mo	V	Nb	Aℓ	B	Ceq.	Pcm	D$_{I\,cal}$
50	0.11	0.25	0.91	0.003	0.0010	0.25	1.04	0.56	0.45	0.050	—	0.056	0.0016	0.526	0.263	5.34
100˙	0.11	0.22	0.90	0.004	0.0006	0.23	2.32	0.58	0.53	0.043	0.012	0.077	0.0013	0.579	0.288	7.94

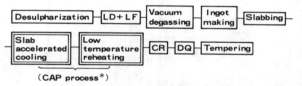

(CAP process*)

∗ CAP process ; Control of Alumiunm Precipitates

Production process of new HT100 steel.

Table XI.   Mechanical Properties of HT100 Steel (39)

Thick-ness (mm)	Treatment (mm)	Location & Direction		Tensile Test				Charpy Impact Test		Austenite grain size (ASTM No.)
				0.2%P.S (kgf/mm²)	T.S (kgf/mm²)	Eℓ (%)	RA (%)	vTrs (℃)	vE₋₅₅ (kgf-m)	
50 (1.0% Ni)	DQ-T	7mm from Surface	L	92.9	99.5	23	67	-84	17.5	8.7
			T	94.2	100.6	22	68	-75	9.3	
		1/4 t	L	92.8	99.5	23	68	-96	16.5	8.7
			T	94.5	100.7	22	68	-91	17.6	
		1/2 t	L	90.5	97.3	23	67	-76	12.5	8.3
			T	90.7	98.1	22	68	-78	12.0	
	Reheated Q-T	Surface	T	91.0	98.7	22	68	-40	5.1	7.6
		1/4 t	T	90.8	98.6	22	68	-70	13.7	7.6
		1/2 t	T	90.2	98.1	23	67	-80	15.9	7.6
	Aim			≧ 90	97/115	—	—	≦-55	≧4.8	—

Table XII. Mechanical Properties of HT100 Steel (39)

Thick-ness (mm)	Treatment (mm)	Location & Direction		Tensile Test				Charpy Impact Test		Austenite grain size (ASTM No.)
				0.2%P.S (kgf/mm²)	T.S (kgf/mm²)	Eℓ (%)	RA (%)	vTrs (°C)	vE₋₅₅ (kgf-m)	
100 (2.3%Ni)	DQT	7mm from Surface	L	104.2	107.0	23	68	−85	17.7	9.5
			T	99.3	104.4	24	67	−70	10.7	
		1/4 t	L	102.1	105.6	23	67	−102	17.7	9.2
			T	97.4	103.7	23	67	−98	18.9	
		1/2 t	L	92.2	98.5	22	64	−68	8.0	8.7
			T	94.3	101.1	24	65	−64	7.8	
	Reheated QT	7mm from Surface	T	92.2	99.0	23	67	−50	6.5	8.4
		1/4 t	T	92.2	99.0	22	64	−70	15.1	8.5
		1/2 t	T	92.3	98.0	22	63	−55	7.9	8.3
100 (3%Ni)	Reheated QT [Multiple Quenching]	7mm from Surface	L	94.0	99.3	22	68	−62	11.7	8.5
			T	93.9	98.3	24	70	−64	12.8	
		1/4 t	L	93.6	98.5	23	68	−80	18.7	9.0
			T	93.7	98.6	24	69	−80	20.1	
		1/2 t	L	93.6	98.7	23	68	−89	19.7	9.0
			T	93.1	97.9	24	69	−90	20.2	
	Aim			≧90	97～115	−	−	≦−60	≧4.8	−

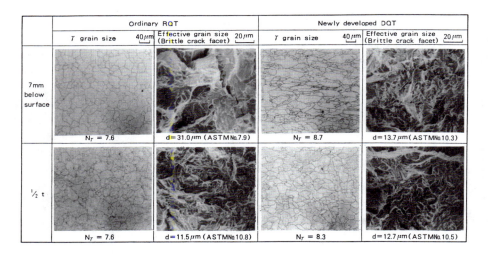

	Ordinary RQT		Newly developed DQT	
	ɤ grain size 40μm	Effective grain size 20μm (Brittle crack facet)	ɤ grain size 40μm	Effective grain size 20μm (Brittle crack facet)
7mm below surface	Nɤ = 7.6	d = 31.0μm (ASTM No.7.9)	Nɤ = 8.7	d = 13.7μm (ASTM No.10.3)
1/2 t	Nɤ = 7.6	d = 11.5μm (ASTM No.10.8)	Nɤ = 8.3	d = 12.7μm (ASTM No.10.5)

Figure 25 - Comparison of ɤ and effective grain size between conventional RQT and newly developed DQT steel (1%Ni, 50mm thick) (39).

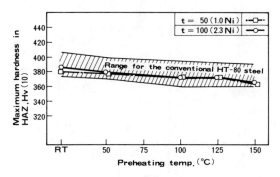

Figure 26 - Maximum hardness in the heat-affected zone (39).

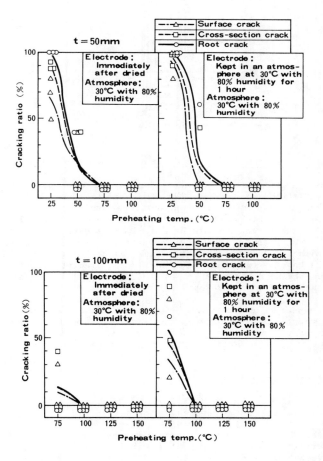

Figure 27 - Results of y-groove restraint weld cracking test (39).

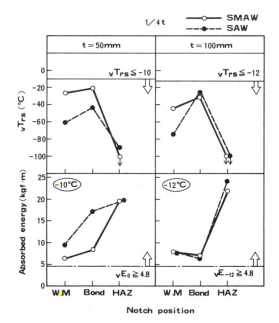

Figure 28 - Charpy test results of welded joints (HI=38 KF/cm) (39).

Development of steel manufacturing technology

As the construction of offshore structures has moved into colder regions, the quality requirements for steel plates have become increasingly severe, necessitating multiple high purification which can meet all of the requirements for ultra-low phosphorus, ultra-low sulphur, low [N], low [O] and low [H]. These requirements have led to the development of new refining techniques and systems. With the enlargement of offshore structures and their installation in deeper waters, steel plates of greater thicknesses and larger sizes have become necessary and the requirements for the soundness of internal quality have become stringent. To meet such requirements, new casting techniques have been developed.

This chapter describes advances in refining and casting techniques and what development has been made in the plate rolling technology to meet the development of thermo-mechanical control process where controlled cooling is done so as to provide products with the designed qualities through rolling alone.

Advances in refining process

To attain multiple high purification of molten steel for plate in the BF-BOF-CC or ingotmaking processes, a system in which high-purity steel is produced on a stable basis by dividing the refining functions heretofore performed in the BOF into several processes has been established.
a) Desulphurizing techniques: The hot metal tapped from the blast furnace normally contains about 0.020% sulphur. When desulphurization is performed in the BOF only, the sulphur level cannot be reduced below 0.010%. However, the sulphur level of the molten steel can be reduced to less than

0.005% by performing preliminary desulphurization of the hot metal. If the molten steel tapped from the BOF is further desulphurized, plates with [S] ≤ 0.003% or [S] ≤ 0.001% can be produced on a regular basis.

b) Dephosphorizing techniques: Normally, the hot metal tapped from the BF contains about 0.100% phosphorus. The phosphorus level of the molten steel is about 0.020% when dephosphorization is performed in the BOF only. This level can be reduced to less than 0.010% by preliminary dephosphorization of the hot metal. If the molten steel tapped from the BOF is further dephosphorized, plates with [P] ≤ 0.005 or [P] ≤ 0.003% can be produced on a stable basis.

Pretreatment of hot metal

a) Torpedo car desulphurization (TDS)
A desulphurizing flux (powder), such as CaC_2, is injected by nitrogen gas into the hot metal in the torpedo car. At present, CaO is also used as a desulphurizing flux, making a contribution to cost reduction.

b) Preliminary dephosphorization of hot metal (ORP) (40)
Dephosphorization is carried out by blowing gaseous oxygen into the hot metal in the torpedo car or ladle and adding an oxide (scale). However, as oxygen is more likely to react with Si in the hot metal, obstructing dephosphorization, it is necessary to perform dephosphorization after desiliconization.
During dephosphorization, CaO and CaF_2 are added so that the P_2O_5 formed by oxidation reactions is stabilized in the slag.

Advances in BOF refining

Multiple oxygen blowing process (LD-OB process)
The multiple oxygen blowing process (LD-OB process) was developed by combining the advantages of the bottom blowing process and the basic oxygen furnace process established as a modern steelmaking process. This process was put into practical operation in the beginning of the 1980s.
In the LD-OB process, 5 ~ 10% of the total blowing oxygen is blown through the duplex nozzle in the bottom to improve and stabilize the efficiency of reaction by stirring the molten steel.

Advances in secondary refining

a) Techniques for desulphurization of molten steel
As sulphur levels cannot be reduced below 0.003% or 0.001% by refining in the BOF alone, desulphurization of molten steel is carried out. For this purpose, the injection method in which a desulphurizing agent is injected into the molten steel, and the ladle refining method using the LF are employed. The former method is suitable for rapid mass treatment and is capable of achieving [S] ≤ 0.002%, while the latter method is suited for the production of ultra-clean steel with [S] ≤ 0.001%

b) Techniques for dephosphorization of molten steel
To reduce the phosphorus level of plate to less than 0.005% or 0.003%, the dephosphorization of molten steel is necessary. For this purpose, several methods, such as tapping dephosphorization method and dephosphorization by LF, have been developed. They are used depending on the purpose.

c) Ca treatment techniques
For prevention of lamellar tear, Ca treatment is performed in addition to extreme reduction in S so as to control the shape of MnS. Two methods have been developed: Ca wire method and injection method. They are used depending on the purpose.
Examples of the combinations of processes from pretreatment of hot metal to secondary refining and the levels of multiple high purification to be achieved by these combinations are shown in Table XIII .

Advances in casting process

As the thickness of continuously cast (CC) slabs have been increased from 200 ~ 250 mm to 300 mm, the production of steel plates up to 35 tons in unit weight from CC slabs has been made possible. For the plates exceeding

Process	Hot metal pre-treatment		BOF [LD-PB]	VSC	Secondary refining			Chemical composition [ppm]				
	TDS	ORP			IP	LF	RH	[P]	[S]	[H]	[O]	[N]
Equipment	CaC_2+N_2	Flux+N_2, O_2	O_2	→Vac	Flux+Ar	Flux	Alloy →Vac					
Function	De-S	De-Si De-P,S	De-C De-P	Deslagging	De-S	De-O De-S Heating	De-H Comp. Adj.					
Treatment pattern I	○	○	○	○	—	○ De-S	○	50	10	1.5	15	30
Treatment pattern II	○	○	○	○	○	—	○	50	10	1.5	—	30
Treatment pattern III	○	○	○	○	—	○ De-P De-S	○	30	10	1.5	15	30

the size range from CC slabs, production by the conventional ingot casting method has been established, making it possible to produce plates of 65 tons max. from 100-ton ingots.

Reducing center segregation of CC slabs   It is considered that the main cause of center segregation is the movement of the unsolidified molten steel in the solidification end zone. The causes of movement of such residual steel are the draft of unsolidified molten steel toward the solidification end due to the solidification contraction and the bulging of the cast slab. The degree of movement of unsolidified molten steel is greatly affected by the solidification structure. If equiaxed dendrites are present in large quantities in the final zone of solidification, the movement of unsolidified molten steel is reduced. Electromagnetic stirring is effective for increasing the quantity of equiaxed dendrites. This stirring significantly reduces the center segregation.

Measures for reduction in inclusions   The main cause of inclusions present in the CC slab is the entrapment of CC powder which is caused mainly by meniscus fluctuations and the entrapment of CC powder by the molten steel flow in the mold.

To prevent such meniscus inclusions and powder entrapment, the mensicus level sensor in the mold and the level control system are employed. Moreover, various techniques, such as preventing powder entrapment by reducing the velocity of molten steel flow using the mold with electromagnetic brake, are under development. Furthermore, it provides a vertical part at the continuous caster for floating separation of inclusions in the caster. This method has already been applied to some casters.

Plate rolling process (including controlled cooling)

Nowadays, the plate rolling process is required to ensure in-line quality by controlled rolling and controlled cooling in addition to the conventional role of rolling plates of designed size and flatness with high efficiency (high yield). This section outlines what development has been made to the plate rolling process to fulfill the conventional role of rolling plates of exact size and flatness and to meet recent requirements for in-line quality assurance, including controlled cooling (41).

Rolling to required size and flatness (42)   When discussing the modern rolling technology, we cannot omit the rolling process control in which the hydraulic AGC is combined with computer control. Table XIV shows the functions of the rolling process control system for 5,500 mm plate mill at Nippon Steel's Oita works.

Table XIV. Summary of Control Functions

	CONTENTS	
AGC	ABS.AGC	to realize the thickness given by computer
	BISRA AGC	traditional type of AGC
	DUAL AGC	to realize the tickness and the wedge given by computer
	FF AGC	to realize the roll gap of FF given by computer
	Super AGC	to realize the roll gap of S-AGC given by computer
	ACC	to realize the roll gap slant of ACC given by computer
	DAT	to realize the roll gap of DAT given by computer
AGC auxiliary functions	mill modulus measuring	mill modulus measuring
	oil contraction compensation	to compensate oil contraction at the entrance of plates
	oil film compensation	to compensate bearing oil film thickness in accordiance with rolling speed
	mill modulus compensation	to compensate mill modulus by plate width and B.U.R. dia. in case of computer down
	roll eccentricity ajustment	to control the roll eccentricity by reference from the roll eccentricity control module
other functions	taperd plates rolling	
	position control zeroing	

The functions given in Table XIV are described in more detail below.

Absolute AGC ... While BISRA AGC is a control method for minimizing thickness deviations in the longitudinal direction, the absolute AGC is a method for control of thickness deviations and securing the aimed thickness in conjunction with a computer. Thickness accuracy has been improved from $1\sigma = 0.12 \sim 0.13$ mm to $0.07 \sim 0.08$ mm by the introduction of this control system.

Dual AGC ... This AGC is designed to improve plate wedge and camber by controlling the screwdown mechanisms at both sides of the rolling mill individually. This AGC system has improved the deviations σ of plate camber from 26 mm to 20 mm.

Super AGC ... This is a kind of feed forward AGC. Super AGC is designed to control thickness deviations so that deviations due to the difference in deformation resistance are not caused after the final pass by converting the difference in deformation resistance due to temperature deviations in the longitudinal direction into thickness deviations in advance.

DAT (Draft Alternation Before Turning) ... As shown in Fig. 29, thickness deviations are given to the slab by the hydraulic AGC system before broadsiding to improve the flat profile after rolling and enhance the plate yield. The yield has been improved by $1.5 \sim 2.0\%$ by this system.

NSC has four plate mills, each having hydraulic AGC.

Controlled rolling. Controlled rolling is defined as a method for improving the mechanical properties of steel through grain refinement by controlling the slab reheating temperature, rolling temperature and rolling

607

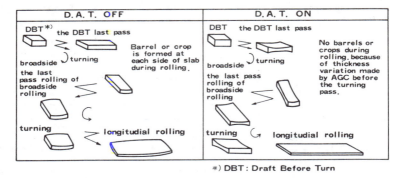

D. A. T. OFF	D. A. T. ON

*) DBT: Draft Before Turn

Figure 29 - Summary of D.A.T. control.

reduction as appropriate. Requirements for rolling have become increasingly severe, including comparatively lower reheating temperature and closer reheating temperature control at the reheating furnace and low-temperature heavy reduction rolling and closer rolling temperature control at the rolling mill in addition to conventional functions.

For the reheating furnace, various measures have been taken, including the adoption of burners which are capable of adjusting the turndown ratio up to 1/10 to permit reheating at a comparatively lower temperature, and the use of computer for reheating combustion control. In the rolling mill, cracks occurred in the screw housings in the past. Accordingly, the mills at NSC have been improved to the most advanced types by improving the portions of the mill where stress is concentrated, and replacing the mill stands in extreme cases. Needless to say, the rolling process has been computerized as described previously.

Controlled cooling (42)   When designing the cooling equipment, due consideration should be paid to the following points:
a) The given process conditions should be satisfied.
b) Uniformity, controllability and reliability of cooling should be ensured.
c) The required flatness of plate should be obtained readily.
d) The construction and running costs should be low.
e) The productivity of the mill should not be adversely affected.

In due consideration of the requirements described above, the closed type cooling equipment which is installed on the rear side of the hot leveller for continuous cooling and can also serve for direct quenching was adopted as NSC's standard type cooling equipment.

As the cooling equipment is installed directly in the rolling line and performs cooling mainly after controlled rolling, the function for controlling the flow rate of cooling water to the edge, front and tail parts was introduced to minimize the effect of uneven cooling of the four sides of the plate during rolling and cooling and to reduce losses due to poor quality and flatness. Control with high accuracy has been realized by extensive application of computers for control of cooling and temperature.

The quantities of plates produced using this cooling equipment and shipped from NSC during the period from the first shipment in February 1983 to the end of September 1986 are shown in Table XV.

608

Table XV. Steel Plates Produced by CLC Process

(from Feb. 1983 to Sep. 1986)

Application	Steel Plates ($\times 10^3$ t)
for Shipbuilding	373.4
for Offshore Structures	104.8
for Line Pipe Stocks	347.4
for Earth movers	24.5
for Others	8.3
Total	858.4

Computerized integrated plate production control system

The production of plates, with the exception of some grades, such as the plate for UO pipe, is small-lot production in many points, such as uses, steel grade, specification and size, and order-oriented production in principle. In most cases, orders for the plates used for a certain project are not repeated. Accordingly, the computerized integrated production control system is indispensable to carry out a series of production cycles, such as grouping of orders received into proper rolling lot in consideration of delivery, rolling in accordance with the specified design conditions (alloy and process), collection and arrangement of production data to determine whether the required properties and delivery have been achieved, and the reflection of the results of production into the next cycle. NSC introduced the EDPS for production control more than twenty years ago. At Kimitsu Plate Mill, the AOL (All-On-Line) system, covering all operations up to the manufacturing process, was introduced during the construction of the mill (in 1968). At Oita Plate Mill which was put into operation in 1977, the system for controlling all operations from reheating and rolling to refining and warehousing was created (43). In a sense, this mill is a fully computerized mill.

These systems have made a great contribution to the improvement of plate quality, cost reduction and delivery.

Subjects for future research and development in connection with heavy plates for offshore structures and related engineering

As the offshore structures for oil and gas production are operated in severe natural environments, the most reliable and economical structures must be constructed in consideration of the operating conditions, production amount, production term, etc. Needless to say, the reliability of these structures is dependent upon optimum design, the selection of suitable materials and excellent workmanship. Development of new materials leads to the design of new types of structures and the adoption of highly effective, stable construction methods. This chapter describes the subjects for future research and development in connection with the heavy plates for structures and related fracture mechanics and welding engineering.

Heavy plates

Recent trend is toward movement of offshore energy development into deeper waters and colder regions. Moreover, the structures have become very large and must be operated under greater loads and at very low temperatures. For heavy plates as the main material of offshore structures, therefore, the development of "steel plates with greater thickness, higher strength and toughness and more excellent weldability" is essential. The aimed strength

609

is 100 kg/mm^2 in terms of tensile strength, and the design temperature is -60 °C. The aimed thickness is about 200 mm. For welding, preheating temperature and heat input are of prime importance.

For the development of such heavy plates, fundamental studies should be continued to establish the following:
a) the relation between the factors affecting the strength and toughness of heavy plate and metallurgical factors,
b) the relation between the factors governing the toughness at the HAZ and metallurgical factors,
c) the factors affecting weld cracking.

Fracture mechanics

Recent progress in fracture mechanics has made a great contribution to the improvement of safety of offshore structures. The results of research in various modes of fracture contribute to the optimum design and operation of structures and provide a basis for selection of optimum materials.

The subjects to be studied in relation to major fracture modes are given below.
Ductile fracture
a) Analysis of buckling under special conditions
Fatigue fracture
a) Effect of plate thickness on crack initiation
b) Quantitative determination of crack propagation and reflection of such determination into design standards
c) Effect of random load
d) Elucidation of corrosion fatigue mechanism and data gathering
Brittle fracture
a) Effect of dynamic load
b) Elucidation of the significance of crack propagation arrest
c) Establishment of reasonable criteria for prevention of crack initiation and propagation

Moreover, studies on HIC, HSCC, SCC, etc. are also very important depending on the environmental and loading conditions to be encountered.

For all modes of fracture described above, the exact prediction of operating conditions of structures to be encountered at site is very important. If structures are designed on the basis of the worst environmental conditions, they will become very uneconomical. If the subjects described above are fully studied and the stochastically proper design is made based on the results of such study, more rational, economical design and material selection will be made possible.

Welding methods and materials

From the standpoints of economy and reliability, fusion welding will continue to be the most suitable method for joining the members of offshore structures. From such standpoints, the subjects to be studied in the future are enumerated below.
Welding methods
a) Narrow-gap welding method with high efficiency and high reliability
b) Welding method with large heat input to ensure high toughness
c) Fully automatic welding method
d) Repair welding method (including under water welding)
e) Application of high-density energy
Welding materials
a) Welding material with high strength and toughness
b) Material for welding with large heat input to ensure high toughness

Prevention of weld cracking

a) Establishment of more rational equation for determination of preheating temperature

The development of simulation model for setting up proper welding conditions at which high productivity and reliability can be ensured is now underway. If this model and data base are completed, real-time welding design will be made possible among researchers, designers and fabricators through the international communication system.

Steel structures are advantageous in many points, including quality assurance and construction term. The quality and economical competitiveness of steel structures must be improved through technological improvements which may enable optimum design and the use of even better materials and construction methods. To achieve the purposes described above, technological development in respective engineering fields and close cooperation among the engineers in the related fields are indispensable.

This symposium has provided a good opportunity to discuss various problems relating to offshore structures and played a significant role in this connection. We put a high appraisal on this symposium and would like to express our sincere gratitude.

References

1. K. Itoh, "The Trend of Development of Steels Used in the Arctic Ocean Field" (Proceedings of the 1st Spilhaus Symposium Arctic Ocean Engineering for the 21st Century. Williamsburg, U.S.A. 14-27 October 1984 Marine Technology Society), 219-234.
2. K. Itoh, P. Jumppanen, W. Sackinger and S. Gowda "Development of New Advanced Materials for Sub-Arctic and Arctic Offshore Structure" (Proceedings of the Int. Offshore and Navigation Conf. and Exhibition, Helsinki, Finland, 27-29 October 1986), 833-868.
3. H. Sekine, "Grain Refinement Through Hot Rolling and Cooling after Rolling" (Proceeding of the Int. Conf. on Thermomechanical Processing of Microalloyed Austenite, Pittsburgh Pennsylvania 17-19 August 1981), 141-161.
4. T. Tanaka, N. Tabata, T. Hatomura and C. Shiga, "Three Stages of the Controlled Process" (Proceedings of an Int. Symposium on HSLA Steels, Washington D.C., 1-3 October 1975), 88-99.
5. I. Kozasu, C. Ouchi, T. Sampei and T. Okita, "Hot Rolling as a High-Temperature Thermo-Mechanical Process", ibid., 100-114.
6. R.A.P. Djaic and J.J. Jonas, "Static Recrystallization of Austenite between Intervals of Hot Working", JISI, 210 (1972), 256-261.
7. C. Ouchi, T. Okita, T. Ishikawa and Y. Ueno, "Hot Deformation Strength of Austenite during Controlled Rolling in a Plate Mill", Trans. ISIJ, 20 (1980) 833-841.
8. A Yoshie, H. Morikawa, Y. Onoe and K. Itoh, "Estimation Model of Austenite Recrystallization Due to Plate Rolling and Its Application to Controlled Rolling" (Paper presented at the TMS-AIME Fall Meeting, Orlando, U.S.A. 5-9 Oct. 1986)
9. T. Okita, C. Ouchi and I. Kozasu, "The Effect of Accelerated Cooling after Controlled Rolling", Tetsu-to-Hagane, 63 (1977) S798.
10. S. Tamukai, Y. Onoe, H. Nakajima, M. Umeno, T. Iwanaga and S. Sasaji, "Production of Extremely Low Carbon Equivalent HT-50", Tetsu-to-Hagane, 67 (1981) S1334.
11. C. Ouchi, T. Okita and S. Yamamoto, "The Effect of Interrupted Accelerated Cooling after Controlled Rolling on Mechanical Properties of Steels", Tetsu-to-Hagane, 67 (1981) 969-978.

12. T. Hashimoto, H. Ohtani, H. Nakagawa, K. Bessho, S. Suzuki and M. Nakamura, "The Mechanical Properties of Controlled Cooled Steel and Its Application", Tetsu-to-Hagane, 68 (1982) A211-A214.
13. T. Utori and M. Ogawa, "Strength and Ductility in a Mixed Ferrite and Martensite Steel Produced by Controlled and Accelerate-cooling-Process", Tetsu-to-Hagane, 68 (1982) A219-A222.
14. M. Machida, S. Kawata, S. Katsumata, H. Kaji and N. Akiyama, "Mechanical Properties and Microstructure of Controlled Cooled Steels", Tetsu-to-Hagane, 68 (1982) A223-A226.
15. C. Shiga, T. Hatomura, K. Amano and T. Enami, "Effect of Cooling Rate and Finishing-cooling Temperature on Properties of the Plates Controlled-rolled and Accelerated-cooled", Tetsu-to-Hagane, 68 (1982) A227-A230.
16. Y. Onoe, H. Morikawa, Y. Sogo and K. Iwanaga, "Effect of Controlled Rolling and Controlled Cooling on the Metallurgical Properties of Plate Steel", Tetsu-to-Hagane, 68 (1982) A231-A234.
17. K. Uchino, Y. Ohno and T. Fujii, "Development of Y.P. 36kgf/mm^2 HT50 for Low Temperature Service under Large Heat Input Welding Condition by Continuous on Line Control Process", Tetsu-to-Hagane, 68 (1982) S1440.
18. M. Nakanishi, S. Watanabe and N. Komatsubara, "Infulences of Rolling Conditions on Hardenability of Direct-quenched Steel Plates", Tetsu-to-Hagane, 69 (1983) S1325.
19. Y. Okamura, S. Yano, T. Inoue, K. Tanabe, K. Kawai and N. Watanabe, "Development of a New 100Kgf/mm^2 Class High Tensile Strength Steel", Tetsu-to-Hagane, 71 (1985) S303.
20. H. Tamehiro, M. Murata, R. Habu and M. Nagumo, "Effect of Composition and Processing Variables on Thermomechanically Processed Niobium-Boron Steel", Tetsu-to-Hagane, 72 (1986) 466-473.
21. H. Morikawa and T. Hasegawa, "Microstructures and Strengthening Factors of Accelerated Cooled Steel" (Paper presented at the Int. Conf. on Accelerated Cooling of Steel, Pittsburgh, U.S.A. 19-21, Aug. 1985)
22. F.B. Pickering, "Physical Metallurgy and Design of Steel", (London, Applied Science Pub. 1978)
23. H. Homma, S. Ohkita, S. Matsuda and K. Yamamoto "Improvement of HAZ Toughness in HSLA Steel by Finely Dispersed Ti-Oxide" (Paper presented at the 67th AWS Covention, Atlanta, U.S.A. 1986)
24. T. Hasegawa, T. Haze and S. Aihara, "Metallurgical Factors Controlling COD Value of 80kgf/mm^2 Steel" (Paper presented at the 112th ISIJ Fall Meeting, Nagoya, Japan, 20-22 Oct. 1986)
25. R. Yamaba, H. Chiba, H. Gokyu and T. Komai, "Metallurgical Factors Controlling HAZ Toughness in HT80" (IIW Doc. 9-1422-86), (Paper presented at the 39th IIW Annual Assembly. Tokyo, 12-19 July, 1986)
26. T. Haze and S. Aihara, "Metallurgical Factors Controlling HAZ Toughness in HT50 Steels", (IIW Doc. 9-1423-86), (Paper presented at the 39th IIW Annual Assembly, Tokyo, 12-19 July, 1986)
27. K. Uchino and Y. Ohno, "A Study of Intercritical HAZ Embrittlement in HT50 for Offshore Structural Use" (Paper to be presented at the 6th OMAE Houston, 1-6 March 1987)
28. S.Aihara and T. Haze, "Micro-mechanical Analysis of the Influence of Metallurgical and Mechanical Inhomogeneity on Cleavage Fracture Initiation" (Paper presented at the Int. Conf. on Fatigue, Corrosion Cracking Fracture Mechanics and Failure Analysis Salt Lake U.S.A. 2-6 Dec. 1985)
29. N. Mori, H. Homma, M. Wakabayashi, S. Ohkita and S. Saito, "Mechanical Properties of Ti-B Weld Metal and Metallurgical Factors controlling Its Excellent Toughness", Seitetsukenkyu no.307, (1982) 104-116.
30. Y. Horii, S. Ohkita, M. Wakabayashi and M. Namura, "Developemnt of Welding Materials for Low Temperature Service" (Paper presented at the International Trends in Welding Research, ASM, Tenessee, May 1986)
31. H. Homma, M. Wakabayashi, S. Ohkita and N. Mori, "Improvement of Mechanical Properties at Welded Joint of the Heavy Plate Steel for Offshore Structure (HT50) Used in Colder Regions", Seitetsukenkyu no.314, (1984)

612

37-43.

32. A. Tsunetomi, Y. Horii, Y. Ogata, T. Koshio and K. Imai, "Development of Welding Technique for High CTOD Value", Seitetsukenkyu no.307, (1982) 44-55.

33. K. Uchino, Y. Ohno, S. Aihara, T. Haze, Y. Hagiwara, K. Horii, Y. Kawashima, Y. Tomita and R. Chijiiwa, "Development of 50kgf/mm² Class Steel for Offshore Structural Use with Superior HAZ Critical CTOD" (Paper presented at the 5th OMAE Tokyo, 13-18 April 1986)

34. Y. Ohno, Y. Okamura, K. Uchino, S. Yano and S. Matsuda, "Development of Low Temperature Use Steel for Large Heat Input Welding" (Proceedings of the Int. Conf. on HSLA, Wollongong, Australia, 20-24 August 1984), 129-133.

35. H. Mimura, Y. Takeshi, "High-Performance Tendon for TLP Manufactured by Quench and Temper" (Paper presented at the 18th Annual OTC Houston, Texas, 5-8 May 1986)

36. S. Yano et al., "A Manufacturing Method of Boron Bearing Low Alloy Steel with High Hardenability, Excellent Strength and Toughness", Japanese Patent No.1102473.

37. S. Yano et al., "Manufacturing Process for Boron Bearing Low-Alloy Steel with Excellent Strength and Toughness", Japanese Patent Application 60-1985.

38. S. Yano, Y. Okamura, A. Toyofuku, K. Ikeda, H. Muraoka and K. Moriyama, "Properties of A Newly Developed 80kgf/mm² Class High Tensile Strength Steel", Seitetsukenkyu no.390, (1982) 110-126.

39. S. Yano, Y. Okamura, K. Tanabe, T. Inoue, Y. Horii, K. Kawai and N. Watanabe, "Development of HT-100 Steel for Heavy Sections through the Application of TMCP" (Paper presented at the 39th IIW Annual Assembly. Tokyo, 12-19 July 1986)

40. K. Sasaki, H. Nakashima, M. Nose, Y. Takasaki and H. Okumura, "A Newly Developed Hot Metal Treatment Has Changed the Idea of Mass Production of Pure Steel" (Paper presented at the 66th Steelmaking Conference of AIME Atlanta, Georgia 1983)

41. Plate Technical Div. Nippon Steel Co., "Steel Plate", Nippon Steel Technical Report, 21 (1983), 83-100.

42. L.J. Fenstermaker, "Low-cost Plate Mill Improvement", Steel Times 214 (6) (1986) 287-290.

43. Y. Miyazaki, M. Kubota, M. Kato, K. Hama, K. Suzuki, I. Degawa, T. Suzuki and Z. Ohba, "Full Computer Control of Plate Mill-Outline of the Plate Mill Control System at Oita Works", Nippon Steel Technical Report. 14 (1979) 120-137.

OPTIMIZATION OF CHEMISTRY AND MANUFACTURING PROCESS

FOR OFFSHORE STRUCTURAL STEEL PLATE FOR LARGE HEAT INPUT WELDING

Y. SAITO [1], Y. NAKANO [2], S. UEDA [2] AND E. KOBAYASHI [3]

1) Plate Research Department, Iron & Steel Research Laboratories, Kawasaki Steel Corporation, Kawasaki-cho, Chiba, 260 Japan.
2) Mizushima Research Department, Iron & Steel Research Laboratories, Kawasaki Steel Corporation, Kurashiki, 712 Japan.
3) Quality Control Department, Mizushima Works, Kawasaki Steel Corporation, Kurashiki, 712 Japan.

Abstract

This paper covers an extensive study on optimization of chemistry and manufacturing process for offshore structural steel plate produced by thermomechanical control process (TMCP) for large heat input welding. This study involves computer simulation, laboratory and production experiments and mechanical test of the steel plate and its welded joints. The paper describes the outline of mathematical models for simulating microstructural changes. The phenomena such as recrystallization, grain growth, carbonitride precipitation and phase transformation in thermomechanical processing are predicted by the models. The effects of microalloying elements, reheating, rolling and cooling conditions on mechanical properties estimated by computer simulation are examined by laboratory experiments. A 460 MPa yield strength steel plate for low temperature use is produced by TMCP on the basis of the herein described fundamental study.

Introduction

High strength steel plates with good low temperature toughness in both base metal and welded joint and with low weld cracking susceptibility are demanded for constructing offshore structures in deep water and low temperature regions with low fabication cost.

Accelerated cooling after controlled rolling has become a major thermomechanical process for producing high strength offshore structural steel plates. One of the most important aspects of accelerated cooling is an appropriate control of reheating condition, rolling schedule and cooling procedure. The optimization of thermomechanical treatment requires precise microstructure control during the process. For this purpose, mathematical models which describe metallurgical phenomena during plate manufacturing process in terms of processing variables have been developed (1).

The following metallurgical factors are simulated by the mathematical models: Formation of austenite, carbonitride dissolution and austenite grain growth in reheating, recrystallization, grain growth, carbonitride precipitation and strain accumulation during rolling and phase transformation during accelerated cooling after rolling. Effects of microalloying elements and manufacturing conditions on mechanical properties estimated by computation are examined by laboratory experiment.

A 460 MPa yield strength (YP460MPa) offshore steel plate for low temperature use which is weldable with a large heat input was produced by the thermomechanical control process (TMCP) on the basis of computer simulation and laboratory experiment.

This paper describes the outline of mathematical models for computer simulation, laboratory experiment for developing a YP460MPa steel plate and the mechanical test results of the plate produced in the factory and its welded joints.

Computer Simulation Models

Computer simulation models consist of two parts; an austenite microstructure control model and a transformation structure control model. The block diagram of the computer simulation model of microstructural changes during plate manufacturing process is shown in Fig. 1.

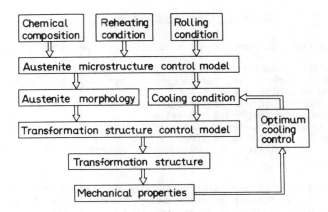

Fig.1   Block diagram for computer simulation of microstructural changes during plate manufacturing process

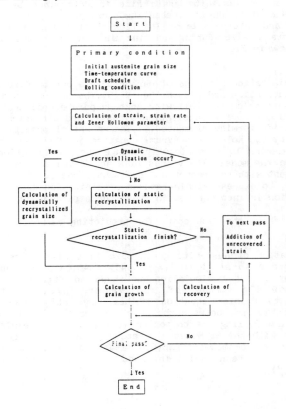

Fig.2   Flow chart for predicting variation of austenite grain size during rolling

Austenite Microstructure Control Model

Microstructural changes in austenite are principally described as follows:
(1) Formation of austenite and grain growth during reheating.
(2) Grain refinement and grain growth by recrystallization in high temperature austenite region.
(3) Dissolution of carbonitrides and coarsening of undissolved carbonitride particles during reheating.
(4) strain-induced precipitation of carbonitrides of microalloyed elements such as Nb during rolling.
(5) Strain accumulation in austenite by deformation in low temperature region.

The variation of austenite grain size during reheating is estimated by simply taking account grain boundary pinning by small particles of undissolved carbonitrides at lower reheating temperatures (2) and impurity drag effect at higher temperatures (3). The kinetics of recrystallization and grain growth is estimated by a modified Sellars' empirical equation (4), in which an austenite grain size is determined by such factors as initial grain size during reheating, reduction per pass, deformation temperature and interpass time. The flow chart for predicting variation of austenite grain size during rolling in the high temperature austenite region is shown in Fig.2.

The precipitation kinetics of carbonitrides can be described by a classical nucleation and growth theory (5).The nucleation and growth rates are the functions of the following unknown parameters: Surface energy of precipitates ç, number of nucleation sites N, and diffusion coefficient D of carbonitride forming elements (5).The amount of precipitation is calculated by the Kolmogorov-Avrami equation (6). The variations of three parameters,ç, N and D, significantly change the precipitation process according to the model. The values of N and D are determined so that the best agreement with the reported experimental results (7) can be obtained. Furthermore, to express strain-induced precipitation kinetics, the effect of deformation is incorporated in these three parameters.

Figure 3 shows the flow chart for predicting carbonitride precipitation kinetics

The variation of austenite grain size in a Si-Mn steel (0.18% C-1.36% Mn steel) and a Nb steel (0.17% C-1.41% Mn-0.031% Nb) during rolling is predicted by computer simulation as shown in Fig. 4 together with the variation of the amount of Nb(C,N) precipitates, the calculated temperature by Fourier equation of heat conduction and the deformation resistance ratio, km, between the Nb steel and the Si-Mn steel measured by composite slab rolling.The composite is produced by welding the Mb steel slab section with the Si-Mn steel section placed in the rolling direction. The increase of the amount of Nb(C,N) leads to the increase of deformation .resistance.This result is in good agreement with the laboratory simulation (8)

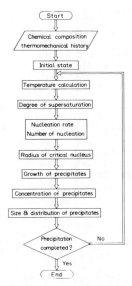

Fig.3 Flow chart for predicting carbonitride precipitation

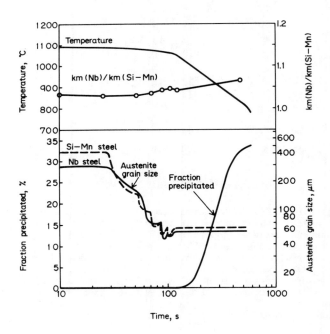

Fig.4   Variation of austenite grain size and amount of Nb(C,N)
          precipitates during rolling

619

The effect of deformation in low temperature austenite region on transformed grain structure is described by accumulated strain in austenite, which is expressed by rolling temperature, interpass time and prestrain (8).
The accumulated strain prior to transformation is an important factor that has an influence on ferrite grain size.
The ferrite grain size dα is estimated by the following equation:

$$\ln d\alpha = 0.92 + 0.44 \ln d\gamma - 0.171 \ln C_R - 0.88 \tanh(10 \, \Delta\varepsilon_t)$$

where dγ, C_R and recrystallized austenite grain size, cooling rate and accumulated strain prior to transformation. The curved surface in Fig. 5 shows the calculated ferrite grain size. The observed value of ferrite grain size as shown by the straight line which is parallel to the Z-axis in Fig. 5 is in good agreement with those calculated.

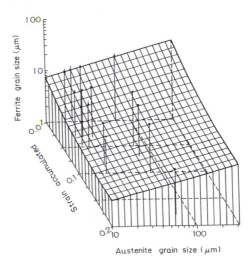

Fig. 5. Comparison of calculated ferrite grain size with observed one

Transformation Structure Control Model

The kinetics of γ to α transformation is calculated by a classical nucleation and growth theory. In order to simplify the calculation the following assumptions are adopted:
(1) Ferrite nucleation is controlled by grain boundary nucleation.
The effect of rolling on phase transformation is taken into consideration in terms of recrystallized austenite grain size, accumulated strain prior to transformation and the amount of carbonitride precipitation.
(2) The growth of ferrite, pearlite and bainite is controlled by the volume diffusion mechanism.
The fraction of transformed phase is calculated by the Kolmogorov-Avrami equation.
(3) The γ to α phase boundary and driving force for transformation are thermodynamically calculated using a quasi-chemical model.

The flow chart for calculation of phase transformation is shown in Fig. 6.

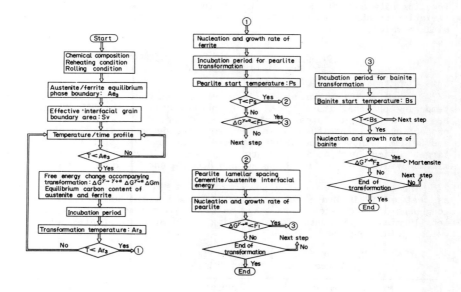

Fig.6. Flow chart for calcuration of phase transformation

An example of computer simulation of phase transformation for a low C and low Mn steel during accelerated cooling is shown in Fig. 7. As shown in the figure, the cooling condition has a small effect on the kinetics of transformation of the low C and low Mn steel.

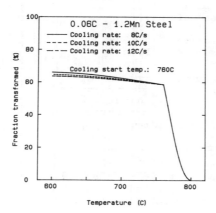

Fig.7. Computer simulation of phase transformation kinetics for low C and low Mn steel.

The mechanical properties of steel plate are estimated using transformed structure obtained by computer simulation. The optimum cooling conditions (cooling rate and finish cooling temperature) are determined so that the variation of strength due to fluctuation in cooling conditions has a minimum value.

Design of Chemistry and Manufacturing Process for 460 MPa Yield Strength Steel Plate

Though a YP460 MPa steel plate can be produced by a quenching and tempering process, its weldability is not so good as the 355 MPa and 415 MPa yield strength steel plates produced by TMCP. Good Charpy impact properties at temperatures down to -60·°C are demanded for the base metal and welded joints of the YP460 MPa steel plate manufactured by TMCP. The target minimum preheat temperature to prevent weld cracking is 0 °C. The minimum yield and tensile strengths are 460 MPa and 590 MPa, respectively.

The concept of chemistry design of YP460 MPa steel plate is the same as that of YP415 MPa steel plate (9), in which the weldability and the toughness of welded joint are taken as key requirements. Table 1 shows the metallurgical approaches for producing high strength plates with good welded joint toughness.

Table 1   Metallurgical approach for improving HAS toughness of large heat input welded joints

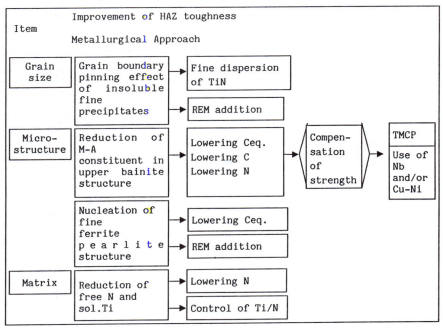

622

The optimum chemistry and manufacturing process are determined by computer simulation and laboratory experiment. Tables 2 and 3 show the range of chemical composition and conditions of thermomechanical treatment, respectively, used in computer simulation and laboratory experiment.

Table 2   Range of chemical composition in computer simulation and laboratory experiment

C	Si	Mn	Cu	Ni	Nb	Ceq	Remark
0.08 ~0.13	0.25 ~0.40	1.30 ~1.50	0.25	0.25	0.015	0.33 ~0.38	REM-Ti Treated

Table 3   Conditions of thermomechanical treatment in computer simulation in computer simulation and laboratory simulation

Reheating temperature	$950°C \sim 1150°C$
Rolling condition Finish rolling temperature	$700°C \sim 780°C$
Cooling condition Cooling rate Finish cooling temperature	$5 \sim 10°C/s$ $300°C \sim 780°C$

The results of the computer simulation are summarized as follows:
(1) In a low C (C<0.08%) and low Mn (Mn<1.2%) steel, the effect of cooling condition on transformed structure is very small.
When heavy controlled rolling (lower reheating temperature, higher deformation in non-recrystallization region and lower finish rolling temperature) is applied to the steel with a lower carbon equivalent (Ceq) value, the yield strength predicted by transformed structure increases with the amounts of C and Mn.
(2) In a high Mn (Mn>1.2%) steel, on the other hand, the effect of cooling conditions on transformed structure is distinctive.
In this case, the yield strength of the steel plate produced in the conditions of high cooling rate and low finish cooling temperature decreases with Mn content
(3) In order to satisfy the target mechanical properties of YP460 MPa steel plate, the optimum C and Mn contents are determined to be 0.1% and 1.30%, respectively.

The effect of reheating temperature on mechanical properties of a 0.1% C and 1.3% Mn steel obtained by laboratory experiment is shown in Fig. 8. The optimum reheating temperature is 1050 °C The transformed structure is ferrite plus bainite.Figure 9 shows the effect of finish cooling temperature on mechanical properties.The finish cooling temperature which satisfies the target mechanical properties ranges from 450 to 500 °C.

623

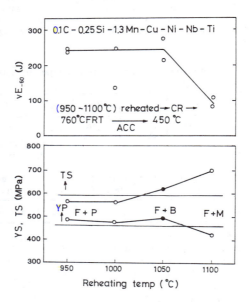

Fig.8.   Effect of reheating temperature on mechanical properties of
         0.1%C-1.3%Mn-Cu-Ni-Nb-Ti Steel

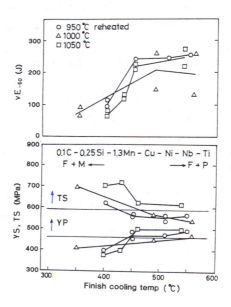

Fig.9.   Effect of finish cooling temperature on mechanical
         properties of 0.1%C-1.3%Mn-Cu-Ni-Nb-Ti Steel

Mechanical Properties of 460 MPa Yield Strength Steel Plate Produced by Thermomechanical Control Process

On the basis of the computer simulation and laboratory experiment described above, a 30 mm thick YP460 MPa steel plate was produced by the production mill and the accelerated cooling device.Table 4 shows the chemical composition of the steel plate..Impurities such as N, P and S were controlled to remain at very low levels. The value of Ceq is 0.35%. Figure 10 shows the manufacturing process. The molten iron was produced by the basic oxygen furnace, treated by vacuum degassing and continuously cast. The steel plate was produced by accelerated cooling after controlled rolling.

Table.4 Chemical composition of steel plate   (wt.%)

C	Si	Mn	P	S	Al	Nb
0.097	0.29	1.33	0.006	0.002	0.027	0.015
Cu	Ni	Ti	REM	N	Ceq	PCM
0.25	0.25	0.009	0.006	0.003	0.35	0.15

Ceq = C + Mn/6 + (V + Mo + Cr)/5 + (Cu + Ni)/15 (%)

PCM = C + Si/30 + (Mn + Cu + Cr)/20 + Ni/60 + Mo/15 + V/5B (%)

Table 5 Mechanical properties of steel plates

Direction	Tensile test*				Charpy impact test			
	YP (MPa)	TS (MPa)	El (%)	RA (%)	Posi-tion	Absorbed energy(J)		50% FATT (°C)
						−60°C	−80°C	
L	505	627	23	−	1/4t	235	225	−125
					1/2t	225	167	−97
T	510	632	23	−	1/4t	157	137	−95
					1/2t	147	98	−75
Z	−	608	−	72	−	−	−	−

Table 5 shows tensile and 2 mm V-notched Charpy impact properties of the steel plate.The tensile and impact properties exceeded the target values. The reduction of area in the through thickness direction was more than 70 %, indicating high resistance against lamellar tearing.

The microstructure of base metal is shown in Fig. 11. Fine ferrite and bainite structures are observed.

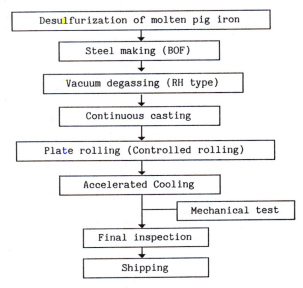

```
┌─────────────────────────────────────┐
│  Desulfurization of molten pig iron  │
└─────────────────────────────────────┘
                  ↓
      ┌─────────────────────────┐
      │    Steel making (BOF)    │
      └─────────────────────────┘
                  ↓
    ┌─────────────────────────────┐
    │  Vacuum degassing (RH type)  │
    └─────────────────────────────┘
                  ↓
      ┌─────────────────────────┐
      │    Continuous casting    │
      └─────────────────────────┘
                  ↓
┌───────────────────────────────────────┐
│  Plate rolling (Controlled rolling)   │
└───────────────────────────────────────┘
                  ↓
     ┌───────────────────────────┐
     │    Accelerated Cooling    │
     └───────────────────────────┘
                  ↓        ┌─────────────────────┐
                  ├────────│   Mechanical test   │
                  ↓        └─────────────────────┘
      ┌─────────────────────────┐
      │    Final inspection     │
      └─────────────────────────┘
                  ↓
      ┌─────────────────────────┐
      │        Shipping         │
      └─────────────────────────┘
```

Fig.10 Manufacturing procedure

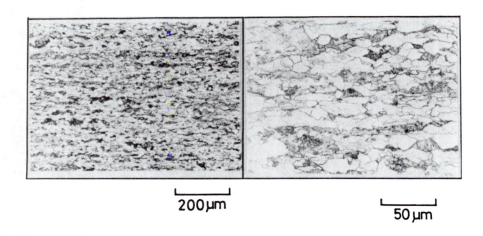

200μm   50μm

Fig.11 Microstructure of YP460MPa steel plate

Table 6   Welding conditions

Welding method	Grove shape	Pass	Elect -rode	Current (A)	Voltage (V)	Speed (cm/min)	Heat imput (KJ/cm)
One- side four- pass SAW	35°	1	L	1000	32	37	52
		2	L	850	38	45	86
			T	800	40	45	
		3	L	900	38	45	91
			T	850	40	45	
		4	L	900	40	40	108
			T	850	42	40	
Multi- Pass SAW	45°		L	550	28	38	50
			T	550	30	38	

The weld cracking susceptibility of the test plate was investigated by the y-groove restraint weld cracking test in accordance with JIS Z3158. No crack was observed when the plate was welded at 0 °C. Therefore, preheat-free welding is possible at temperatures above 0 °C

Table 6 shows the welding conditions. The welding method was submerged arc welding (SAW). One-side four-pass SAW was performed.

The Charpy impact test results of the welded joints made by one-side four-pass and multi-pass SAW were shown in Table 7. The result indicates that all the joints satisfied the target values.

The CTOD test was conducted in accordance with BS5762 on the welded joints. As shown in Table 6, a half K groove was used. The precrack, therefore, was introduced along the straight fusion line. The test result is shown in Table 8. The minimum critical CTOD value for the fusion boundary at -10 °C was 0.46 mm for one-side four-pass SAW and 1.07 mm for multi-pass SAW.

Table 7 Charpy test results

( )Crystallinity(%)

Welding method	Absorbed energy at fusion line (J)		
	−40°C	−60°C	−80°C
One-side four pass SAW	181(27)	74(55)	45(68)
Multi-pass SAW	223(0)	133(53)	81(72)

Table 8 CTOD test result of welded joint at fusion line

Welding method	Temperrature (°C)	Critical CTOD (mm)
One-side four pass SAW	−10	1.52, 0.46
Multi-pass SAW	−10	1.68, 1.07
	−30	2.20, 0.28
	−50	0.19

Conclusion

A 460 MPa yield strength (YP460MPa) offshore structural steel plate for low temperature use was produced through the computer simulation and laboratory experiment, results were applied to design of chemistry and manufacturing process.

The following conclusions are obtained:

(1) Computer simulation technique was successfully applied to thechemistry and manufacturing process.
The plate had sufficient tensile and impact properties
(2) Charpy absorbed energy of the welded joint at -60 °C exceeded 39 J and 78 J for one-side four-pass and multi-pass welded joints, respectively.
(3) The critical CTOD value at -10 °C for the fusion boundary exceeded 0.4mm at -10 °C and 1 mm for one-side four-pass and multi-pass welded joints, respectively.

References

(1) Y. Saito, M. Tanaka, T. Sekine and H. Nishizaki: <u>Proc. of Int'l.   Conf. on High Strength Low Alloy Steel</u>, Wollongong, Australia, 1984, pp. 28
(2) M. Hillert: <u>Acta Metall.</u>, 1965, vol. 13, pp. 227
(3) K. Lucke and K. Detert: <u>Acta Metall.</u>, 1967, vol. 15, pp. 628
(4) C. M. Sellars and J. A. Whiteman: <u>Metall. Sci.</u>, 1979, vol. 13, pp. 187
(5) Y. Saito, M. Kimura, M. Tanaka, T. Sekine, K. Tsubota and T. Tanaka: <u>Kawasaki Steel Technical Report</u>, 1984, no. 9, pp. 12
(6) A. Kolmogorov, <u>Izv. Akad. Nauk SSSR, Ser. Mat</u>, 1937, vol. 1, pp. 355
(7) H. Watanabe: Thesis, Univ. of Michigan, 1975
(8) Y.Saito, T.Enami and T. Tanaka: <u>Trans ISIJ</u>, 1985, vol. 25, pp. 1146
(9) M.Koda, K.Amano, C.Shiga and Y.Nakano: <u>Tetsu to Hagane</u>, 1986, vol. 72, s1157

SUBJECT INDEX

633

AUTHOR INDEX

ANDERSONIAN LIBRARY

★

WITHDRAWN
FROM
LIBRARY
STOCK

★

UNIVERSITY OF STRATHCLYDE